TRANSPORT PROPERTIES OF IONS IN GASES

TRANSPORT PROPERTIES OF IONS IN GASES

Edward A. Mason
Brown University
Providence, Rhode Island

Earl W. McDaniel
Georgia Institute of Technology
Atlanta, Georgia

WILEY

A WILEY-INTERSCIENCE PUBLICATION

JOHN WILEY & SONS

New York / Chichester / Brisbane / Toronto / Singapore

To Ann C. Mason and to Johnny McDaniel,
the grandson with the runaway mobility

Library of Congress Cataloging in Publication Data:

Mason, Edward A. (Edward Allen), 1926–
    Transport properties of ions in gases / Edward A. Mason, Earl W.
McDaniel.
        p.   cm.
    "A Wiley-Interscience publication."
    Includes bibliographies and indexes.
    ISBN 0-471-88385-9
    1. Ionic mobility. 2. Drift mobility. 3. Diffusion.
4. Transport theory. 5. Gases, Kinetic theory of. I. McDaniel,
Earl Wadsworth, 1926– . II. Title.
QC717.5.I6M37 1987
533--dc 19   87-26349
                                                       CIP

Printed in the United States of America

10 9 8 7 6 5 4 3 2 1

PREFACE

Interest in the conduction of electricity through gases stems from 1895 with the discovery of X-rays by Roentgen, and it was soon ascertained that ions and electrons are the charge carriers. Until the 1930s, only crude experimental techniques were available to measure their transport properties in electric fields, but these techniques sufficed for a qualitative mapping of certain important aspects of charged particle transport.

A significant advance in the experimental study of ionic drift velocities was made in the 1930s by Tyndall and his colleagues, who developed powerful new techniques and used them for accurate measurements on alkali ions. On the theoretical side, the 1905 Langevin theory based on the polarization interaction was available; the more general Chapman–Enskog theory, developed circa 1918, was later applied to the calculation of mobilities and diffusion coefficients. However, both of these theories are limited to spherically symmetric ion-molecule combinations and to vanishingly small E/N. (Here E is the intensity of the externally applied electric field, and N is the number density of the gas through which the ions are drifting and diffusing.) The transport theory for gaseous ions underwent a slow development up to the early 1950s, but in 1953 Wannier published a landmark paper that proved to be extremely influential. His paper dealt mainly with the motion of gaseous ions in strong electric fields, and although it did not contain a general solution of the problem, it did give solutions for useful theoretical models and provided much insight into the strong-field problem.

The next important advance came between about 1960 and 1963 on the experimental side with the development of the drift-tube mass spectrometer, which permitted the first direct mobility measurements of mass-identified ions, and which elucidated the complicating effects of chemical reactions between the ions and gas molecules. Second-generation instruments of this type began to come on line in 1966, and by 1970, reliable data on ionic mobilities, and for the first time, on diffusion coefficients, were appearing in copious quantities. The availability of a large amount of data for known species of ions not involved in

v

chemical reactions provided the incentive for the writing of the book *The Mobility and Diffusion of Ions in Gases*, by E. W. McDaniel and E. A. Mason, Wiley, New York (1973). This book provided an exhaustive description of experimental techniques relevant to the study of gaseous ion transport, and a compilation of experimental data on ionic mobilities, drift velocities, and diffusion coefficients. The book also contained a detailed account of the relevant transport theory as it stood in the early 1970s, and an elaborate comparison of theory with experiment.

The main impetus for the present book was provided by new developments in the rigorous kinetic theory of gaseous ion transport that began to take place in about 1975. A new theory has emerged that applies to ion-molecule combinations at any values of the temperature and ionic energy parameter E/N. The present volume presents this theory in considerable detail, and illustrates its use in analyzing experimental data and in extracting spherically symmetric ion-molecule interaction potentials from experimental mobilities obtained as a function of E/N. We have also greatly expanded and updated our previous discussion of experimental techniques, both as applied to the measurement of ionic transport coefficients and of reactions of the ions with neutral molecules. The reaction problem is directly coupled to the transport problem, but we are really concerned only with the interface, and our discussion of reactions is not intended to be comprehensive. The field of ion-neutral reactions has grown from the point in the 1950s when practically nothing was known about them to the point that several volumes are now required for a detailed treatment.

The plan of this volume is as follows: Chapter 1 is largely phenomenological; it introduces concepts and definitions, and describes the ionic transport processes in a qualitative fashion. Chapters 2 and 3 describe experimental techniques for measuring ionic drift velocities and diffusion coefficients. The emphasis is on the drift-tube mass spectrometer, although other instruments, including the flowing afterglow and its variants, are also discussed. Chapter 4 deals with stationary afterglows. Chapters 5 through 8 are essentially theoretical, but also contain extensive comparisons with experimental data. Chapter 5 is a general survey of existing theory; it includes simple physical arguments based on elementary kinetic theory and on equations of momentum and energy balance, and an overview of the more accurate theories based on solutions of the Boltzmann equation. Chapter 6 describes these accurate theories in detail. Chapters 7 and 8 are concerned with applications—Chapter 7 deals with the connection between the ionic transport coefficients and ion-molecule interaction potentials, and Chapter 8 gives a series of brief surveys of applications of the theory to various topics ranging from the use of ion mobility spectrometry in analytical chemistry to the energy relaxation of hot atoms and electrons in gases. Appendix I contains an extensive data index from which the best available experimental results can be tracked down. The volume finishes with two appendices containing compilations of various physical properties and ion-atom model cross sections, useful in connection with Chapter 7.

We have not undertaken to describe the historical development of our subject in systematic detail, although in the text we do not subject it to the procrustean extreme that surely appalls experts when they read this preface. Fuller accounts appear in the books by Loeb, Tyndall, Massey, Huxley and Crompton, and Thomson and Thomson that we cite. We realize that we have failed to give proper credit to many researchers who played an important role in the development of the subject, but we believe that the modern reader may be better served by our moving rapidly to contemporary matters. We also almost entirely avoid discussing situations in which ion-ion interactions are important. Such situations belong in the realm of plasma physics and are treated in books on the subject. We are primarily concerned with unclustered ions.

The audience for whom this book was written includes physicists, chemists, and engineers working in this area, graduate students entering the field, and researchers in gaseous electronics, atomic collisions, the atmospheric sciences, plasma physics, and gas lasers. We assume on the part of the reader a knowledge of elementary kinetic theory and collision theory. On the other hand, we do not assume any detailed knowledge of transport processes—it seems desirable to build from the ground level here.

We wish to thank the Office of Naval Research, the Army Research Office, and the National Science Foundation for their support of our research on the subject of this book through the years. It is also a pleasure to express our appreciation for the useful comments and help of various kinds received from many people. We especially thank Dr. David Burnett, who was not only one of the early pioneers in the rigorous kinetic theory of gases, but who also independently formulated (but unfortunately never published) what we have called the two-temperature theory of ion transport. Dr. R. E. Robson and Dr. L. A. Viehland were also extremely generous in the efforts they made on our behalf. Dr. F. J. Smith very kindly calculated for us many of the cross sections that appear in Appendix II. For reviewing and criticizing various parts of the text we are grateful to D. L. Albritton, M. T. Elford, D. W. Fahey, R. Johnsen, K. Kumar, J. T. Moseley, J. A. Rees, H. R. Skullerud, and D. Smith. Finally, we thank Mrs. Genevieve Goditt and Mrs. Audrey Ralston for their cheerful and efficient work in producing legible text from bad handwriting and ghastly looking equations.

EDWARD A. MASON
EARL W. MCDANIEL

May, 1987
Providence, Rhode Island
Atlanta, Georgia

CONTENTS

1

INTRODUCTION

1-1. QUALITATIVE DESCRIPTION OF THE MOTION OF SLOW IONS IN GASES

Consider a localized collection of ions of a single type in a gas of uniform temperature and total pressure and suppose that the number density n of the ions is low enough to ignore the Coulomb forces of repulsion. As is well known, the ions will become dispersed through the gas by the process of diffusion,* in which there is a net spatial transport of the ions produced by the gradient in their relative concentration. The diffusive flow takes place in the direction opposite that of this gradient and the flow rate is directly proportional to its magnitude. The constant of proportionality is called the (scalar) diffusion coefficient and is denoted by the symbol D. Thus the ionic flux density $\mathbf{J}$ is given by Fick's law of diffusion (Fick, 1855),

$$\mathbf{J} = -D \, \nabla n. \qquad (1\text{-}1\text{-}1)$$

The magnitude of $\mathbf{J}$ equals the number of ions flowing in unit time through unit area normal to the direction of flow. The minus sign indicates that the flow occurs in the direction of decreasing concentration; D is a joint property of the ions and the gas through which they are diffusing and (1-1-1) shows it to be a measure of the transparency of the gas to the diffusing particles. Since the velocity of the diffusive flow $\mathbf{v}$ is given by

$$\mathbf{J} = n\mathbf{v}, \qquad (1\text{-}1\text{-}2)$$

we may also write Fick's law in the form

$$\mathbf{v} = -\frac{D}{n} \nabla n. \qquad (1\text{-}1\text{-}3)$$

*Good general references to the subject of diffusion are Crank (1980) and Hirschfelder et al. (1964).

1

The diffusive flow continues until inequalities in composition have been eliminated by interdiffusion of the ions and gas molecules. This type of flow is separate and distinct from the flow that would result from a nonuniformity of the total pressure.

If a weak uniform electric field is now applied throughout the gas, a steady flow of the ions along the field lines will develop, superimposed on the much faster random motion that leads to diffusion. The velocity of the center of mass of the ion cloud, or equivalently the average velocity of the ions, is called the drift velocity $\mathbf{v_d}$, and this velocity is directly proportional to the electric field intensity $\mathbf{E}$, provided that the field is kept weak. Thus

$$\mathbf{v_d} = K\mathbf{E}, \tag{1-1-4}$$

where the constant of proportionality K is called the (scalar) mobility of the ions; K, like D, is a joint property of the ions and the gas through which the motion occurs.

A simple relation, known as the Einstein equation,* exists between the weak-field mobility and diffusion coefficient. This equation is exact in the limit of vanishing electric field and ion concentration and states that

$$K = \frac{eD}{kT}. \tag{1-1-5}$$

Here e is the ionic charge, k is the Boltzmann constant, and T is the gas temperature. If K is expressed in the usual units, square centimeters per volt per second, D in the usual units, square centimeters per second, and T in kelvin, we have

$$K = 1.1605 \times 10^4 \frac{D}{T}. \tag{1-1-6}$$

(A factor of 299.79 comes into play here because the electrostatic unit of mobility is square centimeters per statvolt per second and 1 statvolt equals 299.79 V.) It is not surprising that K should be directly proportional to D here, since both quantities are a measure of the ease with which the ions can flow through the gas. It is important to point out that (1-1-5) is valid only when the electric field is so weak that the ions are close to being in thermal equilibrium with the gas molecules, that is to say, when "low-field" conditions obtain. Under these conditions the ionic velocity distribution is very nearly Maxwellian. The ionic motion is largely the random thermal motion produced by the heat energy

*This relationship is derived in Section 5-1A. A simple derivation also appears in McDaniel (1964, pp. 490–491). Equation (1-1-5) was first established for ions in gases by Townsend in 1899 (Townsend, 1899). It is frequently referred to as the Nernst–Townsend relation because it had been previously derived (in a different context) by Nernst (1888). Einstein's derivation was published in a paper dealing with the theory of Brownian motion (Einstein, 1905).

of the gas, with a small drift component superimposed in the direction of the applied field.

The constant, steady-state drift velocity of the ions given by (1-1-4) is achieved as a balance between the accelerations in the field direction between collisions with gas molecules and the decelerations that occur during collisions. Since the ionic mass is usually comparable to the molecular mass, only a few collisions are normally required for the ions to attain a steady-state condition after the electric field is applied.

If the electric field intensity is now raised to a level at which the ions acquire an average energy appreciably in excess of the thermal energy of the gas molecules, a number of complications develop. The thermal energy becomes less important, but two large components of motion are produced by the drift field: a directed component along the field lines and a random component representing energy acquired from the drift field but converted into random form by collisions with molecules. The mobility K appearing in (1-1-4) is no longer a constant in general but will usually depend on the ratio of the electric field intensity to the gas number density E/N, which is the parameter that determines the average ionic energy gained from the field in steady-state drift above the energy associated with the thermal motion.* In addition, the energy distribution of the ions becomes distinctly non-Maxwellian, and the diffusion now takes place transverse to the field direction at a rate different from that of the diffusion in the direction of the electric field. The diffusion coefficient† thus becomes a tensor rather than a scalar and has the form

$$\mathbf{D} = \begin{vmatrix} D_T & 0 & 0 \\ 0 & D_T & 0 \\ 0 & 0 & D_L \end{vmatrix}, \tag{1-1-7}$$

where D_T is the (scalar) transverse diffusion coefficient that describes the rate of diffusion in directions perpendicular to $\mathbf{E}$ and D_L is the (scalar) longitudinal diffusion coefficient characterizing diffusion in the field direction (Wannier, 1953). In the "intermediate-field" and "high-field" regions described here the Einstein equation (1-1-5) no longer applies.

It may be of interest to note that the presence of a magnetic field also renders an ionized gas anisotropic and makes both the mobility and the diffusion coefficient assume a tensor form (McDaniel, 1964, pp. 506–512; Seshadri, 1973; Krall and Trivelpiece, 1973; Golant et al., 1980; Stacey, 1981). In this case, however, $\mathbf{D}$ is not in the form indicated by (1-1-7), since two of the off-diagonal components are now different from zero.

---

*The acceleration of the ions to the terminal drift velocity has been treated in detail by Johnsen and Biondi (1972) and Lin et al. (1977).

†The manner in which the diffusion coefficient tensor is used in calculations is indicated in Section 2-6B. It should be noted that the symbols $D_\parallel$ and $D_\perp$ are frequently used in place of D_L and D_T, respectively.

1-2.   PARAMETERS E/N AND E/p

We now attempt to make plausible the statement in Section 1-1 that E/N is the parameter that determines the average ionic energy acquired from the electric field, that is, the "field energy." The electric force on an ion of charge e is eE and the resulting acceleration is eE/m, where m is the mass of the ion. We make the crude assumption that when an ion undergoes a collision it loses, on the average, all the energy it acquired from the field during the preceding free path. Then, if τ denotes the collision period, or mean free time, the velocity acquired just before a collision is $eE\tau/m$. Since $\tau \sim 1/N$, the energy obtained between collisions from the field is thus seen to be proportional to $(E/N)^2$. Rigorous calculations also show E/N to be the parameter that determines the field energy of the ions.

Although E/N is the more fundamental quantity, until recently most experimentalists have reported the results of their measurements in terms of E/p, where p is the gas pressure or, in terms of E/p_0, where p_0 is the "standard pressure," normalized to 0°C. The standard pressure is defined by the equation

$$p_0 = \frac{273.15}{T} p, \qquad (1\text{-}2\text{-}1)$$

where T is the absolute temperature at which the measurement was made. This convention has not been wholly satisfactory, since p_0 has also been used to represent normalizations to temperatures other than 0°C. If we use the parameter E/N, however, there is no ambiguity in comparing experimental results. The conversion relations are

$$\frac{E}{N} = (1.0354 \times T \times 10^{-2}) \frac{E}{p} \qquad (1\text{-}2\text{-}2)$$

or

$$\frac{E}{N} = (2.828) \frac{E}{p_0}, \qquad (1\text{-}2\text{-}3)$$

where E/N is in units of 10^{-17} V-cm^2, T is in kelvin, and E/p or E/p_0 is in volts per centimeter per torr. Huxley, Crompton, and Elford (1966) have suggested that the units of E/N be denoted by the "townsend," or "Td," where $1\text{ Td} = 10^{-17}$ V-cm^2, and this designation is attaining widespread usage. Both E/N and E/p are used in this book.

The field energy is negligible compared with the thermal energy if

$$\left(\frac{M}{m} + \frac{m}{M}\right) eE\lambda \ll kT, \qquad (1\text{-}2\text{-}4)$$

where M and m are the molecular and ionic masses, respectively, and $eE\lambda$ is the energy gained by an ion in moving a mean free path λ in the field direction.* The factor involving the masses accounts for the ability of the ions to store the acquired energy over many collisions if the masses are significantly different. Using the ideal gas law $NkT = p$ and the relationship $\lambda = 1/NQ$, where N is the molecular number density and Q is the ion-molecule collision cross section, we may express the foregoing inequality as $(M/m + m/M)eE \ll pQ$. Taking a singly charged ion moving through the parent gas and making the reasonable assumption that $Q = 50 \times 10^{-16}$ cm^2, we find that the field energy is much less than the thermal energy if $E/p \ll 5 \times 10^{-6}$ (statvolt/cm) per (dyne/cm^2) $\approx$ 2 V/cm-torr. The electric field is said to be "low" when this criterion is satisfied and "high" when the inequality is reversed. It should be noted that a given field in a gas of given density may change from "low" to "high" if the gas temperature is lowered sufficiently.

From what has been said here it may be inferred that the mobility K will be constant, independent of E/p, provided that $E/p \lesssim 2$ V/cm-torr, so that the ionic energy is close to thermal. Actually, theory predicts that K will also be constant at higher ionic energies, provided that the collision frequency does not depend on the energy of the ions. Usually, however, the mobility begins to vary with E/p at the upper end of the low-field region. Under such conditions the concept of mobility loses some of its convenience, but the phenomenological definition of the mobility as the ratio of v_d to E is still useful in comparing experimental data and is used here.

1-3. GENERAL FACTS ABOUT MOBILITIES AND DIFFUSION COEFFICIENTS

The mobility of a given ionic species in a given gas is inversely proportional to the number density of the molecules but relatively insensitive to small changes (a few kelvin) in the gas temperature if the number density is held constant. To facilitate the comparison and use of data a measured mobility K is usually converted to a "standard," or "reduced," mobility, K_0, defined by the equation

$$K_0 = \frac{p}{760} \frac{273.15}{T} K = \frac{p_0}{760} K, \qquad (1\text{-}3\text{-}1)$$

where p is the gas pressure in torr and T is the gas temperature in kelvin

---

*The mass factor in (1-2-4) is introduced on an ad hoc basis to account for the inefficiency of energy transfer in elastic collisions when either M/m or m/M is much greater than unity. A more rigorous discussion of the low-field criterion appears in Section 5-2, culminating in the inequality (5-2-30).

at which the mobility K was obtained.* Under the standard conditions of pressure and temperature (760 torr and 0°C) the gas number density is $2.69 \times 10^{19}/cm^3$. It must be emphasized that the use of (1-3-1) merely provides a standardization or normalization with respect to the molecular number density; the temperature to which the standard mobility actually refers is the temperature of the gas during the measurement. For ions of atmospheric interest in atmospheric gases the standard mobility is of the order of several square centimeters per volt per second. In the modern literature, when a single value is quoted as "the mobility" of an ion in a gas, the value cited is the standard mobility extrapolated to zero field strength.

Ions of the atmospheric gases in their parent gases have diffusion coefficients of the order of 50 cm^2/s at 1 torr pressure and low E/N. The diffusion coefficients usually increase dramatically as E/N is raised above the low-field region. As expected, D is found to vary inversely with the number density of the gas, and diffusion coefficient data are usually presented in the form of the product DN.

Methods of measuring mobilities and diffusion coefficients are described in Chapters 2, 3, and 4. The calculation of these quantities from kinetic theory is discussed in Chapters 5, 6, and 7. Representative data appear throughout the volume, and references to data on several hundred ion-gas combinations are given in Appendix I. The manner in which the standard mobility typically varies with E/N is illustrated in Figs. 7-3-2 through 7-3-6. Fig. 5-2-4 shows the typical variation of diffusion coefficients with E/N.

1-4.   ION-ION INTERACTIONS AND THE EFFECT OF SPACE CHARGE

Under the usual conditions more than one ionic species is present at a given time in an ionized gas. If the density of ionization is low, each species of ions may be considered separately, and the ions of each type drift and diffuse through the gas without interacting appreciably with one another or with members of the other species. For reasons discussed in Chapter 2 this condition must obtain in ion drift experiments if reliable measurements are to be made of the diffusion coefficients (and even the mobilities in certain instances), and it is of interest to establish a criterion for this condition to be satisfied (Wannier, 1953).

We consider the space-charge effect produced mainly by widely separated ions, and we find that its magnitude depends on the dimensions of the apparatus. In one dimension Poisson's equation, $\nabla^2 V = -4\pi\rho$, may be written

---

*All existing mobility theories predict that the standard mobility K_0 should be independent of the gas-number density N except at pressures so high that three-body elastic interactions would be expected to become significant. Elford (1971), however, has reported the observation of a slight dependence of K_0 on N for K^+ ions in He, Ne, Ar, H_2, and N_2 at pressures of the order of 5 torr. McDaniel (1972) and Thomson et al. (1973) have discussed Elford's observations in terms of the clustering of molecules about the K^+ ions, Elford and Milloy (1974) have proposed an explanation in terms of the formation of ion-atom or ion-molecule complexes in orbiting resonant states.

in the form $\partial E/\partial x = 4\pi n e$, where ρ is the charge density, e, the ionic charge, and n, the ionic number density. The criterion for negligible space charge distortion of an applied field E_0 is then

$$n \ll \frac{E_0}{4\pi e L}, \qquad (1\text{-}4\text{-}1)$$

where L is the relevant dimension of the apparatus. If the drift-field intensity E_0 is 2 V/cm and the drift distance L is 10 cm as typical values for mobility and diffusion measurements at low E/N, this inequality predicts significant space charge distortion of the applied field at ion densities of the order of 10^5 cm^{-3}, a value consistent with experimental evidence discussed in Chapter 2. (It may be of interest to point out that in most measurements of mobilities and diffusion coefficients the gas pressure is in the range 10^{-1} to 10 torr, corresponding to a neutral gas-number density of 10^{15} to 10^{17} cm^{-3}.) Larger ion densities may be tolerated in measurements at high field and in apparatus of smaller dimensions.*

We may also consider a second effect of ion-ion interactions produced by random fluctuations of the ionic-number density that may alter a velocity distribution derived on the assumption of a low ionic density. Neighboring ions are more important than remote ions in this connection, since their relative positions can fluctuate more rapidly. The basis of the disturbance is the randomly fluctuating Coulomb force that produces mutual scattering. This random force may be neglected if it is unable to produce a significant deflection in a single mean free path. Since the magnitude of the force is of the order $e^2/d^2 = e^2 n^{2/3}$, where d is the mean ionic spacing, the effect is small if $e^2 n^{2/3} \lambda \ll$ (mean ion energy). In the low-field region the thermal energy

---

*This treatment is valid only for the case of uniformly distributed ions. R. Johnsen (private communication, 1983) has considered the effect of space charge on the spreading of ion *clouds* and obtained some useful results that are reproduced below.

Consider an ion cloud containing S singly charged ions, initially confined to a small volume around a point $r = 0$. According to (1-7-6), at time t the rms displacement from $r = 0$ will be $R = (6Dt)^{1/2}$, where D is the ionic diffusion coefficient, and R will increase further with time as $dR/dt = 3D/R$. Roughly one-half of all of the ions (i.e., $S/2$) will be within the radius R, so the electric field in the radial direction will be

$$E = \frac{S}{2} \frac{e}{4\pi R^2}.$$

The electric field causes an ionic drift with velocity $v = KE$, where K is the ionic mobility. In order that this space-charge drift velocity be small compared with the rate of diffusive spreading, v must be much less than dR/dt. These equations, together with the Einstein relation, give approximately (to about 10%) the condition $S \ll 10^6 R$ (with R in centimeters, S a pure number). Hence for a 10% space-charge repulsion effect, we should have no more than 10^5 ions in a cloud of 1-cm radius. For a smaller cloud, say one of 0.1-cm radius, the number of ions permitted would be only 10^4. Clouds of this size are injected into drift tubes from external ion sources, for example (see Sections 2-5D and 2-5E).

dominates the field energy and the inequality becomes

$$e^2 n^{2/3} \ll pQ. \tag{1-4-2}$$

At high field the relative importance of the thermal and field energies is reversed and the criterion becomes

$$e^2 n^{2/3} \ll eE\left(\frac{M}{m} + \frac{m}{M}\right). \tag{1-4-3}$$

If a pressure of 1 torr and a collision cross section of $50 \times 10^{-16} \, \text{cm}^2$ are assumed, (1-4-2) yields $n \ll 10^{11}$ ions cm^{-3}; (1-4-3) gives similar results.

Throughout most of this book the assumption is made that the ionic-number density is low enough to neglect all ion-ion interactions. This assumption greatly simplifies the mathematical treatment of the ionic motion, for the equation for the velocity distribution function is then linear instead of quadratic (Wannier, 1953).

1-5. IMPORTANCE OF DATA ON IONIC MOBILITIES AND DIFFUSION COEFFICIENTS

Data on ionic mobilities and diffusion coefficients are of both theoretical and practical interest. First of all, experimental values of these quantities, and particularly their dependence on E/N and the gas temperature, can provide information about ion-molecule interaction potentials at greater separation distances than are generally accessible in beam-scattering experiments, as explained in Chapter 7. Second, mobilities are required for the calculation of ion-ion recombination coefficients (Flannery, 1982a, b) and the rate of dispersion of ions in a gas due to mutual repulsion (McDaniel, 1964), pp. 518–520, 575–582). Ionic transport data are also required for the proper analysis of various experiments on chemical reactions between ions and molecules (see Sections 2-4 and 3-1 for discussions of experiments performed with drift tubes). In addition, knowledge of the mobility of an ionic species in a given gas as a function of E/N permits the estimation of the average ionic energy as a function of this parameter (see the second footnote in Section 2-1). Finally, information on both mobilities and diffusion is required for a quantitative understanding of electrical discharges in gases and various atmospheric phenomena (Hirsh and Oskam, 1978; Meek and Craggs, 1978; Massey et al., 1982–1984).

To date, little has been done on the extraction of collision cross sections from experimental data on ionic mobilities and diffusion coefficients. However, at this time, accurate elastic scattering and momentum transfer cross sections can be obtained for atomic ion–atomic gas combinations as a function of impact energy provided that accurate transport properties are known over a wide range of E/N. It is also possible to extract cross sections for symmetric charge transfer from ion-mobility measurements on such systems as Ne^+ in neon gas. It appears likely that with further development of the theory of ionic transport it will become possible to obtain cross sections for inelastic and reactive scattering of ions as well. A much larger effort has gone into the corresponding problem for electrons. A substantial amount of data has been published on electron cross sections derived from electron swarm experiments, but discrepancies have appeared that require a thorough reexamination of the methods used to treat the experimental data. This subject has recently been reviewed by Chutjian and Garscadden (1988).

The transport properties of ions in gases have been studied experimentally since shortly after the discovery of X-rays in 1895 and theoretically since 1903. The first measurements were performed by Thomson, Rutherford, and Townsend at the Cavendish Laboratory of Cambridge University late in the nineteenth century (Thomson and Thomson, 1969). Surveys of the history of experimentation in this field appear in Loeb (1960) and Massey (1971) and are not repeated here; the calculations, however, are reviewed in Chapter 5.

Meaningful direct measurements of ionic diffusion coefficients were first made only in the 1960s and comparatively few data are available. On the other hand, some reliable mobility measurements were made as long ago as the 1930s and a large amount of good data is now on hand. Recent work has indicated, however, that most of the old data and even some of the newer results are either incorrect or refer to ions whose identities were not known. The main reason for this is that in most cases the drifting ions can undergo chemical reactions with the molecules of the gases through which they are moving and thereby change their identities. Techniques for obtaining reliable results, many of them developed only recently, are discussed in Chapters 2, 3, and 4.

1-6. DIFFERENCES IN BEHAVIOR OF IONS AND ELECTRONS

It is appropriate at this time to discuss the differences in behavior of ions and electrons in regard to their drift and diffusion in gases. As might be expected, electrons usually have much higher drift velocities and diffusion coefficients (by orders of magnitude) than ions under given conditions in a given gas. Because of their small mass, electrons are accelerated rapidly by an electric field, and they lose little energy in elastic collisions with molecules (a fraction of the order of m_e/M, where m_e and M are the electronic and molecular masses, respectively.) Therefore, electrons can acquire kinetic energy from an electric field faster than ions, and they can store this energy between collisions to a much greater degree until they reach energies at which inelastic collisions become important. Even with only a weak electric field imposed on the gas through which the electrons are moving, the average electronic energy may be far in excess of the thermal value associated with the gas molecules. Furthermore, the electronic energy distribution is not close to Maxwellian except at extremely low values of E/N.

Other differences between electrons and ions develop in connection with their collision cross sections (Massey and Burhop, 1969; Massey, 1969, 1971; Massey et al., 1974; Hasted 1972). Electronic excitation of atoms and molecules is frequently an important factor in electron collisions even for impact energies of less than 10 eV, and in molecular gases the onset of vibrational and rotational excitation occurs at energies far below 1 eV (Christophorou, 1984; McDaniel, 1989).* These energies are often attained by electrons in situations of common interest. The laboratory-frame thresholds for the corresponding modes of excitation by ions are higher than those for electrons, and the excitation cross sections peak at energies considerably above these thresholds (Bernstein, 1979). Therefore, ions have insufficient energy to produce much excitation under the usual gas kinetic conditions. These considerations tend to make the analysis of electronic motion in gases more difficult than the analysis of ionic motion. A compensating factor is operative, however, because of the relatively small mass of the electron. Since $m_e/M \ll 1$ in any gas, it is possible to make approximations in the analysis of electronic motion that are not valid in the ionic case. These approximations greatly simplify the mathematics and make it possible to calculate accurately the velocity distribution and transport properties of electrons in many gases at high E/N (Huxley and Crompton, 1974). This is not so with ions, as shown in Chapter 5.

On the experimental side certain other important differences appear. Electrons may be produced much more simply than ions by thermionic emission from filaments, by photoemission from surfaces, or by beta decay of radioactive isotopes. Ionic production usually requires the use of much more elaborate apparatus: electron bombardment or photoionization ion sources or an electrical discharge. Furthermore, in an electron-swarm experiment the electronic component of the charge carriers may be easily separated from any ionic component that may be present and no mass analysis is required to interpret the data. In ionic drift and diffusion experiments, on the other hand, mass analysis of the ions is usually essential if unambiguous results are to be obtained. This requirement involves a great complication of the apparatus. On the other hand, electron swarm experiments are usually more sensitive than ion experiments to electric field nonuniformities, contact potentials, and magnetic fields.

A final difference between electron and ion experiments relates to the effects of impurities in the gas being studied. Molecular impurities in an atomic gas can hold the average electronic energy well below the level that would be attained in the pure gas because electrons can lose large fractions of their energy by exciting the rotational and vibrational levels of the molecules. The electronic velocity distribution can be seriously altered in the process. In ionic experiments, however, impurities have little effect on the average ionic energy and velocity distribution. The complication that may develop instead is the production of impurity ions by the reaction of the ions of the main gas with impurity molecules. This is frequently a matter of serious concern.

---

*Another useful, more theoretical reference is Shimamura and Takayanagi (1984).

1-7. SPREADING OF A CLOUD OF IONS BY DIFFUSION THROUGH AN UNBOUNDED GAS

In this section and in Sections 1-8 through 1-11 we concern ourselves with matters related to the spatial dispersion of ions by diffusion through a gas at thermal energy under conditions of low ionization density in which the ions interact only with the gas molecules and not with other ions or electrons. It is assumed that the ions are incapable of undergoing chemical reactions with the gas molecules. The results to be obtained are useful in the analysis of various practical problems and certain types of experiment of interest to us in the present book. In Section 1-12 we consider ambipolar diffusion, which takes place when ions and electrons are both present at high number density, and in Section 2-6 we treat ion-molecule reactions.

First consider a number of ions, S, located at the origin of a one-dimensional coordinate system. If the ions are released at $t = 0$ and allowed to diffuse through a field-free gas filling all space at uniform pressure, the one-dimensional number density of the ions at distance x from the origin at time t is

$$n = \frac{S}{\sqrt{4\pi Dt}}\, e^{-x^2/4Dt}, \tag{1-7-1}$$

where D is the coefficient characterizing the diffusive motion of the ions through the gas.* This equation, as well as (1-1-5), is known as the Einstein relation. At any instant of time a plot of n as a function of x has the shape of a Gaussian error curve. The curve becomes progressively flatter as time elapses. The mean and root-mean-square displacements of the ions from the origin may be calculated from the distribution function in (1-7-1). The results are

$$|\bar{x}| = \frac{1}{S}\int_{-\infty}^{\infty} |x|n\,dx = \frac{2}{S}\int_{0}^{\infty} xn\,dx = \left(\frac{4Dt}{\pi}\right)^{1/2} \tag{1-7-2}$$

and

$$\sqrt{\overline{x^2}} = \left(\frac{1}{S}\int_{-\infty}^{\infty} x^2 n\,dx\right)^{1/2} = \sqrt{2Dt}. \tag{1-7-3}$$

In three dimensions the ionic number density at radius r and time t is

$$n = \frac{S}{(4\pi Dt)^{3/2}}\, e^{-r^2/4Dt}. \tag{1-7-4}$$

---

*Here S is assumed to be small enough that the total pressure will remain essentially constant throughout space.

The mean and root-mean-square displacements are

$$\bar{r} = \left(\frac{16Dt}{\pi}\right)^{1/2} \tag{1-7-5}$$

and

$$\sqrt{\overline{r^2}} = \sqrt{6Dt}, \tag{1-7-6}$$

respectively. In two dimensions

$$\sqrt{\overline{r^2}} = \sqrt{4Dt}. \tag{1-7-7}$$

The equations derived above are useful in the estimation of the average lifetime τ of ions against collision with the walls of a containing vessel. The expressions for the mean displacement indicate that

$$\tau \approx \frac{d^2}{D}, \tag{1-7-8}$$

where d is the relevant dimension of the container. The discussion in Section 1-11 permits the calculation of more accurate values of τ for various geometries. These results are

(a) for an infinitely long rectangular tube of width a and depth b

$$\tau = \left[D\pi^2\left(\frac{1}{a^2} + \frac{1}{b^2}\right)\right]^{-1}; \tag{1-7-9}$$

(b) for an infinitely long cylinder of radius r_0

$$\tau = \frac{1}{D}\left(\frac{r_0}{2.405}\right)^2; \tag{1-7-10}$$

(c) for a sphere of radius r_0

$$\tau = \frac{1}{D}\left(\frac{r_0}{\pi}\right)^2. \tag{1-7-11}$$

To illustrate the use of this concept we may use (1-7-10) to calculate the lifetime of an ion originating on the axis of a tube of 1-cm radius containing nitrogen at 1 torr pressure and room temperature. Taking the diffusion coefficient to be 50 cm^2/s, we find τ to be about 3×10^{-3} s. The total distance traveled during this time is equal to $\bar{v}\tau \approx 160$ cm, where $\bar{v}$ is the mean thermal velocity.

1-8. SPREADING OF AN ION CLOUD DURING ITS DRIFT IN AN ELECTRIC FIELD

It is also of interest to determine the extent of the diffusive spreading of a cloud of ions as it drifts through a gas under the influence of a weak electric field. Let L be the distance of drift during time t, v_d, the drift velocity, E, the electric-field intensity, and V, the potential difference between the extremities of the drift path. The mean displacement of the ions from the center of mass of the moving ion cloud is given by (1-7-2), whereas L is, of course, related to the drift time by the equation $L = v_d t$. Thus

$$\frac{|\bar{x}|}{L} = \left(\frac{4D}{\pi v_d L}\right)^{1/2}. \tag{1-8-1}$$

If we assume a temperature of $0°C$, (1-1-6) shows that $D = K/42.465$. Then, using the relationships $K = v_d/E$ and $E = V/L$, we find that

$$\frac{|\bar{x}|}{L} = \frac{0.173}{\sqrt{V}}. \tag{1-8-2}$$

The ratio of the spread of the ion cloud to the drift distance is thus independent of the diffusion coefficient and mobility and depends only on the total voltage drop experienced by the ions. It should be emphasized that only diffusion effects were considered in the foregoing development. Dispersion due to mutual Coulomb repulsion of the ions was neglected.

1-9. DIFFUSION EQUATION

Let us now consider an ensemble of ions diffusing through an infinite medium that contains no sources or sinks. By definition of the particle flux density $\mathbf{J}$ the net leakage outward through an arbitrarily shaped, imaginary closed surface within the medium is $\int \mathbf{J} \cdot d\mathbf{A}$. Gauss's law shows that this leakage may also be expressed by the integral $\int \mathbf{V} \cdot \mathbf{J} \, dv$, where the integration is performed over the volume bounded by the surface A. Then, if n denotes the number density of the ions, it follows that

$$\int \frac{\partial n}{\partial t} \, dv = -\int \mathbf{V} \cdot \mathbf{J} \, dv$$

or

$$\int \left(\frac{\partial n}{\partial t} + \mathbf{V} \cdot \mathbf{J}\right) dv = 0.$$

Since the choice of A was arbitrary, the integrand must itself vanish. Thus

$$\frac{\partial n}{\partial t} + \mathbf{V} \cdot \mathbf{J} = 0, \tag{1-9-1}$$

which is known as the equation of continuity.

According to Fick's law of diffusion,

$$\mathbf{J} = -D\,\mathbf{V}n. \tag{1-1-1}$$

Therefore,

$$\mathbf{V} \cdot \mathbf{J} = -\mathbf{V} \cdot (D\,\mathbf{V}n) \tag{1-9-2}$$

and the equation of continuity gives

$$\frac{\partial n}{\partial t} = \mathbf{V} \cdot (D\,\mathbf{V}n), \tag{1-9-3}$$

which is known as the time-dependent diffusion equation or Fick's second law. Note that (1-9-3) allows for the possible dependence of D on position [by its dependence on composition (Crank, 1980)]. In Section 2-6 we use an extension of this equation that allows for the drift of the ions in an electric field, chemical reactions, and the tensor nature of the diffusion coefficient at high E/N.

We are now in position to verify the form of the distribution functions given in (1-7-1) and (1-7-4). Direct substitution into (1-9-3) shows that both functions satisfy the diffusion equation.

Let us now suppose that some steady-state distribution of ions $n_0(x, y, z)$ has been established within a gas-filled container. To maintain steady-state conditions ions must be continuously supplied within the gas to replenish losses to the walls by diffusion. This replenishment can be accomplished by continuously ionizing the gas with X-rays or microwaves. Now imagine the ionization source to be abruptly turned off at $t = 0$. If we assume that D is independent of position and separate variables in (1-9-3) by writing

$$n(x, y, z, t) = n_0(x, y, z)T(t), \tag{1-9-4}$$

we obtain an equation for $T(t)$ whose solution is

$$T(t) = e^{-t/\tau}, \tag{1-9-5}$$

where τ is a time constant describing the decay for the kth diffusion mode. Equation (1-9-3) thus yields the time-independent diffusion equation

$$\nabla^2 n_0 + \frac{n_0}{D\tau} = 0. \tag{1-9-6}$$

The solution of (1-9-6) for $n_0(x, y, z)$ is an eigenvalue problem whose solution depends on the geometry of the container and the appropriate boundary conditions.

1-10. BOUNDARY CONDITIONS

Since the diffusion equation is a second-order differential equation, its general solution will contain two arbitrary constants of integration. In the solution of a specific problem the values of these constants are determined by boundary conditions and other physical considerations.

For ions diffusing in a gas-filled container the conditions usually imposed stipulate that the ionic number density must be finite everywhere within the gas and vanish "at" the walls of the container. If this statement is interpreted as meaning that the inward ionic flux density must be zero at the walls so that no ions will be reflected back into the gas on impact, diffusion theory actually requires that the number density vary near the wall in such a way that linear extrapolation would cause it to vanish at a definite distance, d, *beyond* the wall (Fig. 1-10-1). A rather lengthy calculation (McDaniel, 1964, pp. 496–497) shows that for a plane boundary

$$d = \tfrac{2}{3}\lambda, \qquad (1\text{-}10\text{-}1)$$

where λ is the mean free path for elastic scattering of the ions in the gas; d is usually called the linear extrapolation distance and is a measure of the extent by which the dimensions of the container are augmented for the mathematical treatment of the diffusion problem. Slightly different values of d are appropriate for different geometries.

In the study of ions and electrons diffusing in gases the extrapolation distance is negligibly small compared with the container dimensions and ignored in

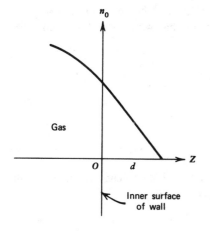

Figure 1-10-1. Linear extrapolation of the ionic number density n_0 past the physical boundary to obtain the extrapolation distance d. This procedure is dictated by the assumption that no reflection occurs at the wall; that is, the ions will become neutralized at the wall on impact if the wall is conducting or stick to the wall if it is an insulator.

calculations; that is, the boundary condition applied is that the ionic or electronic number density go to zero at the inner surface of the containing vessel. In the diffusion of neutrons, on the other hand, d is frequently of significant size and must be taken into account (Glasstone and Edlund, 1952).

1-11. SOLUTION OF THE TIME-INDEPENDENT DIFFUSION EQUATION FOR VARIOUS GEOMETRIES

Here we consider the steady-state diffusion of ions of a single species through a gas of uniform temperature and pressure filling containers of various shapes. In each case the ionic number density is negligible in comparison to the number density of the molecules. No electric field is assumed to be present, so the average ionic energy has the thermal value. Under the conditions described here D will be independent of position. We neglect the extrapolation distance and require n_0 to vanish at the geometrical boundaries of the containers.

A. Infinite Parallel Plates

As the first example, consider a one-dimensional cavity whose walls are the infinite plane parallel plates shown in Fig. 1-11-1. In this simple case the diffusion equation (1-9-6) becomes

$$\frac{d^2 n_0(x)}{dx^2} + \frac{n_0(x)}{D\tau} = 0. \qquad (1\text{-}11\text{-}1)$$

Since $D\tau$ is positive, the solution of (1-11-1) is

$$n_0(x) = A \cos \frac{x}{\sqrt{D\tau}} + B \sin \frac{x}{\sqrt{D\tau}}, \qquad (1\text{-}11\text{-}2)$$

where A and B are constants of integration that must be determined from the boundary conditions and from the requirement that we impose for symmetry

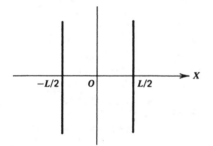

Figure **1-11-1.** One-dimensional cavity with plane parallel walls.

about the midplane. If the width of the cavity is L and the origin of the coordinate system is located at the midplane, the boundary conditions are $n_0(x) = 0$ when $x = \pm L/2$.

The symmetry requirement makes $B = 0$ and the boundary conditions force τ to assume one of the infinite number of values τ_k $(k = 1, 2, 3, \ldots)$ that satisfy the equation

$$\cos \frac{L}{2\sqrt{D\tau_k}} = 0 \quad \text{or} \quad \frac{L}{2\sqrt{D\tau_k}} = (2k - 1)\frac{\pi}{2}. \tag{1-11-3}$$

Now define a quantity Λ_k that represents the characteristic diffusion length for the kth mode of diffusion:

$$\Lambda_k^2 = D\tau_k = \left(\frac{1}{2k - 1}\frac{L}{\pi}\right)^2. \tag{1-11-4}$$

The diffusion length is useful in describing the shape of a cavity in the diffusion process. The solution for the kth mode can then be written

$$n_0(x)_k = A_k \cos \frac{x}{\Lambda_k}. \tag{1-11-5}$$

The function $\cos(x/\Lambda_k)$ assumes negative values in certain regions within the cavity for all modes of diffusion except the lowest, or fundamental, mode corresponding to $k = 1$. Therefore, if we consider each solution singly, we must discard all but the fundamental mode on physical grounds, since the ionic number density can never be negative. Since, however, the diffusion equation is linear, the total solution consists of an infinite number of modes, many of which may be excited simultaneously. Any sum of these modes is then a possible solution, provided the constants A_k have values that prevent the number density from becoming negative. The use of an ionization source that provides uniform ionization throughout the cavity will ensure that the fundamental mode predominates, but in many other experimental arrangements higher modes must be considered.

After the ionization source is abruptly turned off, say at $t = 0$, each diffusion mode decays out with its own characteristic time constant τ_k. The total solution of the time-dependent diffusion problem is thus given by

$$n(x, t) = \sum_{k=1}^{\infty} A_k \left(\cos \frac{x}{\Lambda_k}\right) e^{-t/\tau_k}. \tag{1-11-6}$$

Equation (1-11-3) shows that

$$\frac{\tau_1}{\tau_k} = (2k - 1)^2, \tag{1-11-7}$$

$\tau_1/\tau_2 = 9, \tau_1/\tau_3 = 25, \tau_1/\tau_4 = 49$, etc. Consequently, if higher modes are initially present, they will decay out much faster than the fundamental mode and only this mode will be observable after a time comparable with τ_1.* This fact obviously simplifies the analysis of experiments.

B.   Rectangular Parallelepiped

The next case to be treated is that of a cavity in the form of a rectangular parallelpiped (Fig. 1-11-2). Take the origin of rectangular Cartesian coordinates at the center of the cavity, whose x, y, and z dimensions are a, b, and c, respectively. The time-independent diffusion equation is now

$$\frac{\partial^2 n_0}{\partial x^2} + \frac{\partial^2 n_0}{\partial y^2} + \frac{\partial^2 n_0}{\partial z^2} + \frac{n_0}{D\tau} = 0 \qquad (1\text{-}11\text{-}8)$$

with the boundary conditions that $n_0 = 0$ when $x = \pm a/2$, $y = \pm b/2$, and $z = \pm c/2$. Expressing $n_0(x, y, z)$ as the product of three functions, each of which is a function of only one coordinate,

$$n_0(x, y, z) = X(x)Y(y)Z(z), \qquad (1\text{-}11\text{-}9)$$

we may separate the variables in the diffusion equation and obtain

$$\frac{1}{X}\frac{d^2 X}{dx^2} + \frac{1}{Y}\frac{d^2 Y}{dy^2} + \frac{1}{Z}\frac{d^2 Z}{dz^2} + \frac{1}{D\tau} = 0. \qquad (1\text{-}11\text{-}10)$$

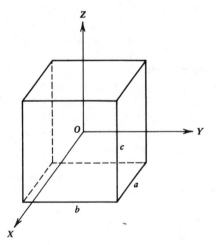

Figure 1-11-2. Cavity with the shape of a rectangular parallelepiped.

*The distance between adjacent regions of high and low particle density decreases as the order of the diffusion mode increases. Hence, in higher modes, the particles need to diffuse over a shorter distance in order to smooth out inequalities in the relative concentration.

Since $D\tau$ is a constant and each of the first three terms is a function of only one variable, we may equate each of them to a separate constant:

$$\frac{1}{X}\frac{d^2 X}{dx^2} = -\alpha^2, \qquad \frac{1}{Y}\frac{d^2 Y}{dy^2} = -\beta^2, \qquad \frac{1}{Z}\frac{d^2 Z}{dz^2} = -\gamma^2. \qquad (1\text{-}11\text{-}11)$$

Equation (1-11-10) shows that

$$\alpha^2 + \beta^2 + \gamma^2 = \frac{1}{D\tau}. \qquad (1\text{-}11\text{-}12)$$

Since there is no essential difference between the x, y, and z directions in this problem and since $\alpha^2 + \beta^2 + \gamma^2$ equals a positive quantity, it follows that α^2, β^2, and γ^2 separately must be positive. Using the boundary conditions and symmetry requirements, we see that the solutions of (1-11-11) are

$$X_i = A_i \cos\frac{(2i-1)\pi x}{a}, \quad Y_j = B_j \cos\frac{(2j-1)\pi y}{b}, \quad Z_k = C_k \cos\frac{(2k-1)\pi z}{c},$$

$$(1\text{-}11\text{-}13)$$

where i, j, and k each may assume any positive integral values. The total solution to the time-dependent problem then has the form

$$n(x, y, z, t) = \sum_{i,j,k=1}^{\infty} G_{ijk} \cos\frac{(2i-1)\pi x}{a} \cos\frac{(2j-1)\pi y}{b} \cos\frac{(2k-1)\pi z}{c} e^{-t/\tau_{ijk}},$$

$$(1\text{-}11\text{-}14)$$

where the three arbitrary constants have been lumped into G_{ijk}. Specification of the mode of diffusion now requires three indices, and corresponding to this triad of indices and this mode of diffusion is a time constant τ_{ijk} given by

$$\frac{1}{\tau_{ijk}} = D\pi^2 \left[\left(\frac{2i-1}{a}\right)^2 + \left(\frac{2j-1}{b}\right)^2 + \left(\frac{2k-1}{c}\right)^2 \right]. \qquad (1\text{-}11\text{-}15)$$

The diffusion length is now given by

$$\Lambda_{ijk}^2 = D\tau_{ijk}. \qquad (1\text{-}11\text{-}16)$$

If the cavity is cubical, $a = b = c$ and

$$\frac{\tau_{111}}{\tau_{211}} = 3.67, \qquad \frac{\tau_{111}}{\tau_{311}} = 9, \qquad \frac{\tau_{111}}{\tau_{411}} = 17.$$

Here the higher modes persist longer in relation to the fundamental mode than in the one-dimensional case and thus their effect is enhanced.

C. Spherical Cavity

Now consider a spherical cavity of radius r_0 (Fig. 1-11-3). For spherical geometry the diffusion equation is

$$\frac{\partial^2 n_0}{\partial r^2} + \frac{2}{r}\frac{\partial n_0}{\partial r} + \frac{1}{r^2 \sin\theta}\frac{\partial}{\partial\theta}\left(\sin\theta\,\frac{\partial n_0}{\partial\theta}\right) + \frac{1}{r^2\sin^2\theta}\frac{\partial^2 n_0}{\partial\phi^2} + \frac{n_0}{D\tau} = 0, \qquad (1\text{-}11\text{-}17)$$

but since there is no preferred direction here we reduce (1-11-17) to

$$\frac{\partial^2 n_0}{dr^2} + \frac{2}{r}\frac{dn_0}{dr} + \frac{n_0}{D\tau} = 0, \qquad (1\text{-}11\text{-}18)$$

discarding in the process all but the fundamental angular mode. To solve this equation easily we put $n_0 = u/r$. We then obtain

$$\frac{d^2 u}{dr^2} + \frac{u}{D\tau} = 0 \qquad (1\text{-}11\text{-}19)$$

whose solution is

$$u = A\cos\frac{r}{\sqrt{D\tau}} + B\sin\frac{r}{\sqrt{D\tau}}, \qquad (1\text{-}11\text{-}20)$$

since $D\tau$ is positive. Thus the solution for n_0 has the form

$$n_0 = \frac{A}{r}\cos\frac{r}{\sqrt{D\tau}} + \frac{B}{r}\sin\frac{r}{\sqrt{D\tau}}, \qquad (1\text{-}11\text{-}21)$$

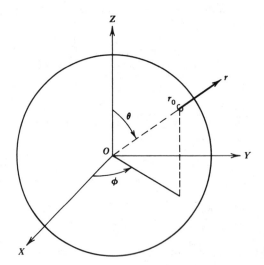

Figure 1-11-3. Spherical cavity.

where evidently A must be zero in order that n_0 may remain finite at the origin. The final time-dependent solution is then

$$n(r, t) = \sum_{k=0}^{\infty} \frac{B_k}{r} \sin \frac{r}{\sqrt{D\tau_k}} e^{-t/\tau_k},\qquad (1\text{-}11\text{-}22)$$

where

$$\frac{r_0}{\sqrt{D\tau_k}} = k\pi \qquad (k = 0, 1, 2, 3, \ldots).\qquad (1\text{-}11\text{-}23)$$

The diffusion length Λ_k is given by the equation

$$\Lambda_k^2 = D\tau_k = \left(\frac{r_0}{\pi k}\right)^2.\qquad (1\text{-}11\text{-}24)$$

D. Cylindrical Cavity

As a final example let us treat a cavity in the form of a right circular cylinder of radius r_0 and height H (Fig. 1-11-4). If we assume symmetry about the axis, there is no dependence on the azimuth angle θ, and the diffusion equation

$$\frac{\partial^2 n_0}{\partial r^2} + \frac{1}{r}\frac{\partial n_0}{\partial r} + \frac{1}{r^2}\frac{\partial^2 n_0}{\partial \theta^2} + \frac{\partial^2 n_0}{\partial z^2} + \frac{n_0}{D\tau} = 0 \qquad (1\text{-}11\text{-}25)$$

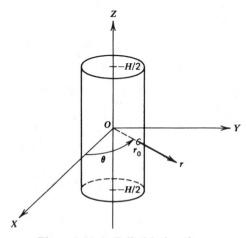

Figure 1-11-4. Cylindrical cavity.

reduces to

$$\frac{\partial^2 n_0}{\partial r^2} + \frac{1}{r}\frac{\partial n_0}{\partial r} + \frac{\partial^2 n_0}{\partial z^2} + \frac{n_0}{D\tau} = 0 \qquad (1\text{-}11\text{-}26)$$

(for the fundamental angular mode). We separate variables by writing

$$n_0(r, z) = R(r)Z(z) \qquad (1\text{-}11\text{-}27)$$

and obtain

$$\frac{1}{R}\left(\frac{d^2 R}{dr^2} + \frac{1}{r}\frac{dR}{dr}\right) + \frac{1}{Z}\frac{d^2 Z}{dz^2} + \frac{1}{D\tau} = 0. \qquad (1\text{-}11\text{-}28)$$

The first term depends only on r and the second only on z, and since $D\tau$ is a constant each of these terms must be equal to a constant. Set

$$\frac{1}{R}\left(\frac{d^2 R}{dr^2} + \frac{1}{r}\frac{dR}{dr}\right) = -\alpha^2 \qquad (1\text{-}11\text{-}29)$$

and

$$\frac{1}{Z}\frac{d^2 Z}{dz^2} = -\beta^2 \qquad (1\text{-}11\text{-}30)$$

so that

$$\alpha^2 + \beta^2 = \frac{1}{D\tau}. \qquad (1\text{-}11\text{-}31)$$

We must now determine whether α^2 and β^2 are positive or negative.

In solving the r equation (1-11-29) it is convenient to make the substitution $r = u/\alpha$. We then obtain the equation

$$u^2 \frac{d^2 R}{du^2} + u\frac{dR}{du} + u^2 R = 0. \qquad (1\text{-}11\text{-}32)$$

Now the general Bessel equation of order n is

$$x^2 \frac{d^2 y}{dx^2} + x\frac{dy}{dx} + (x^2 - n^2)y = 0, \qquad (1\text{-}11\text{-}33)$$

where $(x^2 - n^2)$ is a positive quantity, whereas the modified Bessel equation of order n is

$$x^2 \frac{d^2 y}{dx^2} + x\frac{dy}{dx} - (x^2 - n^2)y = 0. \qquad (1\text{-}11\text{-}34)$$

We see that (1-11-32) is a Bessel equation of order zero, unmodified if α^2 is positive, modified if α^2 is negative. In the first instance the general solution is

$$R = AJ_0(u) + BY_0(u),\qquad\qquad (1\text{-}11\text{-}35)$$

where J_0 and Y_0 are the Bessel functions of the first and second kinds, respectively, of order zero (McLachlan, 1934). In the second instance the solution is

$$R = A'I_0(u) + B'K_0(u),\qquad\qquad (1\text{-}11\text{-}36)$$

where I_0 and K_0 are modified Bessel functions of the first and second kinds, respectively, of order zero. By reference to Fig. 1-11-5 we see that the only satisfactory solution is J_0 and the only possible solution for the r part of our diffusion problem is

$$R(r) = AJ_0(u) = AJ_0(\alpha r).\qquad\qquad (1\text{-}11\text{-}37)$$

Thus α^2 is required to be positive.

By applying the boundary condition that n_0 must vanish at $r = r_0$ we see that $R(r_0) = AJ_0(\alpha r_0) = 0$. The first zero of J_0 occurs at $\alpha r = 2.405$, so that, for the fundamental mode, $\alpha_1 = 2.405/r_0$ and $R(r)$ is given by

$$R(r) = AJ_0\frac{2.405r}{r_0}.\qquad\qquad (1\text{-}11\text{-}38)$$

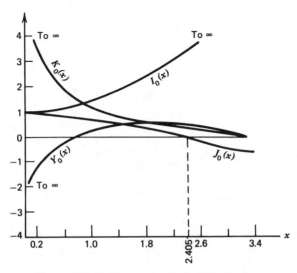

Figure 1-11-5. Zero-order Bessel functions.

Turning now to the z equation

$$\frac{d^2Z}{dz^2} + \beta^2 Z = 0, \tag{1-11-39}$$

we observe that if β^2 is positive the solutions will be $\cos \beta z$ and $\sin \beta z$, whereas if β^2 is negative the solutions are $\cosh \beta z$ and $\sinh \beta h$. Figure 1-11-6 shows that the hyperbolic functions are unacceptable because they would provide a greater ionic number density at the top than at the center of the cavity. The $\sin \beta z$ solution must be discarded because it is an odd function of z. The only remaining possibility is the cosine solution and β^2 must be positive. The solution of the z equation is then

$$Z(z) = C \cos \beta z. \tag{1-11-40}$$

The boundary conditions that n_0 must vanish at $z = \pm H/2$ require that $\beta_1 = \pi/H$ for the fundamental mode.

The time-dependent solution for the lowest mode of diffusion can now be written

$$n(r, z, t) = G_{11} J_0 \frac{2.405r}{r_0} \left(\cos \frac{\pi z}{H} \right) e^{-t/\tau_{11}}, \tag{1-11-41}$$

where

$$\frac{1}{\Lambda_{11}^2} = \frac{1}{D\tau_{11}} = \left(\frac{2.405}{r_0} \right)^2 + \left(\frac{\pi}{H} \right)^2 \tag{1-11-42}$$

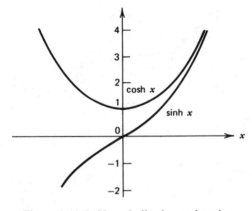

Figure 1-11-6. Hyperbolic sine and cosine.

The total solution which contains the radial higher modes as well as the fundamental is

$$n(r, z, t) = \sum_{i,j=1}^{\infty} G_{ij} J_0(\alpha_i r) \left[\cos \frac{(2j-1)\pi z}{H} \right] e^{-t/\tau_{ij}}. \qquad (1\text{-}11\text{-}43)$$

The diffusion length is given by

$$\frac{1}{\Lambda_{ij}^2} = \frac{1}{D\tau_{ij}} = \alpha_i^2 + \left[\frac{(2j-1)\pi}{H} \right]^2, \qquad (1\text{-}11\text{-}44)$$

where $\alpha_i r_0$ is the ith root of J_0.

1-12. AMBIPOLAR DIFFUSION

We now wish to consider a gas-filled cavity in which both electrons and positive ions are diffusing toward the walls. Usually in studies of such cavities, the interaction between the negative and positive particles can be neglected below ionization densities of about 10^7 to 10^8 cm^{-3}, but above this level space charge effects produced by the Coulomb forces between the electrons and positive ions become important and must be taken into account.

It may be shown that the number density of electrons in a highly ionized gas must approximate the number density of positive ions at each point, provided we are not within about 1 debye length* of a boundary (McDaniel, 1964, App. I; Krall and Trivelpiece, 1973; Seshadri, 1973; Golant et al., 1980; Nicholson, 1983). Any deviation from charge equality produces electrical forces that oppose the charge separation and tend to restore the balance. Because their diffusion coefficient is much higher than that of the ions, the electrons attempt to diffuse more rapidly than the ions toward regions of lower concentration but their motion is impeded by the restraining space charge field thereby created. This same field has the opposite effect on the ions and causes them to diffuse at a faster rate than they would in the absence of the electrons. Both species of charged particles consequently diffuse with the same velocity, and since there is now no difference in the flow of the particles of opposite sign the diffusion is called "ambipolar." The concept of ambipolar diffusion was introduced by Schottky in 1924 in an analysis of the positive column of the glow discharge (Schottky, 1924).

We now derive an expression for the coefficient of ambipolar diffusion. Let n represent the common number density of the electrons and positive ions and v_a, the velocity of ambipolar diffusion. We assume that the gas pressure is high

---

*The debye length is a measure of the distance over which deviations from charge neutrality can occur in an ionized gas. It is directly proportional to the square root of the energy and inversely proportional to the square root of the number density of the charged particles in the ionized gas.

enough for the particles to make frequent collisions. The mobility concept will then be assumed to apply, not only for the ions but for the electrons as well. Let **E** denote the intensity of the electric field established by the charge separation. Since the velocity of diffusion is the same for both species, we have

$$v_a = -\frac{D^+}{n}\frac{dn}{dx} + K^+ E \qquad\qquad (1\text{-}12\text{-}1)$$

and

$$v_a = -\frac{D^-}{n}\frac{dn}{dx} - K^- E \qquad\qquad (1\text{-}12\text{-}2)$$

where K^+ and K^- are the mobilities of the ions and electrons, respectively, and D^+ and D^- are their ordinary or "free" diffusion coefficients. All four coefficients are positive numbers. By eliminating E we obtain

$$v_a = -D_a\frac{1}{n}\frac{dn}{dx} \qquad\qquad (1\text{-}12\text{-}3)$$

where D_a is the coefficient of ambipolar diffusion defined by the equation

$$D_a = \frac{D^+ K^- + D^- K^+}{K^+ + K^-}; \qquad\qquad (1\text{-}12\text{-}4)$$

D_a characterizes the diffusive motion of both species.

If we assume that $K^- \gg K^+$ and $T^- \gg T^+$ and use the Einstein relation

$$\frac{D}{K} = \frac{kT}{e}, \qquad\qquad (1\text{-}1\text{-}5)$$

we find that

$$D_a \approx D^-\frac{K^+}{K^-} = \frac{kT^-}{e}K^+. \qquad\qquad (1\text{-}12\text{-}5)$$

When $T^+ = T^- = T$, on the other hand,

$$D_a \approx 2D^+ = \frac{2kT}{e}K^+. \qquad\qquad (1\text{-}12\text{-}6)$$

The time-dependent diffusion equation for the ambipolar case is

$$\frac{\partial n}{\partial t} = \nabla\cdot(D_a\nabla n). \qquad\qquad (1\text{-}12\text{-}7)$$

If D_a is independent of position and the particle number density is assumed to decay as $e^{-t/\tau}$, the time-independent ambipolar diffusion equation is obtained:

$$\nabla^2 n_0 + \frac{n_0}{D_a \tau} = 0. \tag{1-12-8}$$

This equation is solved for specific problems by the methods in Section 1-11: D_a is given in terms of the decay constant τ, and the appropriate diffusion length Λ, by the equation

$$D_a = \frac{\Lambda^2}{\tau}. \tag{1-12-9}$$

Hence D_a may be evaluated from a determination of the rate of decay of the charged particle density in a cavity after the ionization source has been turned off. Techniques for measuring D_a are discussed in Chapter 4. If the electron, ion, and gas temperatures are equal, (1-12-6) shows that the ionic diffusion coefficient and mobility may be obtained from the measured value of D_a. The zero-field reduced mobility K_0 is related to the ambipolar diffusion coefficient D_a by the equation

$$K_0 = \frac{D_a p}{T^2} 2.086 \times 10^3. \tag{1-12-10}$$

Here p is the pressure in torr and T is the gas temperature in kelvin at which D_a was measured; K_0 is expressed in square centimeters per volt per second and D_a, in square centimeters per second.

REFERENCES

Bernstein, R. B., Ed. (1979). *Atom-Molecule Collision Theory*, Plenum, New York.

Christophorou, L. G., Ed. (1984). *Electron-Molecule Interactions and Their Applications*, 2 vols., Academic, Orlando, Fla.

Chutjian, A. and A. Garscadden (1988). Electron attachment at low electron energies, *Phys. Rep.* in press.

Crank, J. (1980). *The Mathematics of Diffusion*, 2nd ed., Clarendon, Oxford.

Einstein, A. (1905). Über die von der molekular-kinetischen Theorie der Wärme geforderte Bewegung von in ruhenden Flüssigkeiten suspendierten Teilchen., *Ann. Phys. Leipzig* 17, 549–560. Also see Einstein, A. (1908). Elementare Theorie der Brownschen Bewegung, *Z. Elektrochem.* 14, 235–239. English translations of these papers appear in *Investigations on the Theory of the Brownian Movement by Albert Einstein*, R. Fürth, Ed., Dover, New York, 1956, pp. 1–18, 68–85.

Elford, M. T. (1971). An observed pressure dependence of the mobility of potassium ions in gases, *Aust. J. Phys.* **24**, 705–718.

Elford, M. T., and H. B. Milloy (1974). Pressure dependence and end effects in precision ion mobility studies, *Aust. J. Phys.* **27**, 211–225.

Fick, A. (1855). Über Diffusion, *Ann. Phys. Chem.* **94**, 59–86.

Flannery, M. R. (1982a). "Ion-ion recombination in high-pressure plasmas," in *Applied Atomic Collision Physics*, Vol. 3, H. S. W. Massey, E. W. McDaniel, and B. Bederson, Eds., Academic, New York.

Flannery, M. R. (1982b). Theory of ion-ion recombination, *Philos. Trans. R. Soc. London* **A304**, 447–497.

Glasstone, S., and M. C. Edlund (1952). *Elements of Nuclear Reactor Theory*, Van Nostrand, New York.

Golant V. E., A. P. Zhilinsky, and I. E. Sakharov (1980). *Fundamentals of Plasma Physics*, Wiley, New York.

Hasted, J. B. (1972). *Physics of Atomic Collisions*, 2nd ed., American Elsevier, New York.

Hirsch, M. N., and H. J. Oskam, Eds. (1978). *Gaseous Electronics*, Vol. 1, Academic, New York.

Hirschfelder, J. O., C. F. Curtiss, and R. B. Bird (1964). *Molecular Theory of Gases and Liquids*, Wiley, New York.

Huxley, L. G. H., and R. W. Crompton (1974). *The Diffusion and Drift of Electrons in Gases*, Wiley, New York.

Huxley, L. G. H., R. W. Crompton, and M. T. Elford (1966). Use of the parameter E/N, *Br. J. Appl. Phys.* **17**, 1237–1238.

Johnsen, R., and M. A. Biondi (1972). Reaction rates of uranium ions and atoms with O_2 and N_2, *J. Chem. Phys.* **57**, 1975–1979.

Krall, N. A., and A W Trivelpiece (1973). *Principles of Plasma Physics*, McGraw-Hill, New York.

Lin, S. L., L. A. Viehland, E. A. Mason, J. H. Whealton, and J. N. Bardsley (1977). Velocity and energy relaxation of ions in drift tubes, *J. Phys.* **B10**, 3567–3575.

Loeb, L. B. (1960). *Basic Processes of Gaseous Electronics*, 2nd ed., University of California Press, Berkeley, Calif., Chaps. 1 and 2.

McDaniel, E. W. (1964a), *Collision Phenomena in Ionized Gases*, Wiley, New York.

McDaniel, E. W. (1972). The effect of clustering on the measurement of the mobility of potassium ions in nitrogen gas, *Aust. J. Phys.* **25**, 465–467.

McDaniel, E. W. (1989). *Electron and Photon Collisions*, Wiley, New York. A second volume dealing with heavy particle collisions and swarm phenomena is in preparation.

McLachlan N. W. (1934). *Bessel Functions for Engineers*, Clarendon, Oxford.

Massey, H. S. W. (1969). *Electron Collisions with Molecules and Photo-ionization* (2nd ed.), Vol. 2 of *Electronic and Ionic Impact Phenomena*, H. S. W. Massey, E. H. S. Burhop, and H. B. Gilbody, Clarendon, Oxford.

Massey, H. S. W. (1971). *Slow Collisions of Heavy Particles*, 2nd ed., Vol. 3 of *Electronic and Ionic Impact Phenomena*, H. S. W. Massey, E. H. S. Burhop, and H. B. Gilbody, Clarendon, Oxford.

Massey, H. S. W., and E. H. S. Burhop (1969). *Collisions of Electrons with Atoms*, 2nd ed., Vol. 1 of *Electronic and Ionic Impact Phenomena*, H. S. W. Massey, E. H. S. Burhop, and H. B. Gilbody, Clarendon, Oxford.

Massey, H. S. W., and H. B. Gilbody (1974). *Recombination and Fast Collisions of Heavy Particles*, 2nd ed., Vol. 4 of *Electronic and Ionic Impact Phenomena*, H. S. W. Massey, E. H. S. Burhop, and H. B. Gilbody, Clarendon, Oxford.

Massey, H. S. W., E. W. McDaniel, and B. Bederson, Eds. (1982–1984). *Applied Atomic Collision Physics*, 5 Vols., Academic, New York.

Meek, J. M., and J. D. Craggs, Eds. (1978). *Electrical Breakdown of Gases*, Wiley, New York.

Nernst, W. (1888). Zur Kinetik der in Lösung befindlichen Körper: I. Theorie der Diffusion, *Z. Phys. Chem.* **2**, 613–637.

Nicholson, D. R. (1983). *Introduction to Plasma Theory*, Wiley, New York, pp. 1–3, 211–219.

Schottky, W. (1924). Diffusionstheorie der positiven Säule, *Phys. Z.* **25**, 635–640.

Seshadri, S. R. (1973). *Fundamentals of Plasma Physics*, American Elsevier, New York.

Shimamura, I., and K. Takayanagi, Eds. (1984). *Electron-Molecule Collisions*, Plenum, New York.

Stacey, W. M. (1981). *Fusion Plasma Analysis*, Wiley, New York.

Thomson, J. J., and G. P. Thomson (1969). *Conduction of Electricity through Gases*, Vol. 1, Dover, New York (unrevised reprint of 1928 3rd ed.).

Thomson, G. M., J. H. Schummers, D. R. James, E. Graham, I. R. Gatland, M. R. Flannery, and E. W. McDaniel (1973). Mobility, diffusion, and clustering of K^+ ions in gases, *J. Chem. Phys.* **58**, 2402–2411.

Townsend, J. S. (1899). The diffusion of ions into gases, *Philos. Trans. R. Soc. London* **A193**, 129–158.

Wannier, G. H. (1953). Motion of gaseous ions in strong electric fields, *Bell Syst. Tech. J.* **32**, 170–254.

2

MEASUREMENT OF DRIFT VELOCITIES AND LONGITUDINAL DIFFUSION COEFFICIENTS

Direct measurements of drift velocities and longitudinal diffusion coefficients are made with apparatus called drift tubes, descriptions of which are given in Section 2-5. The discussion of drift tubes in this book relates exclusively to their use in the study of ions. Drift tubes are also utilized to investigate the transport properties of electrons in gases and the attachment of electrons to form negative ions, but the techniques in electron studies are somewhat different from those described here. The reader is referred to Huxley and Crompton (1974) for an authoritative discussion of electron drift-tube research.* As we show, the transport of slow ions through gases is closely coupled to the chemical reactions that may occur between the ions and gas molecules, and therefore we must consider in some detail the complicating effects of ion-molecule and charge transfer reactions on drift-tube measurements.† This complication, however, is more than compensated for by the fact that drift tubes may be used for the quantitative study of such reactions, and drift-tube methods of determining ionic rate coefficients are described in this chapter and in Chapter 3. The use of drift tubes for the measurement of transverse diffusion coefficients is treated in Chapter 3. The techniques described there are quite different from those under consideration here.

Drift tubes were first used at the end of the nineteenth century and have been applied to the study of ionic drift velocities almost continuously since that time

---

*It is appropriate to point out that many of the significant advances in electron drift-tube research made during recent years were due to L. G. H. Huxley, R. W. Crompton, and M. T. Elford in Australia and to A. V. Phelps and his colleagues at the Westinghouse Research Laboratories.

†We use the term "ion-molecule reaction" to refer to a heavy particle rearrangement reaction such as $A^+ + BC \rightarrow AB^+ + C$ or $A^- + B + C \rightarrow AB^- + C$. By a "charge transfer reaction" we mean a reaction in which an electron is transferred between the colliding structures, as in $A^+ + B \rightarrow A + B^+$ or $A^- + BC \rightarrow A + BC^-$. Both kinds of reaction are called "chemical." The book edited by Lindinger et al. (1984) is a useful reference on ionic reactions in drift tubes.

31

(Loeb, 1960; Massey, 1971). The early techniques were understandably crude, and few of the data obtained with drift tubes before 1930 are now of other than historical interest. However, noteworthy advances in instrumentation and gas-purification techniques made by Tyndall and his collaborators at Bristol University in England and by Bradbury and Nielsen in the United States during the 1930s led to a significant amount of good mobility data during that decade. Nonetheless, most of the reliable results on hand now are of much more recent vintage. General techniques for obtaining accurate mobilities of known ionic species in gases in which the ions are involved in reactions with the gas molecules became available only in the 1960s with the development of drift-tube mass spectrometers, examples of which are given in Section 2-5. For reasons discussed in this chapter, accurate ionic diffusion coefficients are much more difficult to measure than mobilities, and the first reliable ionic diffusion data for energies above thermal were reported only in the late 1960s. Drift tubes were used to obtain these results. Few meaningful drift-tube measurements of ionic reaction rates were reported before 1960; the widespread use of drift tubes to obtain ionic rate coefficients began only with the advent of drift-tube mass spectrometers. The rates of scores of ion-molecule and charge transfer reactions have now been measured with these instruments.

2-1.   GENERAL CONSIDERATIONS IN DRIFT-TUBE EXPERIMENTS

A conventional ionic drift tube usually consists of an enclosure containing gas, an ion source positioned on the axis of the enclosure, a set of electrodes that establishes a uniform axial electrostatic field along which the ions drift, and a current-measuring collector in the gas at the end of the ionic drift path. The drift field **E** causes the ions of any given molecular composition to "swarm" through the gas with a drift velocity and diffusion rate determined by the nature of the ions and gas molecules, the field strength, and the gas pressure and temperature. For the determination of drift velocities and longitudinal diffusion coefficients the ion source is operated in a repetitive, pulsed mode, and the spectrum of arrival times of the ions at the collector is measured electronically. (Other measurement techniques are required for the determination of transverse diffusion coefficients; they are described in Chapter 3.) A drift-tube mass spectrometer differs from a conventional drift tube by having the ion collector in the gas replaced by a sampling orifice located in the wall at the end of the drift tube, usually on the axis. Ions arriving at the end plate close to the axis pass through the orifice, out of the drift tube, and into an evacuated region that contains a mass spectrometer and an ion detector (usually a pulse-counting electron multiplier). The mass spectrometer can be set to transmit any one of the various ionic species entering it, the other species of ions being rejected in the mass selection process. Thus the arrival-time spectrum can be mapped separately for each kind of ion arriving at the end of the drift tube.

The pressures at which ionic drift-tube experiments have been performed have ranged from about 2.5×10^{-2} torr to above 1 atm. On the assumption of a cross section of 50×10^{-16} cm^2 for collisions of the ions with gas molecules, the ionic mean free path corresponding to a pressure of 2.5×10^{-2} torr is 0.23 cm; the mean free path at 1 atm is 7.4×10^{-6} cm. Most measurements are now made in the pressure range of 7.5×10^{-2} to 10 torr. At pressures lower than the minimum quoted here the ions would make collisions too few for steady-state conditions to be achieved except in apparatus of uncommonly great length. Ion sampling difficulties may appear in drift-tube mass spectrometer measurements at pressures greater than about 10 torr.* The product of the gas pressure p and the drift distance d should be large enough that the ions will travel a negligible fraction of the total distance before energy equilibration in the drift field is achieved. Drift distances of 0.5 to 44 cm have been employed. Because of the large pd products used, each ion makes many collisions with molecules as it drifts the distance d; the average number of collisions per ion usually lies somewhere between 10^2 and 10^7, depending on the value of pd.

Drift-tube measurements are made at ratios of drift field intensity to gas number density (E/N) as low as about 0.3 Td and as high as about 5000. Below $E/N \approx 6$ Td we would expect to be within the "low-field" region (see Section 1-2), in which most of the applications lie and in which the theory of the transport phenomena is easily applied. In a given experiment the lowest value of E/N to which measurements may be extended will depend on intensity considerations, that is, on the ion current reaching the detector and the sensitivity of the detector. The measured ion signal eventually becomes too small for accurate measurements to be made. In many experiments it has proved impossible even to approach the low-field region. The high E/N limit in drift-tube experiments is usually imposed by electrical breakdown within the apparatus. At very high E/N the average ionic energy may attain values of the order of 10 eV† and

---

*By the use of a small exit aperture at the end of the drift tube and fast differential pumping ions may be sampled mass spectrometrically at drift-tube pressures of the order of 1 atm. At high drift-tube pressures, however, molecules may form clusters about the ions in the expanding jet of gas passing through the exit aperture and thereby falsify the mass spectrum (see Milne and Greene, 1967; Hagena and Obert, 1972; Beuhler and Friedman, 1982). Eisele and his colleagues have successfully used drift tube–mass spectrometric techniques to sample and study tropospheric ions at pressures up to 760 torr (see, e.g., Eisele and McDaniel, 1986; Eisele, 1987).

†The equation generally used to calculate the average ionic energy is Wannier's expression

$$\frac{m\overline{v_i^2}}{2} = \frac{mv_d^2}{2} + \frac{Mv_d^2}{2} + \frac{3kT}{2}, \tag{2-1-1}$$

where m and M are the ionic and molecular masses, respectively, $\overline{v_i^2}$ is the mean square of the total ionic velocity, and v_d is the measured drift velocity (Wannier, 1953). This equation was derived for high E/N on the assumption of a constant mean free time scattering model (see Chapter 5). The first term on the right side is the field energy associated with the drift motion of the ion; the second term is the random part of the field energy. The last term represents the thermal energy. This equation illustrates the capacity that light ions in a heavy gas have for storing energy in the form of random inelastic collisions may be important. There is little interest in pushing drift-tube measurements to the highest possible energies, for the data then become difficult to interpret in terms of the transport theory and there appear to be few practical applications in which the results may be used. Indeed, at very high E/N we are more interested in the binary collision properties of the ions than in their transport properties, and ion-molecule collisions in the electron-volt energy range are more accurately studied by beam techniques (Massey et al., 1974; Bowers, 1979; Bernstein, 1979) than by the "swarm" methods described here.

At this point, however, it is of interest to note that conventional single beam-static gas or crossed-beam techniques cannot easily be applied to the study of ionic collisions at laboratory frame ion-beam energies below about 0.5 eV. This limitation develops from the effects of spurious electric fields produced by space charge in the ion beams, contact potentials, and charge accumulation on insulating surfaces. These fields can disperse and deflect slow ion beams and in the process alter the energy of the ions in the beams by unknown amounts.* Drift tubes, on the other hand, are not subject to this limitation if reasonable precautions are observed because in a well-designed experiment the spurious electric fields are much weaker than the applied electric drift field even in the low-field (thermal energy) region. Hence the spurious fields will not seriously

*The recently developed merging-beams method may, however, be used to study many kinds of collisions at impact energies well below 1 eV (Neynaber, 1969, 1980; Gentry et al., 1975; Moseley et al., 1975b; Rundel et al., 1979; Dolder and Peart, 1986). In studies of ion-molecule collisions, a fast beam of ions is merged coaxially in a vacuum with a fast beam of molecules whose velocity is slightly different from that of the ions. Each of the two beams is usually given a laboratory frame kinetic energy in the kiloelectron-volt range so that problems with spurious fields will not appear. The center-of-mass impact energy of the ion-molecule collisions which do occur, however, can be made as low as about 0.15 eV. Furthermore, the spread in the impact energy is much smaller than the laboratory-frame energy spreads of the merging beams and good energy resolution can be obtained.

motion. For ions traveling in a gas composed of molecules of the same mass as that of the ions the ordered and random field energies are equal. For heavy ions in a light gas the random field energy is negligible.

If m and M are expressed in amu, v_d in units of 10^4 cm/s, and T in K, (2-1-1) gives the average ionic energy in eV to be

$$1.036 \times 10^{-4} \frac{m+M}{2} (v_d)^2 + 1.293 \times 10^{-4} T.$$

In those cases in which $m = M$ and resonance charge transfer occurs it is probably better at high field to use the equation

$$\frac{\overline{mv_i^2}}{2} = \frac{\pi}{4} mv_d^2 + \frac{3}{2} kT \tag{2-1-2}$$

(Heimerl et al., 1969), derived from the high-field analysis of Fahr and Müller (1967) for the situation in which resonance charge transfer dominates the picture.

modify the field intentionally employed to produce the drift of the ions, and swarm experiments can be conducted at ionic energies as low as permitted by the thermal contact of the ions with the gas molecules.

It is now appropriate to enumerate some constructional features of drift tubes that should be incorporated into the design if data of the highest quality are to be obtained. First, it is of great importance that the apparatus be of ultrahigh-vacuum construction and that it be bakeable to a suitably high temperature during pump-down before a series of measurements is commenced. Metal gaskets should be used throughout and only insulators of the best vacuum quality should be employed. The vacuum pumps should be isolated from the drift tube by cooled baffles and sorbent or liquid nitrogen traps. These precautions will minimize the presence of outgassed impurities during measurements and prevent the interior surfaces from becoming coated with insulating films of pump fluid. It is frequently advisable to gold-plate the interior metallic surfaces, particularly in the ion-sampling region of a drift-tube mass spectrometer. This step will provide high conductivity surfaces that can be cleaned up as required by baking and minimize the production of spurious electric fields by charge buildup on insulating films. It is important also to shield all the insulators inside the apparatus so that they cannot be "seen" by the ions.

It may be of interest to discuss the different factors that determine the kinds of ions present in a given experiment. The primary* ions produced in the ion source will depend on the gas present in the source and also on the particular ion source used and the conditions under which it is operated; for example, in an electron bombardment ion source containing nitrogen gas both N^+ and N_2^+ ions will be formed if the electron beam energy exceeds 24.2 eV, but only N_2^+ ions will be produced at lower energies because the thresholds for production of N^+ and N_2^+ by electron impact are 24.2 and 15.6 eV, respectively. The primary ions may survive over the entire drift distance or they may be converted in charge transfer or ion-molecule reactions to secondary ions of other types. Whether this happens in a given case depends on the value of pd, which determines the number of collisions the primary ions experience and also perhaps on the pressure itself, since some of the reactions may be the three-body variety with reaction frequencies that vary as p^2. Other determining factors are the values of E/N and T, which characterize the energy of the ion-molecule collisions and thus influence the values of the rate coefficients. If secondary ions are formed, they may also undergo further reactions. Whether or not they do so will depend on the values of pd, p, E/N, and T. In most cases there is more than one kind of ion present under given conditions, and it is often difficult to predict their identities and relative abundances.

Still another matter is that of the states of excitation of the ions. Quantitative information is now available concerning their effect on the transport properties of ions, and some of the discrepancies among experimental data may be due to the ions being in different states. The states in which the primary ions are

---

*Here the word "primary" is used in the sense of "first in time" not "first in importance."

produced will depend on the ion source conditions, and the ability of excited primary ions to retain their excitation energy is governed by their lifetimes against radiative and collisional deexcitation. The excitation states of secondary ions depend on those of the primary ions from which they are produced and the exact nature of the reactions forming them. In most cases the primary ions are probably in the ground electronic state when they enter the drift region from the ion source, but molecular ions may well be vibrationally or rotationally excited. Each case must be examined separately. Of course, what is desired from experiments are mobilities and diffusion coefficients for ions in the ground state or in known states of excitation. Positive ions of the alkalis are particularly interesting in this respect. They may be generated thermionically from coated filaments located within the gas filling the drift tube (McDaniel, 1964), and thermodynamic considerations ensure that they are singly charged and in the ground state. This fact may be partly responsible for the generally good agreement among the mobility data that have been obtained for these ions. Another reason for this good agreement is that the alkali ions do not react to a significant extent with the molecules of many of the gases in which their mobilities have been measured, at least at the temperatures and pressures at which the measurements were made. Hence the arrival-time spectra for these ions usually have a simple shape, and it is an easy matter to infer unambiguous mobilities from them.

2-2.   BASIC ASPECTS OF DRIFT VELOCITY MEASUREMENTS

A.   Experiments on Nonreacting Primary Ions

The accurate determination of ionic drift velocities is a rather simple matter if we are dealing with primary ions that do not react with the gas filling the drift tube and consequently retain their identities during their entire transit through the apparatus. We pulse the ion source repetitively to admit bursts of ions into the drift space and measure the times at which the ions arrive at the detector in relation to the times at which they were gated into the drift space. Data are accumulated over many cycles until a statistically satisfactory spectrum of arrival times is obtained. The average arrival time is computed by a procedure appropriate to the apparatus being used, and the drift velocity is calculated from this quantity and the known drift distance. The drift velocity is a unique function of E/N for a given ion-gas combination at a given gas temperature. The width of the pulses admitting ions into the drift space should be short compared with the average transit time in order to obtain high accuracy. These pulses are typically about 1 μs wide, whereas the transit times are usually of the order of hundreds of microseconds. The pressure of the drift-tube gas should be fairly high (say, in the torr region) to avoid excessive broadening of the arrival-time spectrum due to diffusion and also to permit the pressure to be measured accurately. (As a rule, in drift-tube experiments, the higher the pressure, the more accurately it can be measured in percentages.) Mobilities have been determined with an accuracy of about 1% in the simple cases described here, although the usual accuracy achieved is 4 to 5%.

B.   End Effects and the Shape of the Arrival-Time Spectrum

Although the concept of the measurement discussed above is extremely simple, two matters merit mention before we consider more complicated cases. The first is related to the elimination of end effects arising at the ion source or the detector end of the apparatus like those produced by a nonuniformity of the electric field at the entrance to the drift space or the finite time required for the ions to travel from the drift-tube exit aperture to the detector in a drift-tube mass spectrometer. It is desirable to be able to vary the drift distance by moving the ion source or the detector while leaving everything else unchanged. It is then possible to measure the arrival time for various drift distances and accurately determine the drift velocity from the slope of a plot of arrival time versus drift distance if such a plot has a long linear region. (Note that the extension of the linear region will not necessarily pass through the origin of the plot.) An illustration of end effects is provided by Fig. 2-2-1, which shows mean arrival times for D_3^+ ions in deuterium plotted for various equally spaced ion source positions in experiments by Miller et al. (1968). Use of the data for the high-numbered remote source positions gave the correct drift velocity, whereas the data obtained at ion source positions near the end of the drift path yielded a drift velocity that was too low by 12.6%. The utility of a movable ion source in these experiments is apparent. (Another of its advantages, unrelated to the present discussion, is that it facilitates the study of the approach of the various ion populations to equilibrium in reacting systems and permits the use of an attenuation technique in the measurement of transverse diffusion coefficients and reaction rates. This technique is discussed in Chapter 3.)

The other matter that must be covered here is related to the detailed shape of the arrival-time spectrum. A necessary but not sufficient condition for obtaining drift velocities in which we can have confidence is the close matching of the measured arrival-time spectrum with the shape predicted by a solution of the differential equation governing the drift and diffusion of the ions in the particular apparatus used. The differential equation which applies to the Georgia Tech drift-tube mass spectrometer is presented and solved in Section 2-6. Figure 2-2-2a shows experimental arrival-time spectra obtained with this apparatus at low E/N for "nonreacting"* K^+ ions in nitrogen at room temperature, seven spectra corresponding to seven different ion source positions but equal counting times being superimposed on a single drawing (Moseley et

---

*Actually, K^+ ions can react with N_2 molecules to form $K^+ \cdot N_2$ clusters at room temperature (McDaniel, 1972), but the clustering was inappreciable at the relatively low gas pressure used in obtaining the data shown in Fig. 2-2-2a.

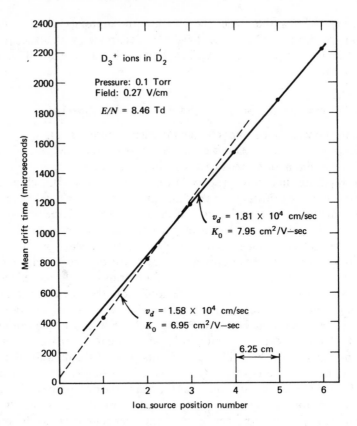

Figure 2-2-1. End effects in drift velocity measurements. The correct drift velocity is given by the solid line. The dashed line yields an erroneous "apparent drift velocity." At a given value of E/N, the drift velocity and standard mobility are related by the equation v_d (in 10^4 cm/s) = 0.02687 × E/N (in Td) × K_0 (in cm²/V-s).

al., 1969a). The solid-line curves represent solutions of the drift-diffusion equation which contain the measured value of the drift velocity and the corresponding value of the diffusion coefficient calculated from the Einstein equation (1-1-5). The solid curves are normalized to match the height of the peak calculated for the shortest drift distance to the corresponding experimental peak. The remaining analytical profiles for the other values of the drift distance are not independently normalized. The spectra illustrate the increasing effects of peak broadening by longitudinal diffusion and intensity loss due to transverse diffusion as the drift distance is increased.

Figure 2-2-2b shows a single experimental arrival-time spectrum for K^+ ions in N_2 obtained at low pressure and a drift distance of 43.77 cm by Thomson et al. (1973). The smooth curve, in whose calculation the experimentally determined values of the mobility and longitudinal diffusion coefficient were used,

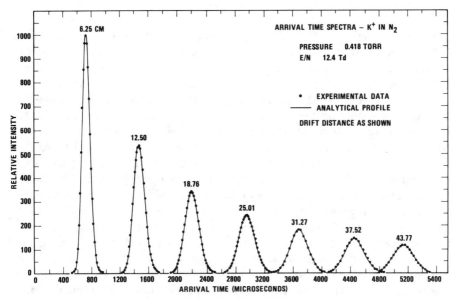

Figure 2-2. **(a)** Arrival-time spectra of "nonreacting" K^+ ions in nitrogen. Recorded at seven ion source positions (dots) compared with the spectra calculated by the analysis of Section 2-6 (solid curve) (Moseley et al., 1969a).

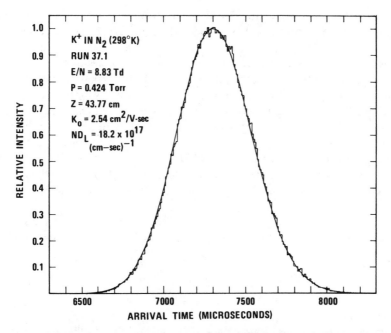

Figure 2-2. **(b)** Arrival-time spectra of "nonreacting" K^+ ions in nitrogen. Recorded for a drift distance of 43.77 cm (histogram) compared with the spectrum calculated by the analysis of Section 2-6 (smooth curve). Reactions were insignificant at the gas pressure used here (Thomson et al., 1973).

gives the prediction of the analysis described in Section 2-6. The agreement between the experimental data and the predictions of the analysis is excellent. Unless this kind of close, quantitative agreement is obtained in a given experiment, there are factors at work that have not been taken into consideration and that might produce faulty results. When good agreement is not achieved, we should probably look first at the question of space charge in the drifting ion swarm. It is widely believed that space charge repulsion will have a negligible effect on the measurement of drift velocities because the effect is usually assumed to be symmetrical about the midplane of the drifting ion cloud perpendicular to the drift-tube axis. Experiments have been performed, however, that indicate that this is not always true. An example of a space-charge effect appeared in drift velocity experiments on hydrogen (Miller et al., 1968) in which an ion of low abundance (H^+) was pushed down the drift tube by the dense space-charge cloud of a much more abundant ion (H_3^+) of lower mobility and the H^+ arrival-time spectrum was seriously distorted in the process. The drift velocity of H^+ inferred by averaging the arrival times was 10% in error until the space-charge effect was reduced to negligible proportions by lowering the total ion current.

C. Experiments on Reacting Systems

The case considered in Section 2-2A, in which measurements are made on a primary ion whose identity is known and which undergoes no reactions in the gas, was the simplest possible. The next simplest case is that in which a known primary ion can be converted to a known secondary species that does not react and thus retains its identity down the drift space. In this situation it is frequently possible to arrange conditions to obtain accurate mobilities for both the primary and secondary ions in essentially the same straightforward manner described in Section 2-2A. An example is provided by the experiments on nitrogen ions in nitrogen gas performed by Moseley and his colleagues (Moseley et al., 1969b). In these experiments two primary ions N^+ and N_2^+ were produced by electron bombardment in the ion source; both can be converted into secondary species by reactions with molecules. The conversion proceeds mainly by the three-body reactions

$$N^+ + 2N_2 \longrightarrow N_3^+ + N_2 \qquad (2\text{-}2\text{-}1)$$

and

$$N_2^+ + 2N_2 \longrightarrow N_4^+ + N_2, \qquad (2\text{-}2\text{-}2)$$

and reconversion to the original ions N^+ and N_2^+ is possible by the reverse reactions only at E/N high enough to permit collisional dissociation of the secondary ions. Below the value of E/N at which dissociation of N_3^+ could occur the drift velocity of N^+ can be measured accurately by working at very low pressures, at which conversion of N^+ to N_3^+ is slow, and at such short drift distances that few of the N^+ ions are converted during drift. Similar techniques permit accurate measurements to be made on the other primary ion N_2^+. To get data on N_3^+ we use high pressures and long drift distances, so that the N^+ ions are converted to N_3^+ in a distance that is short compared with the total drift distance. (Most of the N^+ reacts to form N_3^+ in the ion source.) The same procedure is employed for the other secondary ion N_4^+. For all four ions the experimental arrival-time spectra that were used for the determination of drift velocities closely match the spectra predicted by the solution of the drift-diffusion equation for ions not reacting during their drift. Data for N^+ ions taken at such a low pressure and short drift distance that conversion of N^+ to N_3^+ is negligible are shown in Fig. 2-2-3. (When the pressure and the drift distance are increased to the level at which N^+ is converted to N_3^+ at a significant rate but slow enough that the reactions take place all along the drift tube, the arrival-time spectra shown in Fig. 2-2-4 result. Here there is a ramp on the front of the N_3^+ peak produced by ions that left the ion source as N^+ and traveled part of the way in this form but were converted to N_3^+ before they

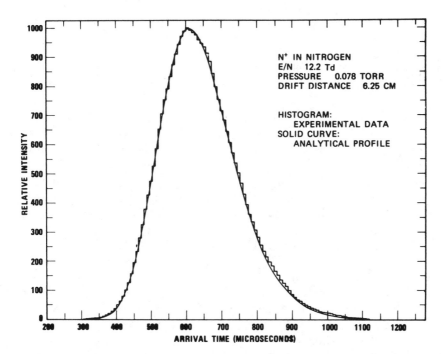

Figure 2-2-3. Experimental arrival time spectrum for N^+ ions in nitrogen when conversion to N_3^+ is negligible compared with the spectrum calculated by the analysis of Section 2-6. In the calculation of the analytical profile a rate coefficient of 2.0×10^{-29} cm^6-s^{-1} was used for the reaction converting N^+ to N_3^+ (2-2-1) (Moseley et al., 1969b).

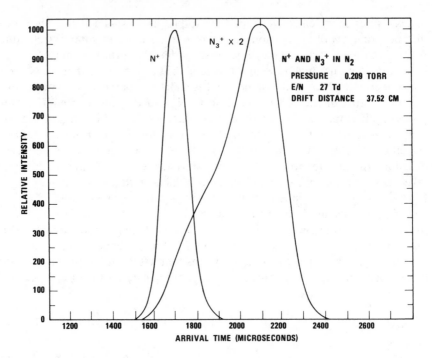

Figure 2-2-4. Comparison of N^+ and N_3^+ arrival-time spectra showing the effect of the formation of N_3^+ from N^+.

arrived at the detector. Data such as these are not used for drift velocity determinations.)

In the nitrogen experiments of Moseley et al., when end effects were eliminated by the use of multiple source positions, mobilities accurate within 3% were obtained for all four species of ions: N^+, N_2^+, N_3^+, and N_4^+. These measurements illustrate the importance of mass selection, high resolution time analysis, and a movable ion source. In the apparatus used the drift distance can be varied during operation from 1 to 44 cm. An important factor is that both the N^+ and N_2^+ ions are converted to secondary form in three-body reactions for which the reaction frequency varies as p^2. Because of this rapid variation with pressure, it proved possible to go from virtually no reaction at all at low pressure to rapid and essentially complete conversion to secondary form at high pressure. This procedure might not be possible in general if the conversion of primary ions to a secondary species proceeded by a two-body reaction for which the reaction frequency varies more slowly with pressure, only as the first power of p.

Frequently we encounter more complicated situations than those we have described. A common case is one in which primary ions are converted to a secondary species, some members of which are reconverted to the primary form

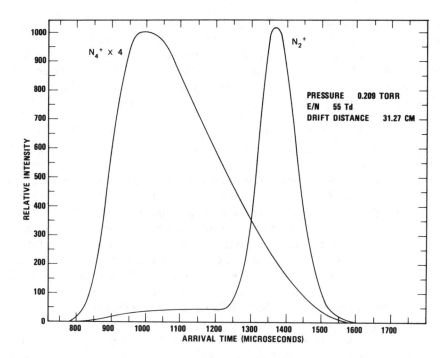

Figure 2-2-5. Comparison of N_2^+ and N_4^+ arrival-time spectra showing the effects of ion-molecule reactions on each profile.

during flight. Figure 2-2-5* is an example that occurs with N_2^+ and N_4^+ ions in nitrogen at room temperature and values of pressure and E/N in the intermediate range (Moseley et al., 1969b). Here N_2^+ is converted to N_4^+ by the "forward" reaction

$$N_2^+ + 2N_2 \longrightarrow N_4^+ + N_2, \qquad (2\text{-}2\text{-}2)$$

and N_4^+ ions are dissociated in collisions with N_2 molecules by the "backward" reaction

$$N_4^+ + N_2 \longrightarrow N_2^+ + 2N_2. \qquad (2\text{-}2\text{-}3)$$

At some values of E/N and pressure these reactions occur so rapidly that the conversion and reconversion processes take place many times during the passage of the ions down the drift tube and a dynamic equilibrium develops

---

*It may be correctly inferred from Fig. 2-2-5 that N_4^+ ions drift through N_2 gas more rapidly than the lighter N_2^+ ions. The reason for this is that N_2^+ ions are retarded in their drift by resonant charge transfer with N_2 molecules.

between the two ionic species. The equilibrium concentrations of N_2^+ and N_4^+ will depend on the value of E/N and the gas temperature and pressure.

In another interesting case the primary ions may be converted to secondary ions which can react with the gas molecules to form a tertiary ionic species. This can happen in hydrogen at room temperature and pressures above about 0.4 torr (Miller et al., 1968) through the sequence of reactions

$$H^+ + 2H_2 \longrightarrow H_3^+ + H_2 \quad \text{and} \quad H_2^+ + H_2 \longrightarrow H_3^+ + H, \qquad (2\text{-}2\text{-}4)$$
$$H_3^+ + 2H_2 \longrightarrow H_5^+ + H_2. \qquad (2\text{-}2\text{-}5)$$

The H_5^+ ions have a low binding energy and are quickly broken up in collisions with H_2 molecules to form H_3^+, even at low E/N and room temperature.

Despite the complexity of the examples described here, it is possible to obtain reliable drift velocities by solving the drift-diffusion equation with reaction terms included and making a multiparameter fit to the experimental arrival-time spectra. In the process the reaction rate coefficients as well as the drift velocities are evaluated. This procedure has been used by several investigators (Beaty and Patterson, 1965; McKnight et al., 1967; see also Takebe et al., 1980, 1981, 1982). Usually, the accuracy achieved in the drift velocity determinations is not as high as in nonreacting systems.

Evidently, a drift-tube mass spectrometer must be employed in a reacting system of appreciable complexity if true drift velocities characteristic of single known species are to be obtained. If a conventional drift tube is used, drift velocities inferred from a given peak in the composite arrival-time spectrum will usually refer to charge carriers that had spent part of their drift time in one ionic form and the remainder in some other, so that what is measured is some unknown kind of average of the true drift velocities of the various species. However, even if a drift-tube mass spectrometer is used for the measurements, we may obtain drift velocities that are badly in error unless the arrival-time spectra are carefully mapped and properly analyzed. Some investigators have determined drift velocities in reacting systems simply by measuring the times at which the ion intensity peaks. Edelson and his colleagues (Edelson et al., 1967) have demonstrated mathematically how this procedure can lead to large errors.

Before terminating this general discussion we should mention one additional advantage offered by drift-tube mass spectrometers that is unrelated to reactions occurring between the ions and gas molecules. It sometimes happens that two primary ions with very nearly equal mobilities are formed in the ion source. A case in point is that of SF_6^- and SF_5^- in SF_6 gas. Certain drift velocity measurements made on this system before the development of drift-tube mass spectrometers revealed only a single peak in the composite arrival-time spectrum, hence indicated the presence of only a single kind of ion (McDaniel and McDowell, 1959; McAfee and Edelson, 1962). When Patterson studied the negative ions of SF_6 with a drift-tube mass spectrometer and obtained separate arrival-time spectra for each ion, it became apparent that two different ionic species were present with closely similar mobilities (Patterson, 1970).

2-3. DETERMINATION OF LONGITUDINAL DIFFUSION COEFFICIENTS

Let us consider again the simple situation in which some particular ionic species is formed entirely within the ion source and travels the length of the drift tube without undergoing reactions with the gas molecules. We would expect that the shape and width of the arrival-time spectrum would depend strongly on the value of the longitudinal diffusion coefficient D_L and that it should prove possible to determine D_L from the shape of the observed spectrum. This is indeed the case, although its determination is much more difficult than that of the drift velocity and its published values date back only to 1968 (Moseley et al., 1968).

To determine D_L accurately we must have a solution of the transport equation describing the drift and diffusion of the ions in the apparatus used for the experiment. Such a solution will provide the expected functional form of the arrival-time spectrum in terms of D_L, the drift velocity v_d, and the transverse diffusion coefficient D_T. The dependence on v_d will be strong, since it determines the length of time the ions spend diffusing in the gas, but, as we have seen in Section 2-2A, the drift velocity can be measured accurately in the simple case considered here and is assumed to be known. The shape of the arrival-time spectrum depends only weakly on the value of D_T, and if D_T is not known from independent measurements we may use its value obtained from the low-field mobility by application of the Einstein equation (1-1-5). Although this value of D_T will apply strictly only in the low-field region, it may be used in determinations of D_L even in the high-field region without introducing a significant error. To obtain slightly greater accuracy we may use the value of D_T calculated for the appropriate E/N from the generalized Einstein relations (Sections 5-2B and 6-4.) Then all the quantities appearing in the analytical expression for the arrival-time spectrum may be considered known except for D_L. We may insert a trial value of D_L into the analytical time profile and adjust this value until the analytical profile and the experimental arrival-time spectrum best correspond when normalized to agree at their points of maximum intensity. In the usual procedure a computer performs a least-squares or least-cubes fit to the experimental data by varying D_L. (It is necessary to shift the analytical profile along the time axis to match up the positions of the peaks if end effects are of significant size in the experiment.)

It is most important in obtaining the experimental arrival-time spectrum that the charge density in the ion swarm be low enough that no detectable space charge expansion of the swarm will occur. A small amount of space charge expansion will not necessarily affect the measurement of v_d to an appreciable extent, but it will materially increase the value of D_L obtained by the fitting procedure described here. As the ion current is increased from an initially low value the peak in the arrival-time spectrum will be noticeably broadened by mutual repulsion long before the measured drift velocity has begun to be seriously affected. A good way to determine whether the ion current is high enough to produce space-charge effects is to compare the measured values of the mobility and the diffusion coefficient at low E/N to see whether the two results are related by the Einstein equation. If the ratio D/K is significantly greater than predicted by this equation, then space charge effects are probably operative and the ion current should be reduced.

The technique we have described for determining D_L may also be applied to the case in which the primary ions formed in the ion source undergo depleting reactions as they drift through the gas, provided that the rate coefficient for the reaction is known and the effect of the reaction is included in the solution of the transport equation (Moseley et al., 1968). If the primary ions are capable of reacting during their drift, the arrival-time spectrum should be mapped at low pressure so that the conversion of the primary ions to secondary form will occur only slowly. Operation at low pressure also offers the advantage of accentuating the effect of diffusion and producing a broad spectrum that can be fitted more accurately with an analytical profile than a narrow spectrum can.

2-4. DETERMINATION OF REACTION RATE COEFFICIENTS FROM ARRIVAL-TIME SPECTRA

It has already been stated that the drift tube offers a useful method of determining the rate coefficients of ion-molecule and charge transfer reactions. The accuracy that can be achieved in such determinations is probably as high as that accessible in any of the alternative methods in use (Bowers, 1979; Franklin, 1972). In favorable cases accuracies of about 5%* can be realized with drift tubes if proper precautions are observed (McDaniel, 1970). The drift tube offers the advantage that data can be obtained not only for thermal energy ions but also for ions of average energy ranging up to several electron volts. The capability of measuring ionic rate coefficients in the suprathermal energy range has excited a great deal of interest, since conventional beam experiments cannot be performed at these energies because of the difficulties present in space charge and stray fields. Merging beams techniques (see Section 2-1) can be applied at suprathermal energies for the study of two-body reactions, but they cannot be utilized for the investigation of reactions of the three-body kind. Hence the drift tube can bridge an important energy gap in the study of many important reactions.

Up to this point in this chapter we have been discussing the methods of determining drift velocities and longitudinal diffusion coefficients from measured arrival-time spectra. It may be appropriate, then, to extend the discussion slightly and show how reaction rates may also be obtained from such spectra. Other drift-tube methods of determining rate coefficients which require the measurement of total ion currents are described in Section 2-5C and Chapter 3.

Let us consider the case in which a primary ion may be converted to a nonreacting secondary species by reactions with the gas molecules during its

---

*For a discussion of experiments in which this degree of accuracy was achieved, see E. Graham (1974), Ph.D. thesis, Georgia Institute of Technology, Atlanta.

drift. The shape of the arrival-time spectrum for the primary ion will not depend strongly on the value of the rate coefficient, although the area under the primary ion spectrum will. The effect of the reaction, however, will be clearly manifested in the shape of the arrival-time spectrum for the secondary ion (provided its drift velocity is not close to that of the primary ion), and often the value of the rate coefficient can be accurately determined from an analysis of this shape. The technique requires the use of a solution of the transport equation for the secondary ion containing a source term for the creation of these ions from the primaries. The rate coefficient for the reaction appears as a parameter in the

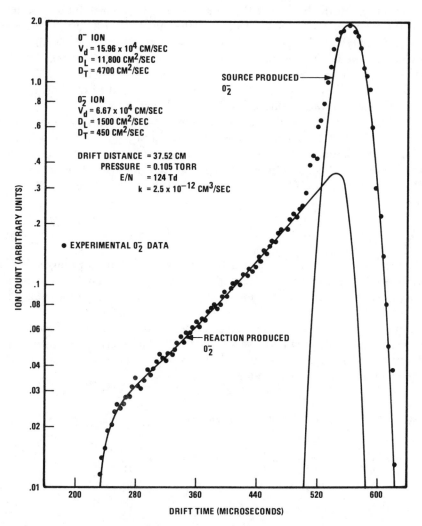

Figure 2-4-1. An O_2^- arrival-time spectrum used to determine the rate coefficient for the reaction $O^- + O_2 \rightarrow O_2^- + O$ (Snuggs et al., 1971a).

solution, and a value for this quantity may be assumed and then varied until the analytical profile matches the experimental arrival-time spectrum for the secondary ion.

Figure 2-4-1 presents the data used to determine the rate coefficient for the two-body reaction $O^- + O_2 \rightarrow O_2^- + O$ at $E/N = 124\,\text{Td}$ and a gas temperature of 301 K (Snuggs et al., 1971b). The dots represent the experimental arrival-time spectrum for the O_2^- ion. The sharp peak at the right corresponds to O_2^- ions that were produced in the ion source and that traveled all the way down the drift tube in this form. The points near the top of the ramp leading to the sharp peak are for O_2^- ions that were formed from O^- early in their drift, whereas those at the bottom of the ramp correspond to ions that underwent a "deathbed conversion" from O^- to O_2^- at the end of the drift, just in time to be detected as O_2^- by the mass spectrometer. The smooth curve is the analytical profile predicted from the solution of the transport equation for the O_2^- ion. The values of the drift velocities and diffusion coefficients used in the analysis are shown on the drawing. Fitting the analytical profile to the experimental data yielded a reaction rate coefficient of $2.5 \times 10^{-12}\,\text{cm}^3/\text{s}$.

The determination of the rate coefficient for a different reaction is now discussed in greater detail to illustrate the strong interplay between the transport properties and reactions of ions in certain drift-tube experiments. The reaction discussed is that of the conversion of CO^+ ions (which we call species A) into the dimer ions $CO^+ \cdot CO$ (species B) in three-body collisions with CO molecules:

$$CO^+ + 2CO \longrightarrow CO^+ \cdot CO + CO. \tag{2-4-1}$$

The measurements were made at 300 K and over the pressure range 0.078 to 0.159 torr. The range of E/N extended from 75 to 150 Td. The drift-tube mass spectrometer described in Section 2-5A was employed in the measurements (Schummers, 1972; Schummers et al., 1973b).

An analytic solution of the transport equation for the product ion B, developed by Gatland (1975) was matched with the experimental arrival-time spectrum for this ion. For the present case in which dissociation of species B is negligible this solution has the form

$$\Phi_B(z, t) = sa \int_0^t du [f_B \delta(t - u) + f_A \alpha_{BA}](\pi r_L^2)^{-1/2}$$

$$\times \left[\frac{2D_{LB}}{r_L^2}(z - r_d) + v_{dB} \right] \exp\left[-\alpha_{BA}(t - u) - \frac{(z - r_d)^2}{r_L^2} \right] \left[1 - \exp\left(-\frac{r_0^2}{r_T^2} \right) \right]. \tag{2-4-2}$$

Here $\Phi_B(z, t)$ is the flux of ions of species B arriving at the detector at time t for a drift distance z. The area of the drift-tube exit aperture is a. Ions of planar density s enter the drift region at time $t = 0$ from the ion source through an aperture of radius r_0. Among these ions a fraction f_A are of species A and a fraction f_B belong to species B. The time of drift spent by an ion in form B is denoted by u. The reaction frequency is $\alpha_{BA} = kN^2$, where k is the rate coefficient; $r_L^2 = 4D_{LA}(t - u) + 4D_{LB}u$ and $r_T^2 = 4D_{TA}(t - u) + 4D_{TB}u$, where D_L and D_T represent longitudinal and transverse diffusion coefficients, respectively. The drift velocity is v_d, and $r_d = v_{dA}(t - u) + v_{dB}u$.

Figure 2-4-2a shows a typical experimental arrival-time spectrum (solid curve) for the product ion $CO^+ \cdot CO$ (species B) obtained at a drift distance of 43.77 cm, a pressure of 0.133 torr, and an E/N of 75 Td. (Since the product ion here has a higher mobility than the parent ion, this spectrum differs markedly in shape from that shown in Fig. 2-4-1, for which the product ion drifts more slowly than the parent.) Also shown as a dashed curve in Fig. 2-4-2a is a plot of the analytic expression (2-4-2) for the same experimental conditions. The drift velocities v_{dA} and v_{dB}, which were used in plotting (2-4-2), were determined in separate measurements, as was D_{LA}. The "modified Wannier equations"* were used to compute D_{LB} and D_{TB} from v_{dB}. These equations, however, cannot be used to calculate D_{LA} and D_{TA} because the diffusion of CO^+ ions in CO is dominated by resonant charge transfer. The best estimate of D_{TA} (given by $ND_{TA} = 17.1 \times 10^{17}$/cm-s) was obtained by an iterative curve-fitting procedure that utilized arrival-time spectra for the product ion mapped at low pressures, where the effects of transverse diffusion are accentuated and D_{TA} can be determined most accurately. With this value of D_{TA} [see (2-4-2)] a value of 1.35×10^{-28} cm^6/s for the rate coefficient k gives the best fit to the experimental arrival-time spectrum mapped at the higher pressure of 0.133 torr, where the effects of the reaction dominate those of transverse diffusion. In the curve-fitting procedure the analytic expression is normalized to agree with the experimental spectrum at its maximum, since the source density s is not accurately known. Two other assumed values for D_{TA} and the corresponding best-fit values of k are indicated on the graph to demonstrate the sensitivity of the evaluation of k to the value of D_{TA} which is used in plotting (2-4-2). These values of D_{TA} are given by $ND_{TA} = 12.8 \times 10^{17}$ (cm-s)$^{-1}$ and $ND_{TA} = 23.5 \times 10^{17}$ (cm-s)$^{-1}$. They produce analytic spectra that do not agree nearly so well with the experimental spectra obtained at low pressures as the value of D_{TA} given by $ND_{TA} = 17.1 \times 10^{17}$ (cm-s)$^{-1}$ (see Fig. 2-4-2b).

---

*The modified Wannier equations are

$$ND_L(E) = ND(0) + \frac{(M + 3.72m)M}{3(M + 1.908m)e} \frac{v_d^3}{E/N},$$

$$ND_T(E) = ND(0) + \frac{(M + m)M}{3(M + 1.908m)e} \frac{v_d^3}{E/N},$$

and may be regarded as a form of the generalized Einstein relations (see McDaniel and Moseley, 1971; Thomson et al., 1973; Whealton and Mason, 1974).

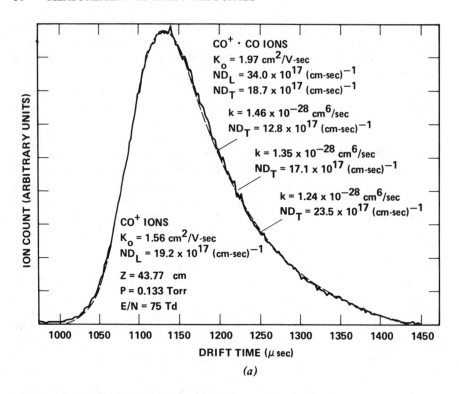

(a)

Figure 2-4-2. Arrival-time spectra of $CO^+ \cdot CO$ ions used to determine the rate coefficient for the reaction $CO^+ + 2CO \rightarrow CO^+ \cdot CO + CO$ (Schummers et al., 1973b).

The measured values of the rate coefficient k for the reaction (2-4-1) range from 1.35×10^{-28} cm^6/s at $E/N = 75$ Td to 1.1×10^{-28} cm^6/s at E/N between 110 and 150 Td. The accuracy of these results is estimated as $\pm 14\%$. No theoretical equation appears to be available for an accurate calculation of the average energy of the reacting CO^+ ions in the intermediate range of E/N explored here. As pointed out in Section 2-1, however, Fahr and Müller (1967) have developed a theory applicable to the high-E/N regime for the case in which resonant charge transfer collisions dominate the transport behavior of the ions, as they do here. Their analysis gives the average energy of the ions derived from the electric field in excess of the thermal values as $\pi m v_d^2/4$, where m is the mass of the ions [see (2-1-2)]. In the absence of a more appropriate expression this result has been used to estimate the average "field energy" of the reacting CO^+ ions above the thermal value in the intermediate range of E/N studied here. The result is that the average "field energy" is estimated as 60% of the thermal value at $E/N = 75$ Td and 184% of the thermal value at $E/N = 150$ Td.

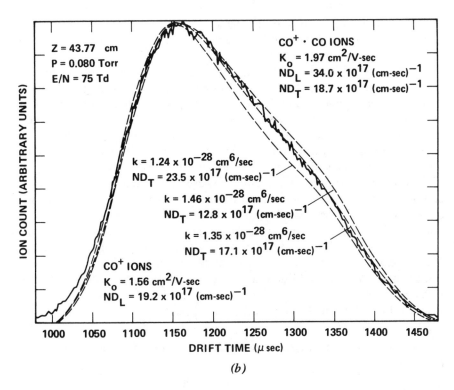

Figure 2-4-2. *Continued.*

2-5.   DESCRIPTIONS OF DRIFT TUBES

Scores of drift tubes, differing greatly in principles of operation and in details of design and construction, have been used for studies of slow ions in gases, and many of them are described in books on atomic collisions and gaseous electronics (Loeb, 1960; Massey, 1971; McDaniel and Mason, 1973; Hasted, 1972). Here we concentrate our attention on drift-tube mass spectrometers in recent use. We also describe several conventional drift tubes and other apparatus that illustrate relevant physical phenomena and experimental techniques.

Our reason for emphasizing drift tubes that incorporate mass analysis is probably apparent at this point in the discussion, but it can be summarized briefly. Even though the identities of the primary ions produced by an ion source are known, reactions of these ions with molecules of the gas filling the drift tube can produce an unexpected assortment of ions, and the interrelationships among the drifting ions can be complicated. Usually it is impossible to disentangle the behavior of the individual species without the use of mass selection and careful analysis of the shapes of the separate arrival-time spectra, and in most cases the use of a drift-tube mass spectrometer is essential if we are to obtain unambiguous results.

Since the early days of mobility research the importance of excluding impurities from drift tubes has been realized because it has long been known that the primary ions produced in the gas intended for study can be efficiently converted to ions of other types in reactions with impurity molecules. Special attention has been paid to polar and highly polarizable impurities (which can produce strong clustering of molecules about the ions) and to impurities of low ionization potential (which can become ionized by charge transfer at the expense of the primary ions). Strangely enough, however, many years elapsed before the complicating effects of reactions with molecules of the *parent* gas were fully appreciated, and it is probably this fact that was responsible for the late appearance of apparatus employing mass analysis. The first drift-tube mass spectrometers were developed in the early 1960s at Georgia Tech (Barnes et al., 1961; McDaniel et al., 1962; Martin et al., 1963) and the Bell Telephone Laboratories (McAfee and Edelson, 1963a, b; Edelson and McAfee, 1964). Both instruments have been replaced by improved apparatus, described below or in McDaniel and Mason (1973, Secs. 2-5D, 2-5E, and 2-5G).

A.   Georgia Tech Low-Pressure Drift-Tube Mass Spectrometer (DTMS)

An experimental facility used at the Georgia Institute of Technology for the measurement of drift velocities, diffusion coefficients, and reaction rates at low pressures was constructed during 1965 and 1966 and is described in a series of papers (Albritton et al., 1968; Miller et al., 1968; Moseley et al., 1968, 1969a, b; Snuggs et al., 1971a, b; Volz et al., 1971; Thomson et al., 1973; Schummers et al., 1973a, b). It consists of a large ultrahigh-vacuum enclosure containing a drift tube and ion-sampling apparatus, plus associated circuitry (see Figs. 2-5-1 and 2-5-2). The gas to be studied is admitted to the drift tube through a servo-controlled leak, and the sample gas flows continuously from the tube through an exit aperture on the axis at the bottom of the tube. The pressure in the drift tube is held constant during operation at some desired value in the range 0.025 to several torr. A pulsed electron-impact ion source is used to create repetitive, short bursts of primary ions at a selected source position on the drift-tube axis. Each burst of ions moves downward out of the source and migrates along the axis of the drift tube under the influence of a uniform electric field produced by electrodes (the drift-field guard rings) inside the tube. When the ions reach the bottom of the drift tube, those close to the axis are swept out through the exit aperture and the core of the emerging jet of ions and gas molecules is cut out by a conical skimmer and allowed to pass into an RF quadrupole mass spectrometer. Ions of a selected charge-to-mass ratio traverse the length of the spectrometer, all other ions being rejected in the mass selection process. The selected ions are then detected individually by a nude electron multiplier

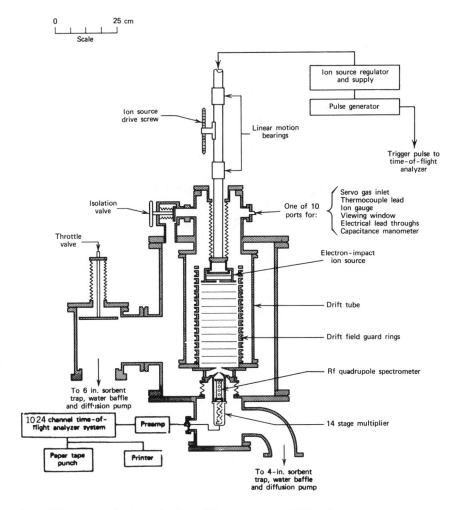

0   25 cm
Scale

Ion source regulator
and supply

Pulse generator

Trigger pulse to
time-of-flight
analyzer

Ion source
drive screw

Linear motion
bearings

Isolation
valve

One of 10
ports for:

Servo gas inlet
Thermocouple lead
Ion gauge
Viewing window
Electrical lead throughs
Capacitance manometer

Throttle
valve

Electron-impact
ion source

Drift tube

Drift field guard rings

Rf quadrupole spectrometer

To 6 in. sorbent
trap, water baffle
and diffusion pump

1024 channel time-of-
flight analyzer system

Preamp

14 stage multiplier

Paper tape
punch

Printer

To 4-in. sorbent
trap, water baffle
and diffusion pump

Figure 2-5-1. Overall schematic view of the low-pressure drift-tube mass spectrometer at the Georgia Institute of Technology. The objects connecting the drift-tube exit aperture plate to the housing of the conical skimmer above the spectrometer are posts which do not significantly impede the action of the pump shown at the left in disposing of the gas flowing from the drift tube. The electron multiplier shown has been replaced by one of the capillary type, located off-axis (McDaniel et al., 1970).

operated as a pulse counter, and the resulting pulses are electronically sorted according to their arrival times by a 1024-channel time-of-flight analyzer. Because of transverse diffusion and geometrical losses, only an extremely small fraction of the ions originally present in each burst reach the detector. A statistically significant histogram of arrival times can, however, be built up by accumulating data from 10^5 to 10^6 ion bursts for a given source position (see

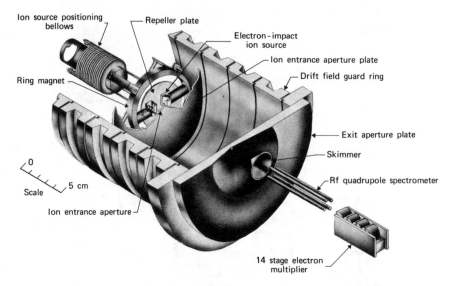

Figure 2-5-2. Isometric view of the drift tube, ion source, and ion sampling apparatus shown in Fig. 2-5-1. The electron multiplier has been replaced by one of the capillary type, located off-axis (McDaniel et al., 1970).

Fig. 2-2-3, for example). This procedure is repeated for various other positions of the source along the axis of the drift tube, following which, the mass spectrometer is tuned successively to other ionic masses and arrival-time spectra are acquired for each type of ion present in the drift tube. Finally, the sequence of measurements is repeated for other values of gas pressure and drift field intensity.

If a given type of primary ion travels all the way from the ion source to the detector without undergoing chemical reactions with the gas molecules, its arrival-time spectrum should consist of a "Gaussian" peak slightly skewed toward later arrival times (see Section 2-6). Typical experimental arrival-time spectra with this expected shape have been shown superimposed in Fig. 2-2-2a for seven different positions of the ion source. Only slight deviations from the peak shape shown in Fig. 2-2-2 are to be expected if the primary species undergoes reactions at a moderate rate as it drifts through the gas, provided that this ion is produced only in the ion source and not, in addition, by reactions along the drift path. Situations occur, however, in which reactions can produce arrival-time spectra of quite different shape for primary ions as well as secondaries. Examples have been displayed in Figs. 2-2-5 and 2-4-1 and are described in the text accompanying them.

The main vacuum chamber shown in Fig. 2-5-1 is constructed of stainless steel and is evacuated by oil diffusion pumps which are separated from the chamber by water-cooled baffles and molecular sieve traps. Metal gaskets provide the vacuum seals, and the chamber is baked at a temperature of 200°C during pumpdown before measurements. The stainless-steel drift tube inside this chamber is heated to 300°C during pumpdown and base vacua below 10^{-9} torr are achieved within it. The background drift-tube pressure after the isolation valve is closed never exceeds 2×10^{-8} torr. Important to the achievement of low base pressures are: (a) the use of organic materials is scrupulously avoided; (b) all insulators are made either of alumina (Al_2O_3) or steatite ($MgO \cdot SiO_2$); and (c) all welds are heliarc welds made from the inside. After the drift tube has been evacuated and filled with the gas to be studied the pressure in the drift tube is measured with a capacitance manometer which has been calibrated by a trapped McLeod gauge. Thermocouples attached to the exterior of three of the drift-field guard rings indicate the temperature inside the drift tube.

Details of the ion source construction are shown in Fig. 2-5-2. The source contains two nonmagnetic stainless-steel boxes, one mounted on each side of a ring magnet which produces a field of about 100 gauss in the magnet gap. Electrons are evaporated from a filament* in the box on the left side and are periodically admitted into the ionization region through a slit in a control plate which is used to gate the passage of the electrons. Because of the magnetic field, the electrons are constrained to move in tight helices. The electron beam has the shape of a narrow ribbon perpendicular to the drift-tube axis, and thus the primary ionization is restricted to a narrow, well-defined region in the gas. After traversing the ionization gap the electrons are collected in the box on the right side of the source. The source frame is maintained at the local equipotential in the drift field, and a suitable potential is applied to the repeller plate at the top of the source to cause the ions formed to move toward the ion entrance aperture plate. A Tyndall gate, not shown, is mounted in the $\frac{3}{4}$-in. hole in this plate. This gate consists of two closely spaced wire meshes mounted perpendicular to the drift-tube axis. Normally a potential is applied between these meshes to prevent the ions from passing through the hole. Periodically, however, this potential is removed to allow the ions to flow through the grid and enter the drift space. The repetition rate of the pulses applied to the control plate and the Tyndall gate is 10^2 to 10^4 s^{-1}, and the width of the pulses is usually less than 1 μs, a negligible time compared with typical drift times. Short pulse widths and an electron beam current of less than 1 μA during each pulse are utilized to restrict the number of ions in each burst to a value that does not produce appreciable space-charge effects in the drift space.

The electron beam in the ion source has a fairly small energy spread (several electron volts), and its energy can be set at any desired value within wide limits. This feature provides closer control on the production of excited and multiply charged ions than is possible with many other kinds of ion sources.

*Thoriated iridium is normally used as the filament material, but in experiments on oxygen a platinum-rhodium filament coated with a $BaZrO_3$-$BaCO_3$-$SrCO_3$ mixture was employed because of its longer lifetime in this reactive gas. This filament, however, has the disadvantage of requiring a great deal of power for its operation and can cause an undesirable temperature gradient along the drift tube.

Both positive and negative ions can be produced by the ion source. Positive ions are created by the ejection of bound electrons into the continuum. In an electronegative gas negative ions can be produced directly by beam electrons in various processes or as the result of capture by gas molecules of electrons ejected from other gas molecules by the ion source beam.

A coated filament is also located inside the ion source on the axis, directly in front of the repeller plate. It can be used to produce positive ions of the alkalis by thermionic emission (see Section 2-1).

The ion source is mounted at the bottom of a long stainless-steel bellows whose length can be varied from outside the apparatus during operation. By changing the length of the bellows we may place the ion source within a few thousandths of an inch at any of 16 positions along the drift-tube axis to provide drift distances that range from 1 to 44 cm.

The drift space which the ions enter after leaving the ion source is bounded by a set of 14 guard rings with 17.5 cm ID. The rings are similar to those used by Crompton et al. (1965) and maintain an axial electric field which is uniform to a fraction of a percent in the region traversed by the ion swarm.* Concealed alumina spacers and dowel pins electrically separate the guard rings and provide alignments to a few thousandths of an inch. All surfaces exposed to the ions here and in other parts of the apparatus are gold-plated to reduce surface potentials. Guard rings of unusually large diameter are used so that the drifting ions will not be able to reach them by transverse diffusion. Thus uncertainties concerning the fate of ions when they strike a surface are avoided, and the mathematical solution of the transport equation describing the ionic motion is considerably simplified because the ions can be assumed to drift and diffuse in a space of unbounded radial extent (see Section 2-6).

Ions leave the drift space through a knife-edged hole, 0.035 cm in diameter, in the exit aperture plate at the bottom of the drift tube. This aperture is similar to a simple molecular beam effusion orifice, and its design minimizes mass discrimination effects in the ionic mass sampling (Parkes, 1971). Strong differential pumping is applied between the drift-tube exit aperture and the skimmer and between the skimmer and the mass spectrometer so that few collisions occur after the ions have passed out of the drift tube. The apparatus may be operated with no guiding or focusing field applied between the drift-tube exit aperture plate and the skimmer so that the ions will not be accelerated until they have passed through the skimmer and entered the analysis region where the pressure is below 10^{-6} torr. This procedure guards against the possibility that weakly bound molecular ions may be dissociated in energetic collisions with gas molecules.† When tests on a given ion in a given gas indicate that it is safe to do

---

*The calculation of the electric field pattern inside the drift tube appears in the thesis of Albritton (1967), which also contains much detailed information concerning the construction of the apparatus described here.

†This dissociation effect has been observed in the sampling of N_4^+ ions from a drift tube containing nitrogen gas (McKnight et al., 1967).

so, however, a drawout potential of 35 V is usually applied to the skimmer to increase the detected ion current and reduce the spread of transit times between the drift tube and the mass spectrometer. The hole in the tip of the skimmer has a diameter of 0.079 cm.

After the ions have passed through the skimmer they are brought to an energy of about 4 eV for analysis in a quadrupole mass filter. Electron multipliers of several different types have been used as pulse counters to detect the ions transmitted through the mass spectrometer. Pulse-counting techniques permit the collection of data even though the signal may be extremely weak. This capability allows the apparatus to be operated at low E/N and with ion currents small enough to prevent space charge effects from appearing. A few counts per second is the practical lower limit on the detector sensitivity. The maximum counting rate that may be used with a dynode-type multiplier (see Fig. 2-5-2) is about 10^5 s^{-1}. With a capillary-type multiplier ("channeltron") the maximum counting rate is about the same, but this kind of detector offers the advantage of requiring less voltage for its operation than the dynode-type. Furthermore, its performance does not deteriorate when it is exposed to air and it is much more convenient to use.

The time interval between the injection of an ion swarm into the drift region and the detection of one of its members is measured and stored by a 1024-channel time-of-flight analyzer. The sweep of the analyzer is triggered by the pulse applied to the Tyndall gate at the upper boundary of the drift space.

The drift tube may be operated over a wide pressure range (about 0.025 to several torr). The lower limit is imposed by the requirement that the product of the gas pressure and drift distance be great enough to permit the ions to make many collisions during their drift. The upper limit arises from the poor performance of the electron bombardment ion source at pressures much above 1 torr. With an ion source of the electrical discharge type measurements with this apparatus could be extended up to drift-tube pressures of about 10 torr before the gas-handling capacity of the pumps would be exceeded.* It is important to be able to operate over a wide range of pressure to determine accurately the pressure dependence of ion-molecule reactions occurring in the drift tube. Furthermore, the reaction pattern in a given gas may make it mandatory to operate at either a high or a low gas pressure to obtain easily interpretable results for a given ionic species in that gas (see Section 2-2C).

The foregoing description of the Georgia Tech drift-tube mass spectrometer applied up to the end of 1971, but in 1972 several important changes were made (Schummers, 1972; Schummers et al., 1973a, b; Thomson et al., 1973). The first change was the removal of the magnet from the ion source to eliminate the fringe field it produced in the top of the ionic drift region. This field had distorted certain arrival-time spectra for secondary ions formed early during the drift of

---

*In measurements made with this apparatus before 1970 the exit aperture at the end of the drift tube had a diameter of 0.079 cm, and it was possible to pressurize the drift tube to only 2 or 3 torr before the pumps were overloaded.

the parent species, although it had not affected any of the published data to a significant extent. With the magnet removed there is sometimes a slight leakage of ions into the drift space when the Tyndall gate is closed, but this leakage has not been a problem.

Next, the conventional cold trap in the gas feed line was replaced by a "refrigerating vapor bath" developed by Puckett et al. (1971). The gas to be admitted to the drift tube can be stored in this device at any temperature between 0 and $-196°C$, a technique that results in some cases in much more efficient trapping of impurities than was possible with the conventional cold trap.

Then, a "multipulsing" scheme was developed which permits the sampling of one ion cloud arriving at the bottom of the drift tube while several other ion swarms are in transit down the tube. This technique is feasible whenever the drift time is much greater than the spread in arrival times of the ions in a given cloud. A major advantage of multipulsing is that with it data can be accumulated at a much faster rate than is possible if only one ion swarm is in transit at a time.

Recent research with this apparatus has centered on measurements of the transport properties of positive alkali ions and negative halogen ions in the noble gases. All of these atomic ion–atomic gas combinations have spherically symmetric interaction potentials, which can be determined from the measurements as functions of E/N (see Chapter 7).

B.   Variable-Temperature Drift-Tube Mass Spectrometers

The Georgia Tech apparatus described in Section 2-5A permits one, but not both of the main classes of measurement. It is operated at a fixed temperature (close to 300 K), and measurements are made over a wide range of E/N, so that the average ionic kinetic energy is varied from thermal (0.04 eV) up to the maximum (typically about 6 eV) that can be attained for the ion-gas combination being studied. Since the ratio of ionic to molecular number densities is of the order of 10^{-10} in most of the measurements made with this apparatus, the drift field does not indirectly heat the gas by directly heating the ions. Collisional excitation of the ions may occur, however, especially in the case of molecular ions at high E/N. The drift-tube gas is always in thermodynamic equilibrium (to an extremely good approximation), but the ions approach equilibrium only as $E/N \to 0$. Until the development of the Viehland–Mason theory of gaseous ion transport starting in the mid-1970s (Chapters 5 and 6), the only rigorous, general theory available was that of Chapman and Enskog (circa 1917), which applied to the case of near-thermodynamic equilibrium and was restricted to spherically symmetric systems. For this reason, fixed-temperature, variable E/N transport data could not be properly interpreted until recently, although accurate data of this type have always been useful in applications. In particular, data outside the low-field region could not be properly utilized in the determination of ion-neutral interaction potentials.

The other main class of experiments involves measurements at extremely low

E/N but at various drift-tube gas temperatures. Here the ions are essentially in thermodynamic equilibrium with the gas at the common temperature prevailing in a given measurement, and the Chapman–Enskog theory has been available since 1917 for the easy and accurate analysis of transport data on spherically symmetric systems. However, as the gas temperature is raised, not only is the translational energy of both the ions and molecules increased, but also the internal state distributions of both kinds of particle altered. At each temperature at which measurements are made, both the ions and the molecules have rotational, vibrational, and electronic distributions characteristic of that temperature. In the case of molecular ions in a molecular gas, the physical system under study may change significantly as the temperature is raised from that of liquid nitrogen, say, to a level hundreds of kelvin higher.* Usually, the differences between T-variation and E/N-variation data are small, but in molecular gases they are often noticeable and can be large enough to be significant (Viehland and Fahey, 1983). The differences can be put to good use, however, since they are indicative of the importance of excited states in the gas population.

Now we can make a detailed comparison between the two types of measurements. The E/N-variation method allows simpler apparatus and a much larger change in the average ion-neutral collision energy. (Typically, as one goes from low field to the highest value of E/N attainable, the average ionic kinetic energy changes by a factor of 150; in the case of Li^+ in He, a factor of 250 has been realized.) A wide range of average impact energy is important for accurate potential determinations. Finally, the Viehland–Mason theory now permits the accurate analysis of data obtained at arbitrary E/N, and if the gas temperature is held fixed, the data refer to the same excited-state distribution of neutrals throughout the entire E/N range covered.

The temperature-variation method offers two strong advantages. First, it permits measurements at gas temperatures far below 300 K that can probe the ion-neutral potential at very large separation distances not accessible at room temperature for any value of E/N. Second, in certain applications such as planetary atmospheres, the gas of interest may be very hot or very cold, and direct measurements at the prevailing temperatures are desirable. The change in the rate coefficient of an ion-molecule reaction with temperature can be extremely dramatic (Schmeltekopf et al., 1968). However, in addition to the inherent complexity of temperature-variation apparatus, this method is severely restricted in its dynamic range—to date, the thermal energy has been varied only through a factor of 12.

---

*In the rigorous kinetic theory of molecular gases, each molecular quantum state is treated as a different chemical species with its own rate equation. For a given collision partner, different molecular states have different interaction potentials or potential surfaces, different differential cross sections, and different transport cross sections. For discussions of the internal energy distributions of molecular ions in drift tubes, and the vibrational excitation and deexcitation of such ions, see Federer et al. (1985) and Lindinger et al. (1984).

1. Canberra Apparatus: The Bradbury–Nielsen Method

This instrument was constructed by Elford and Creaser at the Australian National University during 1966–1968 and subsequently modified to the form shown in Fig. 2-5-3. It permits measurements to be made over a wide range of gas temperature by the use of an internal dewar which surrounds the entire drift section.

The ion sources used are of two types. For alkali-ion measurements (Milloy and Elford, 1975), four coated filaments are used in one assembly so that the ion species emitted can be changed without opening the apparatus. For measurements of mobilities of ions in their parent gas (e.g., Helm, 1975) an ion source using an α-emitting foil (Crompton and Elford, 1973) is used. In both cases the same tubular support is employed. For measurements at 77 K this support is filled with liquid nitrogen to reduce thermal gradients over the electrode system. A radiation shield that is closely linked thermally to the dewar also assists in reducing heat flow into the drift tube.

The ions enter the drift region through a 3.2-mm-diameter hole located in a metal plate at the top of the electrode system. The guard rings (49.5 mm inside diameter) used to produce a uniform electric field over the central region of the drift space are of the "thick ring" type (Crompton et al., 1965).

Two combinations of three electrical shutter grids A, B, and C are used to determine the drift velocity of the ions. Each shutter grid consists of parallel and coplanar Nichrome wires 0.008 cm in diameter spaced 0.04 cm apart. The spacings between the shutters A-B and B-C are 9.0 and 3.0 cm, respectively. All surfaces exposed to the ion swarm are gold plated. The exit plate at the bottom of the drift space is insulated from the dewar and may be used to collect the ion current. A small fraction of this current passes through a knife-edged sampling hole (0.2 mm diameter) in the center of the exit plate and is transmitted via an ion lens to an RF quadrupole mass spectrometer and a particle multiplier, if the ions are of the appropriate mass.

Sine-wave signals are applied between adjacent grid wires of each shutter grid, the mean potentials being those corresponding to a uniform drift field. The signals applied to each shutter grid are in phase and the drift velocity is determined by measuring the transmitted current as a function of the frequency of the sine-wave signal. Maxima in the transmitted ion current are observed when the mean ion transit time is equal to integral multiples of half the period of the sine-wave signal.* The ion current may be measured at the exit plate or, when mass analysis is employed, at the output of the mass spectrometer. A pulse counter or an electrometer is used depending on the magnitude of the ion current transmitted through the quadrupole to the particle multiplier.

The apparatus has been used to measure the mobilities of alkali ions in noble gases (Cassidy and Elford, 1985a, b, 1986) and to investigate mass discrimination effects in sampling ions from drift tubes (Milloy and Elford,

---

*This method of timing the flight of the charge carriers is called the Bradbury–Nielsen method (Bradbury and Nielsen, 1936).

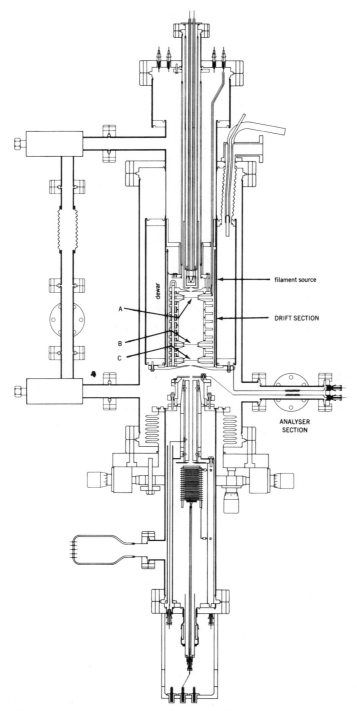

Figure 2-5-3. Canberra drift-tube mass spectrometer. The version shown here differs only slightly from that described by Cassidy and Elford (1985b).

1975). The instrument has also been applied to the study of the mobility and reactions of ions in their parent gases and in particular the dependence of the mobility of atomic ions on the ionic spin state. The first such observation was made by Helm (1975), who measured the mobilities of $Kr^+(^2P_{3/2})$ and $Kr^+(^2P_{1/2})$ in krypton at 295 K and showed that the mobilities differed by about 3% over the E/N range 50 to 85 Td. The study of the mobility of atomic ions in specific spin states was extended to xenon (Helm, 1976a) and neon (Helm and Elford, 1977a).* Mobility measurements were also made using Ar^+ ions in Ar but only one ion species was observed at both 77 and 298 K (Helm and Elford, 1977b). The Canberra DTMS has also been used to study reactions of helium ions in helium at 77 K (Helm, 1976b) and diatomic rare gas ions in their parent gases, helium, neon, argon, and krypton (Helm and Elford, 1978).

2. *Apparatus of Johnsen, Chen, and Biondi*

In this instrument a 15-cm-long drift tube is housed in a large vacuum chamber, which provides both thermal insulation and differential pumping, and also contains an ion analyzer and detector section. One version of the drift tube is equipped with a cooling jacket and also with a heater that may be used to raise the temperature to about 550 K. Since it was not practical to construct a single drift tube capable of operating over the desired range from 77 to about 1500 K, a similar version of this drift tube has been constructed from high-temperature materials (tantalum metal with alumina or sapphire insulators) that is capable of being heated to 1500 K. To date, data have been obtained for temperatures up to only 900 K because of experimental problems arising from chemical reactions of gases with wall materials and electron emission from hot surfaces. Both the high- and low-temperature versions are built in the form of "plug-in units" fitting a common vacuum housing and analyzer section.

In most reaction rate measurements, use is made of the additional-residence-time technique (described in Section 2-5C), in which the ion motion is interrupted by a suitably programmed drift field. The parent ion loss rate, from which the reaction rate coefficient is inferred, is thus measured at the gas temperature, and the ions are in thermal equilibrium with the ambient gas. The apparatus, experimental method, and applications are described by Johnsen et al., (1980a–c).

3. *Georgia Tech High-Pressure, Variable-Temperature Drift-Tube Mass Spectrometer*

This instrument (shown in Fig. 2-5-4) is a modified, improved version of the "Plasma Chromatograph," designed and constructed by the Franklin GNO Corporation. It is capable of operation over the pressure range of about 10 torr up to 2 atm (1500 torr) and at temperatures ranging from about 200 to 675 K. Ions produced by a ^{63}Ni radioactive source drift for either 5 or 10 cm and are

---

*For more recent studies of this type, see Koizumi et al. (1987).

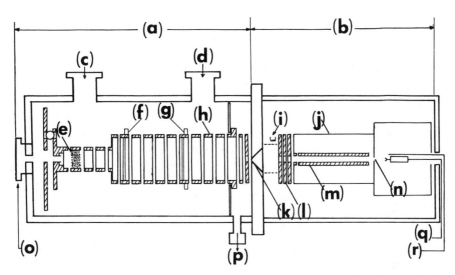

Figure 2-5-4. High-pressure drift-tube mass spectrometer used in the Georgia Tech Laboratory. (a) Drift region, (b) analysis region, (c) gas inlet, (d) pressure gauge port, (e) ^{63}Ni source, (f) grid 1, (g) grid 2, (h) guard ring, (i) filament, (j) quadrupole housing, (k) exit aperture, (l) ion optics, (m) poles, (n) aperture, (o) gas inlet, (p) gas outlet, (q) and (r) electron multiplier terminals. The heating tapes and cooling coils and not shown here.

then gated through either grid 1 or 2 while the other grid remains open. Either of the two identical grids acts as the instantaneous source of the ions, and thus two drift distances are available, making possible a measurement of drift velocity that is independent of end effects. The ions then continue to drift in a uniform electric field produced by 5-cm-ID drift-field electrodes, and finally pass through an exit aperture into a high-vacuum region for focusing, mass selection, and detection by an electron multiplier pulse counter. For drift-tube pressures of 30 torr and 1 atm, exit aperture diameters of 100 and 25 μm, respectively, are used. Neutral molecules passing through the exit aperture can be ionized by electron bombardment and then mass analyzed, if desired.

The drift tube is constructed to meet high-vacuum specifications and is bakeable to 400°C. A 6-in. diffusion pump evacuates the chamber to 10^{-7} torr prior to injection of the gas. Operation at low pressures is limited mainly by the ion source, which emits electrons with a maximum energy of 63 keV. These beta particles are confined to the source region only by collisions with the gas molecules.

Most of the measurements with this apparatus have been made at drift-tube pressures of tens of torr (Ellis et al., 1978; Eisele et al., 1979, 1980, 1981; Perkins et al., 1981, 1983). The instrument is well suited for mobility and reaction rate measurements on many systems, but cannot be used to measure diffusion coefficients because of the high pressures employed. A possible problem that must always be kept in mind concerns the formation of ion clusters in collisions of ions with molecules as the jet of molecules and ions leaves the drift tube and enters the low-pressure analysis region. During the expansion of the jet, the gas temperature drops, and clustering will occur if the drift-tube gas temperature is low enough and if certain species of molecules (notably H_2O) are sufficiently abundant. At present this phenomenon is not understood quantitatively.

An advantage offered by the high-pressure capability of this apparatus is that extremely slow ion-molecule reactions can be studied with it. Another advantage is that weakly bound ions can be formed by thermal collisions of primary ions with the molecules in the source region, and that these ions then survive for study in the drift region. Such ions (H_2SO_4 clusters are good examples) would be difficult to make in a different environment and would probably be dissociated if injected from outside into a drift tube.*

Two disadvantages of the apparatus are: (1) reactions with the drift-tube gas are enhanced because of the high pressures and complications in the drift-tube arrival-time spectra may result; and (2) trace constituents tend to have larger effects than in low-pressure apparatus.

4.   Böhringer and Arnold's Drift Tube with Liquid-Helium-Cooling Capability

A drift-tube mass spectrometer developed at the University of Heidelberg by Böhringer and Arnold (1983, 1986) has an important unique feature—the ability to operate at temperatures as low as 20 K by means of liquid-helium cooling. The drift tube can also be electrically heated to 420 K. The cooling capability requires small dimensions for the drift region (2.9 cm length and 10 cm diameter), and the length is fixed. However, very useful data on ion-molecule reaction rates have been obtained, along with some interesting mobility data. (The Heidelberg laboratory is concerned mainly with the former.) The apparatus is considerably enhanced by a SIFT ion injection source (see Sections 2-5E3 and 2-5E4).

C.   Pittsburgh Reversible Field Apparatus

In the late 1960s, Johnsen and his colleagues at the University of Pittsburgh constructed a drift-tube mass spectrometer that possesses two unusual and especially interesting features. This apparatus was designed primarily for the study of ion-molecule and charge transfer reactions, but it has also been used for some important measurements of mobilities (Heimerl et al., 1969; Johnsen et al., 1970).

The apparatus is shown schematically in Fig. 2-5-5. One of its unusual features relates to the construction of the ion source, which is provided with a separate gas inlet and which communicates with the drift region only through a number of very small holes. This arrangement facilitates the study of ions in a

---

*This problem is mitigated with varying degrees of success by various apparatus that employ SIFT techniques (see Sections 2-5E3 and 2-5E4.

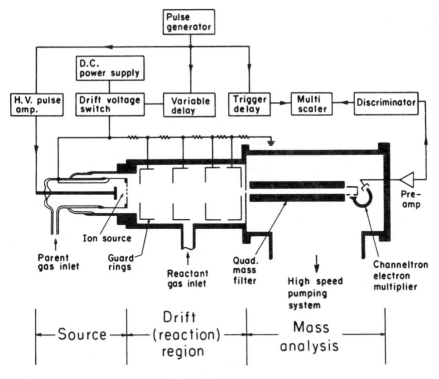

Figure 2-5-5. Original reversible field apparatus (Heimerl et al., 1969).

foreign gas in a manner described below. The other unusual feature of the apparatus is that the electric field in the drift region may be reduced to zero or reversed in direction for a variable length of time during the passage of a burst of ions through the tube. This feature is not utilized in the measurement of drift velocities but has proved to be useful in the study of ionic reactions by a new technique which Johnsen and his associates have called the "additional residence time method." This method is based on the measurement of the loss of parent (i.e., primary) ion current caused by reactions and is discussed at the end of this section.

The ion source in the apparatus employs an electrical discharge between a plane anode and a plane cathode, both of which are made of stainless steel. The cathode is a disk in which a single hole was drilled at its center and six other holes were spaced 60° apart on a 1.3-cm-diameter circle. These seven holes, each of which has a diameter of 0.045 and a thickness of 0.013 cm, permit ions generated in the active discharge to enter the drift region. The disk, 1.9 cm in diameter, which forms the anode is mounted on a threaded assembly that permits variation of the anode-cathode distance from 0 to 1.3 cm by a magnetic coupling to four soft-iron pieces. A high-voltage pulse (1 to 3 kV, 5 to 10 μs long, repeated every 1.3 ms) is applied to the ion source and generates a discharge plasma in the gap between the anode and cathode. The separation distance between the two electrodes is set to optimize the discharge conditions for the particular gas filling used at the time. A negative bias voltage of 50 to 100 V applied to the anode prevents ions from leaving the source during the afterglow period following the active discharge and only a narrow pulse of ions may enter the drift region.

The drift tube is made of stainless steel and is 15.5 cm long. The inside diameter of the guard rings defining the drift field is 2.5 cm. Ions at the end of the drift space are sampled through a hole 0.045 cm in diameter, mass analyzed in an RF quadrupole mass filter, and detected by a pulse-counting Channeltron electron multiplier. The data-handling technique is similar to that described in Section 2-5A. A multichannel analyzer is used to accumulate data over many cycles of operation. A time-of-flight logic unit feeds each ion count into the appropriate channel of a 256-channel computer memory so that the channel address provides a measure of the elapsed time between the discharge pulse in the ion source and registration of an ion. After a sufficient number of cycles have been completed the content of the memory is printed out in digital form to provide the basis of the data analysis.

Ultrahigh-vacuum techniques were employed in the construction of the apparatus and gas-handling system. The entire vacuum system can be baked at 300°C, and a base pressure lower than 10^{-8} torr is usually achieved after bakeout.

In the experiments of Johnsen, Brown, and Biondi with this apparatus (Johnsen et al., 1970) measurements were made of the mobilities of N^+, N_2^+, O^+, and O_2^+ ions in helium. For these measurements a mixture of helium and nitrogen or oxygen was continuously admitted to the ion source through the parent gas inlet at a total pressure of 1 torr, the nitrogen or oxygen being present only to about 1 part in 10^4. Pure helium was admitted to the drift region through its gas inlet to maintain a pressure of 1 torr in this region. Various processes occurring in the ion source led to the production of nitrogen or oxygen ions: direct ionization of the parent molecules by electron impact, ion-molecule and charge transfer reactions between helium ions and the nitrogen or oxygen molecules, and Penning ionization of these molecules by metastable helium atoms produced in the discharge. Arrival-time spectra were mapped for the minority nitrogen or oxygen ions, and in each case the spectrum exhibited an easily identified "source" peak along with a very small tail produced by reactions occurring in the drift region. The source peaks were used to determine the ions' time of flight. Because the gas in the drift region contained a trace of the molecular gas admitted to the ion source an insignificant error in the measured mobilities was produced. In arriving at their final mobility data Johnsen and his associates took into account the time required for an ion to traverse the distance between the sampling plane and the electron multiplier, the difference between the time of arrival of maximum current and the true time of flight, and the effect of spatial nonuniformities in the applied drift field and made appropriate small corrections for these effects.

The reversible-field feature of the apparatus cannot be used to eliminate end effects in drift velocity measurements made with a fixed ion source. Suppose that an ion cloud is released from the source S and allowed to drift partway down the tube to a point A before the field is reversed to bring the cloud back to some point B, from which it then drifts to the exit aperture E under the original field. Only the distance S-E can be directly measured; the distances S-A, A-B, and B-E can only be inferred after the fact from the drift velocity and the times of application of the field in the original, then reversed, and finally again the original directions. These times can be set accurately over a fairly wide range, but end effects could be eliminated only if the overall drift distance S-A-B-E could be determined.

Johnsen et al. measured the rates of reaction of O^+ ions with CO_2, O_2^+ with NO, N^+ with O_2, and N_2^+ with O_2. The parent ions were generated in the manner described above by admitting a mixture of about 10^{-4} torr of oxygen or nitrogen in about 1 torr of helium to the ion source. A small amount of the reactant gas in each case was admitted only into the drift region in a mixture with helium which served as an inert buffer gas to inhibit diffusion and quickly establish energy equilibrium. The pressure in the drift region was maintained slightly lower than that in the ion source so that molecules of the reactant gas could not enter the source. (Diffusion from the drift region into the source is greatly inhibited by the small size of the orifices connecting these regions and the consequent high flow velocity in the opposite direction.) The additional residence-time technique was used for these measurements. In this technique the drift field is turned off or reversed for a selected time interval while the parent ion swarm is near the center of the drift space, thus causing these ions to remain in the drift (reaction) region for an additional residence time Δt. By measuring the decrease in the number of parent ions reaching the mass spectrometer as a function of Δt, first with and then without reactant gas added to the helium in the drift region, the reaction rate could be determined (Heimerl et al., 1969, describe the analysis used.) The net loss of the parent ions due to reaction with the reactant gas during Δt is thus separated from the loss due to transverse diffusion and possible reactions with molecules of the parent gas. The range of E/p covered in these measurements extended from 0 to about 20 V/cm-torr.

When a gas mixture is used in a drift-tube mass spectrometer, as in the experiments described above, a problem may develop in the determination of the composition of the mixture. If the gases have appreciably different molecular weights, they will be pumped out of the drift tube through the exit aperture at significantly different speeds, and the actual composition in the drift tube will differ from that of the mixture admitted to the tube. If this fact is ignored, large errors can result. In the experiments described in this section this factor was taken into account, and the correction that had to be applied typically amounted to about 50%. A detailed analysis of this problem appears in Heimerl's thesis (Heimerl, 1968).

An improved version of this apparatus has been described by Johnsen and Biondi (1973). The improvements include greater length (36 cm between in-

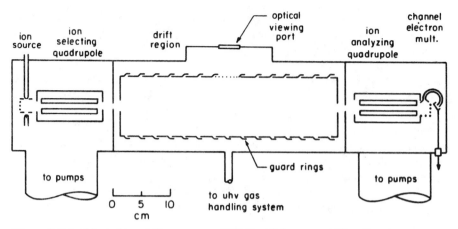

Figure 2-5-6. Selected ion drift apparatus (SIDA) of Johnsen and Biondi (Johnsen et al., 1972).

jection and exit orifices), greater diameter (10 cm), and a differentially pumped electron bombardment ion source. This instrument and a similar one have been used for many important mobility and reaction rate measurements (see, e.g., Johnsen and Biondi, 1972, 1973, 1978, 1979; Graham et al., 1976). The improved instrument discussed here is the same as the SIDA apparatus shown in Fig. 2-5-6, as far as the drift and analysis regions are concerned.

D.   Pittsburgh Selected Ion Drift Apparatus (SIDA)

At least as long ago as the early 1960s it became apparent that considerable advantage would accrue to the use of an external ion source and the injection of a single, selected species of ion into the drift tube. A much wider choice of ion-gas combinations would be made available, and the external ion source would not heat the drift-tube gas. Further, the reduction of the number of ionic species present in the drift tube during a given measurement would simplify the interpretation of the arrival-time spectra in many cases. The successful implementation of this technique, however, requires the use of low injection energies (of the order of tens of eV, or lower) if excitation and dissociation of the ions are to be avoided when the ions enter the drift-tube gas. Early attempts (Bloomfield and Hasted, 1964; Kaneko et al., 1966) required the use of 400-eV injection energies, and higher, and only within the last few years has the injection problem been satisfactorily solved. One of the most successful external ion source–drift tube combinations was constructed at the University of Pittsburgh (Johnsen et al., 1982), and it is chosen for description here.

Ions are produced in an electron-impact ion source (Fig. 2-5-6). A quadrupole mass filter selects ions of the desired mass/charge ratio for injection into the drift region through an orifice of 0.05 cm diameter, typically at an energy of 15 to

20 eV. Breakup of molecular ions in the first energetic collisions with gas molecules has been observed but generally has only a minor disturbing effect. The drift and analysis regions are fairly conventional. The cylindrical drift field electrodes are 10 cm in diameter, and the drift region is 35.3 cm long. Pressures up to 2 torr may be maintained in the drift tube while keeping the differentially pumped mass spectrometer sections at adequately low pressures (several 10^{-5} torr).

The drift tube may be operated in the pulsed mode; that is, ions can be admitted to the drift tube in the form of short pulses, about 10 μs long, repeated at several-millisecond intervals. Each ion pulse contains typically 10^5 to 10^6 ions, of which only a small fraction (1 in 10^5 approximately) is registered by the ion detector. The *total* ion current traversing the drift tube is measured by several electrometers connected to the electrode containing the sampling orifice and the other electrodes at the sampling end of the tube. For some measurements, especially those involving the detection of photons emitted by collision processes in the drift tube, the drift tube may be operated in the continuous mode, and fairly large ion currents (up to 10^{-8} A) have been successfully injected into the drift tube.

The apparatus permits the usual types of drift-tube experiments (e.g., recordings of parent and product ion arrival spectra). In addition, programmed drift fields may be employed for measurements of parent ion loss rates in the presence of reactant gases by means of the additional residence-time method described in Section 2-5C. Because the distance between the injection and sampling orifices cannot be varied, mobility measurements are affected by unavoidable end effects, but with some care data of 3 to 5% accuracy can be obtained.

For spectroscopic observations of optical emissions from ion-molecule or charge transfer reactions the apparatus has been equipped with a sapphire window, through which the central part of the drift region may be viewed with several types of spectroscopic instrumentation, employing either interference filters or grating monochromators in conjunction with low-noise photodetectors. The feasibility of detecting even the feeble radiation emitted by the radiative charge transfer of He^+ ions with neon atoms has been demonstrated recently.

E.   Selected Ion Flow-Drift-Tube (SIFDT) of Albritton, Howorka, and Fahey

We now turn to another formidable apparatus (SIFDT) that incorporates selected ion injection, one, however, that evolved originally from the flowing afterglow (FA), not the drift tube. This difference in origin is reflected in a somewhat different set of problems associated with ion injection, and in certain additional capabilities that can be understood only if one is familiar with the FA techniques. The flowing afterglow has its roots in a remarkably fruitful program of ionic research that was initiated by E. E. Ferguson and his colleagues at what is now known as the Aeronomy Laboratory of the National Oceanic and Atmospheric Administration (NOAA) in Boulder, Colorado. The evolution of SIFDT from the techniques developed there and at the University of Birmingham is discussed below. The rather detailed treatment of the flowing afterglow is justified not only because of the historical connection with SIFDT, but also because so many aspects of the FA research are of general interest in investigations of ionized gases. Further, some important special features of the flowing afterglow can be carried over into SIFDT measurements.

1. Flowing Afterglow (FA) Techniques of Ferguson, Fehsenfeld, and Schmeltekopf

The flowing afterglow techniques developed in the 1960s for ion-molecule reaction studies (Ferguson et al., 1969) differ from the stationary afterglow (SA) techniques to be discussed in Chapter 4 in substituting space resolution for time resolution.* A localized electrical discharge creates a plasma, which is an ionized gas containing both positive and negative charge carriers, the dimensions of the ionized gas being large compared with its debye shielding length. The part of the plasma actually involved in the FA reaction rate measurement is the decaying afterglow plasma that is separated spatially from the active discharge by a fast pumped gas flow. In comparison with the SA, the flowing afterglow offers several important advantages. Of great significance is the fact that the composition of the afterglow can be chemically modified and controlled in space in a way not possible in time. Also, various diagnostic tools such as optical spectrometers and microwave interferometers are more easily used in the FA.

Figure 2-5-7 shows one type of flowing afterglow tube that has been used at NOAA, this one being of Pyrex. Here the discharge electrodes are a large cylindrical cathode and a small wire anode, and the tube is operated in the dc mode. Other tubes have been made of quartz and stainless steel, all of the tubes being about 1 m in length and 8 cm ID. An electron-emitting filament of $BaZrO_3$ is used to attain greater control and wider ranges of the electron density and energy. Both the filament and discharge can be operated in either the pulsed or steady-state mode. In typical operation, helium is pumped through the tube at a pressure of about 0.4 torr, a flow rate of 130 atm-cm^3/s, and a flow velocity of 10^4 cm/s. The helium is exhausted by a large Roots-type blower backed by a large mechanical forepump which obtains a 500-liter/s pumping speed at pressures between about 10 and 10^{-4} torr. The discharge produces about 10^{11} ions/cm^3 and a comparable concentration of triplet helium metastable atoms.

The gas temperature is not raised significantly by the discharge, and electrons produced in the discharge have thermalized by the time they have traveled a few centimeters downstream. The electron density has been measured with both V-

---

*The development of stationary afterglow techniques began decades earlier than the FA (see McDaniel et al. (1970), Sec. 2-3A) and Chapter 4 of this book). The present discussion of FA techniques is an abbreviated and updated version of the treatment prepared by E. E. Ferguson for McDaniel et al. (1970, Sec. 2-3-B).

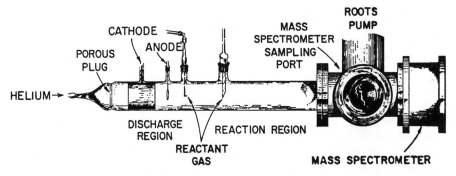

Figure 2-5-7. Pyrex flowing afterglow tube used for ionic reaction studies by Fehsenfeld, Ferguson, and Schemltekopf. Stainless-steel tubes are used almost exclusively today. For a discussion of the use of Langmuir probes in flowing afterglows, see Section 2-5E3 and Smith and Adams (1983) (McDaniel et al., 1970).

(50-GHz) and X-band (9-GHz) microwave interferometers. In addition, the electron density and electron temperature have been measured spectroscopically by relative intensity measurements of Rydberg series extending to very high upper-level principal quantum numbers. Absolute spectroscopic intensity measurements have also been made; they give the absolute population densities of the He-Rydberg states. These densities in turn yield the electron density through the Saha equation. The neutral gas temperature has been determined spectroscopically both by Doppler line width and rotational intensity distribution measurements. Diatomic helium ions are formed from the primary atomic helium ions by the reaction $He^+ + 2He \rightarrow He_2^+ + He$, the He_2^+/He^+ ratio being about 0.1 at the mass spectrometer if very pure helium is used at 0.3 torr. The triplet metastable concentrations have been measured by optical absorption. $He(2^1S)$ singlet metastables are not present in an observable concentration because of their rapid conversion to the triplet metastable state by superelastic electron collisions.

The glass jets shown in Fig. 2-5-7 permit the addition of neutral gases into the flowing helium afterglow plasma, either for the purpose of studying their reactions with the primary discharge ions or for the purpose of producing secondary ions for further reaction studies. This chemical "ion manipulation" will be discussed in detail below. The ion composition of the afterglow is recorded at the end of the reaction region by means of a quadrupole mass spectrometer, followed by a windowless electron multiplier. The decrease in reactant ion current and the increase in product ion current, as a function of concentration of the added neutral reactant gas, constitute the principal measurables of the system. Ion currents at the detector (with direct-current operation) range from 10^{-12} to 10^{-15} A. The sampling hole into the quadrupole is 0.020 in. in diameter, in molybdenum shimstock pieces which have varied from 0.001 to 0.005 in. in thickness. The mass spectrometer chamber is differentially pumped by a 6-in. oil diffusion pump separated from the chamber by a high-efficiency trap. The sampling plate potential is 0.2 V while a drawout potential of about 4.5 V is applied between the plate and the mass spectrometer. Variations in the drawout voltage are found not to affect the measured reaction rates.

The "chemical manipulation" of the helium afterglow to produce the desired ions for reaction studies will now be discussed. Reactions of atomic helium ions can be studied by introducing the neutral reactant into the helium afterglow through one of the glass jets and observing the loss of helium ions as a function of added neutral reactant concentration. The product ions can also be detected and, in addition, ion products of the helium metastable reactions appear. With N_2 addition, the He^+ loss and N^+ increase are due largely to the reaction $He^+ + N_2 \rightarrow He + N + N^+$, whereas the N_2^+ increase is due largely to the slower reaction $He(2^3S) + N_2 \rightarrow He + N_2^+ + e$. Ferguson et al. (1969) discuss the methods of separating product ions into the group produced by an ion (such as He^+) and those produced by a neutral [such as $He(2^3S)$]. Argon is often used in place of helium as a buffer gas and to study argon ion reactions. The large gas flow required precludes routine use of the other rare gases as primary ion sources in the tube because of their cost.

More often the helium or argon afterglow plasma is used to produce a foreign ion whose reactions with neutrals are to be studied. The O^+, N^+, or C^+ ion can be produced by introducing O_2, N_2, or CO downstream from the discharge in a helium afterglow since the cross section for dissociative ionization of these gases by thermal He^+ ions is very large, over 100 $\mathring{A}^2$. Helium triplet metastables react rapidly at thermal energies (cross sections of about 10 $\mathring{A}^2$) with almost every gas to ionize it, so that positive ions of almost all stable molecules can be easily produced by this mechanism. The neon atom is an exception since its ionization energy is greater than the excitation energy of $He(2^3S)$ and not enough internal energy can be transferred to form the Ne^+ ion. No neutral species other than Ne are known to be incapable of ionization by thermal helium triplet metastables.

The FA offers an important advantage in the knowledge and control of the quantum states of the ionic reactants. These ions will almost certainly be in the ground electronic state, either because of the energetics of their formation (e.g., N^+) or, more commonly, because ions (such as O^+) originally produced in excited states are deexcited in superelastic collisions with electrons and collisions with parent gas neutrals before they reach the position where the neutral reactant is added. The situation regarding the vibrational state of molecular ions is much less certain, although in the case of N_2^+ in the helium afterglow, most of the ions are known to be in the ground vibrational state because of the allowed radiative channel from the excited state in which they may have been formed. Of course, the desideratum is a complete knowledge of the full quantum states of both reactants, ion and neutral.

With regard to the knowledge and control of the quantum states of neutral reactants, the latter are usually in their ground electronic, ground vibrational state since they are added to the afterglow without ever entering the discharge upstream. However, the FA permits interesting and important studies of the effects of neutral excitation on the rates of the ion-molecule reactions. For example, Schmeltekopf et al. (1968) excited N_2 vibrationally by a discharge in the side tube before the gas was introduced into the flowing afterglow, determined the vibrational distribution, and deduced the reaction rate constants for He^+ and O^+ as a function of vibrational state. The N_2 vibrational distribution was determined spectroscopically by utilizing electron beam excitation of the vibrationally excited N_2 and analyzing the resulting emission spectrum. The N_2 vibrational temperature was varied from 300 to 6000 K. Such measurements should be possible with many ion-gas systems.

The feature of reactant gas addition downstream from the site of primary ionization (the discharge) permits the study of many ion-neutral combinations with the flowing afterglow that would be inaccessible with other techniques. For example, in a mass spectrometer ion source, a mixture of gases such as H_2 and CO_2 can give rise to a very confusing situation because of concurrent reactions, in this case

$$CO_2^+ + H_2 \rightarrow CO_2H^+ + H$$

and

$$H_2^+ + CO_2 \rightarrow CO_2H^+ + H$$

This confusion does not arise in the FA.*

The FA also demonstrates its versatility by permitting studies of reactions between ions and neutral atoms of unstable species, such as N, O, and H. Nitrogen atoms, for example, are produced by passing N_2 gas through a discharge before admission into the reaction zone, care being taken to ensure that the diameter of the inlet nozzle is large enough that no significant N atom recombination occurs. The N-atom concentration is measured by titrating the N atoms with NO, utilizing the fast reaction $N + NO \rightarrow N_2 + O$. Ferguson et al. (1969) also discuss FA studies made with radicals such as OH and molecules like O_3 and NO_2 that pose special problems in handling and measurement. In addition, they treat the role of impurities in the FA. They argue that impurities do not pose nearly as large a problem, in general, as one might expect provided that certain precautions are observed and checks are made.

Flowing afterglow ion-molecule reaction data can be interpreted in terms of rate coefficients as follows. For a reaction $A^+ + B \rightarrow$ products, the rate coefficient k is defined by

$$-\frac{\partial[A^+]}{\partial t} = k[A^+][B], \qquad (2\text{-}5\text{-}1)$$

---

*Concurrent reactions can be a serious problem in the FA, however, because the gas from which the reactant ions are produced is also present in the afterglow. A good demonstration of this problem appears in attempts to study isotope exchange reactions in a FA. However, the SIFT technique (Section 2-5E) circumvents this problem by excluding the ion source gas from the flow tube.

where the brackets denote concentrations. Over the range of observable A^+ decrease, $[B] \gg [A^+]$, so that $[B]$ can be assumed constant, from the location of the addition jet down the tube to the mass spectrometer port. Equation (2-5-1) then leads to

$$\ln\left(\frac{[A^+]}{[A^+]_0}\right) = -k\tau[B].$$
(2-5-2)

where $[A^+]_0$ is proportional to the A^+ ion signal before B addition. Thus the slope of a logarithmic plot of primary ion current against added neutral reactant concentration, when divided by τ, gives k, Here τ is the ion flow time. Since the absolute neutral reactant concentration must be known in order to obtain k, the neutral flow rate is measured (by flow meters). Note that only the change in the detected current of the reactant ion is required, so possible variation in ion sampling efficiency with ion mass cannot falsify the result.

On pages 14–34 of their review, Ferguson et al. (1969) discuss the analysis* of the flow in their FA apparatus, starting with the simple model of (2-5-1) and (2-5-2). They then consider radial diffusion, nonuniform velocity profile, axial diffusion, axial velocity gradient and slip flow, and inlet effects. If corrections are made for all these effects, the flowing afterglow appears capable of yielding reaction rate coefficients with an accuracy of perhaps 5%. For most of the data on hand, however, only 20 to 30% accuracy is claimed, but this level of accuracy is entirely satisfactory in aeronomy, the field of application of the bulk of the rate coefficients. Most of the other kinds of input data required for atmospheric modeling are known with much less certainty.

Flowing afterglow techniques continue in use at NOAA (Fahey et al., 1982) and elsewhere, and since their introduction in 1963 have been employed in the study of several thousand ion-molecule reactions! Most of the measurements were conducted at room temperature, but variable-temperature apparatus has permitted coverage of the gas temperature range extending from 82 to 900 K.

2. Flow-Drift (FD) Tube of Albritton and McFarland

Ten years after its first use for ionic reaction studies, the flowing afterglow method was extended by Albritton and McFarland at NOAA to allow the measurement of reaction rates, and ion mobilities as well, as functions of E/N and hence average ionic translational energy. The NOAA "flow-drift tube" (McFarland et al., 1973a–c) employs a four-section stainless steel tube 8 cm ID and 125 cm long. Ions are produced by electron impact in the first region, positive ions are separated from negative ions and electrons in the second section, ions of the chosen polarity are periodically bunched by a two-grid shutter in the third section, and then narrow pulses of ions are released at

---

*The most recent analysis of flowing afterglow dynamics was by Adams et al. (1975).

specified times into the drift-reaction section. The latter section contains drift field electrodes similar to those described in Section 2-5A. A buffer gas flow of 80 to 180 atm-cm^3/s (usually helium) is maintained through the tube, where the pressure is typically between 0.19 and 1.25 torr. The apparatus is pumped, and the ions detected, as in the FA apparatus. The FDT is primarily run in a dc mode by leaving the shutters "on." The pulsed mode is used for mobility measurements and some diagnostic studies.

Lindinger and Albritton (1975) have demonstrated the chemical flexibility of the FD tube by making a wide variety of ions in it and measuring the mobility of each species in helium and argon. Lindinger et al., (1975) have also shown the presence of drift-induced vibrational excitation by its effects on rate constants. Perhaps the most important contribution by the FD research was the experimental evidence that the long-standing problem of ion speed distribution has been solved for atomic ions in atomic buffer gases (Albritton et al., 1977; Albritton, 1979). (A detailed discussion of this subject appears in Chapter 8 of this book.)

The flow-drift tube has proved to be a major success. We now turn to the next step in the evolution of SIFDT.

3.   Smith and Adams' Selected Ion Flow Tube (SIFT)

One of the main reasons for the power of the FA and FD techniques is the chemical versatility that they afford, that is, the vast number of ion/gas combinations made available for study. However, Adams and Smith (1976a, b) at the University of Birmingham have taken the matter of primary ion formation in flow tubes yet one step further. A recent version of the part of their apparatus on which we wish to focus our attention is shown in Fig. 2-5-8 (Smith and Adams, 1979); the remainder of the apparatus is similar to the reaction and detection regions of the flowing afterglow tube. Ions are produced in an electrical discharge or electron impact ion source remote from the flow tube. After selection in a quadrupole mass filter, Q, they are injected at low energy through an orifice O into the flow tube. The ions are rapidly thermalized in collisions with molecules of the carrier gas, which is introduced into an annular space around the orifice and enters the flow tube through 12 1-mm-diameter apertures arranged in a circle about the ion beam. These holes direct the gas at high velocity (close to the speed of sound) parallel to the ion "beam" and then down the flow tube. The rapid gas flow causes the pressure in the vicinity of 0 to be much lower than that downstream in the flow tube. A low pressure can be maintained in the mass filter even for a relatively large connecting orifice (typically, 1 mm diameter). The venturi-aspirator action provided by the substantial and rapid gas flow allows useful ion currents to be injected at lab frame energies of about 15 eV, typically; the center-of-mass energy, of course, can be much lower. As stated earlier, a low injection energy is essential if excitation and dissociation are to be avoided.

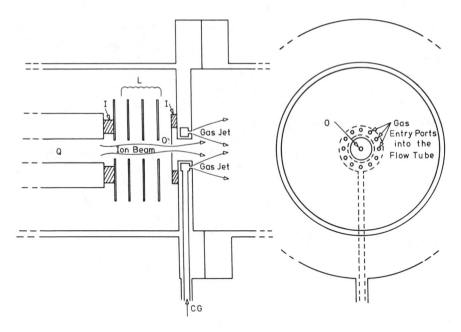

Figure 2-5-8. Elevation and end views of the SIFT ion injector. *Q*: quadrupole mass filter. *I*: insulators. *L*: ion lens. *O*: orifice. *CG*: carrier gas. The dashed lines indicate the tube and cavity through which the carrier gas passes before entry into the flow tube (Smith and Adams, 1979).

It is important to note that only a single species of ion is injected during a given measurement with the SIFT apparatus. The presence of more than one kind of primary ion in the FA and FD apparatus can greatly complicate the analysis of data in certain situations, such as branching ratio measurements (Smith and Adams, 1979). Of course, the fact that charge carriers of only one sign are injected in SIFT means that no plasma is present, in contrast with the FA and FD.

The development of SIFT was a major contribution to the study of ion physics and chemistry. Another unrelated, but important, contribution by the Birmingham group should also be mentioned: techniques for the use of Langmuir probes in flowing afterglows which permit the measurement of rates for electron temperature relaxation, ambipolar diffusion, electron attachment, and electron-ion and ion-ion recombination. In their measurements, Smith and his colleagues use a movable, cylindrical Langmuir probe to determine T_e, n_e, n_+, and n_- along the axis of the afterglow column with a spatial resolution of about 1 mm (Smith and Adams in Lindinger et al., 1984; see also Church and Smith, 1978; Smith and Church, 1976; Smith and Adams, 1982).

Albritton (1978) has compiled data from 1600 ion-neutral reaction measurements made with FA, FD, and SIFT instruments as of the year 1977.

4. *Combination of the FA, FD, and SIFT Techniques to Form the SIFDT*

The SIFDT technique (Fig. 2-5-9) was developed at NOAA as a combination of the techniques discussed above. SIFT furnished the method of producing, selecting, and injecting any single species of ion into a flow tube, the general configuration of which is that of the FA. The technique of incorporating a drift field in the flow tube was provided by the FD.

The emphasis in the NOAA research since its inception has been on the measurement and interpretation of ion-molecule reaction rates of atmospheric significance. As pointed out, a flow tube offers important and unique advantages for rate measurements, not the least of which are associated with choice and preparation of neutral reactants (Schmeltekopf et al., 1968). Mobilities are not required for FA measurements of reaction rates, since the ions flow down the tube at the speed of the gas flow, which is known. If diffusion and mixing are neglected, the rate coefficient is given simply by (2-5-2). In FD and SIFDT, on the other hand, when an electric field is applied down the flow tube, the ions travel faster than the neutral molecules, and the mobility as a function of E/N is required for the rate constant determination. To account for the decreased ionic residence time, we replace the left side of

$$d[A^+]/dt = -k[A^+][B], \qquad (2\text{-}5\text{-}1)$$

by $v\partial[A^+]/\partial z$, where v is the sum of the flow velocity and the drift velocity. The remainder of the analysis is straightforward (McFarland et al., 1973b). The main motivation of the NOAA SIFDT mobility measurements has been to provide

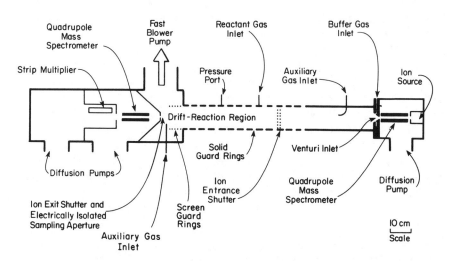

Figure 2-5-9. Selected ion flow-drift-tube mass spectrometer (SIFDT) at NOAA (Rowe et al., 1980).

data for the rate constant determinations, but the intrinsic chemical flexibility of the flow tube technique has prompted mobility measurements for their own sake, especially for ions such as $NO_3^- \cdot HNO_3$ and $NO_3^- \cdot 2HNO_3$ that cannot easily be studied with any other kind of apparatus.

In light of the lengthy description of the FA, FD, and SIFT techniques above, we shall restrict our discussion of SIFDT to a few comments. During a mobility measurement, the "buffer gas" indicated in Fig. 2-5-9 is the single gas present in the flow tube. Typically about 180 STP cm^3/s of this gas is pumped through the tube, with a flow velocity of about 10^4 cm/s and at a pressure of about 0.5 torr. This flow rate is the same as in the FA, and here, also, considerations of expense and corrosive behavior impose some restriction on the gases that may be used: to date, He, Ne, Ar, N_2, O_2, and air have been employed in mobility measurements at NOAA. The pressure in the ion injection chamber is maintained at about 2×10^{-4} torr, and usually 25 to 30 V is applied to inject the ions into the flow tube.* The drift tube section utilizes a double-grid electrical shutter for admitting bursts of ions (≤ 10/s) into the drift region. The apparatus is well suited for mobility and reaction rate measurements, but cannot be used for diffusion studies.

The first paper on the SIFDT apparatus appeared in 1979 (Howorka et al., 1979) and described the first observation of a runaway mobility† (H^+ and D^+ in He; see Chapter 8). Another important early paper dealt with mobility measurements on metastable ions (Rowe et al., 1980). Other experiments are described by Howorka et al. (1980), Fahey et al. (1981a–d), and Viehland and Fahey (1983).

A similar apparatus has been constructed at the University College of Wales and used for a number of mobility measurements (Jones et al., 1981a, b, 1982a, b).

Lindinger and his colleagues at the University of Innsbruck have recently constructed a "slow-flow" selected ion drift tube (SIDT) that utilizes a venturi ion inlet designed so that buffer gas flow rates of only about 1 STP cm^3-s^{-1} can produce a satisfactory jet near the ion injection aperture (Villinger et al., 1984; Lindinger and Smith, 1983). Despite the low gas flow rate, the venturi inlet system permits adequate currents of even weakly bound ions to be injected into the drift region at sufficiently low energies that only a small fraction of the ions are collisionally dissociated. The slow-flow SIDT is proving very useful for the measurement of ion-molecule reaction rate coefficients.

We now turn to two drift tubes without mass spectrometers (Beaty's and Hornbeck's) that illustrate important experimental techniques and that played an important part in the development of the subject of ion mobilities.

---

*Dupeyrat et al. (1982) have made a diagnostic study of venturi inlets for flow tubes. Their results demonstrate significant differences between various inlet designs, and suggest general rules for their use.

†Howorka has discussed this phenomenon in the book by Lindinger et al. (1984).

F.   Beaty's Drift Tube: The Tyndall Four-Gauze Electrical Shutter Method

An extensive and important series of mobility measurements was made by Tyndall and his collaborators at Bristol University during the 1930s with the four-gauze, or four-grid, electrical shutter method developed by Tyndall, Starr, and Powell (Tyndall, 1938; Loeb, 1960). This method has also been used more recently by Beaty (1962), who took advantage of the subsequent advances in electronics and vacuum techniques to improve on the original version of the apparatus. Neither Tyndall nor Beaty employed a mass spectrometer in his apparatus, but both worked with systems that were simple enough to produce results that for the most part could be interpreted in a straightforward manner. The four-gauze method is discussed here in terms of Beaty's apparatus (Fig. 2-5-10): (a) is a schematic representation of the electrode geometry. A typical potential distribution applicable to a portion of the measurement cycle is shown in (b) and the timing sequence, in (c). The two pairs of closely spaced grids form two electrical shutters (Tyndall gates), a distance of about 1 mm separating the

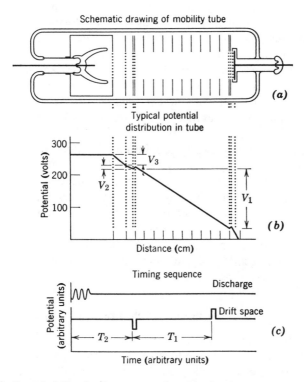

Figure 2-5-10. Beaty's drift tube. Ions are produced in the electrical discharge ion source at the left and are collected by the plane electrode at the right following their drift across the tube. No mass analysis is employed (Beaty, 1962).

grids composing each shutter. The space between the two shutters is the drift space; it has guard rings with holes 1.0 in. in diameter placed 1 cm apart to maintain a uniform electric field. The potential difference across the drift space is labeled V_1. Between the discharge-type ion source and the first shutter is a space in which the ions from the discharge are degraded in energy before their drift velocity is measured. This thermalizing space is divided into two regions by a grid, and, as indicated, the potentials across these two regions are designated V_2 and V_3; V_1, V_2, and V_3 can be varied independently.

The gas in the ion source is ionized periodically for intervals of 7 μs by a gated 20-MHz oscillator. The ions diffuse to the walls under the influence of the space charge fields, and some of the ions pass through the holes at the front of the source and into the thermalizing region. After a time T_2 a negative voltage pulse, which is applied to the entire drift space, opens the first shutter and allows ions to enter the drift space. After another delay T_1 a positive pulse opens the second shutter to admit ions to the collector. The advantage of this arrangement of pulsing is that the electric field is not affected by the pulses anywhere except in the shutters. The output of an electrometer attached to the collector is connected to an X-Y recorder. The electrometer current can be plotted as a function of V_1, V_2, V_3, T_1, or T_2, and if the reaction pattern in the gas is simple the drift velocity can be determined as a function of E/N by a straightforward procedure.

The gas-handling apparatus consists of a Pyrex vacuum system which can be evacuated to a pressure of 10^{-9} torr, a mercury cutoff to disconnect the pump, a bakeable metal valve to admit gas from a cataphoresis purification tube, a mercury manometer, a McLeod gage for reading pressures less than 6 torr, and a liquid nitrogen trap for removing mercury vapor. After sealing the cutoff, and with the ionization gage operating at low emission, it takes several days for the pressure to reach 10^{-6} torr.

Beaty used this apparatus to measure the mobility of positive ions of argon in the parent gas over an E/p_0 range extending from about 1 to 80 V/cm-torr (Beaty, 1962). Pressures of 0.4 to 17 torr were used. Beaty and Patterson, who have also made studies of helium and neon with this apparatus, obtained mobilities and reaction rates for conversion of the atomic to molecular ions (Beaty and Patterson, 1965, 1968). Patterson has constructed a drift-tube mass spectrometer that contains a drift tube similar to the one described here (Patterson, 1970). Recently Williams and Elford (1986) made a study of the action of the shutters in the four-gauze method.

G. Hornbeck's Pulsed Townsend Discharge Apparatus

Another drift tube which makes no provision for mass analysis but which nevertheless has been used in a number of significant experiments on the noble gases is that of Hornbeck (Hornbeck, 1951a, b; Varney, 1952). This apparatus is shown in Fig. 2-5-11. A 0.1-μs burst of photoelectrons is released from a cathode C by ultraviolet light from a spark source operated at a repetition rate of 60 Hz. These electrons are accelerated through the gas at high field and produce a

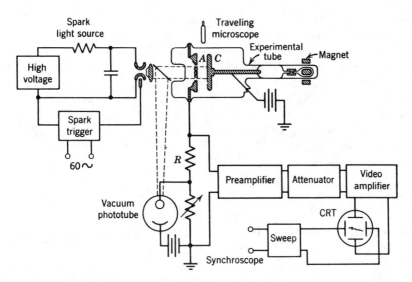

Figure 2-5-11. Hornbeck's pulsed Townsend discharge apparatus. No provision is made for mass analysis of the ions (Hornbeck, 1953a).

Townsend avalanche. The primary and avalanche electrons are collected at the anode A in a time of the order of a few tenths of a microsecond, but the exponential distribution of positive ions formed in the avalanche is swept across to the cathode much more slowly. A voltage transient is developed across the resistor R and displayed on an oscilloscope. The transient consists of a sharp spike, attributable to the photoelectrons and their progeny electrons, and a smaller component produced by the positive ions. The ionic drift velocity is determined by measuring the arrival time of ions formed "at" the anode in the discharge; this time is signaled by a break in the voltage trace on the oscilloscope. The separation between the cathode and anode may be varied by means of an external magnet and measured by a traveling microscope. The gap spacing is typically about 1 cm. Pressures in the range of 0.1 to 30 torr have been used, and ionic drift times of about 2 to 20 μs result. The tube current is of the order of 0.1 μA. Measurements have been made over a range of E/p_0 extending from about 10 to 2000 V/cm-torr.

The Hornbeck technique is particularly useful for obtaining data at high E/p; however, it is intrinsically not well suited to low-field measurements and indeed with this technique it is difficult to reach low values of E/p. Another disadvantage associated with this method is that it does not permit much control over the initial ionization of the gas, since ions are produced by electrons with a wide energy spread and the ionization is not closely confined spatially. The analysis of the ion production and transport is rather involved because of the complicated nature of the physical processes; however, the mathematics is not intractable because the drift-tube geometry and method of generating ions lead to an essentially one-dimensional problem.

The Hornbeck technique has been utilized with mass analysis of the ions arriving at the cathode by McAfee and Edelson in the first drift-tube mass spectrometer built at the Bell Telephone Laboratories (McAfee and Edelson, 1963a). To obtain data at lower E/p than would have otherwise been possible, McAfee and his colleagues subsequently modified their drift tube. They added a grid to confine the Townsend discharge to a smaller volume and thus permit the use of a weaker electric field in the drift region than that required in the discharge region (McAfee et al., 1967).

H. Miscellaneous

We now discuss several additional topics that relate to drift tubes, but not necessarily to their use in the measurement of transport properties of ions.

1. "Continuous Guard-Ring" System and Its Use for Flight-Distance Scanning

Iinuma et al. (1982, 1983) have constructed a new kind of DTMS for the measurement of ionic transport coefficients. They do not employ the conventional set of separate guard rings to establish the drift field. Instead, they use a cylindrical drift tube made of high-purity aluminous porcelain, and uniformly coat its inner surface with an inorganic thick-film resistor cured at a peak temperature of 850°C. Flat metal rings are attached to the ends to serve as terminal electrodes. This "continuous guard-ring" arrangement produces essentially only an axial electric field in the drift region and eliminates the need for insulating spacers and complicated internal wiring. It also permits the use of both the conventional drift time scanning method and a new drift distance scanning method, so that both the conventional arrival time spectra (ATS) and the new flight distance spectra (FDS) can be obtained. In the distance scanning (constant time) method, ion current to a collector at the end of the drift tube is measured while the ion source is slowly moved down the drift tube with a constant delay time between ion gating pulses, and the spatial distribution of the ions is mapped. The solution of the transport equation for the FDS has a considerably simpler mathematical form than does that for the conventional ATS. The solution for the FDS is a function of z with t held constant; the solution for the ATS is a function of t with z held constant (Section 2-6B). In the case of no reactions, the FDS is described by the function

$$\Phi(z) = A(z + C_1)e^{-(z - C_1)^2/C_2}. \qquad (2\text{-}5\text{-}3)$$

Here $\Phi(z)$ is the ion flux on the drift tube axis as a function of the drift distance z, and A, C_1, and C_2 are constants that are closely related to the transport coefficients v_d, D_L, and D_T.

2. Measurements of the Energy Distribution of Drift-Tube Ions

The difficult problem of determining the energy distribution of ions drifting through a gas in steady state at E/N appreciably above zero has long been of interest, and the theoretical and experimental aspects of the subject have been reviewed recently by Rees (1978) and Skullerud (1981).* Until recently, measurements of the energy distribution have been attempted by performing retarding potential measurements on ions sampled from the end of a drift tube (Moruzzi and Harrison, 1974; Ong et al., 1981). This procedure is seriously flawed by the fact that only ions moving nearly parallel to the drift tube axis when leaving the tube are admitted to the retarding potential analyzer. Recently, however, Fhadil et al. (1982) have made a significant improvement by substituting a cylindrical electrostatic momentum analyzer facing the exit aperture of the drift tube and very close to it. It is probably too early to judge the success of this technique. The experimental problems are formidable, and one might do well to rely on theoretical calculations rather than the experimental results at this time. Recent work with the momentum analyzer has been reported by Fhadil et al. (1985); Ong and Hogan (1985); and Hogan and Ong (1986).

A new and promising approach to the measurement of ionic velocity distributions has been introduced by Dressler et al. (1987) at the Joint Institute for Laboratory Astrophysics. They have used single-frequency laser-induced fluorescence to determine Doppler profiles of Ba^+ ions drifting in a helium-filled drift tube. Ionic velocity distributions can be determined from the profile measurements with the laser beam directed both parallel to and perpendicular to the drift field vector.

3. Drift-Tube Studies of Photodetachment and Photodissociation

We now mention an apparatus that is somewhat similar to the Georgia Tech drift-tube mass spectrometer (Section 2-5A) but that possesses an important additional feature. Moseley and his colleagues (Moseley, 1982; Moseley et al., 1975a; Huber et al., 1977; Cosby et al., 1978; see also Hansen et al., 1983) added a laser perpendicular to the drift-tube axis and close to the exit aperture, and by measuring the currents of negative molecular ions emerging from the drift tube with the laser alternately on and off, they have been able to obtain cross sections for photodetachment and photodissociation as a function of the laser frequency. This technique offers the significant advantage of allowing the molecular ions to be collisionally deexcited to the ground vibrational state during their drift before they are exposed to the photons, thus avoiding the usual mixture of different vibrational states. Indeed, by adjusting the drift distance and drift-tube pressure, one may look at the approach to equilibrium in the deexcitation process and measure the rates of deexcitation.

---

*Also see the review by Skullerud and Kuhn in Lindinger et al. (1984).

4. Laser-Induced Fluorescence Studies of Ion Collisional Excitation in a Drift Field

Leone and his colleagues have introduced a new method of investigating the collisional excitation of molecular ions in a flowing afterglow drift tube (Duncan et al., 1983). In their first experiment, N_2^+ ions were brought to a well-defined average kinetic energy in a uniform drift field, and allowed to collide with buffer-gas helium atoms at center-of-mass energies set between 0.039 and 0.054 eV. Rotational, but not vibrational excitation of the N_2^+ ions was observed. The rotational state distribution could be fitted by a Boltzmann temperature corresponding to the CM collision energy. The results indicated that 10 or fewer collisions were required to equilibrate fully the new rotational distribution. Laser-induced fluorescence was used as a direct optical probe of the state distribution of the ions. Similar studies have been made of the vibrational and rotational distributions resulting from nearly thermal charge transfer reactions. A flowing afterglow source was used with a sampling orifice that produces a supersonic expansion (see, e.g., Lin et al., 1985).

5. Study of Tropospheric Ions by Drift-Tube Mass Spectrometric Techniques

In recent years, it has been realized that ions in the troposphere (sea level up to 15 to 20 km) may have environmental significance. Direct sampling and study of these ions is difficult because of their low concentrations and because of the high pressure of the ambient air. However, Eisele has developed a series of instruments at Georgia Tech that do permit direct observations. The latest of his instruments (Eisele, 1987) will be described below, after a few preliminary remarks that will establish the need for some of its unique features.

During the period 1982–1985, Eisele and his colleagues (see, e.g., Eisele and McDaniel, 1986) identified $NH_4^+ \cdot (H_2O)_n$ and $NO_3^- \cdot (HNO_3)_m \cdot (H_2O)_n$ as important constituents of the lower troposphere. Typically, however, $NH_4^+ \cdot (H_2O)_n$ did not account for most of the positive ions observed. Rather, the positive ionic spectrum was complex and variable, and frequently showed core ion peaks at masses 80, 94, and 108 amu; and sometimes peaks at 60 amu and at many masses in the range 150 to 400 amu. Knowledge of the masses of these ions and the fact that they were believed to have high proton affinities did not suffice for a unique chemical identification. Eisele introduced elements into his apparatus to dissociate the complex ions collisionally and thereby reveal the identity of the ionic cores about which the complex ions are formed. This technique is discussed below, with reference to Fig. 2-5-12.

The vertical element at the left of this figure is a "flow-opposed drift tube," with the electric field lines shown for positive ion operation. Air containing ions is drawn in at the top of the column, and the opposing drift field concentrates the sampled ions and raises them to an elevated potential. The ions are extracted from the airflow by a weak transverse electrostatic field, which increases in strength as the ions approach the vacuum entrance aperture. The field is strong

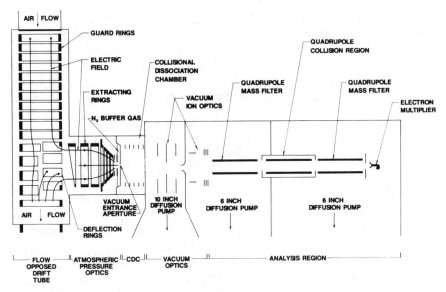

Figure 2-5-12. Eisele's tandem mass spectrometer apparatus for sampling tropospheric ions (Eisele, 1988).

enough (10^3 V/cm) immediately in front of the aperture to move the ions toward the aperture through a flow of dry buffer gas (N_2 or Ar) at the left of the aperture plate. This arrangement permits ions to be sampled without flooding the analysis region with air, which typically contains large concentrations of water vapor. Sampled ions in a dry buffer gas then enter the collisional dissociation chamber (CDC) through the aperture shown, which is 180 μm in diameter for measurements at sea level. This first collision region permits removal of most of the H_2O clusters from the sampled ions by multiple low-energy collisions. Other weakly bound species such as HNO_3 can also be removed in the CDC at E/N values up to 250 Td. The apparatus can be operated with or without the CDC. When it is used, it greatly simplifies the ion spectra and produces a slight increase in ion signal, but causes very little dissociation of the core ions of interest.

Ions leaving the CDC pass through a 15-mm-diameter gridded aperture, are focused onto a skimmer of 10 mm diameter and then enter the analysis region. The first stage of a $\frac{5}{8}$-in. triple quadrupole mass spectrometer (Extrel) transmits only ions of a preselected mass. These ions are then fragmented in the second stage, and the fragment ions are mass identified in the third stage, with 1 amu resolution over the mass range of interest. If all stages of the tandem mass spectrometer are used as indicated above, the detection sensitivity of the apparatus is several hundred ions per cubic centimeter. If, on the other hand, the tandem is operated as a single mass spectrometer, it is possible to mass identify ions present in the sampled air at concentrations of 10 cm^{-3}, provided that the spectrum is simple.

The apparatus is transportable and has been used for field studies at widely separated geographical locations and altitudes. The identification of tropospheric ions and the determination of their abundances is proving valuable in the study of neutral molecules present in the troposphere in trace concentrations. Some of these neutrals, especially those that are very acidic or very alkaline, are known to be important in the environment.

2-6. MATHEMATICAL ANALYSIS OF IONIC MOTION IN DRIFT TUBES

A. Introduction

In Sections 2-2 and 2-3 we discussed in general terms the experimental techniques involved in the determination of ionic drift velocities and longitudinal diffusion coefficients. We saw that a necessary condition for obtaining a drift velocity in which we can have confidence is that the experimental arrival-time spectrum from which the drift velocity is derived must closely resemble the spectrum calculated by solving the transport equation for the ion in question in the apparatus used for the experiment. Furthermore, we saw that a solution of the transport equation is required for the accurate determination of the longitudinal diffusion coefficient. Therefore it is now appropriate to show how the transport equation may be set up and solved for a typical experimental situation existing in the measurement of v_d and D_L with a particular apparatus. The solution we obtain can also be used for the determination of the transverse diffusion coefficient and the rate of chemical reactions between the ions and the molecules filling the drift tube, as we show in Chapter 3.

The apparatus for which the analysis applies is the Georgia Tech drift-tube mass spectrometer described in Section 2-5A. We analyze the space-time behavior of a swarm of primary ions which are diffusing and reacting with molecules as they drift through the apparatus. The case in which no reactions occur may be described by putting the reaction frequency equal to zero in the final solution. We assume that all the primary ions are produced in the ion source and none is derived from other ionic species by reactions along the drift path. This is the usual situation prevailing in the measurement of v_d and D_L.

B. Solution of the Transport Equation for Primary Ions in the Georgia Tech Low-Pressure Drift-Tube Mass Spectrometer (Moseley, 1968; Moseley et al., 1969a)

Consider a population of ions of a single type created at one end of a cylindrically symmetric drift space filled with gas of uniform number density N in which there exists an externally applied uniform electric field $\mathbf{E}$ along the axis. Assume that the number density of the ion swarm $n(\mathbf{r}, t)$ is low enough to make the space-charge field negligible. In the general case in which E/N is not assumed to be small the ionic flux density is given by the equation

$$\mathbf{J}(\mathbf{r}, t) = \mathbf{v_d} n(\mathbf{r}, t) - \mathbf{D} \cdot \nabla n(\mathbf{r}, t) \qquad (2\text{-}6\text{-}1)$$

(cf. Section 1-1). Here $\mathbf{v_d}$ is the drift velocity and $\mathbf{D}$, the diffusion tensor characterizing the motion of the ions through the gas. If we do not allow for the possibility of gain of the ionic species under consideration during the movement down the drift tube but do consider the loss of these ions by chemical reactions with the frequency α, the ion swarm is subject to a continuity equation of the form

$$\frac{\partial n}{\partial t} + \nabla \cdot \mathbf{J} + \alpha n = 0 \qquad (2\text{-}6\text{-}2)$$

[cf. (1-9-1)] or, in light of (2-6-1),

$$\frac{\partial n}{\partial t} - \nabla \cdot \mathbf{D} \cdot \nabla n + \mathbf{v_d} \cdot \nabla n + \alpha n = 0. \qquad (2\text{-}6\text{-}3)$$

(cf. Gatland and McDaniel, 1970). With the explicit form for $\mathbf{D}$ suggested in (1-1-7) and the addition of a source term $\beta(\mathbf{r}, t)$ to represent an input of ions at the beginning of the drift space (2-6-3) becomes, in rectangular coordinates,

$$\frac{\partial n(x, y, z, t)}{\partial t} = D_T\left(\frac{\partial^2 n}{\partial x^2} + \frac{\partial^2 n}{\partial y^2}\right) + D_L\frac{\partial^2 n}{\partial z^2} - v_d\frac{\partial n}{\partial z} - \alpha n + \beta(x, y, z, t).$$

$$(2\text{-}6\text{-}4)$$

In the apparatus under consideration the ions enter the drift space through a circular aperture lying in a plane normal to the drift-tube axis and centered on the axis. If the coordinate system has its origin at the center of this aperture and the electric field along z is in a direction that causes the ions to drift in the positive z direction, the solution to (2-6-4) in unbounded space* is

$$n(x, y, z, t) = \int_{-\infty}^{t} dt' \int_{-\infty}^{\infty} dx' \int_{-\infty}^{\infty} dy' \int_{-\infty}^{\infty} dz' \frac{\beta(x', y', z', t')}{[4\pi(t - t')]^{3/2} D_T D_L^{1/2}}$$

$$\times \exp\left\{ -\alpha(t - t') - \frac{(x - x')^2 + (y - y')^2}{4D_T(t - t')} - \frac{[z - z' - v_d(t - t')]^2}{4D_L(t - t')} \right\}$$

$$(2\text{-}6\text{-}5)$$

*In the Georgia Tech apparatus the drift-field guard rings have such a large inside diameter that few ions ever reach these rings by diffusion. Hence the drift tube may be considered to have an infinite radial extent (Schummers, 1972).

That (2-6-5) is indeed a solution to (2-6-4) has been demonstrated by Moseley (1968). To express (2-6-5) in cylindrical coordinates let

$$x' = r' \cos \theta'; \quad y' = r' \sin \theta'; \quad x = r \cos \theta; \quad y = r \sin \theta. \quad (2\text{-}6\text{-}6)$$

Then

$$n(r, \theta, z, t) = \int_0^\infty r' \, dr' \int_0^{2\pi} d\theta' \int_{-\infty}^\infty dz' \int_{-\infty}^t dt' \frac{\beta(r', \theta', z', t')}{[4\pi(t - t')]^{3/2} D_T D_L^{1/2}}$$

$$\times \exp\left\{ -\alpha(t - t') - \frac{r^2 + r'^2 - 2rr' \cos(\theta - \theta')}{4D_T(t - t')} - \frac{[z - z' - v_d(t - t')]^2}{4D_L(t - t')} \right\}.$$

$$(2\text{-}6\text{-}7)$$

Clearly, if the input β is cylindrically symmetric, the ion number density n will be cylindrically symmetric. Suppose that

$$\beta(r', z', t') = \frac{b}{\pi r_0^2} S(r_0 - r')\delta(z')\delta(t'), \quad (2\text{-}6\text{-}8)$$

where $S(\phi) = 0$ if $\phi < 0$, $S(\phi) = 1$ otherwise. This function describes an axially thin disk source of b ions with uniform surface density and radius r_0, created instantaneously at $t' = 0$ in the plane $z' = 0$. Let $b/\pi r_0^2 = s$, the planar source density. Then

$$n(r, z, t) = \frac{se^{-\alpha t} \exp[-(z - v_d t)^2/4D_L t]}{(4\pi t)^{3/2} D_T D_L^{1/2}}$$

$$\times \int_0^{r_0} r' \, dr' \int_0^{2\pi} d\theta' \exp\left(-\frac{r^2 + r'^2 - 2rr' \cos \theta'}{4D_T t} \right). \quad (2\text{-}6\text{-}9)$$

Now (Gray et al., 1952)

$$I_0(x) = \frac{1}{\pi} \int_0^\pi d\theta \, e^{\pm x \cos \theta} = \sum_{m=0}^\infty \frac{(x/2)^{2m}}{(m!)^2}. \quad (2\text{-}6\text{-}10)$$

Hence

$$n(r, z, t) = \frac{2\pi s e^{-\alpha t} \exp[-(z - v_d t)^2/4D_L t]}{(4\pi t)^{3/2} D_T D_L^{1/2}}$$

$$\times \int_0^{r_0} r' \, dr' \exp\left(-\frac{r^2 + r'^2}{4D_T t} \right) \sum_{m=0}^\infty \frac{1}{(m!)^2} \left(\frac{rr'}{4D_T t} \right)^{2m}. \quad (2\text{-}6\text{-}11)$$

If we let $x = r'^2/4D_T t$, then

$$n(r, z, t) = \frac{s}{(4\pi D_L t)^{1/2}} \exp\left[-\alpha t - \frac{(z - v_d t)^2}{4D_L t} - \frac{r^2}{4D_T t} \right]$$

$$\times \sum_{m=0}^{\infty} \frac{(r^2/4D_T t)^m}{(m!)^2} \int_0^{(r_0^2/4D_T t)} x^m e^{-x} \, dx. \qquad (2\text{-}6\text{-}12)$$

The remaining integral can be done by integrating by parts m times. The result is

$$\int_0^a x^m e^{-x} \, dx = m! \left(1 - e^{-a} \sum_{i=0}^{m} \frac{a^i}{i!} \right). \qquad (2\text{-}6\text{-}13)$$

Substituting in (2-6-12), we obtain

$$n(r, z, t) = \frac{s}{(4\pi D_L t)^{1/2}} \exp\left[-\alpha t - \frac{(z - v_d t)^2}{4D_L t} - \frac{r^2}{4D_T t} \right]$$

$$\times \sum_{m=0}^{\infty} \frac{(r^2/4D_T t)^m}{m!} \left[1 - e^{-r_0^2/4D_T t} \sum_{i=0}^{m} \frac{(r_0^2/4D_T t)^i}{i!} \right] \qquad (2\text{-}6\text{-}14)$$

or

$$n(r, z, t) = \frac{s}{(4\pi D_L t)^{1/2}} \exp\left[-\alpha t - \frac{(z - v_d t)^2}{4D_L t} \right]$$

$$\times \left[1 - \sum_{m=0}^{\infty} \sum_{i=0}^{m} \frac{1}{m!\,i!} \left(\frac{r^2}{4D_T t} \right)^m \left(\frac{r_0^2}{4D_T t} \right)^i \exp\left(-\frac{r_0^2 + r^2}{4D_T t} \right) \right]. \qquad (2\text{-}6\text{-}15)$$

Equation (2-6-15) is an expression for the ion number density at any given time at any point in space for an ion swarm which (a) was instantaneously created with uniform density across an axially thin disk, (b) drifts in unbounded space under the influence of a constant electric field, and (c) possibly undergoes a depleting reaction with the neutral gas molecules.

Since the ions detected in this apparatus are those that exit the drift tube on the axis, the result of interest is the *axial ionic number density*

$$n(0, z, t) = \frac{s e^{-\alpha t}}{(4\pi D_L t)^{1/2}} \left[1 - \exp\left(-\frac{r_0^2}{4D_T t} \right) \right] \exp\left[-\frac{(z - v_d t)^2}{4D_L t} \right], \qquad (2\text{-}6\text{-}16)$$

where s is the initial ion surface density of the delta-function input of ions and r_0 is the radius of the ion entrance aperture.

Two limiting cases are also of interest. For a "point source" in which $r_0 \to 0$ but the total number of ions $b = s(\pi r_0^2)$ in the disk remains finite application of

L'Hospital's rule to the transverse factor in (2-6-15) yields

$$\lim_{r_0 \to 0} \frac{1 - \sum_{m=0}^{\infty} \sum_{i=0}^{m} (1/m!i!) \exp[-(r^2 + r_0^2)/4D_T t](r^2/4D_T t)^m (r_0^2/4D_T t)^i}{r_0^2}$$

$$= \frac{e^{-r^2/4D_T t}}{4D_T t}. \qquad (2\text{-}6\text{-}17)$$

Then the *point source* solution is

$$n(r, z, t) = \frac{b}{(4\pi t)^{3/2} D_T D_L^{1/2}} \exp\left[-\alpha t - \frac{r^2}{4D_T t} - \frac{(z - v_d t)^2}{4D_L t} \right]. \qquad (2\text{-}6\text{-}18)$$

For an *infinite plane source* $r_0 \to \infty$, and now it is the source density s that remains finite. Then from (2-6-5)

$$n(r, z, t) = \frac{s}{(4\pi D_L t)^{1/2}} \exp\left[-\alpha t - \frac{(z - v_d t)^2}{4D_L t} \right]. \qquad (2\text{-}6\text{-}19)$$

Note, as expected, that (2-6-19) is independent of r and the transverse diffusion coefficient does not appear.

The quantity measured experimentally is the flux Φ of ions leaving the drift tube through the exit aperture of area a at a fixed distance z from the source plane:

$$\Phi(0, z, t) = aJ(0, z, t), \qquad (2\text{-}6\text{-}20)$$

where $J(0, z, t)$ is the z component of the ionic flux density in the drift tube, on the axis, at the end of the drift distance z.

The ionic flux density is related to the ionic number density by (2-6-1), and

$$J(0, z, t) = -D_L \frac{\partial n}{\partial z} + v_d n, \qquad (2\text{-}6\text{-}21)$$

where $n(0, z, t)$ is given by (2-6-16). Differentiation and substitution into (2-6-20) gives

$$\Phi(0, z, t) = \frac{a}{2} \left(v_d + \frac{z}{t} \right) n(0, z, t) \qquad (2\text{-}6\text{-}22)$$

or, in full,

$$\Phi(0, z, t) = \frac{sae^{-\alpha t}}{4(\pi D_L t)^{1/2}} \left(v_d + \frac{z}{t} \right)$$

$$\times \left[1 - \exp\left(-\frac{r_0^2}{4D_T t} \right) \right] \exp\left[-\frac{(z - v_d t)^2}{4D_L t} \right]. \qquad (2\text{-}6\text{-}23)$$

In summary, (2-6-23) gives the flux of ions of the single species considered passing through the exit aperture of the drift tube as a function of the time t and drift distance z. All the ions of the single species under consideration are assumed to be introduced from the ion source in periodic delta-function bursts and none is produced by reactions in the drift space. Loss of the ions in reactions producing other species is allowed, however, the rate of loss being described by the frequency α.

Equation (2-6-23) was used to calculate the expected shapes of the arrival-time spectra used for comparison with experimental data in Figs. 2-2-2 and 2-2-3. The foregoing analysis, and the other analyses of the ionic transport equations utilized by the Georgia Tech group, are the work of I. R. Gatland.

C. Other Solutions of Transport Equations for Ions in Drift Tubes

In addition to the analysis given in Section 2-6B, analyses that apply to other apparatus or to different reaction patterns have been published. They are listed below with brief comments concerning the scope of the analysis:

Burch and Geballe (1957). This is a one-dimensional analysis that applies to transient ion currents in a Townsend discharge. The effects of diffusion are not included, but the forward reaction of primary ions to form two kinds of secondary ion is considered.

Edelson and McAfee (1964). This paper discusses the determination of Townsend ionization and electron attachment coefficients and ionic mobilities in pulsed discharge experiments. The analysis is one-dimensional; the effects of diffusion and reactions are considered.

Edelson et al. (1964). In this one-dimensional analysis of a pulsed Townsend discharge ionization, electron attachment, and secondary-electron production at the cathode are considered. The transport of positive and negative ions is treated with and without diffusion. The results are useful in the determination of mobilities.

Frommhold (1964). In this paper electron detachment from negative ions is discussed as a process occurring in avalanches in various gases. Frommhold also considered two interconverting, drifting species in the absence of diffusion and obtained results later derived independently by Edelson et al. (1967).

Beaty and Patterson (1965). The basic analysis is for a cylindrical geometry in which the ion source and collector form the ends of the drift chamber. Drift, diffusion (with $D_L = D_T$), and a depleting reaction to form a secondary species are considered. Both the primary and secondary ions are assumed to have the same temperature. The analysis was applied to experiments on helium to determine the mobilities of the primary and secondary ions and the forward reaction rate.

Barnes (1967). This analysis, basically one-dimensional, deals with drift, diffusion, and depleting reactions. Statistical estimates are developed for the mobility, diffusion coefficient, and reaction rate.

Edelson et al. (1967). This paper contains a one-dimensional analytic solution for a system composed of two kinds of drifting, interconverting ions. Ions of one species are introduced initially by a delta-function source. Diffusion is not included in the model. Numerical solutions, also discussed, contain the effects of diffusion and can be used to determine mobilities and reaction rates.

Edelson (1968). In this analysis the effects of radial diffusion on the collected ion current due to different residence times are determined. The effect of a reduced collector area is also discussed.

Whealton and Woo (1968). This three-dimensional analysis pertains to the determination of reaction rates. The effects produced by the primary and secondary ions with unequal temperatures are discussed.

Woo and Whealton (1969). This analysis, also three-dimensional, considers drift, transverse and longitudinal diffusion, and a forward reaction to a secondary species. The source term corresponds to a disk with arbitrary radial ion density. Expressions are obtained for both the primary and secondary ion densities.

Keller et al. (1970). This paper suggests mathematical modifications to the solution of the Woo–Whealton model which facilitate calculations.

Snuggs (1970). In this thesis a comprehensive analysis is made of an interreacting system of two ion swarms. The geometry is three-dimensional, and drift, transverse and longitudinal diffusion, and reactions are considered. The reaction from each ionic species to the other is taken into account, as well as other depleting reactions. Comparison of experimental data with analytic arrival time spectra permits the determination of mobilities, diffusion coefficients, and reaction rates. Many simplifying assumptions are also considered, and appropriate formulas to aid in data reduction are obtained.

Snuggs et al. (1971a). An analysis is presented for a cylindrical drift tube in which drift, transverse and longitudinal diffusion, and depleting reactions are occurring. The ion source is an axially thin disk and the ion current is measured on the axis of the drift tube. For the primary ion species a closed-form solution for the time-integrated ion current is obtained and used to determine transverse diffusion coefficients and reaction rates by an attenuation method (see Section 3-1). An analytic solution is presented for the arrival-time spectrum of the secondary ions.

Woo and Whealton (1971). This paper deals mainly with the error introduced by neglecting transverse diffusion when calculating reaction rates from arrival-time spectra. Three-dimensional geometry was used and various ion source configurations were considered. Longitudinal diffusion was neglected and only a forward reaction was included.

Schummers (1972). This thesis describes an analysis of the drift, longitudinal and transverse diffusion, and interreaction of two ion species in cylindrical geometry. The ion source is an axially thin disk and the ion current is evaluated on the axis. This analysis can be used for both ionic species to determine mobilities, diffusion coefficients, and forward-backward reaction rates from arrival-time spectra.

Gatland (1975). This paper describes a Green's function solution of the ion transport equations. This type of solution is particularly suitable when the drift-tube boundaries have only a slight effect on the observed ions. Solutions are worked out for a variety of reaction patterns. Ionic drift, longitudinal and transverse diffusion, and reactions are treated exactly. Boundary effects can be included as perturbations. Some of the analysis summarized here is reproduced in the thesis by Schummers (1972).

In closing this section it is important to point out that even when no analytic solution is available we may always resort to numerical methods to solve the relevant transport equations.

REFERENCES

Adams, N. G., and D. Smith (1976a). The selected ion flow tube (SIFT): A technique for studying thermal energy ion-neutral reactions, *Int. J. Mass Spectrom. Ion Phys.* **21**, 349–359.

Adams, N. G., and D. Smith (1976b). Product ion distributions for some ion-molecule reactions, *J. Phys.* **B9**, 1439–1451.

Adams, N. G., M. J. Church, and D. Smith (1975). An experimental and theoretical investigation of the dynamics of a flowing afterglow plasma, *J. Phys.* **D8**, 1409–1422.

Albritton, D. L. (1967). Ph.D. thesis, Georgia Institute of Technology, Atlanta.

Albritton, D. L. (1978). Ion-neutral reaction-rate constants measured in flow reactors through 1977, *At. Data Nucl. Data Tables* **22**, 1–101.

Albritton, D. L. (1979). Energy dependences of ion-neutral reactions studied in drift tubes, in *Kinetics of Ion-Molecule Reactions*, P. Ausloos, Ed., Plenum, New York, pp. 119–142.

Albritton, D. L., T. M. Miller, D. W. Martin, and E. W. McDaniel (1968). Mobilities of mass-identified H_3^+ and H^+ ions in hydrogen, *Phys. Rev.* **171**, 94–102.

Albritton, D. L., I. Dotan, W. Lindinger, M. McFarland, J. Tellinghuisen, and F. C. Fehsenfeld (1977). Effects of ion speed distributions in flow-drift tube studies of ion-neutral reactions, *J. Chem. Phys.* **66**, 410–421.

Barnes, W. S. (1967). Method of analysis of ion swarm experiments, *Phys. Fluids* **10**, 1941–1952.

Barnes, W. S., D. W. Martin, and E. W. McDaniel (1961). Mass spectrographic identification of the ion observed in hydrogen mobility experiments, *Phys. Rev. Lett.* **6**, 110–111.

Beaty, E. C. (1962), Mobilities of positive ions in argon, *Proc. 5th Int. Conf. Ionization Phenomena Gases*, Munich, 1961, Vol. **1**, North-Holland, Amsterdam, pp. 183–191.

Beaty, E. C., and P. L. Patterson (1965). Mobilities and reaction rates of ions in helium, *Phys. Rev.* **137**, 346–357.

Beaty, E. C., and P. L. Patterson (1968). Mobilities and reaction rates of neon ions in neon, *Phys. Rev.* **170**, 116–121.

Bernstein, R. B., Ed. (1979). *Atom-Molecule Collision Theory, A Guide for the Experimentalist*, Plenum, New York.

Beuhler, R. J., and L. Friedman (1982). A study of the formation of high molecular weight water cluster ions ($m/e < 59,000$) in expansion of ionized gas mixtures, *J. Chem. Phys.* **77**, 2549–2557.

Bloomfield, C. H., and J. B. Hasted (1964). New technique for the study of ion-atom interchange, *Discuss. Faraday Soc.* **37**, 176–184.

Böhringer, H., and F. Arnold (1983). Studies of ion/molecule reactions, ion mobilities, and their temperature dependence to very low temperatures using a liquid-helium-cooled ion drift tube, *Int. J. Mass Spectrom. Ion Phys.* **49**, 61–83.

Böhringer, H., and F. Arnold (1986). Temperature and pressure dependence of the reaction of He^+ ions with H_2, *J. Chem. Phys.* **84**, 1459–1462.

Bowers, M. T., Ed. (1979). *Gas Phase Ion Chemistry*, Vols. 1 and 2, Academic, New York (Vol. **3**, 1984).

Bradbury, N. E., and R. A. Nielsen (1936). Absolute values of the electron mobility in hydrogen, *Phys. Rev.* **49**, 388–393.

Burch, D. S., and R. Geballe (1957). Clustering of negative ions in oxygen, *Phys. Rev.* **106**, 188–190.

Cassidy, R. A., and M. T. Elford (1985a). The reaction rate coefficient for the clustering of He to Li^+ at low center of mass energies, *Aust. J. Phys.* **38**, 577–585.

Cassidy, R. A., and M. T. Elford (1985b). The mobility of Li^+ ions in helium and argon, *Aust. J. Phys.* **38**, 587–601.

Cassidy, R. A., and M. T. Elford (1986). The mobility of K^+ ions in helium and argon, *Aust. J. Phys.* **39**, 25–33.

Church, M. J., and D. Smith (1978). Ionic recombination of atomic and molecular ions in flowing afterglow plasmas, *J. Phys.* **D11**, 2199–2206.

Cosby, P. C., G. P. Smith, and J. T. Moseley (1978). Photodissociation and photodetachment of negative ions: IV. Hydrates of O_3^-, *J. Chem. Phys.* **69**, 2779–2781.

Crompton, R. W., and M. T. Elford (1973). The drift velocity of electrons in oxygen at 293 K, *Aust. J. Phys.* **26**, 771–782.

Crompton, R. W., M. T. Elford, and J. Gascoigne (1965). Precision measurements of the Townsend energy ratio for electron swarms in highly uniform electric fields, *Aust. J. Phys.* **18**, 409–436.

Dolder, K. T., and B. Peart (1986). Electron-ion and ion-ion collisions with intersecting beams, *Adv. At. Mol. Phys.* **22**, 197–241.

Dressler, R. A., H. Meyer, A. O. Langford, V. M. Bierbaum, and S. R. Leone (1987). Direct observation of Ba^+ velocity distributions in a drift tube using single-frequency laser-induced fluorescence. *J. Chem Phys.* **87**, 5578–5579.

Duncan, M. A., V. M. Bierbaum, G. B. Ellison, and S. R. Leone (1983). Laser-induced fluorescence studies of ion collisional excitation in a drift field: Rotational excitation of N_2^+ in helium, *J. Chem. Phys.* **79**, 5448–5456.

Dupeyrat, G., B. R. Rowe, D. W. Fahey, and D. L. Albritton (1982). Diagnostic studies of venturi-inlets for flow reactors, *Int. J. Mass Spectrom. Ion Phys.* **44**, 1–18.

Edelson, D. (1968). Radial diffusion in drift tube experiments, *J. Appl. Phys.* **39**, 3497–3498.

Edelson, D., and K. B. McAfee (1964). Improved pulsed Townsend experiment, *Rev. Sci. Instrum.* **35**, 187–194.

Edelson, D., J. A. Morrison, and K. B. McAfee (1964). Ion distributions in a pulsed Townsend discharge, *J. Appl. Phys.* **35**, 1682–1690.

Edelson, D., J. A. Morrison, L. G. McKnight, and D. P. Sipler (1967). Interpretation of ion-mobility experiments in reacting systems, *Phys. Rev.* **164**, 71–75.

Ellis, H. W., F. L. Eisele, and E. W. McDaniel (1978). Temperature dependent mobilities of negative ions in N_2 and O_2, *J. Chem. Phys.* **69**, 4710–4711.

Eisele, F. L. (1988). First tandem mass spectrometric measurements of tropospheric ions, *J. Geophys. Res.* **93**, 716–724.

Eisele, F. L., and E. W. McDaniel (1986). Mass spectrometric study of tropospheric ions in the Northeastern and Southwestern United States, *J. Geophys. Res.* **91**, 5183–5188.

Eisele, F. L., H. W. Ellis, and E. W. McDaniel (1979). Temperature dependent mobilities: O_2^+ and NO^+ in N_2, *J. Chem. Phys.* **70**, 5924–5925.

Eisele, F. L., M. D. Perkins, and E. W. McDaniel (1980). Mobilities of NO_2^-, NO_3^-, and CO_3^- in N_2 over the temperature range 217–675 K, *J. Chem. Phys.* **73**, 2517–2518.

Eisele, F. L., M. D. Perkins, and E. W. McDaniel (1981). Measurement of the mobilities of Cl^-, $NO_2^- \cdot H_2O$, $NO_3^- \cdot H_2O$, $CO_3^- \cdot H_2O$, and $CO_4^- \cdot H_2O$ in N_2 as a function of temperature, *J. Chem. Phys.* **75**, 2473–2475.

Fahey, D. W., I. Dotan, F. C. Fehsenfeld, D. L. Albritton, and L. A. Viehland (1981a). Energy dependence of the rate constant of the reaction $N^+ + NO$ at collision energies 0.04 to 2.5 eV, *J. Chem. Phys.* **74**, 3320–3323.

Fahey, D. W., F. C. Fehsenfeld, and D. L. Albritton (1981b). Mobilities of N^+ ions in helium and argon, *J. Chem. Phys.* **74**, 2080–2081.

Fahey, D. W., F. C. Fehsenfeld, and E. E. Ferguson (1981c). Rate constant for the reaction $C^+ + CO_2$ at collision energies 0.04 to 2.5 eV, *Geophys. Res. Lett.* **8**, 1115–1117.

Fahey, D. W., F. C. Fehsenfeld, E. E. Ferguson, and L. A. Viehland (1981d). Reactions of Si^+ with H_2O and O_2 and SiO^+ with H_2 and D_2, *J. Chem. Phys.* **75**, 669–674.

Fahey, D. W., H. Böhringer, F. C. Fehsenfeld, and E. E. Ferguson (1982). Reaction rate constants for $O_2^-(H_2O)_n$ ions $n = 0$ to 4, with O_3, NO, SO_2, and CO_2, *J. Chem. Phys.* **76**, 1799–1805.

Fahr, H., and K. G. Müller (1967). Ionenbewegung unter dem Einfluss von Umladungsstössen, *Z. Phys.* **200**, 343–365.

Federer, W., H. Ramler, H. Villinger, and W. Lindinger (1985). Vibrational temperature of O_2^+ and N_2^+ drifting at elevated E/N in helium, *Phys. Rev. Lett.* **54**, 540–543.

Ferguson, E. E., F. C. Fehsenfeld, and A. L. Schmeltekopf (1969). Flowing afterglow measurements of ion-neutral reactions, in *Advances in Atomic and Molecular Physics*, D. R. Bates and I. Estermann, Eds., Vol. 5, Academic, New York, pp. 1–56.

Fhadil, H. A., D. Mathur, and J. B. Hasted (1982). Mobilities of O^+, O^+*, and O_2^{2+} in He and Ar from ion energy distribution measurements in an injected-ion drift tube, *J. Phys.* **B15**, 1443–1453.

Fhadil, H. A., A. T. Numan, T. Shuttleworth, and J. B. Hasted (1985). Energy distributions of atomic ions drifting in rare gases, *Int. J. Mass Spectrom. Ion Phys.* **65**, 307–319.

Franklin, J. L., Ed. (1972). *Ion-Molecule Reactions*, Plenum, New York.

Frommhold, L. (1964). Über verzögerte Elektronen in Elektronenlawinen, insbesondere in Sauerstoff und Luft, durch Bildung und Zerfall negativer Ionen (O^-), *Fortschr. Phys.* **12**, 597–643.

Gatland, I. R. (1975). Analysis for ion drift tube experiments, in *Case Studies in Atomic Collision Physics*, E. W. McDaniel and M. R. C. McDowell, Eds., Vol. 4, North-Holland, Amsterdam, pp. 369–437.

Gatland, I. R., and E. W. McDaniel (1970). Reply to "Longitudinal diffusion coefficients misnamed," by R. N. Varney, *Phys. Rev. Lett.* **25**, 1603–1605.

Gentry, W. R., D. J. McClure, and C. H. Douglass (1975). Merged beams at Minnesota, *Rev. Sci. Instrum.* **46**, 367–375.

Graham, E., M. A. Biondi, and R. Johnsen (1976). Spectroscopic studies of the charge transfer reaction $He^+ + Hg \rightarrow He + (Hg^+)^*$ at thermal energy, *Phys. Rev.* **A13**, 965–968.

Gray, A., G. B. Mathews, and T. M. MacRoberts (1952). *Bessel Functions*, 2nd ed., Macmillan, London, pp. 20, 46.

Hagena, O. F., and W. Obert (1972). Cluster formation in expanding supersonic jets, *J. Chem. Phys.* **56**, 1793–1802.

Hansen, J. C., C. H. Kuo, F. J. Grieman, and J. T. Moseley (1983). High resolution absorption spectroscopy of N_2^+ $X(v'' = 0) \rightarrow A(v' = 4)$ using charge exchange detection, *J. Chem. Phys.* **79**, 1111–1115.

Hasted, J. B. (1972). *Physics of Atomic Collisions*, 2nd ed., American Elsevier, New York.

Heimerl, J. M. (1968). Ph.D. thesis, University of Pittsburgh, Pittsburgh, Pa.

Heimerl, J. M., R. Johnsen, and M. A. Biondi (1969). Ion-molecule reactions, $He^+ + O_2$ and $He^+ + N_2$, at thermal energies and above, *J. Chem. Phys.* **51**, 5041–5048.

Helm, H. (1975). The mobilities of $Kr^+(^2P_{3/2})$ and $Kr^+(^2P_{1/2})$ in krypton at 295 K, *Chem. Phys. Lett.* **36**, 97–99.

Helm, H. (1976a). The mobilities of atomic kyrpton and xenon ions in the $^2P_{1/2}$ and $^2P_{3/2}$ state in their parent gas, *J. Phys.* **B9**, 2931–2943.

Helm, H. (1976b). The mobilities and equilibrium reactions of helium ions in helium at 77 K, *J. Phys.* **B9**, 1171–1189.

Helm, H., and M. T. Elford (1977a). The influence of fine-structure splitting on the mobility of atomic neon ions in neon, *J. Phys.* **B10**, 983–991.

Helm, H., and M. T. Elford (1977b). The mobility of Ar^+ ions in argon and the effect of spin-orbit coupling, *J. Phys.* **B10**, 3849–3851.

Helm, H., and M. T. Elford (1978). Mobilities and reactions of diatomic rare-gas ions in their parent gases helium, neon, argon and krypton, *J. Phys.* **B11**, 3939–3950.

Hogan, M. J., and P. P. Ong (1986). Comparison of theoretical analytic ion velocity distribution functions with experimental measurements, *J. Phys.* **D19**, 2123–2128.

Hornbeck, J. A. (1951a). Microsecond transient currents in the pulsed Townsend discharge, *Phys. Rev.* **83**, 374–379.

Hornbeck, J. A. (1951b). The drift velocities of molecular and atomic ions in helium, neon, and argon, *Phys. Rev.* **84**, 615–620.

Howorka, F., F. C. Fehsenfeld, and D. L. Albritton (1979). H$^+$ and D$^+$ ions in He: Observations of a runaway mobility, *J. Phys.* **B12**, 4189–4197.

Howorka, F., I. Dotan, F. C. Fehsenfeld, and D. L. Albritton (1980). Kinetic energy dependence of the branching ratios of the reactions of N$^+$ ions with O$_2$, *J. Chem. Phys.* **73**, 758–764.

Huber, B. A., P. C. Cosby, J. R. Peterson, and J. T. Moseley (1977). Photodetachment and de-excitation of excited NO$_2{}^-$, *J. Chem. Phys.* **66**, 4520–4526.

Huxley, L. G. H., and R. W. Crompton (1974). *The Diffusion and Drift of Electrons in Gases*, Wiley, New York.

Iinuma, K., M. Takebe, Y. Satoh, and K. Seto (1982). Design of a continuous guard ring and its application to swarm experiments, *Rev. Sci. Instrum.* **53**, 845–850.

Iinuma, K., M. Takebe, Y. Satoh, and K. Seto (1983). Measurements of mobilities and longitudinal diffusion coefficients by a flight-distance scanning method: Comparison with the flight-time scanning method, *J. Chem. Phys.* **79**, 3906–3910.

Johnsen, R., and M. A. Biondi (1972). Mobilities of uranium and mercury ions in helium, *J. Chem. Phys.* **57**, 5292–5295.

Johnsen, R., and M. A. Biondi (1973). Measurements of the O$^+$ + N$_2$ and O$^+$ + O$_2$ reaction rates from 300 K to 2 eV, *J. Chem. Phys.* **59**, 3504–3509.

Johnsen, R., and M. A. Biondi (1978). Mobilities of doubly charged rare gas ions in their parent gases, *Phys. Rev.* **A18**, 989–995.

Johnsen, R., and M. A. Biondi (1979). Mobilities of singly and doubly charged rare gas ions in helium and in neon, *Phys. Rev.* **A20**, 221–223.

Johnsen, R., M. A. Biondi, and M. Hayashi (1982). Mobilities of ground state and metastable O$^+$, O$_2{}^+$, O^{2+}, and O$_2{}^{2+}$ ions in helium and in neon, *J. Chem. Phys.* **77**, 2545–2548.

Johnsen, R., H. L. Brown, and M. A. Biondi (1970). Ion-molecule reactions involving N$_2{}^+$, N$^+$, O$_2{}^+$, and O$^+$ ions from 300°K to about 1 eV, *J. Chem. Phys.* **52**, 5080–5084.

Johnsen, R., A. Chen, and M. A. Biondi (1980a). Dissociative charge transfer of He$^+$ ions with H$_2$ and D$_2$ molecules from 78 K to 330 K, *J. Chem. Phys.* **72**, 3085–3088.

Johnsen, R., A. Chen, and M. A. Biondi (1980b). Three-body association reactions of He$^+$, Ne$^+$, and Ar$^+$ ions in their parent gases from 78 K to 300 K, *J. Chem. Phys.* **73**, 1717–1720.

Johnsen, R., A. Chen, and M. A. Biondi (1980c). Association reactions of C$^+$ ions with H$_2$ and D$_2$ molecules at 78 K and 300 K, *J. Chem. Phys.* **73**, 3166–3169.

Jones, T. T. C., J. D. C. Jones, and K. Birkinshaw (1981a). Gaseous ion mobility measurements: Ne$^+$ and Xe$^+$ in argon, *Chem. Phys. Lett.* **82**, 377–379.

Jones, T. T. C., J. Villinger, D. G. Lister, M. Tichy, K. Birkinshaw, and N. D. Twiddy (1981b). The energy dependence of some neon-ion-neutral reaction rate coefficients investigated in a flow-drift tube experiment, *J. Phys.* **B14**, 2719–2729.

Jones, T. T. C., J. D. C. Jones, K. Birkinshaw, and N. D. Twiddy (1982a). Gaseous ion mobility measurements: Ne$^+$ and Kr$^+$ in helium at 294 K, *Chem. Phys. Lett.* **86**, 503–505.

Jones, T. T. C., J. D. C. Jones, K. Birkinshaw, and N. D. Twiddy (1982b). Gaseous ion mobility measurement: N$_2$H$^+$ in helium and argon at 295 K, *Chem. Phys. Lett.* **89**, 442–445.

Kaneko, Y., L. R. Megill, and J. B. Hasted (1966). Study of inelastic collisions by drifting ions, *J. Chem. Phys.* **45**, 3741–3751.

Keller, G. E., M. R. Sullivan, and M. D. Kregel (1970). Transport model for converting charged species in drift tubes, *Phys. Rev.* **A1**, 1556–1558.

Koizumi, T., T. Tsurugai, and I. Ogawa (1987). State dependence of mobilities for Kr^{++} in Kr at 88 K, *J. Phys. Soc. Jpn.* **56**, 17–20.

Lin, G.-H., J. Maier, and S. R. Leone (1985). Nascent vibrational and rotational distributions from the charge transfer reaction $Ar^+ + CO \rightarrow CO^+ + Ar$ at near thermal energies, *J. Chem. Phys.* **82**, 5527–5535.

Lindinger, W., and D. L. Albritton (1975). Mobilities of various mass-identified positive ions in helium and argon, *J. Chem. Phys.* **62**, 3517–3522.

Lindinger, W., and D. Smith (1983). Influence of translational and internal energy on ion-neutral reactions, in *Kinetics and Energetics of Small Transient Species*, A. Fontijn, Ed., Academic, New York.

Lindinger, W., M. McFarland, F. C. Fehsenfeld, D. L. Albritton, A. L. Schmeltekopf, and E. E. Ferguson (1975). Translational and internal energy dependences of some ion-molecule reactions, *J. Chem. Phys.* **63**, 2175–2181.

Lindinger, W., T. D. Märk, and F. Howorka, eds. (1984). *Swarms of Ions and Electrons in Gases*, Springer-Verlag, Vienna. Includes: Historical development and present state of swarm research, by E. W. McDaniel, p. 1; Velocity distribution functions of atomic ions in drift tubes, by H. R. Skullerud and S. Kuhn, p. 13; Internal-energy distribution of molecular ions in drift tubes, by L. A. Viehland, p. 27; Determination of ion-atom potentials from mobility experiments, by I. R. Gatland, p. 44; Transverse ion diffusion in gases, by E. Märk and T. D. Märk, p. 60; Runaway mobilities of ions in helium, by F. Howorka, p. 87; Theory of ion-molecule collisions at (1 eV–5 keV)/amu," by M. R. Flannery, p. 103; Vibrational excitation and de-excitation, and charge transfer of molecular ions in drift tubes, by E. E. Ferguson, p. 126; Kinetic and internal energy effects on ion-neutral reactions, by W. Lindinger, p. 146; Cluster ion association reactions: Thermochemistry and relationship to kinetics, by A. W. Castleman, Jr., and R. G. Keesee, p. 167; Temperature dependences of positive-ion molecule reactions, by N. G. Adams and D. Smith, p. 194; Reactions of negative ions, by A. A. Viggiano and J. F. Paulson, p. 218; Studies of plasma reaction processes using a flowing-afterglow/Langmuir probe apparatus, by D. Smith and N. G. Adams, p. 284.

Loeb, L. B. (1960). *Basic Processes of Gaseous Electronics*, 2nd ed., University of California Press, Berkeley, Calif., Chap. 1.

McAfee, K. B., and D. Edelson (1962). *Bull. Am. Phys. Soc.* **7**, 135.

McAfee, K. B., and D. Edelson (1963a). Drift velocities of atomic and molecular ions in nitrogen, *Proc. 6th Int. Conf. Ionization Phenomena Gases* (Paris), Vol. 1, SERMA, Paris, pp. 299–301.

McAfee, K. B., and D. Edelson (1963b). Identification and mobility of ions in a Townsend discharge by time-resolved mass spectrometry, *Proc. Phys. Soc. London* **81**, 382–384.

McAfee, K. B., D. Sipler, and D. Edelson (1967). Mobilities and reactions of ions in argon, *Phys. Rev.* **160**, 130–135.

McDaniel, E. W. (1964). *Collision Phenomena in Ionized Gases*, Wiley, New York, pp. 683–684.

McDaniel, E. W. (1970). Possible sources of large error in determinations of ion-molecule reaction rates with drift tube mass spectrometers, *J. Chem. Phys.* **52**, 3931–3935.

McDaniel, E. W. (1972). The effect of clustering on the measurement of the mobility of potassium ions in nitrogen gas, *Aust. J. Phys.* **25**, 465–467.

McDaniel, E. W., and M. R. C. McDowell (1959). Low-field mobilities of the negative ions in oxygen, sulfur hexafluoride, sulfur dioxide, and hydrogen chloride, *Phys. Rev.* **114**, 1028–1037.

McDaniel, E. W., and E. A. Mason (1973). *The Mobility and Diffusion of Ions in Gases*, Wiley, New York.

McDaniel, E. W., and J. T. Moseley (1971). Tests of the Wannier expressions for diffusion coefficients of gaseous ions in electric fields, *Phys. Rev.* **A3**, 1040–1044.

McDaniel, E. W., D. W. Martin, and W. S. Barnes (1962). Drift tube mass spectrometer for studies of low-energy ion-molecule reactions, *Rev. Sci. Instrum.* **33**, 2–7.

McDaniel, E. W., V. Čermák, A. Dalgarno, E. E. Ferguson, and L. Friedman (1970). *Ion-Molecule Reactions*, Wiley, New York.

McFarland, M., D. L. Albritton, F. C. Fehsenfeld, E. E. Ferguson, and A. L. Schmeltekopf (1973a, b, c). A flow-drift technique for ion mobility and ion-molecule reaction rate constant measurements: 1. Apparatus and mobility measurements, 2. The positive ion reactions of N^+, O^+, and N_2^+ with O_2 and O^+ with N_2 from thermal to about 2 eV, 3. The negative ion reactions of O^- with H_2, D_2, CO, and NO, *J. Chem. Phys.* **59**, 6610–6619, 6620–6628, 6629–6635.

McKnight, L. G., K. B. McAfee, and D. P. Sipler (1967). Low-field drift velocities and reactions of nitrogen ions in nitrogen, *Phys. Rev.* **164**, 62–70.

Martin, D. W., W. S. Barnes, G. E. Keller, D. S. Harmer, and E. W. McDaniel (1963). Mobilities of mass-identified ions in nitrogen, *Proc. 6th Int. Conf. Ionization Phenomena Gases*, Paris, Vol. **1**, SERMA, Paris, pp. 295–297.

Massey, H. S. W. (1971). *Slow Collisions of Heavy Particles*, 2nd ed., Vol. 3 of *Electronic and Ionic Impact Phenomena*, H. S. W. Massey, E. H. S. Burhop, and H. B. Gilbody, Clarendon, Oxford, Chap. 19.

Massey, H. S. W., and H. B. Gilbody (1973). *Recombination and Fast Collisions of Heavy Particles*, 2nd ed., Vol. 4 of *Electronic and Ionic Impact Phenomena*, H. S. W. Massey, E. H. S. Burhop, and H. B. Gilbody, Clarendon, Oxford.

Miller, T. M., J. T. Moseley, D. W. Martin, and E. W. McDaniel (1968). Reactions of H^+ in H_2 and D^+ in D_2: Mobilities of hydrogen and alkali ions in H_2 and D_2 gases, *Phys. Rev.* **173**, 115–123.

Milloy, H. B., and M. T. Elford (1975). Mass discrimination in ion sampling from drift tubes, *Int. J. Mass Spectrom. Ion Phys.* **18**, 21–31.

Milne, T. A., and F. T. Greene (1967). Mass spectrometric observations of argon clusters in nozzle beams, *J. Chem. Phys.* **47**, 4095–4101.

Moruzzi, J. L., and L. Harrison (1974). Energy distributions of O^- and O_2^- ions produced in drift tubes, *Int. J. Mass Spectrom. Ion Phys.* **13**, 163–171.

Moseley, J. T. (1968). Ph.D. thesis, Georgia Institute of Technology, Atlanta.

Moseley, J. T. (1982). Determination of ion molecular potential curves using photodissociative processes, in *Applied Atomic Collision Physics*, Vol. 5, H. S. W. Massey, E. W. McDaniel, and B. Bederson, Eds., Academic, New York, pp. 269–283.

Moseley, J. T., R. M. Snuggs, D. W. Martin, and E. W. McDaniel (1968). Longitudinal and transverse diffusion coefficients of mass-identified N^+ and N_2^+ ions in nitrogen, *Phys. Rev. Lett.* **21**, 873–875.

Moseley, J. T., I. R. Gatland, D. W. Martin, and E. W. McDaniel (1969a). Measurement of transport properties of ions in gases: Results for K$^+$ ions in N$_2$, *Phys. Rev.* **178**, 234–239.

Moseley, J. T., R. M. Snuggs, D. W. Martin, and E. W. McDaniel (1969b). Mobilities, diffusion coefficients, and reaction rates of mass-identified nitrogen ions in nitrogen, *Phys. Rev.* **178**, 240–248.

Moseley, J. T., P. C. Cosby, R. A. Bennett, and J. R. Peterson (1975a). Photodissociation and photodetachment of molecular negative ions: 1. Ions formed in CO$_2$/H$_2$O mixtures, *J. Chem. Phys.* **62**, 4826–4834.

Moseley, J. T., R. E. Olson, and J. R. Peterson (1975b). Ion-ion mutual neutralization, *Case Stud. At. Phys.* **5**, 1–45.

Neynaber, R. H. (1969). Experiments with Merging Beams, in *Advances in Atomic and Molecular Physics*, Vol. 5, D. R. Bates and I. Estermann, Eds., Academic, New York, pp. 57–108.

Neynaber, R. H. (1980). Merging-beams experiments with excited atoms, in *Electronic and Atomic Collisions*, N. Oda and K. Takayanagi, Eds., North-Holland, Amsterdam, pp. 287–300.

Ong, P. P., and M. J. Hogan (1985). Velocity distributions of He$^+$, Ne$^+$, and Ar$^+$ in their parent gases, *J. Phys.* **B18**, 1897–1906.

Ong, P. P., D. Mathur, M. H. Khatri, J. B. Hasted, and M. Hamdan (1981). Energy distribution of CO$^+$ ions drifting in He, Ne, and Ar, *J. Phys.* **D14**, 633–641.

Parkes, D. A. (1971). Electron attachment and negative ion-molecule reactions in pure oxygen, *Trans. Faraday Soc.* **67**, 711–729.

Patterson, P. L. (1970). Mobilities of negative ions in SF$_6$, *J. Chem. Phys.* **53**, 696–704.

Perkins, M. D., F. L. Eisele, and E. W. McDaniel (1981). Temperature dependent mobilities: NO$_2^-$, NO$_3^-$, CO$_3^-$, CO$_4^-$, and O$_2^+$ in O$_2$, *J. Chem. Phys.* **74**, 4206–4207.

Perkins, M. D., R. D. Chelf, F. L. Eisele, and E. W. McDaniel (1983). Temperature dependent mobilities of NH$_4^+$ and Br$^-$ ions in N$_2$ and O$_2$, *J. Chem. Phys.* 79, 5207–5208.

Puckett, L. J., M. W. Teague, and D. G. McCoy (1971). Refrigerating vapor bath, *Rev. Sci. Instrum.* **42**, 580–583.

Rees, J. A. (1978). Fundamental processes in the electrical breakdown of gases, in *Electrical Breakdown of Gases*, J. M. Meek and J. D. Craggs, Eds., Wiley, New York, pp. 1–128.

Rowe, B. R., D. W. Fahey, F. C. Fehsenfeld, and D. L. Albritton (1980). Rate constants for the reactions of metastable O^{+*} ions with N$_2$ and O$_2$ at collision energies 0.04 to 0.2 eV and the mobilities of these ions at 300 K, *J. Chem. Phys.* **73**, 194–205.

Rundel, R. D., D. E. Nitz, K. A. Smith, M. W. Geis, and R. F. Stebbings (1979). Resonant charge transfer in He$^+$–He collisions studied with the merging-beams technique, *Phys. Rev.* **A19**, 33–42.

Schmeltekopf, A. L., E. E. Ferguson, and F. C. Fehsenfeld (1968). Afterglow studies of the reactions of He$^+$, He (2^{3}S), and O$^+$ with vibrationally excited N$_2$, *J. Chem. Phys.* **48**, 2966–2973.

Schummers, J. H. (1972). Ph.D. thesis, Georgia Institute of Technology, Atlanta.

Schummers, J. H., G. M. Thomson, D. R. James, I. R. Gatland, and E. W. McDaniel (1973a). Mobilities and longitudinal diffusion coefficients of mass-identified positive ions in carbon monoxide gas, *Phys. Rev.* **A7**, 683–688.

Schummers, J. H., G. M. Thomson, D. R. James, E. Graham, I. R. Gatland, and E. W. McDaniel (1973b). Measurement of the rate coefficient of the reaction $CO^+ + 2CO \rightarrow CO^+ \cdot CO + CO$ in a drift tube mass spectrometer, *Phys. Rev.* **A7**, 689–693.

Skullerud, H. R. (1981). Ion energy distribution functions, in *Electron and Ion Swarms*, L. G. Christophorou, Ed., Pergamon, Elmsford, N.Y., pp. 157–163.

Smith, D., and N. G. Adams (1979). Recent advances in flow tubes: Measurement of ion-molecule rate coefficients and product distributions, in *Gas Phase Ion Chemistry*, Vol. 1, M. T. Bowers, Ed., Academic, New York, pp. 1–44.

Smith, D., and N. G. Adams (1983). Studies of ion-ion recombination using flowing afterglow plasmas, in *Proc. NATO Adv. Study Inst. Ion-Ion Electron-Ion Collisions*, F. Brouillard and J. Wm. McGowan, Eds., Plenum, New York.

Smith, D., and M. J. Church (1976). Binary ion-ion recombination coefficients determined in a flowing afterglow plasma, *Int. J. Mass Spectrom. Ion Phys.* **19**, 185–200.

Snuggs, R. M. (1970). Ph.D. thesis, Georgia Institute of Technology, Atlanta.

Snuggs, R. M., D. J. Volz, I. R. Gatland, J. H. Schummers, D. W. Martin, and E. W. McDaniel (1971a). Ion-molecule reactions between O^- and O_2 at thermal energies and above, *Phys. Rev.* **A3**, 487–493.

Snuggs, R. M., D. J. Volz, J. H. Schummers, D. W. Martin, and E. W. McDaniel (1971b). Mobilities and longitudinal diffusion coefficients of mass-identified potassium ions and positive and negative oxygen ions in oxygen, *Phys. Rev.* **A3**, 477–487.

Takebe, M., Y. Satoh, K. Iinuma, and K. Seto (1980). Mobilities and longitudinal diffusion coefficients for K^+ ions in nitrogen and argon, *J. Chem. Phys.* **73**, 4071–4076.

Takebe, M., Y. Satoh, K. Iinuma, and K. Seto (1981). Direct measurement of mobility for cluster ions $K^+(H_2O)$ in N_2 at room temperature, *J. Chem. Phys.* **75**, 5028–5030.

Takebe, M., Y. Satoh, K. Iinuma, and K. Seto (1982). Mobilities and longitudinal diffusion coefficients for Li^+ ions in Ar, Kr, and Xe at room temperature, *J. Chem. Phys.* **76**, 2672–2674.

Thomson, G. M., J. H. Schummers, D. R. James, E. Graham, I. R. Gatland, M. R. Flannery, and E. W. McDaniel (1973). Mobility, diffusion, and clustering of K^+ ions in gases, *J. Chem. Phys.* **58**, 2402–2411.

Tyndall, A. M. (1938). *The Mobility of Positive Ions in Gases*, Cambridge University Press, Cambridge.

Varney, R. N. (1952). Drift velocities of ions in krypton and xenon, *Phys. Rev.* **88**, 362–364.

Viehland, L. A., and D. W. Fahey (1983). The mobilities of NO_3^-, NO_2^-, NO^+, and Cl^- in N_2: A measurement of inelastic energy loss, *J. Chem. Phys.* **78**, 435–441.

Villinger, H., J. H. Futrell, A. Saxer, R. Richter, and W. Lindinger (1984). An evaluation of the role of internal energy and translational energy in the endothermic proton transfer reaction of N_2H^+ with Kr, *J. Chem. Phys.* **80**, 2543–2547.

Volz, D. J., J. H. Schummers, R. D. Laser, D. W. Martin, and E. W. McDaniel (1971). Mobilities and longitudinal diffusion coefficients of mass-identified potassium ions and positive nitric oxide ions in nitric oxide, *Phys. Rev.* **A4**, 1106–1109.

Wannier, G. H. (1953). Motion of gaseous ions in strong electric fields, *Bell Syst. Tech. J.* **32**, 170–254.

Whealton, J. H., and E. A. Mason (1974). Transport coefficients of gaseous ions in electric fields, *Ann. Phys.* **84**, 8–38.

Whealton, J. H., and S. B. Woo (1968). Anomalous ion-molecule reaction rates in drift tubes, *Phys. Rev. Lett.* **20**, 1137–1141.

Williams, O. M., and M. T. Elford (1986). The dependence of measured drift velocity on shutter open time in the four gauze time-of-flight method, *Aust. J. Phys.* **39**, 225–235.

Woo, S. B., and J. H. Whealton (1969). Transport model for converting charged species in drift tubes, *Phys. Rev.* **180**, 314–319. Errata published *Phys. Rev.* **A1**, 1558 (1970).

Woo, S. B., and J. H. Whealton (1971). Determination of reaction rates in drift tubes, *Phys. Rev.* **A4**, 1046–1051.

Note added in proof:

Important extensions of his applications of lasers to the study of ions in drift tubes that were discussed in Sections 2-5-H-2 and 2-5-H-4 are described by S. R. Leone (1989). Laser probing of ion collisions in drift fields: state excitation, velocity distributions, and alignment effects, in *Gas Phase Bimolecular Reactions*, M. N. R. Ashfold and J. E. Baggott, Eds., Royal Society of Chemistry, London. *See also* R. N. Zare (1988) *Angular Momentum*, Wiley, New York.

3

MEASUREMENT OF TRANSVERSE DIFFUSION COEFFICIENTS*

Measurements of D_T are performed in drift tubes but by techniques different from those used to determine v_d and D_L and discussed in Chapter 2. Two main experimental approaches have been utilized in measurements of D_T to date—the attenuation method and the Townsend method. Both are described in this chapter. The remarks made in Section 2-1 concerning desirable features of drift-tube construction apply with equal force here but are not repeated. Likewise it is hardly necessary to repeat the earlier arguments (Sections 2-2B and 2-3) concerning the requirement of low current densities and negligible space charge if accurate data are to be obtained.

3-1. ATTENUATION METHOD

This method was introduced by the Georgia Tech group (Miller et al., 1968; Moseley et al., 1969a, b) and is described in terms of their drift-tube mass spectrometer (see Section 2-5A). It is convenient to divide the discussion into two parts: the first applies to cases in which the ions being investigated do not react chemically with the gas filling the drift tube; the second to cases in which reactions occur.

A. No Reactions

Let us suppose that we are dealing with an ionic species produced only in the ion source in bursts that are negligibly short compared with the drift time. Then the flux $\Phi(0, z, t)$ of the ions passing through the exit aperture of the drift tube is

*Much of the ground covered here is also covered in the review by Märk and Märk (1984).

given as a function of the time t and the drift distance z by (2-6-23). Integrating this expression over time gives

$$I = \int_0^\infty \Phi(0, z, t) \, dt, \qquad (3\text{-}1\text{-}1)$$

the total number of ions flowing through the exit aperture following each pulse of the ion source. Analytically, I may be expressed as

$$I(z) = \frac{sa \exp(zv_d/2D_L)}{4D_L^{1/2}} \left\{ \left[2D_L^{1/2} + \frac{v_d}{(\alpha + v_d^2/4D_L)^{1/2}} \right] \exp\left[\frac{-z}{D_L^{1/2}} \left(\frac{v_d^2}{4D_L} + \alpha \right)^{1/2} \right] \right.$$

$$- \left[\frac{z}{(z^2/4D_L + r_0^2/4D_T)^{1/2}} + \frac{v_d}{(\alpha + v_d^2/4D_L)^{1/2}} \right]$$

$$\left. \exp\left[-2 \left(\frac{z^2}{4D_L} + \frac{r_0^2}{4D_T} \right)^{1/2} \left(\frac{v_d^2}{4D_L} + \alpha \right)^{1/2} \right] \right\} \qquad (3\text{-}1\text{-}2)$$

(Snuggs, 1970). In the case under discussion here the ions are assumed to be incapable of reacting with the drift-tube gas; therefore the reaction frequency α equals zero. The drift velocity v_d and the longitudinal diffusion coefficient D_L are assumed to be known from the measurements described in Chapter 2.* The area of the drift-tube exit aperture a and the radius of the ion entrance aperture r_0 are also known quantities. The planar source density s depends on the operating conditions of the ion source. It now becomes apparent that if all the operating parameters of the apparatus are maintained constant except the ion source position I is a function of the drift distance z alone and (3-1-2) can be used to determine the value of the transverse diffusion coefficient. The technique is to measure the ionic current reaching the detector for various source positions, assume a value of D_T in (3-1-2), and adjust the assumed value of D_T until the best agreement between the analytical and experimental variations of I with z is achieved. During the measurements the ion source may be operated in either the pulsed or the dc mode, but in either one the source output must be maintained constant at all source positions.

Figure 3-1-1 shows typical data used in the determination of D_T for K^+ ions in nitrogen under conditions in which reactions are negligible. The symbols indicate the experimental decrease in count rate as the drift distance is increased at five different drift-tube pressures. (As the drift distance is increased, the

---

*Actually, the value of D_T which is obtained by the analysis described here is insensitive to the value of D_L which is used. Likewise, as stated in Section 2-3, the transverse diffusion only very weakly affects the determination of D_L, and the two diffusion coefficients may be obtained essentially independently of one another. In experimental determinations of D_T, if D_L has not been measured as a function of E/N, the low-field value D obtained from the mobility by the Einstein equation may be used for D_L without significant error. A better procedure would be to use the appropriate generalized Einstein relation to calculate D_L as a function of E/N (see Sections 5-2B and 6-4).

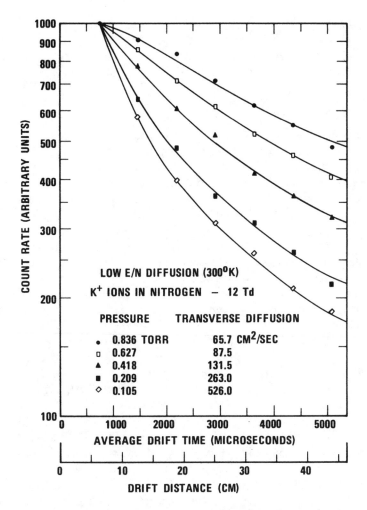

Figure 3-1-1. The experimental decrease in count rate as the drift distance is increased, compared with the predictions of (3-1-2) for K^+ ions in N_2 (Moseley et al., 1969a).

enhanced effect of transverse diffusion causes a smaller fraction of the ions to leave the drift tube through the exit aperture and reach the detector.) The smooth curves in each case are plots of I vs. z obtained from (3-1-2), with the value of D_T that gives the best fit with the experiment being used at each pressure. Since the absolute intensity of the ion swarm entering the drift tube and the ion sampling efficiency at the end of the drift tube are not known, the curves are normalized to agree with the experimental points obtained at the shortest drift distance. It will be noticed that the measured values of D_T vary as $1/N$, where N is the drift-tube gas number density, as should be the case. The experimental results obtained for D_T at low E/N are in agreement with the value predicted by the Einstein equation from the measured value of K.

Even for the simple situation described here the accuracy achieved (typically, 10 to 20%) is not so high as would be hoped for, due in large measure to the difficulty in meeting the requirement that the ion source output be held constant as the source is moved along the axis of the drift tube.

B. Depleting Reactions

Let us continue to restrict our attention to a species of ions created only in the ion source but now allow for the possibility of the destruction of these ions by chemical reactions with the gas molecules. Then (3-1-2) applies if some appropriate nonzero value is given to α, the frequency of the reactions depleting the primary ion population. At a given gas temperature α is a function of E/N and the gas pressure. The evaluation of D_T is considerably complicated by the presence of reactions because now both the reactions and the transverse diffusion contribute to the decrease in counting rate for the primary ion as the drift distance is increased. The problem is to separate the two effects, and we find that in the process of doing so we can evaluate the reaction frequency α (and rate coefficient k) as well as the transverse diffusion coefficient. Let us assume that the reaction of the ions with the gas molecules proceeds through a single channel, as is frequently the case, and that α refers to a single type of reaction. We take advantage of the fact that D_T varies as $1/N$, whereas α varies as kN in a two-body reaction or as kN^2 in a three-body reaction; this enables us to emphasize the effect of diffusion by operating at low pressures and the effect of reactions by working at high pressures. For small N the effect of D_T on $I(z)$ is much greater than the effect of α; for large N the situation is reversed.

We begin by measuring $I(z)$ at low E/N, where D_T can be calculated from the measured value of the mobility by the Einstein equation (1-1-5). Then α, the only unknown, may be evaluated directly at low E/N (thermal energy) by assuming a value for this quantity and varying it in (3-1-2) until the best fit is obtained with the experimental attenuation data. After this is done E/N is increased slightly and a first approximation for D_T at this higher E/N is made by assuming that it has its low E/N value. This first approximation for D_T is used with high-pressure experimental data to obtain a first approximation for α at the higher value of E/N. The first approximation for α is then used with low-pressure experimental data to obtain a second approximation for D_T, which in turn can be used to determine a second approximation for α from the high-pressure data, and so on until stationary values of D_T and α are obtained for this higher value of E/N. E/N is increased to successively higher values, and the procedure described above is used to evaluate D_T and α as functions of E/N up to the highest value of this parameter that can be reached in the experiment. From the pressure dependence of the measured values of α the order of the reaction can be determined and the reaction rate coefficient evaluated.

Figure 3-1-2 shows typical attenuation data for N^+ ions in N_2 which were used to determine D_T at 300 K and $E/N = 54$ Td. The dots represent experimental points; the curves are plots of (3-1-2) for various assumed values of D_T.

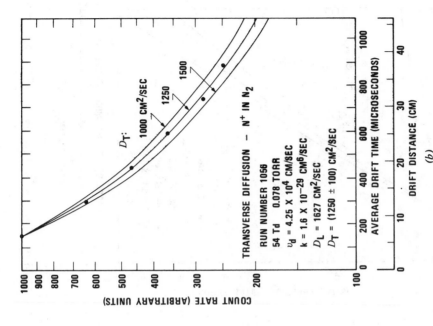

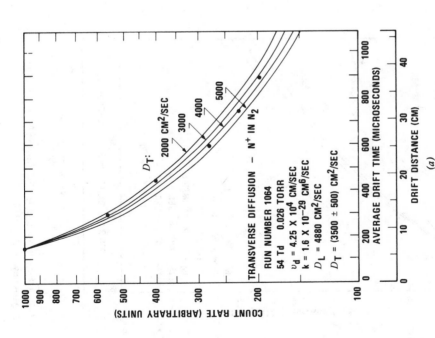

Figure 3-1-2. Attenuation data for N^+ ions in N_2 used to determine D_T at $E/N = 54$ Td and $T = 300$ K. (*a*) Pressure of 0.026 torr; (*b*) pressure of 0.078 torr.

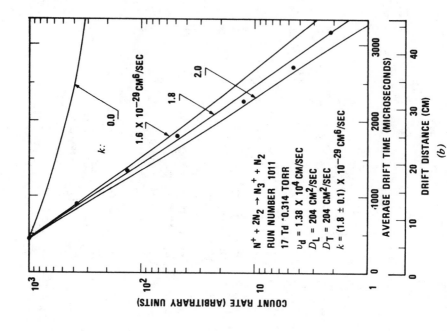

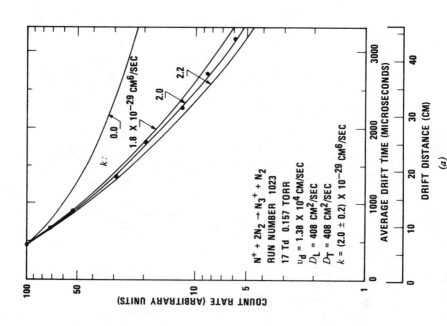

Figure 3-1-3. Attenuation data for N^+ ions in N_2 used to determine k at $E/N = 17$ Td and $T = 300$ K: (a) Pressure of 0.157 torr; (b) pressure of 0.314 torr (Moseley et al., 1969b).

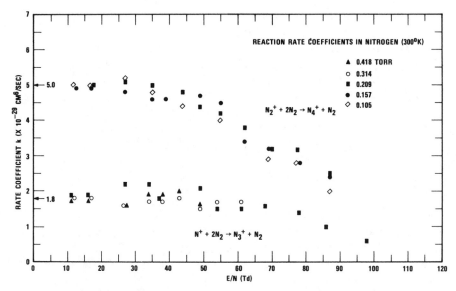

Figure 3-1-4. Rate coefficients as functions of E/N for the indicated N^+ and N_2^+ reactions in N_2 at 300 K (Moseley et al., 1969b).

The final values of D_T derived from these data are (3500 ± 500) cm^2/s at 0.026 torr and (1250 ± 100) cm^2/s at 0.078 torr.

N^+ ions react with nitrogen molecules according to the reaction $N^+ + 2N_2 \rightarrow N_3^+ + N_2$. Attenuation data used to determine the rate coefficient for this reaction at $E/N = 17$ Td and $T = 300$ K are shown in Fig. 3-1-3, in which the experimental points are indicated by dots and plots of (3-1-2) by curves. The rate coefficients for the N^+ reaction and for the $N_2^+ + 2N_2 \rightarrow N_4^+ + N_2$ reaction as well are shown in Fig. 3-1-4. The attenuation method was utilized to obtain these data. The accuracy of the low-field determinations of k was estimated to be about 11%.

3-2.   TOWNSEND METHOD

This method represents the application to ions of techniques developed many years ago by Townsend for measurements of the transport properties of slow electrons in gases (McDaniel, 1964; Huxley and Crompton, 1974). Judging by the sparsity of the published data on ions obtained by the Townsend method and the generally poor agreement among them, we conclude that reliable ionic measurements based on the Townsend technique are considerably more difficult to make than the corresponding electron measurements, which have produced a large volume of accurate results on a variety of gases. This situation is due mainly to the greater difficulty of producing satisfactory inputs of ions to the diffusion cell and to the continually vexing problems associated with chemical reactions between the ions and molecules of the gas through which they are moving. The Townsend method does not provide values of D_T directly but rather the ratio of the transverse diffusion coefficient to the mobility D_T/K; D_T can be obtained, however, at a given E/N if the mobility is separately determined at the same E/N for the same ionic species. The earliest published data on D_T/K for ions obtained by the Townsend method are those of Llewellyn-Jones (1935) for unselected argon ions in argon: subsequently, little work of this type was done until the 1960s. In this section we describe three variants of the Townsend method used recently for measurements of D_T/K for ions.

A.   Unselected Ions: Movable Ion Collector

The apparatus of Fleming et al. (1969a, b) at the University of Liverpool is described here to illustrate this method. It consists of a cylindrical chamber containing a dc ion source on the axis, a drift region in which the ions are allowed to come into energy equilibrium at the chosen value of E/N, an entrance aperture plate through which the ions then enter the diffusion region, and a collector assembly mounted perpendicular to the axis at the far end of the diffusion region. It is assumed that only a single, nonreacting species of ion is created by the ion source. A uniform axial electric field is provided in both the drift and diffusion regions by sets of thick cylindrical electrodes of the type shown in Fig. 2-5-2. The ion collector assembly (Fig. 3-2-1) consists of a pair of semicircular electrodes surrounded by a grounded electrode and mounted on a sheet of flat plate glass. The semicircular electrodes are separated from one another by a 0.0025-cm gap and from the grounded electrode by a 0.01-cm gap. The entire collector assembly is movable as a unit in a direction perpendicular to the axis of the electrode system, its position being adjusted by an external screw

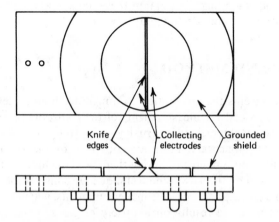

Figure 3-2-1. Ion collector assembly in the transverse diffusion apparatus of Fleming, Tunnicliffe, and Rees (Fleming et al., 1969a).

drive. The entrance aperture which allows the ions to pass from the drift region into the diffusion region is a slit 1 cm long and 0.01 cm wide, aligned parallel with the gap between the semicircular collecting electrodes. The distance between the entrance aperture plate and the collector assembly is 10 cm.

The ion current to each of the two collector electrodes is measured as a function of the off-axis displacement of the collector assembly with a pair of electrometer amplifiers. A curve of the ratio of the two currents versus the displacement is then plotted, and values of D_T/K are obtained from the analysis of this curve. Skullerud (1966a,b) has also constructed an apparatus similar to that of Fleming and his associates and applied it to measurements of D_T/K.

Evidently the success of this method hinges on the assumption that only a single known ionic species is present in the apparatus, and the technique cannot be employed with confidence when the ion source may produce more than one kind of ion or when reactions may occur between the ions and the gas molecules. By comparison the attenuation method described in Section 3-1 is not subject to this limitation.

B.   Drift-Selected Ions: Stationary Source and Collector

This method was developed in an attempt to overcome the problems associated with the production of a multiplicity of ionic species in the ion source and can be described in terms of the apparatus at the University College of Swansea (Dutton et al., 1966; Dutton and Howells, 1968). The apparatus (Fig. 3-2-2) consists of two sections. The first is a four-gauze electrical shutter drift tube similar to that described in Section 2-5F and is shown schematically at the left of the plate J. Ions are created in the glow discharge ion source S, and their drift velocities are determined by measuring the ion currents through the system as

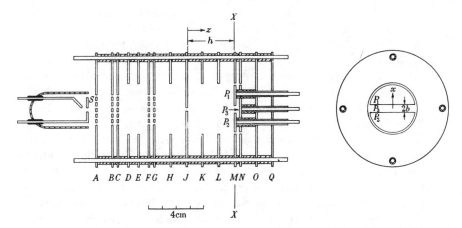

Figure 3-2-2. Swansea apparatus used for determinations of the drift velocity and transverse diffusion coefficient (Dutton et al., 1966).

functions of the frequency of the pulses applied to the shutters B-C and F-G.*
The drift section is also used to select the desired species of ion for admission
through a slit in J to the second section, which consists of a Townsend-type
transverse diffusion cell used for the determination of D_T/K for the selected
species. The collector assembly at the end of this cell consists of two D-shaped
electrodes, P_1 and P_2, and a circular electrode, P_3. The ratio D_T/K at a given
value of E/N is determined from the measured value of the ratio R of ion current
received at P_3 to the total ion current collected by P_1, P_2, and P_3. Obviously
reliable measurements for the selected species can be made only if its mobility is
significantly different from those of the other species and if the selected ions do
not react with the gas during their passage through the diffusion cell. Even then
the utility of the value of D_T/K obtained depends on the unambiguous
identification of the selected species through its measured mobility.

In the work of Dutton and his associates diffusion in the field direction was
ignored in the analysis of the data, but consideration of a more accurate
expression for R containing an explicit dependence on D_L indicated that this
procedure would probably lead to an error of the order of only 1%. A more
serious difficulty came to light when the measured values of D_T/K at low E/N
were found to be in significant disagreement with the predictions of the Einstein
equation (1-1-5). This discrepancy was attributed to the distortion of the electric
field in the vicinity of the collecting electrodes due to the finite thickness
(0.06 cm) of P_1 and P_2. A rather large empirical correction was applied to the
data at all values of E/N to correct for this effect.

C. Apparatus of Gray and Rees

Features of both instruments described in Sections 3-2A and 3-2B have been
combined by Gray and Rees (1972) at the University of Liverpool into the
apparatus shown in Figs. 3-2-3 and 3-2-4. Ions are produced in an electron
bombardment ion source at the top of the apparatus and pushed downward into
the drift region, which has a length of 6 cm. This region can be operated as a
Bradbury–Nielsen drift tube (see Section 2-5B) when electrical shutters are
inserted at the top and bottom of the drift space. Hence the apparatus can
provide either unselected ions or drift-selected ions for admission through the
source electrode into the diffusion region. The orifice in the source electrode is a
2.5 cm × 0.02 cm slit perpendicular to the plane of drawing in Fig. 3-2-3. The
length of the diffusion region is 10 cm. The collector assembly at the bottom of
this region consists of two semicircular disk electrodes shown in Fig. 3-2-4, and
in one of these electrodes there is a small hole leading to a differentially pumped
quadrupole mass spectrometer. The source electrode is mounted on ruby balls
2 mm in diameter and can be moved perpendicular to the axis of the apparatus
along a line in the plane of Fig. 3-2-3. All components above the source electrode
are rigidly connected to this electrode and move with it. The ion currents to each

---

*This method of measuring the drift velocity is similar to that described in Section 2-5B1.

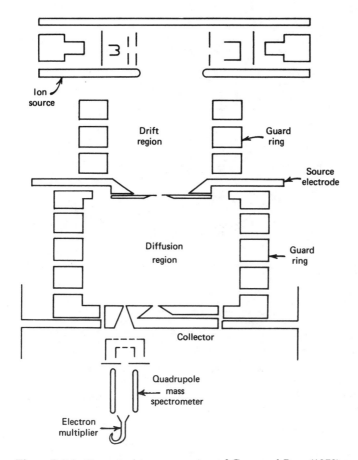

Figure 3-2-3. Townsend-type apparatus of Gray and Rees (1972).

of the collector electrodes are measured as a function of the lateral displacement of the slit in the source electrode. In addition, the mass spectrometer scans the diffusing stream of ions as it is translated by moving the source electrode. Analysis of the collector and mass spectrometer currents, when combined with drift velocity data, will yield reliable values of D_T (Rees, 1974; Alger et al., 1978; Rees and Alger, 1979). The greatest accuracy is achieved for the simplest physical situation, wherein only a single nonreacting ionic species enters the diffusion region through the source slit. Good accuracy can also be obtained for the case of two species of ions already in reaction equilibrium when they enter the diffusion chamber—here the fractional population of each species remains constant during the diffusion.

Another apparatus with an ion source capable of transverse movement has been recently developed by Sejkora et al. (1984). It utilizes alpha-particle ionization, an insulated exit aperture plate, and a quadrupole ion filter and

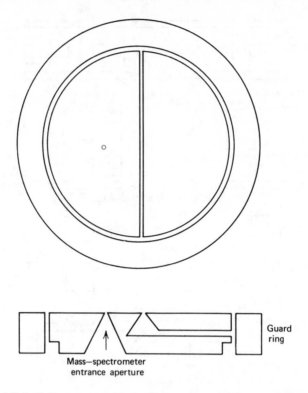

Guard
ring

Mass—spectrometer
entrance aperture

Figure 3-2-4. Collector assembly in the apparatus of Gray and Rees (1972).

channeltron multiplier for ion identification and detection. In measurements on argon ions in Ar, it was possible to obtain D_T/K data over the E/N range 30 to 290 Td. This apparatus has been described in the review by Märk and Märk (1984). The method used with this instrument to obtain transverse diffusion coefficients is referred to as the "radial ion distribution" method; it is a variant on the Townsend technique.

REFERENCES

Alger, S. R., T. Stefánsson, and J. A. Rees (1978). Measurements of the lateral diffusion of O_2^+ ions in oxygen, N_3^+ and N_4^+ ions in nitrogen, and $CO^+ \cdot CO$ ions in carbon monoxide, *J. Phys.* **B11**, 3289–3297.

Dutton, J., and P. Howells (1968). The motion of oxygen ions in oxygen, *J. Phys.* **B1**, 1160–1170.

Dutton, J., F. Llewellyn Jones, W. D. Rees, and E. M. Williams (1966). Drift and diffusion of ions in hydrogen, *Philos. Trans. R. Soc. London* **A259**, 339–354.

Fleming, I. A., R. J. Tunnicliffe, and J. A. Rees (1969a). The drift and diffusion of potassium ions in nitrogen, *J. Phys.* **D2**, 551–556.

Fleming, I. A., R. J. Tunnicliffe, and J. A. Rees (1969b). Concerning determinations of the drift velocity and lateral diffusion of positive ions in hydrogen, *J. Phys.* **B2**, 780–789.

Gray, D. R., and J. A. Rees (1972). The lateral diffusion of mass identified positive ions in oxygen, *J. Phys.* **B5**, 1048–1055.

Huxley, L. G. H., and R. W. Crompton (1974). *The Diffusion and Drift of Electrons in Gases*, Wiley, New York.

Llewellyn-Jones, F. (1935). The energy of agitation of positive ions in argon, *Proc. Phys. Soc.* **47**, 74–85.

McDaniel, E. W. (1964). *Collision Phenomena in Ionized Gases*, Wiley, New York, pp. 524–528.

Märk, E., and T. D. Märk (1984). Transverse ion diffusion in gases, in *Swarms of Ions and Electrons in Gases*, W. Lindinger, T. D. Märk, and F. Howorka, Eds., Springer-Verlag, Berlin, pp. 60–86.

Miller, T. M., J. T. Moseley, D. W. Martin, and E. W. McDaniel (1968). Reactions of H^+ in H_2 and D^+ in D_2: Mobilities of hydrogen and alkali ions in H_2 and D_2 gases, *Phys. Rev.* **173**, 115–123.

Moseley, J. T., I. R. Gatland, D. W. Martin, and E. W. McDaniel (1969a). Measurement of transport properties of ions in gases: Results for K^+ ions in N_2, *Phys. Rev.* **178**, 234–239.

Moseley, J. T., R. M. Snuggs, D. W. Martin, and E. W. McDaniel (1969b). Mobilities, diffusion coefficients, and reaction rates of mass-identified nitrogen ions in gases, *Phys. Rev.* **178**, 240–248.

Rees, J. A. (1974). Transport properties of ions in electronegative gases, *Vacuum* **24**, 603–607.

Rees, J. A., and S. R. Alger (1979). Transport properties of mass-identified ions in oxygen, nitrogen, carbon dioxide, and carbon monoxide, *Proc. Inst. Electr. Eng.* **126**, 356–360.

Sejkora, G., P. Girstmair, H. C. Bryant, and T. D. Märk (1984). Transverse diffusion of Ar^+ and Ar^{2+} in Ar, *Phys. Rev.* **A29**, 3379–3387.

Skullerud, H. R. (1966a). *Tech. Rep. GDL 66-1*, Gassutladningslaboratoriet, Institutt for Teknisk Fysikk, Norges Tekniske Högskole, Trondheim, Norway.

Skullerud, H. R. (1966b). *Proc. 7th. Int. Conf. Phenomena Ionized Gases*, Belgrade, 1965, Vol. 1, Gradevinska Knjiga, Belgrade, p. 50.

Snuggs, R. M. (1970). Ph.D. thesis, Georgia Institute of Technology, Atlanta.

4

STATIONARY AFTERGLOW
TECHNIQUES

The main theme in this chapter is the determination of zero-field ionic diffusion coefficients from observations on a decaying plasma* during the "afterglow" period, although the evaluation of other quantities from such studies is briefly discussed. The afterglow is the time regime following the removal of the ionizing source that produces the plasma during which the gas remains ionized to an appreciable extent. Except for times late in the afterglow, the ionization densities are much higher than those in the drift-tube measurements described in Chapters 2 and 3 and high enough to make the attractive forces between the electrons and positive ions important. Under these conditions the diffusion is ambipolar (see Section 1-12) and the diffusion coefficient measured is the coefficient of electron-ion ambipolar diffusion D_a.† The afterglow experiments discussed here are made with no applied electric field and usually under conditions in which the ion and electron temperatures equal the gas temperature. Under these conditions the ordinary, or free, diffusion coefficient for the positive D^+ equals one-half of D_a [see (1-12-6)]; this relationship permits the determination of D^+ from the measured value of the ambipolar diffusion coefficient.

Afterglow experiments designed to provide information on atomic collisions and transport phenomena are of two basically different types—stationary and flowing. In a stationary afterglow experiment, the only kind treated in detail here, we observe the temporal variation of the charged-particle number densities

---

*By a "plasma" we mean an ionized gas whose dimensions are large compared with its debye shielding length. The differences between a plasma and an ordinary ionized gas are discussed in detail by McDaniel (1964), App. I.

†In an electronegative gas such as nitric oxide the electrons produced in the ionization process may be captured to form negative ions before they diffuse to the chamber walls. Hence late in the afterglow the ambipolar diffusion may actually involve negative and positive ions rather than electrons and positive ions. This type of ion-ion ambipolar diffusion has been observed and is discussed in Section 4-1C.

following an ionizing pulse in a gas that is either at rest or flowing with a negligible speed. Measurements on stationary afterglows have provided momentum transfer cross sections for electrons, electron attachment coefficients, recombination coefficients, and rate coefficients for chemical reactions between ions and molecules as well as ambipolar diffusion coefficients (McDaniel, 1964, pp. 120–122, 514–518, 605–607; Biondi, 1968; Oskam, 1969; McDaniel et al., 1970). Flowing afterglow experiments yield mainly information on chemical reactions (Ferguson et al., 1969; McDaniel et al., 1970; Franklin, 1972). In such experiments we sample the reaction products in a rapidly flowing field-free gas at some location downstream from the point at which the reactant was introduced into the flow and substitute a spatial measurement for a temporal measurement. This kind of measurement has been extraordinarily fruitful in studies of ion-molecule and charge transfer reactions, but it is limited to the study of thermal energy processes. Data on suprathermal reactions may be obtained by using a combination of flowing afterglow and drift-tube techniques in which the ions are heated by a drift field during their flow. The flowing afterglow technique and its variants that utilize drift fields were discussed in Section 2-5E.

At present stationary afterglow techniques can provide zero-field ionic diffusion coefficients with an accuracy of about 8 to 20%. Much better accuracy ($\frac{1}{2}$ to 4%) can usually be obtained in the evaluation of these coefficients if we resort to the indirect method of calculating them from experimental zero-field mobilities by use of the Einstein equation (1-1-5). Nonetheless, stationary afterglow methods are of considerable interest and warrant a rather detailed treatment in this book. We discuss three quite different techniques, all of which have produced important results.

4-1. TECHNIQUE OF LINEBERGER AND PUCKETT

Lineberger and Puckett constructed a stationary afterglow apparatus at the Aberdeen Proving Ground and applied it to the study of diffusion and reactions in nitric oxide (Lineberger and Puckett, 1969a). They subsequently used the apparatus for some interesting experiments on reactions of positive and negative ions in gas mixtures (Lineberger and Puckett, 1969b; Puckett and Lineberger, 1970; Puckett and Teague, 1971a, b; Puckett et al., 1971), but we shall confine our attention to the work reported in their first paper.

A. Apparatus

The apparatus of Lineberger and Puckett is shown schematically in Fig. 4-1-1. The afterglow cavity is a bakeable, gold-plated, stainless-steel cylinder, 18 in. in diameter and 36 in. long, which is sealed by metal gaskets. When the cavity is filled with nitric oxide, NO^+ ions are produced in periodic bursts by the krypton discharge lamp. Ions of this type and other species formed in secondary

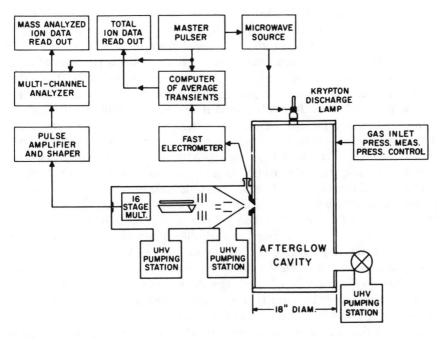

Figure 4-1-1. Schematic diagram of the stationary afterglow apparatus of Lineberger and Puckett (1969a).

processes diffuse to the wall and are mass spectrometrically sampled as a function of time through the orifice shown halfway along the wall of the cavity. The cavity and the differential pumping regions shown at the left are pumped by oil diffusion pumps equipped with water-cooled baffles and sorbent traps. The cavity can be isolated from its pumping station by means of an all-metal, bakeable valve. Typical base pressures within the cavity following a 48-h 200°C bake are about 10^{-9} torr, with a rate of rise less than 10^{-10} torr/s. Gas is admitted into the cavity through a servo-driven variable leak valve coupled to a feedback control system that regulates the cavity pressure to a value set by the capacitance manometer used to measure the pressure.

The discharge lamp, which is microwave-powered, produces the primary NO^+ ions by photoionization of the nitric oxide molecules with krypton resonance radiation (123.6 and 116.5 nm). The photons have insufficient energy to form any other ionic species in NO. The lamp body is quartz with a MgF_2 window 1 mm thick which transmits approximately 50% of the resonance radiation. The intensity of the light inside the afterglow cavity is about 10^{15} photons/s. With this level of radiation the initial charged-particle number densities are low enough that electron-positive ion recombination makes a negligible contribution to the decay of the plasma, and this fact greatly simplifies the analysis of the data. In the midplane of the cavity, where the ions diffusing to the wall are sampled, the radial distribution of the light is approximately a truncated cosine distribution that fills about 80% of the chamber. This distribution is similar to a zeroth-order Bessel function J_0 (see Section 1-11D), and the ions quickly stabilize into a fundamental mode spatial distribution.

The sampling orifice, which has a 0.60-mm diameter, is contained in a disk 6 cm in diameter, machined to match the contour of the cavity wall. The orifice disk is insulated from the wall so that it can be electrically biased and the transient current to the disk is measured with a computer of average transients. This feature permits tests to be made to determine the optimum potential that should be applied to the sampling disk to produce a count rate of mass-analyzed ions directly proportional to their number density in the cavity throughout the plasma decay. Lineberger and Puckett consider this matter in great detail and present data to show the pronounced effect of orifice potentials on the measured time constants of the afterglow decay. Most of their measurements are made with attractive potentials of several tens of millivolts applied to the orifice plate.

The relatively high pressures used in the afterglow cavity (up to about 300 millitorr) make advisable the use of differential pumping in the ion sampling apparatus. As Fig. 4-1-1 shows, two stages of differential pumping are provided, each affording a pressure reduction by a factor of about 10^3. These two stages are separated by an insulated conical skimmer with a hole 1 mm in diameter in its tip. The skimmer is normally operated at an attractive potential of 25 to 300 V to enhance the ion sampling efficiency and to reduce mass discrimination in the sampling. Checks are made to ensure that molecular ions are not dissociated in energetic collisions during the sampling process. The mass spectrometer is a General Electric monopole spectrometer, modified to permit operation with ions formed at ground potential. Those ions that pass through the spectrometer are counted with a 16-stage Ag-Mg electron multiplier whose output pulses are sorted by a 1024-channel time analyzer. A master pulser simultaneously triggers the discharge lamp, the time analyzer, and the computer of average transients. Repetitive pulsing of the apparatus permits the accumulation of a sufficient number of counts in each time channel to provide a statistically significant history of the plasma decay.

B. Data Analysis

In the analysis of their data Lineberger and Puckett assume fundamental mode diffusion and negligible recombination losses and demonstrate the validity of these assumptions except for times early in the afterglow. Here we reproduce their analysis. Let us suppose that electron-ion ambipolar diffusion of NO^+ and a reaction with NO molecules having a frequency v_l are the only significant loss processes for NO^+ ions. Then, if there are no sources of NO^+ during the afterglow, the continuity equation for the NO^+ ions is

$$\frac{\partial}{\partial t}[NO^+(r, t)] = D_a \nabla^2[NO^+(r, t)] - v_l[NO^+(r, t)], \qquad (4\text{-}1\text{-}1)$$

where the brackets [ ] denote number densities and D_a is the NO^+ electron-ion ambipolar diffusion coefficient. For a long cylindrical cavity of radius R and an initial NO^+ distribution given by

$$[NO^+(r,0)] = [NO^+(0,0)]J_0\left(2.405\frac{r}{R}\right) \qquad (4\text{-}1\text{-}2)$$

the solution of (4-1-1) is

$$[NO^+(r,t)] = [NO^+(0,0)]J_0\left(2.405\frac{r}{R}\right)\exp\left[-\left(\frac{D_ap}{\Lambda^2 p}+v_l\right)t\right], \quad (4\text{-}1\text{-}3)$$

where Λ is the characteristic diffusion length of the cavity, p is the gas pressure, and $[NO^+(0,0)]$ is the initial axial number density. (The reader may find it helpful to review Section 1-11D at this point.)

In the Lineberger-Puckett apparatus the direct physical observable is the count rate at the mass spectrometer, not the ionic volume number density. If we assume that the NO^+ count rate $CR(NO^+)$ is proportional to the NO^+ wall-current density and that the wall-current density is simply a diffusion current, driven by the ion density gradient, then

$$CR(NO^+) \propto D_a[NO^+(0,0)]\exp\left[-\left(\frac{D_ap}{\Lambda^2 p}+v_l\right)t\right]$$

$$\propto D_a[NO^+(0,0)]e^{-vt}, \qquad (4\text{-}1\text{-}4)$$

where v is the total NO^+ loss frequency. Thus, under the assumptions made above, the NO^+ count rate is directly proportional to the NO^+ number density in the cavity. This simple proportionality does not necessarily hold, however, in the event of a time-dependent spatial distribution, surface charging effects, or the use of a drawout voltage on the orifice plate. Lineberger and Puckett discuss experimental tests made to check the validity of each of their assumptions.

It is now apparent that both D_a and v_l may be determined if v is measured as a function of the pressure. The data presented in the next section indicate that NO^+ is lost through the reaction

$$NO^+ + 2NO \longrightarrow NO^+ \cdot NO + NO, \qquad (4\text{-}1\text{-}5)$$

where v_l and the reaction rate coefficient k are related by the equation

$$v_l = k[NO]^2. \qquad (4\text{-}1\text{-}6)$$

Hence the p^{-1} (diffusion) and p^2 (reaction) contributions to v can be separated and D_a and k may be evaluated individually.

C. Experimental Results

At room temperature and over the pressure range of 10 to 200 millitorr the dominant positive ions present in pure NO are NO^+ and the dimer $NO^+ \cdot NO$, whereas NO_2^- dominates the negative ion spectrum by a factor of about 100. (At pressures above 200 millitorr hydrated impurity ions begin to appear in significant amounts.) Figure 4-1-2 shows the NO^+ count rate as a function of time at 20 millitorr pressure. The data were accumulated over 10^4 light pulses and extend over more than six decades of afterglow decay. The maximum ionic number density at the center of the cavity, while not directly measured, is estimated to be about $10^7 \, cm^{-3}$. The decay between about 90 and 160 ms is associated with electron-positive ion ambipolar diffusion and is exponential, as expected. The loss frequency during this period, $v = 56 \, s^{-1}$, is the reciprocal time constant for this period. The break in the NO^+ count rate near 170 msec is due to the sudden transition from electron-positive ion ambipolar diffusion domination (characterized by a coefficient that we label $D_{+,e}$) to domination by negative ion-positive ion ambipolar diffusion (characterized by $D_{+,-}$). These data represent the first reported observation of such a transition.* The structure in the count rate in the vicinity of the transition is the result of the decay of the electron and ion densities to a level too small to sustain the space-charge field that produces the electron-positive ion ambipolar diffusion and of the release of negative ions whose diffusion (unlike that of the positive ions) is opposed by the electron-positive ion space charge field. The diffusion of negative ions influences

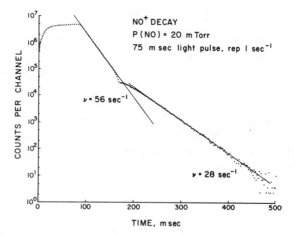

Figure 4-1-2. Typical NO^+ decay curve at 20 millitorr pressure in the experiments of Lineberger and Puckett (1969a).

*The transition from electron-ion to ion-ion afterglow plasma observed in the SA was subsequently studied and used to great effect in the FA (Smith and Church, 1976).

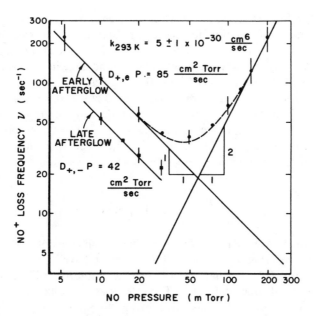

Figure 4-1-3. Variation of the NO^+ loss frequency in pure NO as a function of NO pressure. The closed circles represent early (electron-ion) afterglow data, the closed squares, late (ion-ion) afterglow data. The dashed line is the sum of the asymptotic p^{-1} and p^2 lines (Lineberger and Puckett, 1969a).

the positive ion wall current by the establishment of a weak positive ion-negative ion space-charge field through which positive and negative ions are coupled in ambipolar diffusion.

The experimental data for the pressure variation of the loss frequency of NO^+ ions are displayed in Fig. 4-1-3 and were obtained with orifice disk potentials varied between zero and 100 mV negative with respect to the cavity wall. Over the pressure range of 5 to 200 millitorr the NO^+ loss frequency may be expressed as

$$v = \frac{a}{p} + bp^2, \qquad (4\text{-}1\text{-}7)$$

which implies that the dominant NO^+ loss processes are diffusion and a three-body reaction between NO^+ ions and NO molecules. Since the only significant secondary ion observed in this pressure range was the dimer, the reaction occurring is that described by (4-1-5). The rate coefficient for this reaction was determined from b and found to be $(5 \pm 1) \times 10^{-30}$ cm^6/s.

Since it was determined that electron-ion ambipolar diffusion was the process operative during the early afterglow between 90 and 160 ms, the observed diffusion loss frequency, together with the diffusion length of the cavity,* provides enough information for the determination of the electron-ion ambipolar diffusion coefficient D_a. The result obtained is $D_a p = 85$ cm²-torr/s. Since the electron, ion, and gas temperatures are the same in this experiment, the free diffusion coefficient for NO^+ ions, D^+, is given by $D_a/2$.

Decay rates for NO^+ ions were also measured late in the afterglow (after the transition at 170 ms) to obtain information on the ion-ion ambipolar diffusion. The analysis presented above is applicable here if the ion-ion ambipolar diffusion coefficient $D_{+,-}$ is substituted for D_a. Lineberger and Puckett have shown that if the ion and gas temperatures are assumed to be equal, then

$$D_{+,-} = 2\frac{D^+ D^-}{D^+ + D^-}, \tag{4-1-9}$$

where D^+ and D^- are the free diffusion coefficients for the positive and negative ions, respectively. It is apparent that $D_{+,-}$ must have a value intermediate between D^+ and D^- and that if $D^+ = D^-$ the ambipolar diffusion process is effectively one of free diffusion. Three species of diffusing ions—NO^+, $NO^+ \cdot NO$, and NO_2^-, the last two being dominant—appear late in the afterglow. All three species have approximately the same free diffusion coefficients,† in contrast with the early afterglow in which the dominant charge carriers (NO^+ and electrons) have vastly different free diffusion coefficients. Hence in the late afterglow the ion-ion ambipolar diffusion coefficient will be nearly equal to the free diffusion coefficient for each of the three ionic species. Then at low pressures, where the only significant loss process is through diffusion, the late afterglow loss frequency (which is the same for all three types of ion) should be one-half the early afterglow loss frequency for NO^+ if in the early afterglow the electron and ion temperatures equal the gas temperature. That this is indeed the case is verified in Fig. 4-1-3, which shows that in the late afterglow $D_{+,-} p = 42$ cm²-torr/s. (Here the plus sign refers equally well to NO^+ and to $NO^+ \cdot NO$ and the minus sign refers to NO_2^-.) This is an important observation, for it shows that the electrons are not preferentially heated in the afterglow.

Lineberger and Puckett also sampled the $NO^+ \cdot NO$ and NO_2^- ions as a function of time following the light pulse. Early afterglow measurements of the $NO^+ \cdot NO$ decay frequency gave a $D_a p$ product of 84 cm²-torr/s to describe the

---

*The diffusion length used in the data analysis is that for the lowest order radial and axial modes of the cavity,

$$\Lambda^2 = \left[\left(\frac{2.405}{R}\right)^2 + \left(\frac{\pi}{L}\right)^2\right]^{-1} = 78.5 \text{ cm}^2, \tag{4-1-8}$$

where R and L are, respectively, the cavity radius and length in centimeters.

†The NO^+ ion has a free diffusion coefficient as low as that of the other much heavier ions because of the retarding effect of resonance charge transfer it experiences.

electron-ion ambipolar diffusion for this species; the estimated uncertainty was $\pm 15\%$. The uncertainty in the determination of the electron-ion ambipolar diffusion coefficient quoted above for NO^+ was estimated to be slightly lower—$\pm 12\%$. About one-fourth of this total uncertainty was associated with the fact that variations in the potential applied to the sampling orifice affected the results, and it was impossible to determine precisely the proper potential to use.

4-2. TECHNIQUE OF SMITH AND HIS COLLEAGUES

David Smith and his colleagues at the University of Birmingham have developed some useful stationary afterglow techniques and applied them to an extensive study of momentum transfer, ambipolar diffusion, chemical reactions, and electron-ion recombination in the noble gases, oxygen, and nitrogen. Their apparatus differs from that of Lineberger and Puckett, described in Section 4-1-A, in several important respects. Their afterglow cavity is much smaller and is made of glass. The initial ionization is produced by a pulsed RF discharge, and Langmuir probes (Chen, 1965; Smith and Plumb, 1972; Smith et al., 1974) are used, as well as mass spectrometric sampling of the ions, to obtain information on the decaying plasma. Furthermore, the apparatus may be operated over a wide range of temperature.

A. Apparatus

A typical experimental tube used by Smith and his associates is shown in Figs. 4-2-1 and 4-2-2. It is a Pyrex cylinder which has an internal diameter of 11.0 cm and an active length of 18.0 cm. Energy to ionize the gas in the tube is supplied as repetitive 10 μs pulses of 10-MHz power coupled capacitively to the gas by two external sleeve electrodes, the power in each pulse normally being about 20 kW. This amount of power produces an initial electron density of the order of 10^{10} cm^{-3}, as indicated by Langmuir probe measurements (Smith et al., 1968). Electrical contact with the plasma is provided by two plane circular nickel electrodes mounted inside the tube. These electrodes, which also serve in conjunction with the glass wall to define the plasma boundary, are positioned normal to the cylinder axis with a separation of 18.0 cm so that the characteristic diffusion length for the fundamental mode is $\Lambda = 2.11$ cm ($\Lambda^2 = 4.44$ cm^2). The ion sampling orifice is a hole 0.10 mm in diameter in a Nilo-K alloy disk 0.05 mm thick which is sealed to the cylindrical wall midway between the internal disk electrodes. The sampling disk is maintained at the same potential as the internal electrodes. Ions effusing from the plasma are mass-selected by a quadrupole mass spectrometer and detected by an electron multiplier. The variation of the ion current with time in the afterglow is observed by displaying the amplified signal on an oscilloscope (Fig. 4-2-3).

The entire vacuum system with the exception of the pumping lines is baked for several hours at 350°C before experiments are performed and the metal

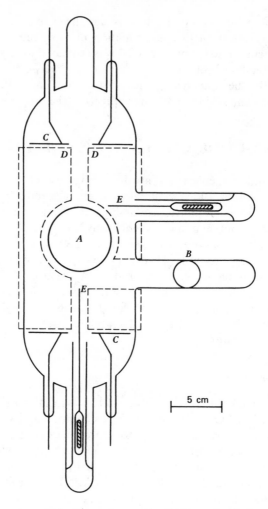

Figure 4-2-1. Top view of afterglow tube used by Smith et al. *A* is the mass spectrometer sidearm; *B*, the pumping and gas admittance sidearm; *C*, the nickel internal electrodes; *D*, the external sleeve electrodes; and *E*, the Langmuir probes.

electrodes are outgassed by induction heating. The concentration of impurities is further reduced by flushing the discharge tube with a sample of the gas to be used and running a discharge in this gas for several minutes to displace residual impurity atoms absorbed on the interior surfaces. This procedure results in residual gas pressures of the order of 5×10^{-7} torr, the residual gas consisting mainly of atoms of the gas used to flush the tube. When appropriate, the gas used in the experiment is purified by cataphoresis before being admitted to the tube.

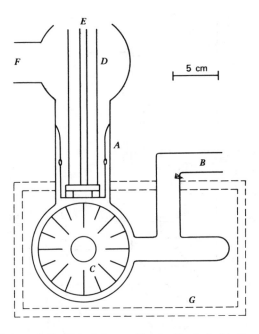

Figure 4-2-2. End view of afterglow tube used by Smith et al. A is the mass spectrometer sidearm; B, the pumping and gas admittance sidearm; C, the nickel internal electrode with slits for uniform RF induction heating; D, the quadrupole mass spectrometer; E, to electron multiplier; F, to differential pumping system; and G, oven or cooling bath.

B.   Studies of Ambipolar Diffusion and Chemical Reactions

For measurements of ambipolar diffusion and reaction rates the ion decay rates in the afterglow are measured with the mass spectrometer as a function of the gas pressure under the assumption that the sampled ion current is directly proportional to the ion number density in the body of the plasma throughout the observation (Smith and Cromey, 1968; Smith and Copsey, 1968; Smith and Fouracre, 1968; Smith et al., 1970). Such measurements are described here in terms of experiments on argon (Smith et al., 1972a) in which the atomic ion is converted to the molecular ion in the three-body reaction

$$\text{Ar}^+ + 2\text{Ar} \longrightarrow \text{Ar}_2^+ + \text{Ar}. \qquad (4\text{-}2\text{-}1)$$

Studies of the atomic ion are facilitated because the loss of this ion by recombination with electrons occurs at a negligibly slow rate compared with that due to diffusion and reaction. Hence the rate of loss of the Ar^+ ions is governed by the equation

$$\frac{\partial n}{\partial t} = D_a \nabla^2 n - vn, \qquad (4\text{-}2\text{-}2)$$

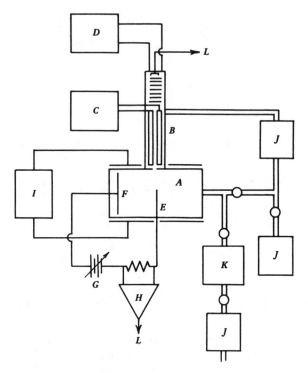

Figure 4-2-3. Block diagram of apparatus of Smith et al. A is the afterglow tube; B, the mass spectrometer sidearm; C, the quadrupole mass spectrometer power supply; D, the electron multiplier power supply; E, the Langmuir probe; F, the internal electrode; G, the probe bias; H, the probe amplifier; I, the pulsed RF supply; J, the vacuum pumps; K, the gas-handling system; and L, to oscilloscope input.

where n is the atomic ion number density, D_a is the electron-ion ambipolar diffusion coefficient for Ar^+, and v is the frequency of the reaction (4-2-1). Observations are made late enough in the afterglow that diffusion occurs almost entirely in the fundamental mode. The solution of (4-2-2) for fundamental mode diffusion indicates an exponential decrease in n with time; the decay constant λ is given by

$$\lambda = \frac{D_a}{\Lambda^2} + \beta p_0^2, \qquad (4\text{-}2\text{-}3)$$

where Λ is the fundamental-mode characteristic diffusion length, β is the rate coefficient for (4-2-1), and p_0 is the gas pressure reduced to the standard temperature of 273 K; β and v are related by the equation

$$v = \beta p_0^2. \qquad (4\text{-}2\text{-}4)$$

Since D_a is inversely proportional to p_0, Smith and his colleagues find it convenient to express (4-2-3) as

$$\lambda p_0 = \frac{D_a p_0}{\Lambda^2} + \beta p_0^3. \tag{4-2-5}$$

Their method of data analysis is somewhat different from that of Lineberger and Puckett (Section 4-1B).

Experimental data on the variation of λp_0 with reduced pressure p_0 are shown in Fig. 4-2-4. The curve drawn through the experimental points is flat in region B; this region corresponds to domination of the Ar^+ ion decay rate by ambipolar diffusion. The ambipolar diffusion coefficient is obtained by solving (4-2-5) for D_a by using the value of λp_0 obtained in region B and neglecting the term βp_0^3. At higher pressures in region C diffusion is inhibited, and the reaction (4-2-1) begins to dominate the decay; β is determined from the slope of a plot of λp_0 vs. p_0^3 by using the data of region C and its extension to still higher pressures. In region A the phenomenon of "diffusion cooling," first observed and discussed by Biondi (1954), is in evidence. This term describes the situation in which the average energy of the electrons is reduced by the diffusion of the faster electrons to the walls. In a low-pressure argon afterglow the thermal contact between the

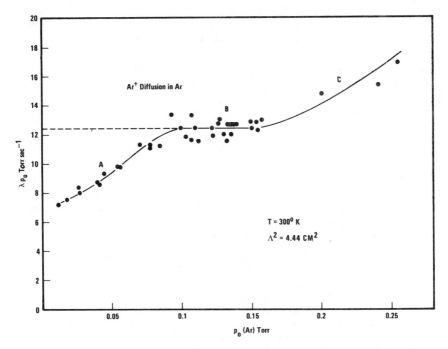

Figure 4-2-4. λp_0 as a function of p_0 for pure argon. A is the diffusion cooled region; B, normal ambipolar diffusion; and C, the onset of reaction (Smith et al., 1972a).

electrons and the gas atoms is poor, and the rapid diffusion loss of the faster component of the electrons causes a large reduction of the electron temperature with respect to the gas temperature.

The ambipolar diffusion coefficient of the molecular ions Ar_2^+ is somewhat more difficult to obtain than that of the atomic ions because Ar_2^+ undergoes recombination with electrons at a rate orders of magnitude more rapid than Ar^+ (McDaniel, 1964, Chap. 12); D_a for Ar_2^+, however, may be determined from observations of the Ar_2^+ decay rate late in the afterglow when the electron density and the recombination rate are small.

In their work on the noble gases Smith and his colleagues consider several effects not yet mentioned here and show that it is unlikely that they affected their experiments significantly. In particular, they discuss the production of free electrons in collisions between pairs of metastable atoms, the heating of electrons in superelastic collisions with metastables, and the heating of the gas atoms by the discharge.

More recent studies of reactions in the Birmingham stationary afterglow apparatus have been reported by Plumb et al. (1972), Adams et al. (1972), Smith et al. (1972b), Dean et al. (1974).

C.   Studies of Electron-Ion Recombination

The Langmuir probes shown in Figs. 4-2-1 and 4-2-2 may be used to measure electron number densities and electron temperatures and have been employed to determine coefficients of dissociative electron-ion recombination for molecular ions (Smith and Goodall, 1968; Smith et al., 1970). If the measurements are made under conditions in which some nonreacting molecular ion is the dominant species, the rate of change of the electron number density n_e may be described by the equation

$$\frac{\partial n_e}{\partial t} = D_a \nabla^2 n_e - \alpha n_e^2, \tag{4-2-6}$$

where D_a is the ambipolar diffusion coefficient of the molecular ions and α is the two-body coefficient that describes dissociative recombination between these ions and electrons. (The latter coefficient is defined such that the number of recombination events per unit volume and unit time is $\alpha n_e n_+$ but here $n_+ \approx n_e$.) If the gas pressure is high enough that diffusion has a small effect and if the ionization density is also sufficiently high, the afterglow may be said to be recombination controlled.* Under these conditions the approximate solution to

---

*In a very important paper, which has been summarized by McDaniel (1964), Chap. 12, Gray and Kerr (1962) considered in detail the effects of diffusion and recombination in spheres and infinitely long cylinders and derived explicit criteria for domination of the plasma decay by each process. One important conclusion gained from these studies is that linearity of the plot of reciprocal electron number density versus time does not in itself imply that a meaningful value for α can be deduced from the slope of the plot; linearity can be observed even when the decay is controlled by diffusion. The Gray-Kerr criteria should be carefully heeded in any stationary afterglow experiment. The reader is also referred to a paper by Oskam (1958) for an analysis of plasma decay in infinite parallel plate geometry.

(4-2-6) is

$$\frac{1}{n_e(t)} = \frac{1}{n_e(0)} + \alpha t, \qquad (4\text{-}2\text{-}7)$$

and observations of the electron density as a function of time in the recombination-controlled afterglow can yield the recombination coefficient.

4-3. MICROWAVE TECHNIQUES

Microwave techniques for studying atomic collisions and transport phenomena were introduced by M. A. Biondi and S. C. Brown at the Massachusetts Institute of Technology during the late 1940s. These techniques have undergone continuous development since that time, particularly by Biondi and his co-workers at the Westinghouse Research Laboratories and the University of Pittsburgh, and have yielded a great deal of extremely important information. They have been especially valuable in the investigation of electron-ion recombination and ambipolar diffusion, and more has been learned about the former phenomenon by microwave techniques than by any other method. Because of the vast literature on microwave studies of plasmas, we cite here only a few of the recent papers by Biondi and his colleagues: Frommhold et al. (1968), Weller and Biondi (1968), Kasner and Biondi (1968), Mehr and Biondi (1968, 1969), Philbrick et al. (1969), Frommhold and Biondi (1969), Leu et al. (1973), Huang et al. (1975), Shiu et al. (1977), and Whitaker et al. (1981a, b). J. L. Dulaney, M. A. Biondi, and R. Johnson (1987). References to Biondi's earlier work and the work of other investigators are listed in reviews (Biondi, 1968, 1976, 1982; Oskam, 1969).

The microwave studies of interest here involve the use of small cavities whose dimensions are typically a few centimeters. The gas or gas mixture to be investigated is contained in the cavity at a pressure that might be as low as a fraction of a torr for diffusion studies or as high as tens of torr in recombination measurements. Frequently, provision is made for varying the cavity temperature over a wide range. The gas is ionized periodically, usually by short pulses of microwave radiation, and the decrease in electron number density n_e is measured as a function of time in the decaying afterglow by a microwave probing signal. This measurement, at any given time, is accomplished by determining the shift in the resonant frequency of the cavity produced by the presence of the free electrons (McDaniel, 1964, Chap. 12). Since the sensitivity of the probing technique is not great enough to allow determination of electron number densities below about $10^7 \, \text{cm}^{-3}$, the quantity measured in diffusion studies is the coefficient of ambipolar diffusion D_a, as in the other stationary afterglow experiments discussed in this chapter. In some studies the ions diffusing to the walls are sampled mass spectrometrically as a function of time in the afterglow to ascertain the identities and relative abundances of the ions present. In addition, the spectral distribution and temporal variation of the light emitted by the plasma are sometimes measured. Such measurements have proved useful in studies of dissociative recombination (Frommhold and Biondi, 1969). In another useful technique that has been employed in recombination experiments a separate microwave input is applied to heat the electrons during the afterglow so that the recombination coefficient can be obtained as a function of the electron temperature.

In the determination of ambipolar diffusion coefficients by microwave techniques it is important, of course, to ensure that the plasma decay is dominated by diffusion and not affected appreciably by electron-ion recombination. If the decay is diffusion-dominated and fundamental mode diffusion has been established, a plot of n_e versus time on a semilogarithmic scale should give a straight line and D_a may be obtained from its slope. Meaningful results will be obtained in general only if a single type of positive ion is dominant and the electrons are not subject to appreciable attachment to form negative ions. Techniques for obtaining both D_a and the electron attachment coefficient when attachment occurs in a three-body collision are discussed by Weller and Biondi (1968).

Microwave techniques have been used to study a feature of diffusion which is of general interest in this book and should be mentioned at this point. The discussion in Section 1-11 indicates that the higher diffusion modes decay much faster than the fundamental mode and that even if higher modes are strongly excited by breaking down a gas in an asymmetric fashion the fundamental mode will eventually dominate the diffusion. This fact is illustrated in Fig. 4-3-1, which shows data obtained by Persson and Brown (1955) in a microwave afterglow experiment. The two plots shown here correspond to different discharge

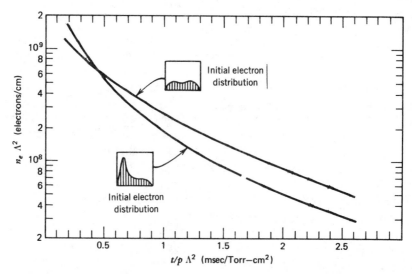

Figure 4-3-1. Influence of the initial spatial distribution on the decay of the electron number density n_e (Persson and Brown, 1955).

conditions and different initial combinations of diffusion modes. Note that the curves become straight and parallel to one another at large t, each having the slope corresponding to the fundamental mode. The inset drawings are qualitative measures of the initial spatial electron density at the start of the decay period obtained by scanning the discharge with a photomultiplier and slit system and displaying the signal on an oscilloscope.

We must now make special mention of the recent paper by Dulaney, Biondi, and Johnsen (1987) that deals with a redetermination of the electron temperature dependence of the dissociative recombination of electrons with NO^+ ions. The previous measurement performed by Huang et al. (1975) in Biondi's laboratory yielded the result $\alpha \sim T_e^{-0.37}$ at high electron temperatures, in conflict with the trapped-ion result $\alpha \sim T_e^{-0.85}$ of Walls and Dunn (1974). Dulaney et al. (1987) found that the electron temperature achieved by microwave heating in the experiments of Huang et al. was lower than calculated because of the neglect of the effects of inelastic collisions of electrons with minority NO molecules in the NO-rare gas mixtures used. In addition, their assumption of spatially uniform electron temperatures was not justified. Dulaney et al. (1987) used an improved data analysis and obtained results consistent with those of Walls and Dunn. The problems revealed by Dulaney and his colleagues are shown to have significantly affected some, but not all, of Biondi's other measurements of dissociative recombination rate coefficients.

REFERENCES

Adams, N. G., A. G. Dean, and D. Smith (1972). Thermal energy reactions of rare gas atomic ions with molecular oxygen and nitrogen, *Int. J. Mass Spectrom. Ion Phys.* **10**, 63–76.

Biondi, M. A. (1954). Diffusion cooling of electrons in ionized gases, *Phys. Rev.* **93**, 1136–1140.

Biondi, M. A. (1968). Afterglow experiments: Atomic collisons of electrons, ions and excited atoms, in *Methods of Experimental Physics*, Vol. 7B, B. Bederson and W. L. Fite, Eds., Academic, New York, pp. 78–124.

Biondi, M. A. (1976). Recombination, in *Principles of Laser Plasmas*, G. Bekefi, Ed., Wiley, New York.

Biondi, M. A. (1982). Electron-ion recombination in gas lasers, in *Applied Atomic Collision Physics*, H. S. W. Massey, E. W. McDaniel, and B. Bederson, Eds., Vol. 3, *Gas Lasers*, E. W. McDaniel and W. L. Nighan, Eds. Academic, New York.

Chen, F. F. (1965). Electric probes, in *Plasma Diagnostic Techniques*, R. H. Huddlestone and S. L. Leonard, Eds., Academic, New York.

Dean, A. G., D. Smith, and N. G. Adams (1974). Observations of electron temperature relaxation rates in rare gas afterglow plasmas, *J. Phys.* **B7**, 644–656.

Dulaney, J. L., M. A. Biondi, and R. Johnsen (1987). Electron temperature dependence of the recombination of electrons with NO^+ ions. *Phys. Rev. A.* **36**, 1342–1350.

Ferguson, E. E., F. C. Fehsenfeld, and A. L. Schmeltekopf (1969). Flowing afterglow measurements of ion-neutral reactions, in *Advances in Atomic and Molecular Physics*, Vol. 5, D. R. Bates and I. Estermann, Eds., Academic, New York, pp. 1–56.

Franklin, J. L., Ed. (1972). *Ion-Molecule Reactions*, Plenum, New York.

Frommhold, L., M. A. Biondi, and F. J. Mehr (1968). Electron temperature dependence of electron-ion recombination in neon, *Phys. Rev.* **165**, 44–52.

Frommhold, L., and M. A. Biondi (1969). Interferometric study of dissociative recombination radiation in neon and argon afterglows, *Phys. Rev.* **185**, 244–252.

Gray, E. P., and D. E. Kerr (1962). The diffusion equation with a quadratic loss term applied to electron-ion volume recombination in a plasma, *Ann. Phys. (N.Y.)* **17**, 276–300.

Huang, C.-M., M. A. Biondi, and R. Johnsen (1975). Variation of electron-NO^+-ion recombination coefficient with electron temperature, *Phys. Rev.* **A11**, 901–905.

Kasner, W. H., and M. A. Biondi (1968). Temperature dependence of the electron-O_2^+-ion recombination coefficient, *Phys. Rev.* **174**, 139–144.

Leu, M. T., M. A. Biondi, and R. Johnsen (1973). Measurements of recombination of electrons with H_3^+ and H_5^+ ions, *Phys. Rev.* **A8**, 413–419.

Lineberger, W. C., and L. J. Puckett (1969a). Positive ions in nitric oxide afterglow, *Phys. Rev.* **186**, 116–127.

Lineberger, W. C., and L. J. Puckett (1969b). Hydrated positive ions in nitric-oxide-water afterglows, *Phys. Rev.* **187**, 286–291.

McDaniel, E. W. (1964). *Collision Phenomena in Ionized Gases*, Wiley, New York.

McDaniel, E. W., V. Čermák, A. Dalgarno, E. E. Ferguson, and L. Friedman (1970), *Ion-Molecule Reactions*, Wiley, New York.

Mehr, F. J., and M. A. Biondi (1968). Electron-temperature dependence of electron-ion recombination in argon, *Phys. Rev.* **176**, 322–326.

Mehr, F. J., and M. A. Biondi (1969). Electron temperature dependence of recombination of O_2^+ and N_2^+ ions with electrons, *Phys. Rev.* **181**, 264–271.

Oskam, H. J. (1958). Microwave investigation of disintegrating gaseous discharge plasmas, *Philips Res. Rep.* **13**, 335–457.

Oskam, H. J. (1969). Recombination of rare gas ions with electrons, in *Case Studies in Atomic Collision Physics*, Vol. 1, E. W. McDaniel and M. R. C. McDowell, Eds., North-Holland, Amsterdam, pp. 463–523.

Persson, K. B., and S. C. Brown (1955). Electron loss process in the hydrogen afterglow, *Phys. Rev.* **100**, 729–733.

Philbrick, J., F. J. Mehr, and M. A. Biondi (1969). Electron temperature dependence of recombination of Ne_2^+ ions with electrons, *Phys. Rev.* **181**, 271–274.

Plumb, I. C., D. Smith, and N. G. Adams (1972). Formation and loss of O_2^+ and O_4^+ ions in krypton-oxygen afterglow plasmas, *J. Phys.* **B5**, 1762–1772.

Puckett, L. J., and W. C. Lineberger (1970). Negative-ion reactions in $NO-H_2O$ mixtures, *Phys. Rev.* **A1**, 1635–1641.

Puckett, L. J., and M. W. Teague (1971a). Production of $H_3O^+ \cdot nH_2O$ from NO^+ precursor in $NO-H_2O$ gas mixtures, *J. Chem. Phys.* **54**, 2564–2572.

Puckett, L. J., and M. W. Teague (1971b). Ion-molecule reactions in $NO-NH_3$ gas mixtures, *J. Chem. Phys.* **54**, 4860–4863.

Puckett, L. J., M. D. Kregel, and M. W. Teague (1971). A new technique for the measurement of electron attachment in afterglows, *Phys. Rev.* **4**, 1659–1666.

Shiu, Y.-J., M. A. Biondi, and D. P. Sipler (1977). Dissociative recombination in xenon: Variation of the total rate coefficient and excited-state production with electron temperature, *Phys. Rev.* **A15**, 494–498.

Smith, D., and M. J. Church (1976). Binary ion-ion recombination coefficients determined in a flowing afterglow plasma, *Int. J. Mass Spectrom. Ion Phys.* **19**, 185–200.

Smith, D., and M. J. Copsey (1968). Investigations of the helium afterglow: I. Mass spectrometric observations, *J. Phys.* **B1**, 650–659.

Smith, D., and P. R. Cromey (1968). Conversion rates and ion mobilities in pure neon and argon afterglow plasmas, *J. Phys.* **B1**, 638–649.

Smith, D., and R. A. Fouracre (1968). The temperature dependence of the reaction rate coefficients of O^+ ions with molecular oxygen and nitrogen, *Planet. Space Sci.* **16**, 243–252.

Smith, D., and C. V. Goodall (1968). The dissociative recombination coefficient of O_2^+ ions with electrons in the temperature range of 180°–630°K, *Planet. Space Sci.* **16**, 1177–1188.

Smith, D., and I. C. Plumb (1972). An appraisal of the single Langmuir probe technique in the study of afterglow plasmas, *J. Phys.* **D5**, 1226–1238.

Smith, D., C. V. Goodall, and M. J. Copsey (1968). Investigations of the helium afterglow: 2. Langmuir probe observations, *J. Phys.* **B1**, 660–668.

Smith, D., C. V. Goodall, N. G. Adams, and A. G. Dean (1970). Ion- and electron-density decay rates in afterglow plasmas of argon and argon–oxygen mixtures, *J. Phys.* **B3**, 34–44.

Smith, D., A. G. Dean, and N. G. Adams (1972a). Diffusion cooling in neon, argon and krypton afterglow plasmas, *Z. Phys.* **253**, 191–199.

Smith, D., A. G. Dean, and I. C. Plumb (1972b). Three-body conversion reactions in pure rare gases, *J. Phys.* **B5**, 2134–2142.

Smith, D., A. G. Dean, and N. G. Adams (1974). Space-charge fields in afterglow plasmas, *J. Phys.* **D7**, 1944–1962.

Walls, F. L., and G. H. Dunn (1974). Measurement of total cross sections for electron recombination with NO^+ and O_2^+ using ion storage techniques. *J. Geophys. Res.* **79**, 1911–1915.

Weller, C. S., and M. A. Biondi (1968). Recombination, attachment, and ambipolar diffusion of electrons in photo-ionized NO afterglow, *Phys. Rev.* **172**, 198–206.

Whitaker, M., M. A. Biondi, and R. Johnsen (1981a). Electron-temperature dependence of dissociative recombination of electrons with $CO^+ \cdot (CO)_n$-series ions, *Phys. Rev.* **A23**, 1481–1485.

Whitaker, M., M. A. Biondi, and R. Johnsen (1981b). Electron-temperature dependence of dissociative recombination of electrons with $N_2^+ \cdot N_2$ dimer ions, *Phys. Rev.* **A24**, 743–745.

5

KINETIC THEORY OF MOBILITY AND DIFFUSION

The purpose of the theory outlined in this chapter is to give a molecular description of the various phenomena described in the preceding chapters. That is, we wish to relate the phenomenological coefficients K and D to the properties of the ions and the neutral molecules, and to their interaction. Like so many kinetic problems in physics, this problem comes down to finding appropriate solutions of the Boltzmann equation, which describes the velocity distribution function of the ions. The Boltzmann equation is formidable: almost no nontrivial exact solutions are known after more than 100 years of effort. As a result, discussions of the Boltzmann equation can become quite technical and elaborate from a mathematical viewpoint. We try in this chapter to avoid as much as possible this kind of discussion, whereby the physics is often obscured by the mathematics, and adopt a more workaday approach. First, we wring as much as possible out of dimensional arguments, symmetry, and simple physical considerations of momentum and energy balance. Most of the essential physics can be brought out in this way, although of course most of the numerical factors will be inaccurate. Second, when considerations of accuracy and quantitative agreement with experiment force us to attack the Boltzmann equation, we will adopt only one main approach, a moment method in matrix form. This approach has two major virtues: low-order approximations turn out to have obvious connections to the simple physical arguments given previously, and most actual numerical computations of ion mobility and diffusion have used some version of a moment method. Third, we will only indicate other approaches to solving the Boltzmann equation, and their relations to the moment method and to each other. Details are not given here, but their lack can be made up by reference to two remarkably complete and scholarly reviews, by Kumar et al. (1980) and by Kumar (1984).

5-1.   DEFINITIONS AND GENERAL RESULTS

In this section we extract as much information as possible from a few general characteristics of mobility and diffusion. In particular, we obtain a relation between K and D on the basis of the assumption that the system is near equilibrium (weak fields and small gradients). The density dependence of K and D can next be deduced from the assumption that the rate of transport is controlled entirely by binary collisions between ions and neutral molecules. The binary collision assumption also can be used to show that the field and density dependence are not independent, but occur together as the variable E/N. Symmetry considerations show that K and D must be even functions of E/N. Finally, we invoke all these assumptions (quasi-equilibrium, binary collisions, and symmetry) to discuss the behavior of the higher nonlinear transport coefficients that may be invoked when the mobility becomes non-Ohmic and the diffusion becomes non-Fickian.

The phenomenonlogical definitions of the scalars K and D that have been given in Section 1-1 will suffice until we have to deal with the higher nonlinear coefficients. However, we should remark parenthetically that great simplifications are introduced by the condition that the ions are present in only trace amounts. This allows us to escape many of the painful situations that arise in trying to describe diffusion in a general gas mixture of arbitrary composition, such as the impossibility of simultaneously having zero net flux and zero total pressure gradient (Mason and Marrero, 1970).

A.   Simple Nernst–Townsend–Einstein Relation

As was mentioned in Section 1-1, a simple relation exists between K and D, which holds quite generally as long as both the field and the concentration gradient are small, so that the system is near equilibrium. This relation is not restricted to gases. Its equivalent was first obtained by Nernst (1888), who used the concept of osmotic pressure as the driving force in connection with solutions of electrolytes. The relation was later derived independently for the special case of ions in gases by Townsend (1899), who took as his starting point the fundamental paper on kinetic theory by Maxwell (1867). The name of Einstein (1905, 1908) later became attached to the relation through his work on Brownian motion, in which he estimated K by means of the Stokes formula for the viscous drag on a sphere by a continuum fluid. Various combinations of all three names are used in connection with the relation.

A simple, but essentially rigorous, derivation of the Nernst–Townsend–Einstein relation can be given that is valid not only for dilute gases, but also for dense gases, liquids, and isotropic solids. For weak fields and small gradients we can take the flux of ions to be linear in both the field and the gradient, combine equations (1-1-1)–(1-1-4), and write

$$\mathbf{J} = nK\mathbf{E} - D\nabla n. \qquad (5\text{-}1\text{-}1)$$

This equation is assumed to hold generally, so we can pick any special case that is convenient for finding a relation between K and D. We choose the case of equilibrium, for which $\mathbf{J} = 0$ and for which ∇n can be found from equilibrium statistical mechanics. The physical meaning of this choice is that the electric field causes a gradient of ion concentration, and at equilibrium the diffusion down this gradient exactly balances the forced flow caused by the electric field. At equilibrium, the ion distribution in space is given by a Boltzmann exponential,

$$n = n_0 \exp(+e\mathbf{E} \cdot \mathbf{r}/kT), \tag{5-1-2}$$

the positive sign in the exponential occurring because of the sign conventions on e, $\mathbf{E}$, and $\mathbf{r}$. Differentiation yields

$$\frac{1}{n}\nabla n = \frac{e}{kT}\mathbf{E}. \tag{5-1-3}$$

Substituting this expression into (5-1-1) and setting $\mathbf{J} = 0$, we obtain the desired relation,

$$K = eD/kT. \tag{5-1-4}$$

The requirement of a linear transport equation, (5-1-1), and of quasi-equilibrium, means that this relation holds only in the low-field region. Other than that, it is quite general.

For dilute gases the Nernst–Townsend–Einstein relation will follow, as it must, from the solution of the Boltzmann equation in the low-field region, but the foregoing derivation shows the relation to be much more general than the Boltzmann equation, which describes only binary-collision processes. Thus the relation should hold for all degrees of approximation in constructing solutions of the Boltzmann equation, not just in the first approximation.

B. Density and Field Dependence

The fact that mobility and diffusion are controlled by binary collisions in dilute gases allows further conclusions to be drawn without detailed calculations. At very low fields, where the ion energy is entirely thermal, both K and D are inversely proportional to N (recall that $n \ll N$). The reason is that the rate of transport is controlled entirely by collisions, and the frequency of collisions increases directly with N if the temperature is held constant. Thus $\mathbf{J}$ is inversely proportional to N, and so also are K and D, according to (5-1-1). This argument, despite its seeming casualness, gives an exact result because only the counting of collisions is essential, not their specific nature, and collision frequency is exactly proportional to N in the binary collision regime. This is a standard result of classical kinetic theory (Cowling, 1950, Chap. 6).

The foregoing argument fails at higher fields, however, where the average ion speed depends on E as well as T. To see this, suppose that N is changed while E and T are held fixed. Then the distance traveled between collisions (the mean free path) will be changed, and this is the mean distance over which an ion is accelerated. The mean speed is thereby changed, and this affects the collision frequency because fast ions make more collisions than do slow ions. Thus K is not simply inversely proportional to N at constant E. But we can use the insight obtained in arriving at this apparently unpleasant result to draw an important exact conclusion that holds at all field strengths. This is the fact, already mentioned in Section 1-1, that all dependence on E occurs through the variable E/N rather than through E itself. The proof is simple. If we double E, we double the acceleration an ion receives between collisions, and the ion arrives at its collision more quickly and with a higher speed. But if we also double N, we halve the distance traveled (binary collision assumption) and the ion arrives at its next collision in the same time as originally, and with the same speed. Thus the effect of any change in E can be nullified by a corresponding change in N. The conclusion is that v_d, K, D_L, and D_T can depend on E only through the ratio E/N.

Once the E/N dependence is understood, we can go back to the original argument on density dependence and see that it holds if E/N and T are taken as the variables rather than E and T. The general dependence on density and field is therefore as follows: at fixed T, the quantities v_d, NK, ND_L, and ND_T depend only on E/N. This result depends only on the binary-collision condition, and not on any details concerning the nature of the collisions. The latter determines the precise dependence of these quantities on E/N, and the field dependence of K is an important probe of ion-molecule forces, as will be discussed later in detail.

A short digression is in order here as a reminder that different conventions are used for the density dependences of different transport coefficients when reporting experimental results. It is customary to report ion mobility as a standard mobility, K_0, which is the measured mobility converted to the mobility at standard gas *density* on the supposition that K varies inversely with N, at the measured values of T and E/N. Ion diffusion coefficients, however, are usually reported as the products ND_L and ND_T at the measured values of T and E/N. In contrast, diffusion coefficients of neutral gases are customarily reported at a standard *pressure* of 1 atm at the temperature of measurement, on the asssumption that D varies inversely with p. The choice of standard density or standard pressure involves a trivial but annoying factor of T, since $p = nkT$.

A final generalization, which depends only on symmetry, can be made about the dependence of K_0, ND_L, and ND_T on E/N, namely that they are all even functions of E/N—that is, they do not change sign if E changes sign. Again, the proof is simple. Suppose we change our coordinate system around by looking at the apparatus from the other side, so that the coordinate of the field direction changes from z to $-z$. Then E becomes $-E$, and v_d becomes $-v_d$. But $K = v_d/E$ does not change sign because the minus signs cancel, nor should it, since we have changed nothing physically by just viewing the apparatus differently. Thus if we expand K_0 as a power series in E/N, only even powers will appear:

$$K_0(E) = K_0(0)[1 + \alpha_2(E/N)^2 + \alpha_4(E/N)^4 + \cdots]. \qquad (5\text{-}1\text{-}5a)$$

Note that the coefficients α_2, α_4, etc. are independent of N. Similar results hold for ND_L and ND_T,

$$ND_L(E) = ND(0)[1 + d_2^L(E/N)^2 + d_4^L(E/N)^4 + \cdots], \qquad (5\text{-}1\text{-}5b)$$

$$ND_T(E) = ND(0)[1 + d_2^T(E/N)^2 + d_4^T(E/N)^4 + \cdots]. \qquad (5\text{-}1\text{-}5c)$$

This symmetry in E holds for any isotropic medium, not just dilute gases, but the explicit dependences on N depend on the binary-collision assumption. The expansions themselves, however, do not converge well, and are useful only up to fields of moderate strength (Whealton and Mason, 1974).

C. Generalized Nernst–Einstein Relations for Higher-Order Transport Coefficients

It is an interesting exercise to see how the foregoing arguments involving quasi-equilibrium, binary collisions, and inversion symmetry can be used to obtain relations among higher-order transport coefficients (Weinert and Mason, 1980; Hope et al., 1981). These coefficients arise when E and ∇n are not small enough for Ohm's law and Fick's law to hold accurately; examples are the α_i, d_i^L, and d_i^T in (5-1-5). To describe such cases we simply add terms with higher powers of E and higher derivatives of n to the equation for J. No terms like $(\nabla n)^2$ occur, however, because of the assumption of trace ion concentration. In all of this there is an implicit assumption that the field is homogeneous, inasmuch as no derivatives of E occur. It is convenient to write the flux equation initially with the dependence on E concealed in the coefficients, as follows:

$$\mathbf{J} = nK\mathbf{E} - \mathbf{D}^{(2)} \cdot \nabla n + \mathbf{Q}^{(3)} : \nabla\nabla n - \mathbf{R}^{(4)} \vdots \nabla\nabla\nabla n + \cdots. \qquad (5\text{-}1\text{-}6a)$$

This tensor notation is rather standard (see, e.g., Ferziger and Kaper, 1972, App. A, pp. 491–500). In any case, the results are written out explicitly as needed in what follows. The mobility is still a scalar, but the diffusion coefficient $\mathbf{D}^{(2)}$ is a second-order tensor, $\mathbf{Q}^{(3)}$ a third-order tensor, and $\mathbf{R}^{(4)}$ a fourth-order tensor. It was pointed out in Section 1-1 that $\mathbf{D}^{(2)}$ has only two independent components, usually expressed as D_T and D_L. This is a consequence of rotational symmetry, which can be used to show that $\mathbf{Q}^{(3)}$ has in general only three independent components and $\mathbf{R}^{(4)}$ only five (Whealton and Mason, 1974). A quick alternative way to deduce the structure of these tensors is to recognize that the only nonscalar quantities left in the problem after ∇n is removed are the electric field $\mathbf{E}$ and the unit (second-order) tensor $\mathbf{I}$. Any other tensors introduced in the description must therefore be constructed from these (Robson, 1982, personal communication). The only possibilities are as follows:

$$\mathbf{D}^{(2)} = d_0\mathsf{I} + d_2\mathbf{EE}, \tag{5-1-6b}$$

$$\mathbf{Q}^{(3)} = q_1\mathsf{I}\mathbf{E} + q_1'\mathbf{E}\mathsf{I} + q_3\mathbf{EEE}, \tag{5-1-6c}$$

$$\mathbf{R}^{(4)} = r_0\mathsf{II} + r_2\mathsf{I}\mathbf{EE} + r_2'\mathbf{E}\mathsf{I}\mathbf{E} + r_2''\mathbf{EE}\mathsf{I} + r_4\mathbf{EEEE}, \tag{5-1-6d}$$

where the coefficients d_i, q_i, and r_i are scalar functions of E^2. (The subscripts on the coefficients keep track of the powers of $\mathbf{E}$.) The number of independent components is obvious in this construction.

We next expand the transport coefficients in powers of $\mathbf{E}$. Inversion symmetry requires that K and $\mathbf{D}^{(2)}$ be even functions of E, as before, and further that $\mathbf{Q}^{(3)}$ be an odd function of E, and $\mathbf{R}^{(4)}$ an even function. This symmetry is also apparent in the expressions (5-1-6). The assumption that the field is homogeneous suppresses the appearance of some of the components of $\mathbf{Q}^{(3)}$ and $\mathbf{R}^{(4)}$, but it is easy to let the angle between $\mathbf{E}$ and ∇n be arbitrary. It is also easy to exhibit the dependence on N explicitly (a consequence of the binary-collision condition); to shorten the notation a little, we let

$$\mathscr{E} \equiv \mathbf{E}/N. \tag{5-1-7}$$

Indicating by superscripts E the components parallel to $\mathbf{E}$, and by superscripts n the components parallel to ∇n, we obtain

$$\mathbf{D}^{(2)} = D(0)[\mathsf{I}(1 + d_2^n\mathscr{E}^2 + d_4^n\mathscr{E}^4 + \cdots) + \mathscr{E}\mathscr{E}(d_2^E + d_4^E\mathscr{E}^2 + \cdots)], \tag{5-1-8a}$$

$$\mathbf{Q}^{(3)} = N[\mathsf{I}\mathscr{E}(q_1^n + q_3^n\mathscr{E}^2 + \cdots) + \mathscr{E}\mathsf{I}(q_1^E + q_3^E\mathscr{E}^2 + \cdots)], \tag{5-1-8b}$$

$$\mathbf{R}^{(4)} = R(0)\mathsf{II}(1 + \cdots), \tag{5-1-8c}$$

where I is the unit second-order tensor. The expansion for K is the same as (5-1-5a). Using these expansions, we can write out the gradient terms in the flux equation explicitly:

$$\mathbf{D}^{(2)} \cdot \nabla n = D(0)[\nabla n + d_2^n\mathscr{E}^2\,\nabla n + d_2^E\mathscr{E}(\mathscr{E}\cdot\nabla n) + \cdots], \tag{5-1-9a}$$

$$\mathbf{Q}^{(3)}:\nabla\nabla n = N[q_1^n(\mathscr{E}\cdot\nabla)\nabla n + q_1^E\mathscr{E}(\nabla^2 n) + \cdots], \tag{5-1-9b}$$

$$R^{(4)} \vdots \nabla\nabla\nabla n = R(0)(\nabla\cdot\nabla)\nabla n + \cdots. \tag{5-1-9c}$$

The next step is to use the quasi-equilibrium condition in the form of the Boltzmann exponential (5-1-2), to find relations between the gradients of n and the powers of $\mathscr{E}$, just as we did in obtaining the simple Nernst–Townsend–Einstein relation, but a question arises. In the earlier derivation we used a linear flux expression (5-1-1) that referred to a system deviating only slightly from equilibrium, but for the ion distribution in space we used the equilibrium

Boltzmann exponential. If we extend the flux equation to allow for higher-order deviations from equilibrium, perhaps we should expect to have to take into account at least first-order deviations from the Boltzmann exponential? In general this expectation is justified, because a strong field, for example, could locally increase the ion density enough to cause appreciable ion-ion interactions. What saves the result in this case, however, is the explicit assumption of trace concentration of ions, plus the implicit assumption that the background medium of neutral molecules is not affected by the field. We therefore proceed to differentiate the equilibrium Boltzmann exponential (5-1-2) repeatedly, to obtain

$$\nabla n = nN(e/kT)\mathscr{E}, \tag{5-1-10a}$$

$$\nabla\nabla n = nN^2(e/kT)^2\mathscr{E}\mathscr{E}, \tag{5-1-10b}$$

$$\nabla\nabla\nabla n = nN^3(e/kT)^2\mathscr{E}\mathscr{E}\mathscr{E}, \text{ etc.} \tag{5-1-10c}$$

Substituting these expressions back into (5-1-9), setting $\mathbf{J} = 0$ in (5-1-6), and equating the coefficients of different powers of $\mathscr{E}$ separately to zero, we obtain

$$K(0) - D(0)(e/kT)) = 0, \tag{5-1-11a}$$

$$K(0)\alpha_2 - D(0)(d_2^n + d_2^E)(e/kT)$$
$$+ N^2(q_1^n + q_1^E)(e/kT)^2 - N^2 R(0)(e/kT)^3 = 0, \quad \text{etc.} \tag{5-1-11b}$$

The first relation is the simple Nernst–Townsend–Einstein relation, and (5-1-11b) is the first higher-order relation. The procedure for obtaining even higher-order relations is clear, but it is doubtful that any of them have any practical use, because of the great experimental difficulty of measuring the non-Fickian transport coefficients.

The explicit dependence of the higher transport coefficients on N can be seen directly from (5-1-11b). We have noted in Section 5-1B that $K(0)$ and $D(0)$ vary inversely with N, and that α_2, d_2^T and d_2^L are independent of N; a similar result holds for d_2^n and d_2^E. In order for (5-1-11b) to hold at all densities, it is clear that q_1^n, q_1^E, and $R(0)$ must be proportional to N^{-3}.

The foregoing results show that there is a whole hierarchy of relations among the higher-order transport coefficients, analogous to the well-known Nernst–Townsend–Einstein relation between the two linear coefficients. The main restriction is that the ions be present in only trace concentration, a restriction that does not apply to the two linear coefficients. The relations are otherwise general, and apply to liquids as well as to gases, for instance. Only the density dependence is restricted to dilute gases, because it depends on the binary-collision condition.

The generalized Nernst–Einstein relations derived in this section should not be confused with some apparently similar relations frequently used for gaseous ions at high electric fields, which are usually called generalized Einstein relations (GER) and which are discussed later in detail. The GER consider only Fickian diffusion, but E may be large, so that K, D_L, and D_T depend on E. The GER are approximate expressions that give $D_L(E)$ and $D_T(E)$ in terms of $K(E)$ and its field derivatives; no expansions in powers of E are involved. The essential difference is that the present generalized Nernst–Einstein relations are exact relations among field-independent higher-order transport coefficients, whereas the GER are approximate relations among field-dependent linear (Fickian) coefficients.

5-2.   SIMPLE PHYSICAL ARGUMENTS AND ELEMENTARY DERIVATIONS

In Section 5-1 we exhibited just those exact results that can be obtained from considerations of quasi-equilibrium, binary collisions, and symmetry. Much further information can be obtained with only modest effort if we are willing to make a few approximations, which is the subject of this section. For high field strengths we must give up quasi-equilibrium and expansions in powers of E/N, but of course we do not tamper with symmetry. This leaves binary collisions as the centerpiece of the discussion, but now we must be concerned with some details about collisions, having already extracted everything possible from their mere existence. It is over these details that approximations occur, as well as in connection with such time-honored tricks as replacing the average of a product by the product of the averages. We thereby find, to a quite respectable order of accuracy, a number of scaling rules and other relations among macroscopically observable quantities. These include the following: an expression for the mean ion energy in terms of T, v_d, m, and M, which then leads to criteria for low-field and high-field behavior; the dependence of the mobility on m and M at all values of T and E/N; a rule for scaling the separate dependence of K on T and on E/N into a single relation; and the dependence of K, D_L, and D_T on composition in a mixture of neutral gases. To a somewhat poorer accuracy, we find how K depends on T and E/N in terms of the ion–neutral interaction, and also find expressions for D_L and D_T in terms of K for all values of T and E/N.

The approximation scheme we adopt is based on considerations of momentum and energy balance, in which the momentum and energy gained by the ion from the field must at steady state be balanced by losses through collisions with the neutral molecules. There is nothing approximate about the physics behind these ideas, but approximations enter when we try to calculate the collisional losses. The scheme has come to be called *momentum-transfer theory* because of the role played by momentum balance. Energy balance was omitted from the name for historical reasons: the theory was first used to discuss the diffusion of neutral gases and the mobility of ions at very low field strengths, where the energy can always be taken to be entirely thermal. Although momentum-transfer theory is less well known than mean-free-path theory, which is the standard version of elementary kinetic theory appearing in most textbooks, it is equally venerable. Both theories were introduced in the same paper by Maxwell (1860). Momentum-transfer theory has many advantages over mean-free-path theory for the discussion of ion transport, and the two theories are compared near the end of this section. The final part of this section is devoted to some comments about the magnitude and energy dependence of the momentum-transfer or diffusion cross section that appears in the theory.

A. Momentum-Transfer Theory for Mobility

Consider first the momentum balance for the ions. The momentum gained from the field is easily calculated exactly. The force on an ion is eE, and by Newton's second law of motion this must be exactly equal to the momentum gained from the field in unit time. The momentum lost by collisions is much harder to calculate, and we shall be satisfied with approximating it as the average momentum loss per collision, multiplied by the average number of collisions per unit time that produce such an average momentum loss, and summed over all possible collisions. The relative momentum of an ion-neutral collision is $\mu \mathbf{v}_r$, where $\mu = mM/(m + M)$ is the reduced mass and $\mathbf{v}_r$ is the relative velocity. It is easy to show that in one collision the momentum transferred to a gas molecule has a component parallel to $\mathbf{v}_r$ of (Present, 1958, p. 136)

$$\delta(\mu \mathbf{v}_r)_{\|} = \mu \mathbf{v}_r (1 - \cos \theta), \qquad (5\text{-}2\text{-}1)$$

where θ is the relative deflection angle of the collision. If we average over many collisions, we expect all the random components of $\mathbf{v}_r$ to average to zero and only the drift velocity to contribute. The average momentum communicated to the gas per collision is therefore approximately

$$\mu v_d (1 - \cos \theta). \qquad (5\text{-}2\text{-}2)$$

But the average number of collisions an ion makes per unit time having a deflection angle between θ and $\theta + d\theta$ is

$$N \bar{v}_r 2\pi \sigma(\theta, \bar{v}_r) \sin \theta \, d\theta, \qquad (5\text{-}2\text{-}3)$$

where $\bar{v}_r$ is the mean relative speed and $\sigma(\theta, \bar{v}_r)$ is the differential cross section for scattering through an angle θ at speed $\bar{v}_r$. This expression is really just the definition of $\sigma(\theta, \bar{v}_r)$. Combining (5-2-2) and (5-2-3) and adding up all possible collisions by integrating over all deflection angles, we obtain

$$\text{momentum loss} = \mu v_d N \bar{v}_r 2\pi \int_0^\pi (1 - \cos \theta) \sigma(\theta, \bar{v}_r) \sin \theta \, d\theta. \qquad (5\text{-}2\text{-}4)$$

It is convenient to compress the notation a bit at this point by defining a momentum-transfer or diffusion cross section Q_D,

$$Q_D(\bar{\varepsilon}) \equiv 2\pi \int_0^\pi (1 - \cos \theta)\sigma(\theta, \bar{v}_r) \sin \theta \, d\theta, \qquad (5\text{-}2\text{-}5)$$

where

$$\bar{\varepsilon} \equiv \tfrac{1}{2}\mu \overline{v_r^2} \qquad (5\text{-}2\text{-}6)$$

is the mean relative energy of collision. It is also convenient to define an ion-neutral collision frequency as

$$\nu(\bar{\varepsilon}) \equiv N\bar{v}_r Q_D(\bar{\varepsilon}). \qquad (5\text{-}2\text{-}7)$$

Note that $\nu(\bar{\varepsilon})$ has the dimensions of s^{-1} and is directly proportional to N. The momentum-balance equation is therefore

$$eE = \mu v_d \nu(\bar{\varepsilon}). \qquad (5\text{-}2\text{-}8)$$

This is the formula for v_d, and hence K, that we want, except that we do not know $\bar{\varepsilon}$ or $\bar{v}_r$. It is consistent with our level of approximation not to worry about the difference between $\bar{v}_r^2$ and $\overline{v_r^2}$, and the latter is

$$\overline{v_r^2} = \overline{(\mathbf{v} - \mathbf{V})^2} = \overline{v^2} + \overline{V^2}, \qquad (5\text{-}2\text{-}9)$$

where $\mathbf{v}$ is the ion velocity and $\mathbf{V}$ is the neutral velocity. The second equality in (5-2-9) follows because the cross term averages to zero. We know that $\overline{V^2}$ is entirely thermal,

$$\tfrac{1}{2}M\overline{V^2} = \tfrac{3}{2}kT. \qquad (5\text{-}2\text{-}10)$$

If we were concerned only with low fields, $\overline{v^2}$ would also be entirely thermal and our calculation would be finished. But at high fields $\overline{v^2}$ has both thermal and field components. To find $\overline{v^2}$ for fields of arbitrary strength we must turn to the energy-balance equation.

It might be thought that at high fields, where the thermal component of $\overline{v^2}$ is negligible, we could take $\overline{v^2} = v_d^2$, but this is incorrect. The reason is that collisions do not merely absorb some of the ion energy, they also randomize some of it, so that $\overline{v^2}$ consists of a part visible as drift motion, namely v_d^2, plus a random part. We need a more careful calculation of energy balance.

The average work done by the electric field on an ion per unit time is eEv_d; this is therefore the average energy gained from the field per unit time. The average energy lost per collision is obviously

$$\tfrac{1}{2}m\overline{v^2} - \tfrac{1}{2}m\overline{v'^2},$$

where v' is the ion speed after collision. We can approximate the total energy loss per unit time by multiplying this by the collision frequency, so that the energy-balance equation is

$$\tfrac{1}{2}m(\overline{v^2} - \overline{v'^2})v(\bar{\varepsilon}) = eEv_d. \tag{5-2-11}$$

The problem remaining is to find $\mathbf{v}'$ in terms of $\mathbf{v}$ and $\mathbf{V}$, square the expression, and average over collisions. This requires only a straightforward use of the laws of conservation of momentum and energy in a single collision, but involves some algebra and the use of center-of-mass coordinates. The reader who is uninterested in such details can skip directly to (5-2-19). The center-of-mass and relative velocities are, respectively,

$$\mathbf{v}_{cm} = \frac{m\mathbf{v} + M\mathbf{V}}{m + M}, \tag{5-2-12}$$

$$\mathbf{v}_r = \mathbf{v} - \mathbf{V}. \tag{5-2-13}$$

In terms of these quantities, the ion velocities before and after collision are easily found to be

$$\mathbf{v} = \mathbf{v}_{cm} + \frac{M}{m + M}\mathbf{v}_r, \tag{5-2-14}$$

$$\mathbf{v}' = \mathbf{v}'_{cm} + \frac{M}{m + M}\mathbf{v}'_r. \tag{5-2-15}$$

Conservation of momentum requires that $\mathbf{v}'_{cm} = \mathbf{v}_{cm}$, and conservation of energy (elastic collisions) requires that $v'^2_r = v^2_r$. Squaring and subtracting (5-2-14) and (5-2-15), we then obtain

$$v^2 - v'^2 = \frac{2M}{m + M}\mathbf{v}_{cm}\cdot(\mathbf{v}_r - \mathbf{v}'_r). \tag{5-2-16}$$

We now average over all collisions and make an assumption. We assume that the average involving $\mathbf{v}'_r$ vanishes. This is equivalent to assuming that in any particular collision, the ion is just as likely to be scattered in one direction as in another, as viewed in a coordinate system moving with the center of mass of the colliding pair. This assumption cannot be strictly correct unless the ions and neutrals collide like rigid spheres, but it is a reasonable approximation and saves much algebra at this stage. We thus obtain

$$\overline{v^2} - \overline{v'^2} = \frac{2M}{m + M}(\overline{\mathbf{v}_{cm} \cdot \mathbf{v}_r}).$$ (5-2-17)

To evaluate the right-hand side of this expression, we substitute for $\mathbf{v}_{cm}$ and $\mathbf{v}_r$ from (5-2-12) and (5-2-13) and obtain

$$\mathbf{v}_{cm} \cdot \mathbf{v}_r = \frac{mv^2 - MV^2 + (M - m)\mathbf{v} \cdot \mathbf{V}}{m + M}.$$ (5-2-18)

The last term of this expression will vanish on averaging over collisions because the neutral gas is stationary, so that $\bar{\mathbf{V}} = 0$. Substituting back into (5-2-17) and multiplying by the collision frequency, we obtain the energy-balance equation,

$$eEv_d = \frac{mM}{(m + M)^2}(\overline{mv^2} - M\overline{V^2})\nu(\bar{\varepsilon}).$$ (5-2-19)

We can now combine the equations of momentum and energy balance, (5-2-8) and (5-2-19), to find expressions for the mean ion energy and the mobility. If we eliminate $eE/\nu(\bar{\varepsilon})$ between them, we find the ion energy to be

$$\tfrac{1}{2}m\overline{v^2} = \tfrac{1}{2}M\overline{V^2} + \tfrac{1}{2}mv_d^2 + \tfrac{1}{2}Mv_d^2.$$ (5-2-20)

This surprisingly simple formula turns out to be quite accurate, as we shall see later. It was first shrewdly deduced by Wannier (1951, 1953) on the basis of some calculations for a special model, and is actually rather general. The interpretation of the three terms on the right-hand side is also simple. The first term is equal to $3kT/2$ according to (5-2-10), and is the thermal energy acquired by collisions with the gas. The second term is obviously that part of the energy from the field that is visible as the drift motion. The third term therefore represents the random part of the field energy. As might be expected, light ions in a heavy gas $(m \ll M)$ have most of their field energy as random motion, since the heavy neutrals are extremely effective in deflecting the ions but absorb little energy in recoil. Heavy ions in a light gas $(m \gg M)$, however, have most of their energy as drift motion, for the light neutrals are ineffective either in deflecting the ions or in absorbing their energy. For $m = M$ the ion energy is equally divided between drift and random components.

The mean relative energy follows immediately from (5-2-20) by substitution into (5-2-6) and (5-2-9),

$$\bar{\varepsilon} = \tfrac{1}{2}\mu(\overline{v^2} + \overline{V^2}) = \tfrac{1}{2}M\overline{V^2} + \tfrac{1}{2}Mv_d^2.$$ (5-2-21)

It is useful to think of $\bar{\varepsilon}$ in terms of an effective ion temperature, T_{eff}, defined as

$$\tfrac{3}{2}kT_{\text{eff}} \equiv \bar{\varepsilon} = \tfrac{3}{2}kT + \tfrac{1}{2}Mv_d^2. \tag{5-2-22}$$

Thus T_{eff} represents the total random energy of the ions, which consists of a thermal part and a field part.

We get an expression for the mobility by solving (5-2-8) for v_d and using (5-2-22) for $\bar{\varepsilon}$ or T_{eff},

$$K \equiv \frac{v_d}{E} = \frac{e}{\mu v(\bar{\varepsilon})} = \frac{e}{N}\left(\frac{1}{3\mu k T_{\text{eff}}}\right)^{1/2}\frac{1}{Q_D(T_{\text{eff}})}. \tag{5-2-23}$$

This constitutes, along with (5-2-22), the complete momentum-transfer theory for ion mobility in a single gas, obtained from momentum and energy balance at the expense of a few approximations and a little algebra. It obviously has some defects, because we have been casual about taking averages, and we can check this immediately by passing to the low-field limit. Here the ion energy is entirely thermal, the exact Nernst–Townsend–Einstein relation applies, and our result is therefore equivalent to the accurate Chapman–Enskog result for the diffusion coefficient (Chapman and Cowling, 1970, Sec. 9.81; Hirschfelder et al., 1964, Sec. 8.2; Ferziger and Kaper, 1972, Sec. 7.3). We find that the numerical factor in (5-2-23) of $3^{-1/2} = 0.577$ should be $3(2\pi)^{1/2}/16 = 0.470$ in a first Chapman–Enskog approximation, an error of about 20%. This error is fortuitously reduced somewhat when higher Chapman–Enskog approximations are considered. Moreover, instead of a momentum-transfer cross section taken at a mean relative energy of $\bar{\varepsilon} = 3kT_{\text{eff}}/2$, we should have a cross section averaged over a distribution of relative energies; that is, in place of $Q_D(T_{\text{eff}})$ there should appear the quantity

$$\bar{\Omega}^{(1,1)}(T) = \tfrac{1}{2}(kT)^{-3}\int_0^\infty Q_D(\varepsilon)\exp(-\varepsilon/kT)\varepsilon^2\,d\varepsilon. \tag{5-2-24}$$

The distinction disappears for rigid spheres, but can be of some importance for other interactions.

Nevertheless, the foregoing derivation gives all the important physical features of ion mobility correctly, and it is worthwhile to examine these in a little detail. We begin by checking that the results conform to the general features elucidated in Section 5-1. First, we note that K is inversely proportional to N at fixed $\bar{\varepsilon}$ or T_{eff}. This dependence appeared already in the momentum-balance equation, (5-2-8), where we can see that it arose directly from a binary-collision assumption, as advertised. Second, we should therefore expect $\bar{\varepsilon}$ or T_{eff} to depend on the two variables T and E/N, so that the field dependence of K is through E/N rather than through E itself. From (5-2-22) we see that T_{eff} depends on T and v_d, and from (5-2-23) we see that v_d is equal to E/N multiplied by some function of T_{eff}. Thus there is a complicated relation connecting T_{eff} and E/N via

v_d, which we cannot even hope to solve unless the functional form of $Q_D(T_{\text{eff}})$ is known, but we do see that the relation involves only E/N, and not E and N separately. Thus v_d, NK, and T_{eff} depend only on E/N at fixed T, as required.

We can also see from (5-2-22) and (5-2-23) that the symmetry is given correctly. Suppose that we change v_d to $-v_d$. Then T_{eff} remains the same because it depends on v_d^2, but E/N must change sign according to (5-2-23). Hence v_d and E/N change sign together, so that K does not change sign and must be an even function of E/N. We can even carry out the expansion as a power series in $(E/N)^2$ by expanding T_{eff} around T, assuming that $Mv_d^2/3kT$ is a small quantity. From (5-2-22) and (5-2-23), respectively, we write

$$T_{\text{eff}} - T = Mv_d^2/3k = (M/3k)(NK)^2(E/N)^2$$

$$= \frac{M}{\mu}\left(\frac{e}{3kTQ_D(T)}\right)^2\left(\frac{E}{N}\right)^2 + \cdots, \tag{5-2-25}$$

and we expand $Q_D(T_{\text{eff}})$ in a Taylor series around $Q_D(T)$,

$$Q_D(T_{\text{eff}}) = Q_D(T) + \frac{dQ_D(T)}{dT}(T_{\text{eff}} - T) + \cdots$$

$$= Q_D(T)\left[1 + \frac{d \ln Q_D(T)}{dT}\frac{Mv_d^2}{3k} + \cdots\right]. \tag{5-2-26}$$

Substituting these expressions back into (5-2-23), expanding, and collecting terms, we obtain

$$K(E) = K(0)\left[1 - \frac{m+M}{m}\left(\frac{1}{2} + \frac{d \ln Q_D}{d \ln T}\right)\left(\frac{e}{3kTQ_D}\right)^2\left(\frac{E}{N}\right)^2 + \cdots\right]. \tag{5-2-27}$$

This is exactly the expansion predicted in (5-1-5a), but with an explicit expression (approximate, of course) for the coefficient α_2. Comparison with accurate kinetic-theory calculations (Mason and Schamp, 1958) shows that α_2 is of this form, at least in a low-order approximation, but that the numerical coefficient is not very accurate.

Having checked that the momentum-transfer calculation has not violated any general principles, we now look for some new features. An obvious one is a criterion for the low-field behavior of K, or for the onset of nonlinear behavior in v_d, as a function of E/N, which was referred to back in Section 1-2. We find a criterion immediately in the expression (5-2-22) for $\bar{\varepsilon}$ or T_{eff}: we can expect low-field behavior whenever the random field energy is much less than the thermal energy,

$$\tfrac{1}{2}Mv_d^2 \ll \tfrac{3}{2}kT. \tag{5-2-28}$$

This can be put into a criterion in terms of E/N by substituting $v_d = KE$ and using (5-2-23) for K. However, we have already done this algebra in deriving the expansion in (5-2-27) above, and (5-2-28) is the same as saying that the second term in brackets in (5-2-27) is small compared to unity. This leads to the criterion

$$\frac{E}{N} \ll \left(\frac{m}{m+M}\right)^{1/2} \frac{3kTQ_D}{e} \left(\frac{1}{2} + \frac{d \ln Q_D}{d \ln T}\right)^{-1/2}. \qquad (5\text{-}2\text{-}29)$$

For a quick numerical estimate, we can ignore the temperature derivative of Q_D (this is a rigid-sphere approximation), take $T = 300 \text{ K}$, and obtain

$$\frac{E}{N}(\text{Td}) \ll 0.78 \left(\frac{m}{m+M}\right)^{1/2} Q_D(\text{Å}^2), \qquad (5\text{-}2\text{-}30)$$

with $1 \text{ Td} = 10^{-17} \text{ V-cm}^2$ and $1 \text{ Å}^2 = 10^{-16} \text{ cm}^2$. There is nothing in the argument restricting the mass ratio, and the criterion applies to electrons as well as to heavy ions (Robson and Mason, 1982). Clearly, what may be a low field for a heavy ion may be a very high field for an electron in the same gas.

The foregoing arguments are concerned with an expansion valid when the field energy is small compared to the thermal energy, but an analogous expansion in the opposite extreme is also possible, in which the small quantity is taken to be $3kT/Mv_d^2$ rather than its reciprocal. Such an expansion might be useful if a solution for the high-field case were available, for example based on the idea that the velocity distribution function of the neutral gas could be taken to be a delta function at high fields (Wannier, 1951, 1953; Skullerud and Forsth, 1979). The expansion proceeds as before, and we obtain a so-called "cold-gas expansion":

$$K(E,T) = K(E,0)\left[1 - \frac{3}{2}\left(\frac{m}{m+M}\right)^{1/2}\left(\frac{1}{2} + \frac{d \ln Q_D}{d \ln (E/N)}\right)\frac{Q_D NkT}{eE} + \cdots\right]. \qquad (5\text{-}2\text{-}31)$$

Accurate kinetic-theory calculations (Skullerud and Forsth, 1979) give an expansion of this form, but the numerical coefficient in (5-2-31) is, not unexpectedly, inaccurate. The convergence of this expansion is particularly good when m/M is small (Skullerud and Forsth, 1979).

A word of caution is in order concerning the high-field limit, by which we mean $3kT/Mv_d^2 \to 0$. This is most easily attained in a theoretical calculation by taking $T \to 0$ and keeping E/N and v_d nonzero. An added benefit is that the distribution function of the neutral gas can then be taken as a delta function. Strictly speaking, this is a "cold-gas" limit, and is obviously an impractical limit for an experiment. Instead, the experimenter keeps T conveniently nonzero and makes E/N very large; usually this causes no difficulty. However, in a mathematical sense the limits $T \to 0$ and $E/N \to \infty$ are not necessarily the same, and even in a real physical sense there can be important differences, depending on how Q_D varies with ε. The reason is that there is another energy scale in the problem in addition to kT and Mv_d^2, namely a scale concerning the ion-neutral potential energy. In particular, if Q_D decreases too rapidly at large ε, no steady state may be possible at large E/N, and the ions continue to accelerate and "run away." Thus the $E/N \rightarrow \infty$ limit does not exist in this case, although the $T \rightarrow 0$ limit does. This is not just an academic point, since H^+ and D^+ ions have been found to undergo runaway in He gas at large E/N (Lin et al., 1979a; Howorka et al., 1979). A similar difficulty occurs when the limit $m/M \rightarrow \infty$ is taken. Then there is still a cold-gas, low-field limit, but the high-field limit may not exist because of runaway (Burnett, 1982, personal communication). Here the relevant third energy scale is mv_d^2.

We now examine more quantitatively the predictions of (5-2-22) and (5-2-23) for mass, temperature, and field dependence. To isolate the mass dependence, we can imagine experiments with isotopic variations of ions and neutrals, such as H^+/D^+ or $^3He/^4He$. If E/N is kept small enough for the ion energy to be essentially thermal, then (5-2-23) shows that $\mu^{1/2}K_0$ vs. T should give a single curve for all isotopic combinations, assuming that the ion-neutral potential is independent of nuclear mass. In fact, (5-2-23) shows that $\mu^{1/2}K_0$ vs. T_{eff} should give a single curve even if E/N is not kept small. However, it may be more useful to think in terms of K_0 vs. E/N at fixed T, which is the usual form in which experimental results are reported. We thus need to know the mass dependence of T_{eff} in terms of E/N, which we can find by the following argument. From (5-2-22) we see that $T_{\text{eff}} - T$ is proportional to $Mv_d^2 = M(KE)^2$; this means that $M(E/N)^2K_0^2$ depends only on T_{eff} if T is fixed. But from (5-2-23) we see that $\mu^{1/2}K_0$ depends only on T_{eff} for a particular ion-neutral interaction. We can therefore write,

$$M(E/N)^2K_0^2 = f_1(T_{\text{eff}}) \qquad \text{for } T \text{ fixed},$$

$$\mu^{1/2}K_0 = f_2(T_{\text{eff}}) \qquad \text{for } V(r) \text{ fixed},$$

from which it follows by simple algebra that

$$(M/\mu)^{1/2}(E/N) = f_1^{1/2}/f_2 = F(T_{\text{eff}}).$$

This means that there is a one-to-one relation between T_{eff} and the variable $(M/\mu)^{1/2}(E/N)$. If a plot of $\mu^{1/2}K_0$ vs. T_{eff} gives a single curve for isotopic variations of both ions and neutrals, then a plot of $\mu^{1/2}K_0$ vs. $(M/\mu)^{1/2}(E/N)$ should also give a single curve if T is fixed. The latter may sometimes be a more convenient way of expressing experimental data. We shall see that this scaling prediction is accurately followed, using H^+ and D^+ in He as an example (Figs. 6-1-2 and 8-3-2).

We next check the accuracy of the formula for the mean ion energy, as given in (5-2-20), by comparison with known accurate results obtained for special models. For the special case of a cold gas (negligible thermal energy of both neutrals and ions) and an inverse-power repulsive potential between ions and neutrals, $V(r) = C_n/r^n$, Skullerud (1976) was able to extend a special moment method devised by Wannier (1951, 1953), and thereby to calculate some accurate values of velocity moments and diffusion coefficients. We need not be concerned with any details at this point; all we want to do is use the results to check the accuracy of the momentum-transfer formula for the mean ion energy. According to this formula, the following ratio should be unity when the thermal energy is negligible:

$$\frac{\overline{mv^2}}{(m + M)v_d^2} \approx 1. \tag{5-2-32}$$

Values of this ratio are shown in Table 5-2-1 for a number of values of the mass ratio m/M and for four values of the potential index n. (The value $n = \infty$ corresponds to a rigid-sphere interaction.) The results for the extreme mass ratios of $m/M = 0$ and ∞ were not obtained by the moment method, but by special tricks suitable for these limits; again the details do not matter at this point. The cold gas limit can usually be approximated in practice by using very high fields at room temperature.

Two points stand out on examination of Table 5-2-1. The first is the remarkable overall accuracy of the mean energy formula—the maximum

Table 5-2-1 Accuracy of the Wannier Formula (5-2-20) for the Mean Ion Energy in a Cold Gas ($T = 0$) for Some Ion-Neutral Potentials of the Form $V(r) = C_n/r^n$ at Various Values of the Ion/Neutral Mass Ratio, m/M (Skullerud, 1976)

	$\overline{mv^2}/(m + M)v_d^2$			
m/M	$n = 4$	$n = 8$	$n = 12$	$n = \infty$
0	1	1.0202	1.0324	1.0606
0.1	1	1.00	1.01	1.029
0.2	1	0.987	0.990	1.003
0.5	1	0.955	0.946	0.935
0.8	1	0.941	0.928	0.903
1.0	1	0.938	0.923	0.895
1.5	1	0.940	0.925	0.896
2.0	1	0.946	0.932	0.906
3.0	1	0.957	0.946	0.925
4.0	1	0.965	0.956	0.939
∞	1	1.0000	1.0000	1.0000
$\langle|\text{dev}|\rangle$	0.0%	3.6%	4.5%	6.3%

deviation amounts only to about 10%. We may suspect that this is at least partly due to the fact that we obtained the formula by eliminating the collision frequency $\nu(\bar{\varepsilon})$ between the equations for momentum and energy balance; in other words, we eliminated the crudest part of the calculation, leaving a rather accurate relation. The second point about Table 5-2-1 is that the results are exact for $n = 4$. This case corresponds to a constant collision frequency independent of $\bar{\varepsilon}$, which introduces many simplifications into kinetic theory. It happens that both the momentum-balance and the energy-balance equations are exact in this case, and accurate versions of kinetic theory often start from this point as a first approximation.

The case of constant collision frequency is the famous Maxwell model, which occupies a prominent place in the history of kinetic theory. Maxwell (1867) found that great simplifications occurred with this model, and that many exact results could be obtained. It plays a role in kinetic theory analogous to that of the hydrogen atom in atomic physics: it is the model on which all new calculation schemes are tested, and it is the starting point for many kinds of successive approximations.

Finally, we examine the accuracy of the expression (5-2-23) for the mobility. Without even bothering about numerical factors or about the precise behavior of Q_D, we notice that (5-2-23) predicts a remarkable scaling rule. Instead of being a function of the two independent variables T and E/N, the mobility and drift velocity depend on only one combined variable, an effective temperature T_{eff} given by (5-2-22), which is important enough to bear repetition:

$$\tfrac{3}{2}kT_{\text{eff}} = \tfrac{3}{2}kT + \tfrac{1}{2}Mv_d^2 = \tfrac{3}{2}kT + \tfrac{1}{2}M(NK)^2(E/N)^2. \tag{5-2-33}$$

It is easy to find T_{eff} for real systems because it is directly obtained from experimental quantities. Thus measurements at fixed T and variable E/N can be made to coincide with independent measurements in which E/N is fixed (usually at zero) and T is varied. This is a much more important scaling rule than the mass scaling rule discussed above, because it shows how to replace a difficult experimental operation (variation of temperature) with an easier one (variation of voltage). The rule is easily tested, because only experimental quantities are involved. Figure 5-2-1 shows such a test for K^+ ions in He, Ne, and Ar, where it can be seen that the scaling holds within the experimental error. The same test applied to electrons in He, Ne, and Ar is shown in Fig. 5-2-2. Here some small discrepancies can be seen on close examination, but the overall agreement is still remarkably good. Such discrepancies can also be found in other systems, and suggest that the deficiencies of (5-2-23) for K are not entirely in purely numerical factors. This is indeed the case, but it takes a full-scale assault on the Boltzmann equation to do anything about it.

It is worth emphasizing that this simple scaling in terms of T_{eff} depends on the assumption that all ion-neutral collisions are elastic. Some remarks on the results with molecular species are given in Section 2-5B, and a full kinetic-theory treatment appears in Chapter 6.

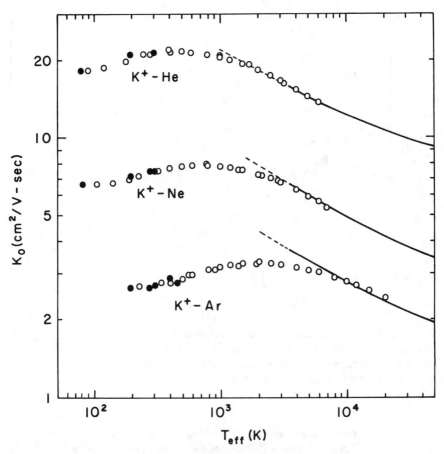

Figure 5-2-1. Scaling rule test of K_o as a function of T_{eff} for K^+ ions in He, Ne, and Ar. The filled circles are from zero-field measurements as a function of temperature, and the open circles are from measurements as a function of E/N at fixed temperature. Data are from Hoselitz (1941), Creaser (1969, 1974), and the survey of Ellis et al. (1976). The curves at high temperatures are zero-field mobilities calculated from the results of ion-beam scattering measurements (Inouye and Kita, 1972; Amdur et al., 1972); the dashed portions represent extrapolations to lower energies.

To check the numerical accuracy of (5-2-23) for the mobility, we again use known accurate results for special models. It is not worth going into much detail on this score, since we are already prepared to expect errors of the order of 20%. Results for the inverse-power repulsive potential are shown in Table 5-2-2 for a few values of the repulsion index n and of m/M, for both the zero-field case (Mason, 1957a, b) and the high-field (cold-gas) case (Skullerud, 1976). There seem to be no surprises, but it is interesting that the simple formula is so accurate at high fields, although it is probably fortuitous. It is also interesting that the error is independent of m/M for $n = 4$, and independent of n for

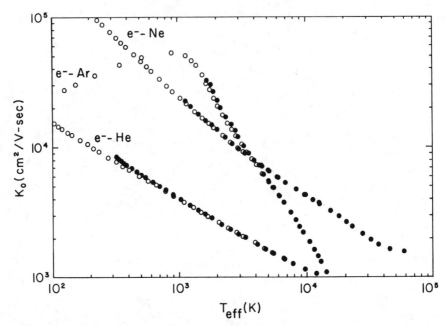

Figure 5-2-2. Scaling rule test of K_o as a function of T_{eff} for electrons in He, Ne, and Ar. The filled circles are from measurements as a function of E/N at $T = 293$ K, and the open circles from measurements at $T = 77$ K for He and Ne and $T = 90$ K for Ar. Data are from Huxley and Crompton (1974, Chap. 14).

Table 5-2-2 Accuracy of the Simple Momentum-Transfer Formula (5-2-23) for the Mobility, at Low Fields and at High Fields (Cold Gas) for Some Ion-Neutral Potentials of the Form $V(r) = C_n/r^n$ at Various Values of the Ion/Neutral Mass Ratio, m/M (Mason, 1957a, b; Skullerud, 1976)

	K(approx.)/K(accurate)			
m/M	$n = 4$	$n = 8$	$n = 12$	$n = \infty$
Low Field				
0	1.23	1.19	1.16	1.08
1	1.23	1.22	1.22	1.21
∞	1.23	1.23	1.23	1.23
High Field				
0	1.00	1.06	1.08	1.11
1	1.00	1.02	1.03	1.04
∞	1.00	1.00	1.00	1.00

$m/M = \infty$. This suggests that the theory might turn out to be simple for both the Maxwell model and the heavy-ion limit (often called the Rayleigh model). This is indeed the case, as we shall see later.

Just to make sure that nothing bizarre happens to K at intermediate field strengths, it is nice to have an accurate test case that can be treated at all field strengths. Three soluble models are known: the Maxwell model (constant collision frequency), the Rayleigh model ($m \gg M$), and the Lorentz model ($m \ll M$). The first two behave too simply to be good test cases, so we select the Lorentz model. The mobility for this model with a rigid-sphere interaction can be found by a fairly simple numerical integration (Hahn and Mason, 1972), giving results valid at all values of T and E/N. The comparison is facilitated by defining dimensionless field strength, drift velocity, and mobility as follows:

$$\mathscr{E}^* \equiv \frac{3\pi^{1/2}}{16kT}\left(\frac{m+M}{m}\right)^{1/2}\frac{eE}{N\pi d^2}, \tag{5-2-34}$$

$$v_d^* \equiv (M/2kT)^{1/2}v_d, \tag{5-2-35}$$

$$K^* \equiv v_d^*/\mathscr{E}^*, \tag{5-2-36}$$

where d is the diameter of the rigid-sphere potential. In terms of these quantities, the momentum-transfer formula (5-2-23) for the mobility becomes

$$\frac{3(6\pi)^{1/2}}{16}v_d^*(1 + \tfrac{2}{3}v_d^{*2})^{1/2} = \mathscr{E}^*. \tag{5-2-37}$$

The comparison with the accurate results of Hahn and Mason is shown in Fig. 5-2-3, where it can be seen that nothing dramatic happens at intermediate values of $\mathscr{E}^*$.

Let us now summarize the results we have obtained for the mobility by simple arguments on momentum and energy balance. A few of them are absolutely accurate, such as that NK is an even function only of E/N at fixed T. These are the general results that follow from the existence of a binary-collision mechanism and of spatial inversion symmetry; any theory that does not yield them is clearly faulty. Three results are more special, but have quite respectable accuracy, as follows:

1. Mean ion energy formula (5-2-20) in terms of m, M, and v_d. The accuracy (see Table 5-2-1) is presumably due to the elimination of those parts of the calculation that refer specifically to ion-neutral collisions, so that only a relation involving measurable quantities remains.
2. Scaling rule for the mass dependence of K: $\mu^{1/2}K_0$ vs. $(M/\mu)^{1/2}(E/N)$ is a single curve for isotopic variations of ions and neutrals at fixed T.
3. Scaling rule for the dependence of K on T and E/N: the dependence is on only one variable, not two, namely T_{eff} as given by (5-2-22) or (5-2-33). This rule works both for heavy ions (Fig. 5-2-1) and for electrons (Fig. 5-2-2).

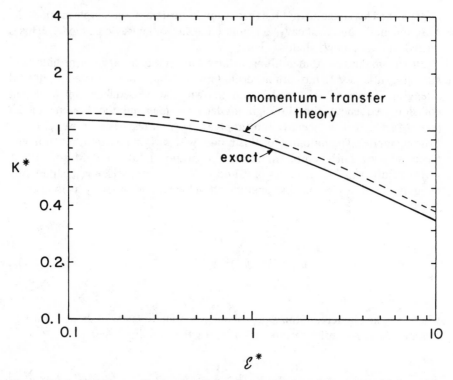

Figure 5-2-3. Mobility vs. field strength for the rigid-sphere Lorentz model ($m \ll M$): momentum-transfer formula (5-2-23) compared with accurate results from numerical integration (Hahn and Mason, 1972). The dimensionless mobility K^* and field strength $\mathscr{E}^*$ are defined by (5-2-34)–(5-2-36).

Both of these scaling rules can be condensed into a single rule if desired: for a given system the variations of isotopic mass, of T, and of E/N can be given as a single relation between $\mu^{1/2} K_0$ and T_{eff}.

Finally, the most specific result of all is the calculation of K as a function of T and E/N, given only the masses and the ion-neutral interaction potential. Here we achieve only a so-so accuracy of the order of 20% (Table 5-2-2 and Fig. 5-2-3). At this point we omit a lot of details about the behavior of K with the nature of the ion-neutral potential, which we save for the accurate kinetic theory. The aim is merely to give an impression of the sort of accuracy obtainable with the simplest version of momentum-transfer theory.

We conclude with a confession about momentum-transfer theory: a lot of the foregoing impressive accuracy with such a simple calculation is really being wise after the event. Elementary versions of kinetic theory are notoriously tricky, giving good results in some cases and disastrous ones in others. It is almost necessary to have an accurate theory available to tell which parts of an elementary theory are good and which are bad. It happens that momentum-transfer theory does very well, especially for ion mobility, but it would not be so easy to tell this a priori. Moreover, many versions of elementary kinetic theory are virtually impossible to improve in a systematic way, leading H. A. Kramers (1949) to remark on "those unimprovable speculations of which the kinetic theory of gases affords such ghastly examples." Mean-free-path theory is a classic example. Momentum-transfer theory again happens to do well in this regard, being the first approximation of a systematic moment solution of the Boltzmann equation, but this only becomes clear in retrospect.

We turn now to the calculation of ion diffusion coefficients, and then to the problem of mixtures.

B. Ion Diffusion: Generalized Einstein Relations

Instead of trying to develop a momentum-transfer theory of ion diffusion from scratch, we use the shortcut of first seeking a relation between diffusion and mobility, much like the simple Nernst–Townsend–Einstein relation that is valid at zero field strength. We thereby avoid repeating most of the preceding calculations on mobility, and can concentrate on just those parts that are new. We can expect diffusion to be more trouble theoretically than mobility. For one thing, the diffusion coefficient is a tensor, and for another it varies much more strongly with field strength. Over a range of field strength in which K varies by less than a factor of 2, D_L and D_T may vary by a factor of nearly 10^2.

We proceed as before by setting up (approximate) equations for momentum and energy balance. For mobility, the external force is obviously supplied solely by the electric field. For diffusion, the external force is regarded as coming from the gradient of partial pressure of the diffusing species (Present, 1958, Secs. 4-2 and 8-3). In place of (5-2-8) we then obtain the momentum-balance equation,

$$ne\mathbf{E} - \nabla \cdot \mathbf{p}_i = n\mu\bar{\mathbf{v}}\nu(\bar{\varepsilon}), \tag{5-2-38}$$

where $\mathbf{p}_i$ is the partial pressure of the ions. This vector-tensor notation is explained by Ferziger and Kaper (1972, App. A). The reason for taking $\mathbf{p}_i$ to be a tensor is easily seen. If we do a simple calculation of the normal component of ion momentum crossing a reference plane, the same calculation as done in elementary kinetic theory, we find the diagonal components of the pressure to be

$$p_{xx} = nm\overline{v_x^2}, \qquad p_{yy} = nm\overline{v_y^2}, \qquad p_{zz} = nm(\overline{v_z^2} - \bar{v}_z^2), \tag{5-2-39}$$

where the z-axis is chosen to point in the direction of $\mathbf{E}$. The off-diagonal components of $\mathbf{p}_i$, due to $\overline{v_x v_y}$, etc., must be zero by symmetry, because $\mathbf{E}$ is the only vector quantity in the system. That is, a uniform electric field cannot cause stresses in an isotropic system. However, in the presence of a strong electric field, p_{zz} will in general not be equal to p_{xx} and p_{yy}. The structure of $\mathbf{p}_i$ can be neatly summarized by defining an ion temperature tensor (diagonal), as follows:

$$\mathbf{p}_i = nk\mathbf{T}_i, \tag{5-2-40}$$

where

$$kT_T \equiv m\overline{v_x^2} = m\overline{v_y^2}, \qquad kT_L \equiv m(\overline{v_z^2} - \bar{v}_z^2). \tag{5-2-41}$$

Notice that in (5-2-38) we may not disregard the effect of the gradient on $\bar{\mathbf{v}}$ and $\bar{\varepsilon}$; to do so would be to discard diffusion entirely. However, $\bar{\mathbf{v}} = \mathbf{v}_d$ in the spatially homogeneous case (no gradients).

The energy balance proceeds as in the spatially homogeneous case. The work done on the ions is now not simply $neEv_d$, but

$$(ne\mathbf{E} - \nabla \cdot \mathbf{p}_i) \cdot \bar{\mathbf{v}}. \tag{5-2-42}$$

The rest of the calculation is the same, and we find the mean relative energy to be

$$\bar{\varepsilon} = \tfrac{3}{2}kT + \tfrac{1}{2}M\bar{v}^2, \tag{5-2-43}$$

where $\bar{v}$ is *not* the same as v_d, which we consider as referring to the spatially homogeneous case.

It remains only to define $\mathbf{D}$, and then we will have taken care of all the physics involved. We define $\mathbf{D}$ in terms of the excess ion flux over that occurring in the spatially homogeneous case (due to the field only),

$$n(\bar{\mathbf{v}} - \mathbf{v}_d) \equiv -\mathbf{D} \cdot \nabla n. \tag{5-2-44}$$

Before proceeding with the algebra, let us summarize the physical assumptions involved. These are as follows: momentum-transfer treatment of $\nabla \cdot \mathbf{p}_i$ as an external force; the structure of $\mathbf{p}_i$; and the definition of $\mathbf{D}$. We will also assume that $\nabla \cdot \mathbf{p}_i$ and ∇n are small quantities in what follows, since we are concerned here only with linear (Fickian) diffusion.

We can get an inkling of the sort of result to expect by examining (5-2-38). Notice that $\nabla \cdot \mathbf{p}_i$ appears in the same way that $\mathbf{E}$ does, so that the addition of a gradient is in some sense equivalent to a variation of $\mathbf{E}$. We might therefore expect $\mathbf{D}$ to be related to the differential mobility, dv_d/dE, and this turns out to be the case. To show this, we expand about the spatially homogeneous reference case, which we denote by a superscript $^\circ$, and keep only first-order terms because the gradient is assumed small. We first subtract the spatially homogeneous result from (5-2-38), obtaining

$$-\nabla \cdot \mathbf{p}_i = n\mu\bar{\mathbf{v}}\nu(\bar{\varepsilon}) - n\mu\mathbf{v}_d\nu(\bar{\varepsilon}^\circ)$$

$$= n\mu(\bar{\mathbf{v}} - \mathbf{v}_d)\nu(\bar{\varepsilon}^\circ) + n\mu\bar{\mathbf{v}}[\nu(\bar{\varepsilon}) - \nu(\bar{\varepsilon}^\circ)], \tag{5-2-45}$$

where $\bar{\mathbf{v}}^\circ \equiv \mathbf{v}_d$, and then expand the last term,

$$v(\bar{\varepsilon}) - v(\bar{\varepsilon}^\circ) = \frac{dv(\bar{\varepsilon}^\circ)}{d\bar{\varepsilon}^\circ}(\bar{\varepsilon} - \bar{\varepsilon}^\circ) + \cdots, \tag{5-2-46}$$

$$\bar{\varepsilon} - \bar{\varepsilon}^\circ = \tfrac{1}{2}M(\bar{v}^2 - v_d^2) = Mv_d(\bar{v} - v_d) + \cdots. \tag{5-2-47}$$

We now note that $\bar{\varepsilon} - \bar{\varepsilon}^\circ$ behaves rather differently depending on whether the gradient is parallel or perpendicular to the electric field. If the gradient is parallel, then $\bar{\varepsilon} - \bar{\varepsilon}^\circ$ is a small quantity of first order and must be retained. But if the gradient is perpendicular, to first order v_d is simply rotated through a small angle to become $\bar{v}$, and their magnitudes differ only in second order. Thus $\bar{\varepsilon} - \bar{\varepsilon}^\circ$ is a small quantity of second order and should be dropped in this case. It is thus better to treat the perpendicular (transverse) and parallel (longitudinal) cases separately from here on.

For the transverse case we have, from (5-2-44),

$$n(\bar{\mathbf{v}} - \mathbf{v}_d)_x = -D_T(\partial n/\partial x), \tag{5-2-48}$$

and we can also write

$$-(\mathbf{V}\cdot\mathbf{p}_i)_x = -kT_T(\partial n/\partial x). \tag{5-2-49}$$

Substituting these into (5-2-45) and canceling terms, we obtain

$$kT_T = D_T\mu v(\bar{\varepsilon}^\circ). \tag{5-2-50}$$

But from the spatially homogeneous equation we have $\mu v(\bar{\varepsilon}^\circ) = e/K$, so that our final result is

$$D_T = (kT_T/e)K. \tag{5-2-51}$$

This is the first generalized Einstein relation.

For the longitudinal case we must keep all of (5-2-45), which then becomes

$$-kT_L\frac{\partial n}{\partial z} = n\mu(\bar{v} - v_d)v(\bar{\varepsilon}^\circ) + n\mu v_d\frac{dv(\bar{\varepsilon}^\circ)}{d\bar{\varepsilon}^\circ}Mv_d(\bar{v} - v_d) + \cdots. \tag{5-2-52}$$

Substituting for $(\bar{v} - v_d)$ from (5-2-44) and canceling terms, we obtain

$$kT_L = D_L\mu\left[v(\bar{\varepsilon}^\circ) + Mv_d^2\frac{dv(\bar{\varepsilon}^\circ)}{d\bar{\varepsilon}^\circ}\right]. \tag{5-2-53}$$

The factor in brackets can be related to dv_d/dE by differentiating both sides of the spatially homogeneous equation, $eE = \mu v_d v(\bar{\varepsilon}^\circ)$, with respect to E:

$$e = \mu \frac{dv_d}{dE} v(\bar{\varepsilon}^\circ) + \mu v_d \frac{dv(\bar{\varepsilon}^\circ)}{d\bar{\varepsilon}^\circ} \frac{d\bar{\varepsilon}^\circ}{dv_d} \frac{dv_d}{dE},$$

$$e = \mu \frac{dv_d}{dE} \left[v(\bar{\varepsilon}^\circ) + M v_d^2 \frac{dv(\bar{\varepsilon}^\circ)}{d\bar{\varepsilon}^\circ} \right]. \tag{5-2-54}$$

Comparing (5-2-53) and (5-2-54), we obtain our final result,

$$D_L = \frac{kT_L}{e} \frac{dv_d}{dE} = \frac{kT_L}{e} K \left(1 + \frac{d \ln K}{d \ln E} \right), \tag{5-2-55}$$

where the last step results from differentiating $v_d = KE$. This is the second generalized Einstein relation (GER).

Both GER can be written in similar form, if desired, as follows:

$$D_{T,L} = \frac{kT_{T,L}}{e} \left(\frac{dv_d}{dE} \right)_{T,L}. \tag{5-2-56}$$

It is necessary to remember that $(dv_d/dE)_T$ means adding a small field increment dE_T perpendicular to the original field. This merely rotates both E and v_d through the same small angle without changing the magnitude of either (to first order), so that

$$\left(\frac{dv_d}{dE} \right)_T = \frac{v_d}{E} = K. \tag{5-2-57}$$

Notice that this is the same argument by which we dropped the term $\bar{\varepsilon} - \bar{\varepsilon}^\circ$ in (5-2-45) for the transverse case, which of course is no accident. At any rate, the conjecture that **D** might be related to the differential mobility is confirmed.

The GER of (5-2-51) and (5-2-55)—or combined as in (5-2-56)—turn out to be rather general, just like the Wannier ion energy formula (5-2-20). This generality seems plausible when we note that the crudest part of the calculation has been eliminated from the GER, namely that referring specifically to the ion-neutral collisions. The GER were first conjectured by Wannier (1952, 1953) on the basis of his calculations for the Maxwell model; the use of the differential mobility was a conjecture because it happens to be equal to the mobility itself at all field strengths for the Maxwell model. However, the two differ by a factor of 2 for rigid spheres at high fields. The case for the differential mobility was strengthened by a comment of Klots and Nelson (1970), and particularly by a thermodynamic derivation by Robson (1972). The use of nonequilibrium thermodynamics would seem to limit the range of validity of the GER to weak fields. However, Wannier (1973) suggested, on the strength of a heuristic derivation based on the Langevin equation, that the GER were at least approximately valid at all field strengths. There have been a number of further attempts to give a sound theoretical basis to the GER, since they seem to work quite well in practice; most of these derivations have been based on approximate solutions of the Boltzmann equation. The most thorough treatment along these lines shows that even higher approximations add only small correction terms to the GER as given above (Waldman and Mason, 1981).

The derivation given above is based largely on the ideas of Robson (1972) and of Whealton et al. (1974). It contains what seems to be the minimum input needed to obtain the desired result, and thereby indicates that the GER are generally valid over a wide range. Momentum-transfer theory has also been used to estimate some of the small corrections to the GER (Robson, 1976a), but this topic is best left to the accurate kinetic theory.

To check the numerical accuracy of the GER, we again use the known accurate results of Skullerud (1976) for special models at high field strengths (i.e., in the cold-gas limit). No check is necessary at low fields because in this case the GER obviously revert to the simple Nernst–Townsend–Einstein relation ($T_T = T_L = T$, and $dv_d/dE = v_d/E$), which is exact. Tables 5-2-3 and 5-2-4 list the following ratios:

$$eD_T/kT_T K \quad \text{and} \quad eD_L/[kT_L(dv_d/dE)], \qquad (5\text{-}2\text{-}58)$$

Table 5-2-3 Accuracy of the Generalized Einstein Relation (5-2-51) for D_T in a Cold Gas ($T = 0$) for Some Ion-Neutral Potentials of the Form $V(r) = C_n/r^n$ at Various Values of the Ion/Neutral Mass Ratio, m/M (Skullerud, 1976)

	$eD_T/kT_T K$			
m/M	$n = 4$	$n = 8$	$n = 12$	$n = \infty$
0	1	1.044	1.072	1.144
0.1	1	1.00	1.06	1.052
0.2	1	0.97	1.000	1.003
0.5	1	0.957	0.996	0.954
0.8	1	0.959	0.957	0.951
1.0	1	0.961	0.957	0.952
1.5	1	0.972	0.961	0.959
2.0	1	0.971	0.965	0.961
3.0	1	0.983	0.982	0.973
4.0	1	0.987	0.986	0.978
∞	1	1.0000	1.0000	1.0000
$\langle |dev| \rangle$	0.0%	2.6%	3.0%	4.3%

Table 5-2-4 Accuracy of the Generalized Einstein
Relation (5-2-55) for D_L in a Cold Gas ($T = 0$) for
Some Ion-Neutral Potentials of the Form $V(r) = C_n/r^n$
at Various Values of the Ion/Neutral Mass Ratio, m/M
(Skullerud, 1976)

m/M	\multicolumn{4}{c}{$eD_L/[kT_L(dv_d/dE)]$}
	$n = 4$	$n = 8$	$n = 12$	$n = \infty$
0	1	1.035	1.091	1.124
0.1	1	0.96	0.99	1.021
0.2	1	0.94	0.949	0.941
0.5	1	0.902	0.869	0.832
0.8	1	0.909	0.885	0.833
1.0	1	0.914	0.895	0.846
1.5	1	0.929	0.915	0.881
2.0	1	0.946	0.932	0.905
3.0	1	0.960	0.950	0.933
4.0	1	0.971	0.961	0.953
∞	1	1.0000	1.0000	1.0000
$\langle|dev|\rangle$	0.0%	5.5%	6.8%	9.3%

which should be unity if the GER were exact. The results are quite good, although not as good as those for the mean ion energy.

The success of the GER is unfortunately somewhat illusory, because the ion temperatures are seldom known independently. All that we know at this point is that their sum must give the correct total mean ion energy,

$$2kT_T + kT_L = m(\overline{v_x^2} + \overline{v_y^2} + \overline{v_z^2}) - mv_d^2$$
$$\approx 3kT + Mv_d^2 \approx 3kT_{\text{eff}}. \tag{5-2-59}$$

We must therefore attempt to calculate how the mean ion energy is partitioned into transverse and longitudinal components, in order for the GER to be generally useful. However, we can already obtain one useful result for the special case of $m \ll M$ (e.g., electrons). Because a collision merely rotates the ion velocity without changing its magnitude, we would expect that $\overline{v_x^2} = \overline{v_y^2} = \overline{v_z^2}$, or $T_L = T_T$. Thus the ratio D_L/D_T can be found from the quantity $(d \ln K/d \ln E)$, according to (5-2-51) and (5-2-55); this result is of considerable practical value for electrons.

C. Ion Diffusion: Partitioning of Ion Energy

We need another energy-balance equation involving either $\overline{v_z^2}$ or $\overline{v_x^2} = \overline{v_y^2}$ in order to calculate the energy partitioning. We need find only $\overline{v_z^2}$, since

$\overline{v_x^2} + \overline{v_y^2} + \overline{v_z^2} = \overline{v^2}$, and we already have an expression for $\overline{v^2}$ in (5-2-20). It is convenient to consider the group $(3\overline{v_z^2} - \overline{v^2})$; this is zero in the absence of a field, and will presumably be proportional to eEv_d. We therefore write another energy-balance equation as follows:

$$\tfrac{1}{2}m[(3\overline{v_z^2} - \overline{v^2}) - (3\overline{v_z'^2} - \overline{v'^2})]v(\bar{\varepsilon}) = 2eEv_d. \qquad (5\text{-}2\text{-}60)$$

We can see the need for the factor of 2 on the right-hand side by considering the extreme case of very heavy ions in a cold light gas ($T \to 0$), for which $\overline{v_z^2} \approx \overline{v^2}$ and $\overline{v_z'^2} \approx \overline{v'^2}$. Consistency with the energy-balance equation (5-2-11) for $\overline{v^2}$ then requires a factor of 2.

The problem now is to find $\overline{v_z'^2}$ in terms of $\overline{v_z^2}$, after which only algebra should be necessary. As before, this requires some manipulation in center-of-mass coordinates, and the reader who is uninterested in these details can skip to (5-2-69). We begin by following exactly the same steps as given in (5-2-12) through (5-2-18) of Section 5-2A, but use only the z-components instead of the full velocities. In place of (5-2-17) and (5-2-18) we then obtain

$$\overline{v_z^2} - \overline{v_z'^2} = \frac{2M}{(m+M)^2}(m\overline{v_z^2} - M\overline{V_z^2}) + \left(\frac{M}{m+M}\right)^2 (\overline{v_{rz}^2} - \overline{v_{rz}'^2}), \qquad (5\text{-}2\text{-}61)$$

where v_{rz} is the z-component of the relative velocity, $\mathbf{v}_r$. If we wrote down the corresponding equations for the x- and y-components and added the three expressions, the last term would vanish because conservation of energy in a single collision requires that $v_r'^2 = v_r^2$. However, it is *not* true that $v_{rz}'^2 = v_{rz}^2$. The problem now is reduced to finding v_{rz}' in terms of v_{rz}.

The advantage of center-of-mass coordinates is that an elastic collision does not change the magnitude of $\mathbf{v}_r$, but simply rotates it through some angle. We can describe this in terms of the customary polar angles θ and ϕ in a coordinate system in which $\mathbf{v}_r$ points along the z-axis, so that

$$\mathbf{v}_r' = v_r(\mathbf{i} \sin\theta \cos\phi + \mathbf{j} \sin\theta \sin\phi + \mathbf{k} \cos\theta), \qquad (5\text{-}2\text{-}62)$$

where $\mathbf{i}, \mathbf{j}, \mathbf{k}$ are the unit vectors along the axes. We now suppose that $\mathbf{v}_r$ makes an angle ψ with the real z-axis defined by the electric field, and find the components of $\mathbf{v}_r$ and $\mathbf{v}_r'$ along this real z-axis (which without loss of generality can be taken to lie in the $\mathbf{i}$–$\mathbf{k}$ plane):

$$v_{rz} = v_r \cos\psi, \qquad (5\text{-}2\text{-}63)$$

$$v_{rz}' = v_r(\sin\theta \cos\phi \sin\psi + \cos\theta \cos\psi). \qquad (5\text{-}2\text{-}64)$$

We now square these expressions, average over all collisions, and make two approximations. The first is to take the average of a product equal to the product of the averages. The second is to assume, as we did before, that the scattering is isotropic in the center-of-mass system. This means that we take, for three-dimensional scattering,

$$\overline{\sin^2\theta} = \tfrac{2}{3}, \quad \overline{\cos^2\theta} = \tfrac{1}{3}, \quad \overline{\cos\phi} = 0, \quad \overline{\cos^2\phi} = \tfrac{1}{2}. \qquad (5\text{-}2\text{-}65)$$

The result is fairly simple,

$$\overline{v_{rz}^2} - \overline{v_{rz}'^2} = \overline{v_r^2}(\overline{\cos^2\psi} - \tfrac{1}{3}) = \overline{v_{rz}^2} - \tfrac{1}{3}\overline{v_r^2}. \qquad (5\text{-}2\text{-}66)$$

Substituting $\overline{v_r^2} = \overline{v^2} + \overline{V^2}$ and $\overline{v_{rz}^2} = \overline{v_z^2} + \overline{V_z^2}$ from (5-2-9), we obtain

$$\overline{v_{rz}^2} - \overline{v_{rz}'^2} = \overline{v_z^2} - \tfrac{1}{3}\overline{v^2}, \qquad (5\text{-}2\text{-}67)$$

since $\overline{V_z^2} = \overline{V^2}/3$. Then (5-2-61) is

$$\overline{v_z^2} - \overline{v_z'^2} = \left(\frac{2M}{m+M}\right)^2 (m\overline{v_z^2} - M\overline{V_z^2}) + \left(\frac{M}{m+M}\right)^2 (\overline{v_z^2} - \tfrac{1}{3}\overline{v^2}), \qquad (5\text{-}2\text{-}68)$$

and the energy-balance equation (5-2-60) becomes

$$\frac{mM(2m+M)}{4(m+M)^2}(3\overline{v_z^2} - \overline{v^2})\nu(\bar{\varepsilon}) = eEv_d. \qquad (5\text{-}2\text{-}69)$$

This is the desired result, analogous to equation (5-2-19) for the overall energy balance.

Only straightforward algebra is now needed to find the ion temperatures. We eliminate $eEv_d/\nu(\bar{\varepsilon})$ between the two energy-balance equations, (5-2-19) and (5-2-69), and after some manipulation obtain

$$kT_L \equiv m(\overline{v_z^2} - v_d^2) = kT + \frac{1}{3}\left(\frac{4m+M}{2m+M}\right)Mv_d^2. \qquad (5\text{-}2\text{-}70a)$$

The transverse temperature is then found by difference, using (5-2-59),

$$kT_T \equiv m\overline{v_x^2} = m\overline{v_y^2} = kT + \frac{1}{3}\left(\frac{m+M}{2m+M}\right)Mv_d^2. \qquad (5\text{-}2\text{-}70b)$$

These equations predict that T_L and T_T are equal for light ions in a heavy gas, as we would expect. For heavy ions in a light gas they predict that T_L is four times larger than T_T at high fields. This is qualitatively in the direction expected, but for quantitative conclusions we must refer to accurate calculations.

We again refer to the accurate results of Skullerud (1976) for special models in the cold gas limit in order to test the expressions for T_T and T_L for numerical accuracy. The results are given in Tables 5-2-5 and 5-2-6. We should expect somewhat worse agreement than for the total ion energy, inasmuch as we are asking for more detailed information, and have been rather casual about treating averages. Even so, the results are rather disappointing. For heavy ions in a light gas the discrepancies range up to a factor of 2. Here the simple momentum-transfer theory has let us down rather badly, but we can nevertheless make a little progress in understanding why. There are two main candidates for the discrepancies: the momentum and energy balance equations themselves, and the isotropic averaging over scattering. We can partly apportion the blame by examining the results for the $n = 4$ case (the Maxwell model). Here we know that the balance equations are correct, so all the discrepancies must be due to the isotropic averaging, since an r^{-4} repulsive potential does not scatter isotropically. Tables 5-2-5 and 5-2-6 indicate that improving the averaging over scattering angles would improve the agreement substantially, but that the likely improvement is still far from the required factor of 2, especially when we recall that the scattering *is* isotropic for the $n = \infty$ case (rigid spheres). Unfortunately, it is not clear how to fix the remaining discrepancy by modifying the momentum and energy balance equations.

In the face of such difficulties, there are two main responses. One is to give up on the simple theory and try to develop a more accurate theory, usually by some approximate solution of the Boltzmann equation. This procedure was followed

Table 5-2-5 Accuracy of the Relation (5-2-70b) for the Transverse Ion Temperature, T_T, in a Cold Gas ($T = 0$) for Some Ion-Neutral Potentials of the Form $V(r) = C_n/r^n$ at Various Values of the Ion/Neutral Mass Ratio, m/M (Skullerud, 1976)

	$kT_T \left[\dfrac{1}{3}\left(\dfrac{m + M}{2m + M} \right) Mv_d^2 \right]^{-1}$			
m/M	$n = 4$	$n = 8$	$n = 12$	$n = \infty$
0	1.0000	1.0202	1.0324	1.0606
0.1	1.063	1.05	1.05	1.067
0.2	1.113	1.082	1.075	1.068
0.5	1.216	1.128	1.100	1.044
0.8	1.280	1.153	1.109	1.023
1.0	1.310	1.161	1.115	1.004
1.5	1.363	1.181	1.118	0.984
2.0	1.397	1.200	1.135	0.975
3.0	1.438	1.228	1.155	0.982
4.0	1.461	1.247	1.161	1.004
∞	1.551	1.301	1.210	1.0000

Table 5-2-6 Accuracy of the Relation (5-2-70a) for the Longitudinal Ion Temperature, T_L, in a Cold Gas ($T = 0$) for Some Ion-Neutral Potentials of the Form $V(r) = C_n/r^n$ at Various Values of the Ion/Neutral Mass Ratio, m/M (Skullerud, 1976)

$kT_L\left[\dfrac{1}{3}\left(\dfrac{4m+M}{2m+M}\right)Mv_d^2\right]^{-1}$			
m/M	$n = 4$	$n = 8$	$n = 12$	$n = \infty$
0	1.0000	1.0202	1.0324	1.0606
0.1	0.901	0.93	0.95	0.982
0.2	0.849	0.856	0.873	0.917
0.5	0.784	0.738	0.740	0.760
0.8	0.760	0.674	0.665	0.659
1.0	0.752	0.650	0.634	0.621
1.5	0.741	0.617	0.593	0.569
2.0	0.735	0.600	0.575	0.547
3.0	0.731	0.583	0.556	0.527
4.0	0.729	0.577	0.551	0.510
∞	0.725	0.566	0.537	0.5000

by Viehland et al. (1974), who found corrections proportional to $(d \ln K/d \ln E)$ for T_{eff}, T_T, and T_L. This work will be discussed later in connection with the accurate kinetic theory. Another approach is to solve a few crucial test cases, look for trends, and set up a semiempirical parameterization scheme. Skullerud (1976) has followed such a procedure, which we briefly describe. For the general Maxwell model of constant collision frequency but unspecified angular scattering pattern, Wannier (1953) showed that T_T and T_L were given exactly by

$$kT_{T,L} = kT + \zeta_{T,L}Mv_d^2, \tag{5-2-71a}$$

where

$$\zeta_T = \frac{(m + M)[Q^{(2)}/Q^{(1)}]}{4m + 3M[Q^{(2)}/Q^{(1)}]}, \tag{5-2-71b}$$

$$\zeta_L = \frac{4m - (2m - M)[Q^{(2)}/Q^{(1)}]}{4m + 3M[Q^{(2)}/Q^{(1)}]}, \tag{5-2-71c}$$

$$Q^{(l)} = 2\pi \int_0^\pi (1 - \cos^l\theta)\sigma(\theta, \varepsilon) \sin \theta \, d\theta, \tag{5-2-71d}$$

Notice that ζ_T and ζ_L reduce to the previous expressions of (5-2-70) on setting

$Q^{(2)}/Q^{(1)} = \frac{2}{3}$, the value appropriate for isotropic scattering (rigid spheres). Notice also that the relation $2\zeta_T + \zeta_L = 1$ always holds. The cross section $Q^{(1)}$ is the same as Q_D of the momentum-transfer theory. Using numerical results from Monte Carlo calculations and from special moment methods, Skullerud (1973a, 1976) investigated the accuracy of (5-2-71) for cases in which the collision frequency was not constant, but $Q^{(2)}/Q^{(1)}$ was of course still calculable. He found that the results for T_T were reasonable, but that those for T_L were essentially useless. He then investigated the special case of heavy ions in a light gas ($m/M \to \infty$, the worst case in Table 5-2-6), for which Wannier (1953) had shown how to obtain exact solutions for T_T and T_L in the cold-gas limit. Skullerud (1976) carried the solution through explicitly for inverse-power potentials, for which $v(\varepsilon) \propto \varepsilon^{\gamma/2}$, where $\gamma = 1 - (4/n)$. He found that the expression for ζ_T was essentially unchanged, but that for ζ_L was divided by the factor $(1 + \gamma)$. He also noted that the field dependence of the mobility in the cold-gas limit is

$$K' \equiv \frac{d \ln K_0}{d \ln (E/N)} = -\frac{\gamma}{1 + \gamma} = -\frac{1 - (4/n)}{2 - (4/n)}, \qquad (5\text{-}2\text{-}72)$$

a result derived later in Section 5-2F (essentially by dimensional analysis). Thus ζ_L can be corrected by multiplication by $(1 + K')$. For smaller m/M the correction needed is also smaller (see Table 5-2-6), so that a correction factor of the form $(1 + \beta_L K')$ might be appropriate, where β_L depends primarily on m/M and varies between 0 and 1. This correction is also of the general form suggested by the work of Viehland et al. (1974), and subsequent more elaborate calculations have shown that the correction is quite successful (Waldman and Mason, 1981; Waldman et al., 1982). The numerical values of β_L as a function of m/M must of course be determined empirically (Skullerud, 1976). The final results can thus be written as

$$kT_{T,L} = kT + \zeta_{T,L} M v_d^2 (1 + \beta_{T,L} K'), \qquad (5\text{-}2\text{-}73a)$$

with $\beta_T = 0$ and β_L given as a numerical table. In addition, Skullerud recommended the value $Q^{(2)}/Q^{(1)} = 0.85$ for use in ζ_T and ζ_L. Some numerical results are shown in Tables 5-2-7 and 5-2-8 for the same examples as in Tables 5-2-5 and 5-2-6. There is a marked improvement, and the agreement would be even better if the proper values of $Q^{(2)}/Q^{(1)}$ had been used instead of the fixed value of 0.85. For convenience in quick calculations, the numerical values of β_L can be represented fairly well in terms of the variable $m/(m + M)$ by a straight line plus a parabola,

$$\beta_L \approx \frac{m}{m + M} + \frac{mM}{(m + M)^2}. \qquad (5\text{-}2\text{-}73b)$$

Notice that acceptance of (5-2-73a) for T_T and T_L implies a modification in the

Table 5-2-7 Accuracy of the Modified Relation (5-2-73) for T_T, for the Same Systems as in Table 5-2-5[a]

	$kT_T[\zeta_T M v_d^2(1 + \beta_T K')]^{-1}$			
m/M	$n = 4$	$n = 8$	$n = 12$	$n = \infty$
0	1.0000	1.0202	1.0324	1.0606
0.1	1.025	1.01	1.01	1.029
0.2	1.044	1.015	1.008	1.002
0.5	1.085	1.006	0.981	0.931
0.8	1.110	1.000	0.962	0.887
1.0	1.122	0.994	0.952	0.859
1.5	1.143	0.990	0.938	0.825
2.0	1.156	0.993	0.939	0.807
3.0	1.172	1.001	0.941	0.800
4.0	1.181	1.008	0.938	0.812
∞	1.216	1.021	0.949	0.784
$\langle|dev|\rangle$	11.4%	0.9%	4.1%	12.6%

[a] $\beta_T = 0$ and ζ_T is given by (5-2-71b) with $Q^{(2)}/Q^{(1)} = 0.85$.

Table 5-2-8 Accuracy of the Modified Relation (5-2-73) for T_L for the Same Systems as in Table 5-2-6[a]

	$kT_L[\zeta_L M v_d^2(1 + \beta_L K')]^{-1}$			
m/M	$n = 4$	$n = 8$	$n = 12$	$n = \infty$
0	1.0000	1.0202	1.0324	1.0606
0.1	0.957	1.00	1.03	1.070
0.2	0.931	0.991	1.022	1.093
0.5	0.892	0.988	1.026	1.115
0.8	0.875	0.986	1.029	1.116
1.0	0.868	0.988	1.028	1.121
1.5	0.859	0.985	1.024	1.119
2.0	0.854	0.986	1.031	1.134
3.0	0.849	0.987	1.035	1.155
4.0	0.847	0.985	1.040	1.139
∞	0.840	0.985	1.038	1.159
$\langle|dev|\rangle$	11.2%	1.3%	3.0%	11.7%

[a] β_L is given numerically by Skullerud (1976), and ζ_L is given by (5-2-71c) with $Q^{(2)}/Q^{(1)} = 0.85$.

formula for the total ion energy, $\frac{1}{2}m\overline{v^2}$, and for T_{eff}:

$$\frac{1}{2}m\overline{v^2} = \frac{3}{2}kT + \frac{1}{2}(m + M)v_d^2(1 + \beta K'), \tag{5-2-74a}$$

$$\frac{3}{2}kT_{\text{eff}} = \frac{1}{2}\mu(\overline{v^2} + \overline{V^2}) = \frac{3}{2}kT + \frac{1}{2}Mv_d^2(1 + \beta K'), \tag{5-2-74b}$$

$$\beta = \frac{M}{m + M}(2\zeta_T\beta_T + \zeta_L\beta_L). \tag{5-2-74c}$$

The accuracy of this formula for $\frac{1}{2}m\overline{v^2}$ is tested in Table 5-2-9 for the same systems as shown in Table 5-2-1. The result is a distinct improvement in an already good formula, and the agreement is nearly perfect, except when $m \ll M$.

Still further improvements are possible, both in the ion temperatures and in the GER themselves, but are best discussed in connection with the accurate kinetic theory in Chapter 6.

We have now succeeded in deriving approximate relations involving only experimentally accessible quantities. One momentum-balance equation, two energy-balance equations, and some semiempirical modifications of T_T and T_L were needed. The GER allow the diffusion coefficients to be calculated entirely from mobility measurements. In particular, D_T is found from (5-2-51) and D_L from (5-2-55), plus (5-2-73) for T_T and T_L. This is an important result

Table 5-2-9 Accuracy of the Modified Relation (5-2-74) for the Mean Ion Energy for the Same Systems as in Table 5-2-1[a]

	$\overline{mv^2}[(m + M)v_d^2(1 + \beta K')]^{-1}$			
m/M	$n = 4$	$n = 8$	$n = 12$	$n = \infty$
0	1	1.020	1.0324	1.0606
0.1	1	1.01	1.02	1.038
0.2	1	1.004	1.011	1.030
0.5	1	0.999	0.999	1.001
0.8	1	0.996	0.994	0.985
1.0	1	0.995	0.992	0.980
1.5	1	0.995	0.990	0.977
2.0	1	0.996	0.992	0.981
3.0	1	0.999	0.996	0.987
4.0	1	1.000	0.997	0.991
∞	1	1.0000	1.0000	1.0000
$\langle|\text{dev}|\rangle$	0.0%	0.5%	0.9%	2.1%

[a]The values of $\beta_{T,L}$ and $\zeta_{T,L}$ used to determine β are those used in Tables 5-2-7 and 5-2-8.

because mobilities are much easier to measure than are diffusion coefficients. Figure 5-2-4 shows a test of these relations for K$^+$ ions in He, Ne, and Ar. The overall agreement is rather good for such a simple theory. The discrepancies are partly due to inaccuracies in the calculated ion temperatures and partly due to inaccuracies in the GER themselves. A major theoretical effort is needed to secure improvement, and is described in Section 6-4.

Two final comments on ion diffusion coefficients are suggested by Fig. 5-2-4. First, the dependence on E/N is much stronger than is the case for mobility; most of this can be attributed to T_T and T_L, and the dependence thus goes roughly as $(E/N)^2$. Second, the use of logarithmic scales to accommodate this strong dependence obviously helps to conceal discrepancies between theory and experiment, as well as structure in the curves, which can be detected only by close inspection. We shall later show a convenient way to present diffusion results that removes the main quadratic dependence on E/N and reveals the structure (Section 6-2B3).

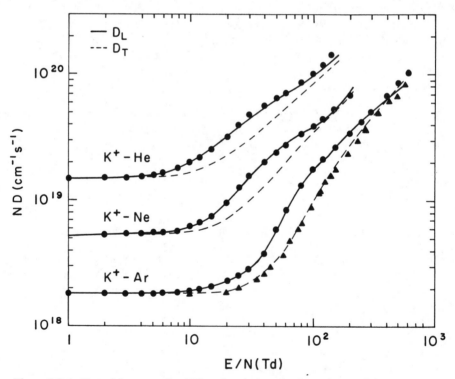

Figure 5-2-4. Test of the generalized Einstein relations for D_L and D_T and the expressions for T_L and T_T in terms of experimental measurements. The curves are calculated from mobility data at 300 K via formulas (5-2-51), (5-2-53), and (5-2-73). The circles are experimental measurements of D_L from the survey of Ellis et al. (1978), and the triangles are from measurements of D_T for K$^+$ in Ar by Skullerud (1972, 1973b).

D. Mixtures

Here the aim is to calculate the transport coefficients of an ion in a gas mixture, assuming that the transport coefficients are known in each of the single pure components. We do not need to worry about mixtures of ions because our basic assumption of trace ion concentration assures that the ions act independently, so that a mixture of ions behaves like a simple superposition of the behaviors of the separate ions. We should expect to obtain good results from even an approximate theory in a case like this, since we ask the theory only to act as an interpolation formula and not to furnish any absolute magnitudes. With a little care we should be able to eliminate most or all of those features where approximate theories characteristically go wrong, such as numerical factors, detailed dependence on T and E/N, etc. Momentum-transfer theory is especially suited for the description of even multicomponent mixtures, because the assumption of independent binary collisions allows the equations of momentum and energy balance to be written down in a very straightforward way. It has thus been extensively exploited for describing the composition dependence of ion transport coefficients (Mason and Hahn, 1972; Whealton and Mason, 1972; Milloy and Robson, 1973; Whealton et al., 1974).

1. Mobility

The total momentum transferred per ion to a gas mixture is just the sum of the momenta transferred to each species in the mixture. However, the collision frequency of the ion with species j in the mixture is *not* the same as it would be in pure gas j at the same number density and the same values of T and E/N, because the mean relative collision energy is different in the mixture than in a pure gas. This is not so surprising, since the drift velocity itself is different in the two cases. To keep matters straight, we will denote quantities in the mixture by the symbol $\langle \cdots \rangle_{\text{mix}}$ where necessary. The momentum-balance equation is then

$$eE = \langle v_d \rangle_{\text{mix}} \sum_j \mu_j N_j \langle \bar{v}_{rj} \rangle_{\text{mix}} Q_{Dj}(\langle \bar{\varepsilon}_j \rangle_{\text{mix}})$$

$$= \langle v_d \rangle_{\text{mix}} \sum_j x_j \mu_j \langle v_j(\bar{\varepsilon}_j) \rangle_{\text{mix}}, \qquad (5\text{-}2\text{-}75)$$

where $x_j \equiv N_j/N$ is the mole fraction of species j, $\mu_j = mM_j/(m + M_j)$ is the reduced mass, and $\langle v_j(\bar{\varepsilon}_j) \rangle_{\text{mix}}$ is the collision frequency of the ion with species j in *the mixture*, normalized to the total number density N. This is the generalization to mixtures of equations (5-2-4)–(5-2-8) for single gases. Comparing (5-2-75) with the corresponding expression for a single gas, we can write the mobility in the mixture, K_{mix}, in terms of the mobilities K_j in the pure components at the same total number density,

$$\frac{1}{K_{\text{mix}}} = \sum_j \frac{x_j}{K_j} \frac{\langle v_j(\bar{\varepsilon}_j) \rangle_{\text{mix}}}{v_j(\bar{\varepsilon}_j)}. \qquad (5\text{-}2\text{-}76)$$

To evaluate the ratios $\langle v_j(\bar{\varepsilon}_j)\rangle_{\mathrm{mix}}/v_j(\bar{\varepsilon}_j)$ we need to use the energy-balance equation, but we can first obtain the low-field result immediately.

At low fields the mean collision energies are all thermal and the ratios of collision frequencies all become unity, so that (5-2-76) reduces to

$$\frac{1}{K_{\mathrm{mix}}(0)} = \sum_j \frac{x_j}{K_j(0)}. \qquad (5\text{-}2\text{-}77)$$

This result is known as Blanc's law (Blanc, 1908). It also holds for the diffusion coefficient at low fields. Physically, Blanc's law is just a statement about the additivity of cross sections in the binary collision regime, or the independence of successive collision events. It should therefore be quite accurate, since it is independent of all the approximate details of momentum-transfer theory, or indeed of any other reasonable elementary theory. So great is the confidence in Blanc's law that deviations are often taken as experimental evidence that an ion changes its chemical identity through clustering or complex formation as the gas composition is varied (Loeb, 1960, pp. 129ff; Biondi and Chanin, 1961). However, nonnegligible kinetic-theory deviations are also possible, although these appear only in higher approximations of accurate kinetic theory, and will accordingly be dealt with later (Section 6-1G).

The equation for energy balance is also straightforward to write down; we take the total energy transferred per ion to the gas mixture to be the sum of the energies transferred to each species. In place of (5-2-19) for a single gas we then obtain

$$eE\langle v_d\rangle_{\mathrm{mix}} = \sum_j x_j \frac{mM_j}{(m+M_j)^2}(m\langle \overline{v^2}\rangle_{\mathrm{mix}} - M_j\overline{V_j^2})\langle v_j(\bar{\varepsilon}_j)\rangle_{\mathrm{mix}}. \qquad (5\text{-}2\text{-}78)$$

Combining this with the momentum-balance equation (5-2-75), we find an expression for the mean ion energy

$$m\langle \overline{v^2}\rangle_{\mathrm{mix}}\sum_j x_j \frac{M_j\langle v_j\rangle_{\mathrm{mix}}}{(m+M_j)^2} = \sum_j x_j \frac{M_j\langle v_j\rangle_{\mathrm{mix}}}{(m+M_j)^2}[M_j\overline{V_j^2} + (m+M_j)\langle v_d\rangle_{\mathrm{mix}}^2],$$

$$(5\text{-}2\text{-}79)$$

where we have shortened the notation for the collision frequency to $\langle v_j\rangle_{\mathrm{mix}}$. This is the mixture analog of the Wannier formula (5-2-20), but is obviously more complicated. The analogy can be made more obvious by noting that $M_j\overline{V_j^2} = 3kT$ and by defining a suitable mean mass of the gas mixture, whereby (5-2-79) becomes

$$\tfrac{1}{2}m\langle \overline{v^2}\rangle_{\mathrm{mix}} = \tfrac{3}{2}kT + \tfrac{1}{2}m\langle v_d\rangle_{\mathrm{mix}}^2 + \tfrac{1}{2}\langle M\rangle_{\mathrm{mix}}\langle v_d\rangle_{\mathrm{mix}}^2, \qquad (5\text{-}2\text{-}80)$$

where

$$\langle M \rangle_{\mathrm{mix}} \equiv \sum_j \omega_j M_j \Big/ \sum_j \omega_j, \tag{5-2-81}$$

$$\omega_j \equiv x_j M_j \langle v_j \rangle_{\mathrm{mix}}/(m + M_j)^2. \tag{5-2-82}$$

To evaluate the ratios $\langle v_j \rangle_{\mathrm{mix}}/v_j$ we need the mean relative energy between the ion and species j in the mixture, which is

$$\langle \bar{\varepsilon}_j \rangle_{\mathrm{mix}} = \tfrac{1}{2}\mu_j \langle \overline{v_{rj}^2} \rangle_{\mathrm{mix}} = \tfrac{1}{2}\mu_j \langle \overline{v^2 + V_j^2} \rangle_{\mathrm{mix}}. \tag{5-2-83}$$

Substituting for $\langle \overline{v^2} \rangle_{\mathrm{mix}}$ from (5-2-80) and remembering that $M_j \overline{V_j^2} = 3kT$, we convert this expression to

$$\langle \bar{\varepsilon}_j \rangle_{\mathrm{mix}} = \frac{3}{2}kT + \frac{1}{2}\left(\frac{m + \langle M \rangle_{\mathrm{mix}}}{m + M_j} \right) M_j \langle v_d \rangle_{\mathrm{mix}}^2, \tag{5-2-84}$$

which corresponds to the formula (5-2-22) for T_{eff} in a single gas. Instead of a single effective temperature representing the total random ion energy (thermal plus random field energy), there is now a different effective temperature for the ions with respect to each species in the gas mixtures.

The problem is now in principle solved: K_{mix} is given by (5-2-76), with $\langle \bar{\varepsilon}_j \rangle_{\mathrm{mix}}$ given by (5-2-84), and $\langle M \rangle_{\mathrm{mix}}$ by (5-2-81) and (5-2-82). Unfortunately, the result for K_{mix} is only an implicit relation because the unknown $\langle v_j \rangle_{\mathrm{mix}}$ occur in the formulas for both K_{mix} and $\langle M \rangle_{\mathrm{mix}}$, and solutions must usually be found by iteration. Moreover, it is not sufficient to know the values of the K_j at just the same value of T and E/N at which we want to find K_{mix}; they must be known over at least a small range of E/N, because the value of $\langle \bar{\varepsilon}_j \rangle_{\mathrm{mix}}$ is not the same as $\bar{\varepsilon}_j$ for the ion in single gas j. The inherent complexity of the mixture problem lies in this fact. It is not the fault of momentum-transfer theory that the final result is complicated, and we can even expect the result from accurate kinetic theory to be more complicated.

Just to get some idea of the deviations from Blanc's law to be expected at high fields, we consider some special cases. At one extreme we can take the Maxwell model, in which collision frequencies are constants. All the ratios $\langle v_j \rangle_{\mathrm{mix}}/v_j$ then become unity, and the results collapse to Blanc's law, which holds at all field strengths.

At another extreme we can take the case of constant cross sections (rigid-sphere assumption), with the field strong enough for the thermal energy to be negligible compared to the random field energies. The relative speed is then the same as the ion speed, and after some algebra the mobility formula simplifies to

$$\frac{1}{K_{\mathrm{mix}}^2} = \sum_j \frac{x_j}{K_j^2}\left(\frac{m + \langle M \rangle_{\mathrm{mix}}}{m + M_j} \right)^{1/2}. \tag{5-2-85}$$

Notice that it is the *squares* of the mobilities that occur here. There is also much cancellation of constant factors in the ω_j used to calculate $\langle M \rangle_{\text{mix}}$ from (5-2-81); leaving out such factors, we can write the ω_j as

$$\omega_j = \frac{x_j}{K_j^2}\left(\frac{m}{m + M_j}\right)^{3/2}. \tag{5-2-86}$$

Some simulated results are shown in Fig. 5-2-5 for K^+ ions in He + Ar mixtures, assuming rigid-sphere collisions. The mobilities in the single gases correspond to the real systems at $E/N = 150$ Td and 300 K (Ellis et al., 1976). Under these conditions, the value of T_{eff} for K^+ in He is 5500 K, and for K^+ in Ar is 3000 K, and these values change appreciably in the mixture as the composition is varied, according to the expression (5-2-84) for $\langle \bar{\varepsilon}_j \rangle_{\text{mix}}$. The figure shows that fairly large deviations from Blanc's law are predicted on the basis of rigid-sphere collisions. The assumption of rigid-sphere collisions always gives deviations in the direction shown.

We expect the behavior of most real systems to be less extreme than that of rigid spheres, and some systems are known to give deviations from Blanc's law in the direction opposite to that predicted for rigid-sphere collisions. The assumption of such collisions is thus likely to give poor predictions for real systems, and we must fall back on the full set of implicit relations, (5-2-76), (5-2-81), (5-2-82), and (5-2-84). These are inconvenient to use, and it would be a great advantage if the deviations from Blanc's law could be expressed entirely in terms of the behavior of the ions in the single gases. Milloy and Robson (1973) have shown how this can be done, at least in a first approximation, by expanding the collision frequencies in the mixture in terms of the collision frequencies in the single gases. This is the same sort of expansion we have already used several times in this chapter, most recently in connection with the generalized Einstein relations. We write

$$\langle v_j \rangle_{\text{mix}} = v_j + \frac{dv_j}{d\bar{\varepsilon}_j}(\langle \bar{\varepsilon}_j \rangle_{\text{mix}} - \bar{\varepsilon}_j) + \cdots, \tag{5-2-87}$$

$$\langle \bar{\varepsilon}_j \rangle_{\text{mix}} - \bar{\varepsilon}_j = \frac{1}{2}\left(\frac{m + \langle M \rangle_{\text{mix}}}{m + M_j}\right)M_j \langle v_d \rangle_{\text{mix}}^2 - \tfrac{1}{2}M_j v_{dj}^2. \tag{5-2-88}$$

The derivative is readily evaluated from the expression for the mobility in a single gas, $\mu_j K_j v_j = e$, yielding

$$\frac{d \ln v_j}{d\bar{\varepsilon}_j} = -\frac{d \ln K_j}{dE}\frac{dE}{dv_{dj}}\frac{dv_{dj}}{d\bar{\varepsilon}_j} = -\frac{1}{M_j K_j^2 E^2}\frac{K_j'}{1 + K_j'}, \tag{5-2-89}$$

where K_j' is the logarithmic field derivative of K_j, as defined in (5-2-72). In the expression for $\langle \bar{\varepsilon}_j \rangle_{\text{mix}} - \bar{\varepsilon}_j$, we need evaluate $\langle M \rangle_{\text{mix}}$ and $\langle v_d \rangle_{\text{mix}}$ only to first order, in which we take $\langle v_j \rangle_{\text{mix}} \approx v_j$, because the whole expression occurs only

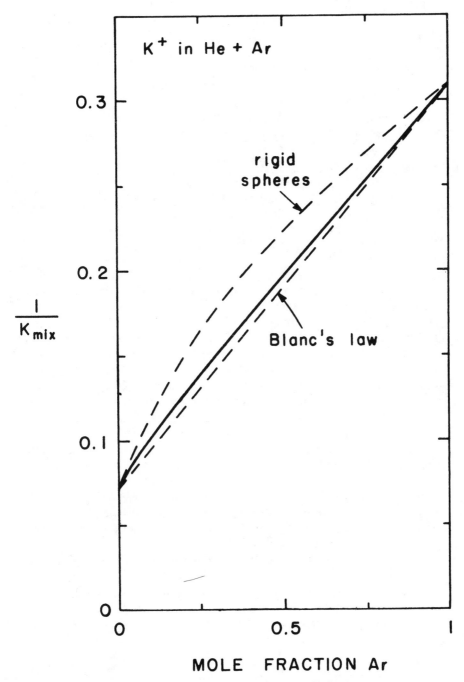

Figure 5-2-5. Predicted deviations from Blanc's law for the mobility of K^+ ions in He + Ar mixtures at $E/N = 150$ Td and $T = 300$ K. The upper dashed curve is based on the assumption of rigid-sphere collisions. The solid curve is that given by (5-2-91).

in a correction term. For $\langle v_d \rangle_{\text{mix}}$, or K_{mix}, we therefore use Blanc's law, and for $\langle M \rangle_{\text{mix}}$ we obtain, after a little algebra,

$$m + \langle M \rangle_{\text{mix}} \approx \left[K_{\text{mix}} \sum_i \frac{x_i}{(m + M_i)K_i} \right]^{-1}. \tag{5-2-90}$$

Substituting these results back into (5-2-76) for K_{mix}, we obtain

$$\frac{1}{K_{\text{mix}}} = \sum_j \frac{x_j}{K_j} \left[1 + \frac{1}{2} \left(\frac{K_j'}{1 + K_j'} \right)(1 - \delta_j) \right], \tag{5-2-91}$$

$$\delta_j^{-1} = (m + M_j)K_j^2 \left(\sum_i \frac{x_i}{K_i} \right) \left[\sum_i \frac{x_i}{(m + M_i)K_i} \right] \approx \frac{\bar{\varepsilon}_j - \frac{3}{2}kT}{\langle \bar{\varepsilon}_j \rangle_{\text{mix}} - \frac{3}{2}kT} \tag{5.2.92}$$

Notice that this result reduces to Blanc's law if $K_j' = 0$, as is the case for the Maxwell model. Notice also that the original need to know the K_j over a range of E/N now appears through the derivatives K_j'.

Probably (5-2-91) is not to be trusted if any of the K_j' is large, which is fortunately rare for real systems. As an example, we show the calculated results for K^+ ions in He + Ar mixtures in Fig. 5-2-5, taking the data for the single gases from the compilation of Ellis et al. (1976). The deviations from Blanc's law are much less than for rigid-sphere collisions, largely because of a small value of K_j' for K^+ in Ar, but are still noticeable. There are very few measurements on mobilities in mixtures at high fields with which to test (5-2-91). Unfortunately, most of the systems that have been studied have happened to show only small deviations from Blanc's law (Biondi and Chanin, 1961; Milloy and Robson, 1973).

Iinuma et al. (1987) have tested (5-2-91) using simulated data obtained from accurate moment solutions of the Boltzmann equation (Section 5-3). They found that (5-2-91) can fail disastrously in many cases, and traced the failure to the expansion (5-2-87) for $\langle v_j \rangle_{\text{mix}}$. This expansion can be very inaccurate when $\langle \bar{\varepsilon}_j \rangle_{\text{mix}}$ is much different from $\bar{\varepsilon}_j$. In such cases the full set of implicit equations for K_{mix}, namely (5-2-76), (5-2-81), (5-2-82), and (5-2-84), should be solved by iteration. Fortunately, the approximate relation (5-2-92) for δ_j, which relates $\langle \bar{\varepsilon}_j \rangle_{\text{mix}}$ to $\bar{\varepsilon}_j$, is fairly accurate and leads to rapid convergence when it is used as the starting point for the iteration. The relation (5-2-92) also allows a quick estimate of the likely accuracy of (5-2-91), which is easy to use, without having to carry out a full iteration: (5-2-91) will fail whenever δ_j as calculated from (5-2-92) is not small compared to unity.

2. Diffusion

A check on the derivation in Section 5-2B shows that the generalized Einstein relations apply to gas mixtures as well as to single gases. The momentum-balance equation (5-2-38) must be replaced by a summation, as in (5-2-75), and the neutral mass M must be replaced by the mean mass $\langle M \rangle_{\text{mix}}$ in the energy equations, but the derivation otherwise proceeds essentially unchanged (Whealton et al., 1974).

We next need the transverse and longitudinal ion temperatures in the mixture. The energy-balance equation (5-2-69) becomes, for a mixture,

$$\sum_j x_j \frac{mM_j(2m + M_j)}{4(m + M_j)^2} (3\langle \overline{v_z^2} \rangle_{\text{mix}} - \langle \overline{v^2} \rangle_{\text{mix}}) \langle v_j \rangle_{\text{mix}} = eE\langle v_d \rangle_{\text{mix}}. \qquad (5\text{-}2\text{-}93)$$

Combining this with the other energy-balance equation, (5-2-78), we find expressions for the ion temperatures that correspond to the expressions (5-2-70) for a single gas, with M replaced by $\langle M \rangle_{\text{mix}}$ and v_d replaced by $\langle v_d \rangle_{\text{mix}}$. Unfortunately, we have no reason to expect these expressions to be any more accurate than the single-gas results (Tables 5-2-5 and 5-2-6). The best we can do at this point is to suggest an obvious extension of Skullerud's formula (5-2-73) to a mixture, and write

$$\langle kT_{T,L} \rangle_{\text{mix}} = kT + \langle \zeta_{T,L} \rangle_{\text{mix}} \langle M \rangle_{\text{mix}} \langle v_d \rangle_{\text{mix}}^2 (1 + \langle \beta_{T,L} \rangle_{\text{mix}} K'_{\text{mix}}), \qquad (5\text{-}2\text{-}94)$$

$$\langle \zeta_T \rangle_{\text{mix}} \approx \frac{(0.85)(m + \langle M \rangle_{\text{mix}})}{4m + 3(0.85)\langle M \rangle_{\text{mix}}}, \qquad (5\text{-}2\text{-}95a)$$

$$\langle \zeta_L \rangle_{\text{mix}} \approx \frac{4m - (0.85)(2m - \langle M \rangle_{\text{mix}})}{4m + 3(0.85)\langle M \rangle_{\text{mix}}}, \qquad (5\text{-}2\text{-}95b)$$

$$\langle \beta_T \rangle_{\text{mix}} \approx 0, \qquad (5\text{-}2\text{-}96a)$$

$$\langle \beta_L \rangle_{\text{mix}} \approx \frac{m}{m + \langle M \rangle_{\text{mix}}} + \frac{m\langle M \rangle_{\text{mix}}}{(m + \langle M \rangle_{\text{mix}})^2}. \qquad (5\text{-}2\text{-}96b)$$

However, we can hope that any errors thereby introduced will tend to cancel, because the final result will be used only as an interpolation formula to predict composition dependence.

We can now find expressions for $\langle D_T \rangle_{\text{mix}}$ and $\langle D_L \rangle_{\text{mix}}$ by using the GER and substituting back into the expression (5-2-76) for K_{mix},

$$\frac{\langle T_{T,L} \rangle_{\text{mix}}}{\langle D_{T,L} \rangle_{\text{mix}}} (1 + \gamma_{T,L} K'_{\text{mix}}) = \sum_j x_j \frac{(T_{T,L})_j}{(D_{T,L})_j} \frac{\langle v_j \rangle_{\text{mix}}}{v_j} (1 + \gamma_{T,L} K'_j), \qquad (5\text{-}2\text{-}97)$$

where $\gamma_T = 0$ and $\gamma_L = 1$. This is the final answer, but it is an implicit formula for $\langle D_{T,L} \rangle_{\text{mix}}$, just like (5-2-76) for K_{mix}. The important new feature is the appearance of the ion temperatures, which are not the same in the mixture and in the pure components at the same value of E/N. As a consequence of this, (5-2-97) does *not* reduce to Blanc's law for the Maxwell model, except at such low fields that all the temperatures are thermal, even though all the ratios $\langle v_j \rangle_{\text{mix}}/v_j$ are unity and all the K'_j are zero. Some numerical examples of the deviations

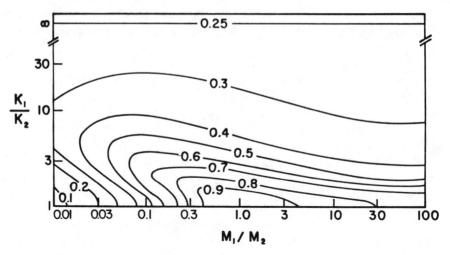

Figure 5-2-6. Ratio of $\langle D_L \rangle_{\text{mix}}$(Blanc) to $\langle D_L \rangle_{\text{mix}}$ from (5-2-98) at high fields for an equimolar Maxwell-model binary gas mixture, shown as contours in a K_1/K_2 vs. M_1/M_2 plane, with $m = M_2$. Notice that the deviations from Blanc's law are always in the same direction. The corresponding ratio for K_{mix} is always unity for the Maxwell model.

from Blanc's law for the Maxwell model are shown in Fig. 5-2-6; very large deviations are possible.

The ratios $\langle v_j \rangle_{\text{mix}}/v_j$ in (5-2-97) can obviously be evaluated as for the mobility, in order to obtain an expression for $\langle D_{T,L} \rangle_{\text{mix}}$ that can be expressed in terms of the behavior of the ion in the single gases. But we must still deal with $\langle T_{T,L} \rangle_{\text{mix}}$ and K'_{mix}. The latter can be dealt with by differentiating (5-2-91). It is consistent with the expansion used for $\langle v_j \rangle_{\text{mix}}/v_j$, however, to drop second derivatives and powers of K'_j in this differentiation, so that we obtain

$$K'_{\text{mix}} \approx \sum_j (x_j K'_j/K_j)/\sum_j (x_j/K_j). \qquad (5\text{-}2\text{-}98)$$

The problem of $\langle T_{T,L} \rangle_{\text{mix}}$ is the problem of $\langle M \rangle_{\text{mix}}$, since all the other parts of the expressions (5-2-94) and (5-2-96) can now be evaluated. This, in turn, is a question of the ω_j in the definition (5-2-81) of $\langle M \rangle_{\text{mix}}$; these involve the ratios $\langle v_j \rangle_{\text{mix}}/v_j$ again, and here we must keep the correction term. Omitting constant factors that cancel in the calculation of $\langle M \rangle_{\text{mix}}$, we obtain

$$\omega_j \approx \frac{x_j}{(m + M_j)K_j}\left[1 + \frac{1}{2}\left(\frac{K'_j}{1 + K'_j}\right)(1 - \delta_j)\right], \qquad (5\text{-}2\text{-}99)$$

where δ_j is given by (5-2-92). The ratio $\langle v_j \rangle_{\text{mix}}/v_j$ is just the factor in brackets in (5-2-99); it should be calculated by iteration unless $\delta_j < 1$, as in the case of K_{mix}.

This completes the recipe for calculating $\langle D_{T,L} \rangle_{\text{mix}}$ from the behavior of the ion in the single component gases. We need to know $(D_{T,L})_j$, K_j, and K'_j; even the $(D_{T,L})_j$ can be dispensed with, since they can be calculated from K_j, K_j' by the GER. Although the expressions to be used are rather lengthy, only algebra is involved. Short of going to full implicit expressions, which are inconvenient to use, the only improvement that accurate kinetic theory has to offer over the present results is somewhat better formulas for the $T_{T,L}$. These are given later (Section 6-2A2b). Unfortunately, no experimental data exist at the present time on ion diffusion coefficients in mixtures at high fields (that is, above normal energies). Simulated data have been calculated theoretically by Iinuma et al. (1987) to test these formulas. The results are similar to those for K_{mix}; that is, (5-2-99) will fail unless $\delta_j < 1$. However, even when a full iteration calculation is carried out, the results for $\langle D_{T,L}\rangle_{\mathrm{mix}}$ are quantitatively less satisfactory than those for K_{mix}. The fault probably lies with the approximate calculation of $\langle T_{T,L}\rangle_{\mathrm{mix}}$.

Let us summarize the results for the mobility and diffusion of ions in gas mixtures, according to simple arguments on momentum and energy balance. The calculation procedure is straightforward, but the results are implicit equations that may need to be solved by iteration. This complexity is inherent in the problem: the mean relative ion energy depends strongly on the composition of the neutral gas. It is possible to obtain explicit expressions by carrying out Taylor expansions, but these are inaccurate unless $\delta_j < 1$. The full formulas work surprisingly well for K_{mix} but less well for $\langle D_{T,L}\rangle_{\mathrm{mix}}$, probably owing to the calculation of the ion temperatures. These conclusions apply only to systems undergoing elastic collisions.

E. Comparison of Momentum-Transfer Theory with Free-Path and Free-Flight Theories

Although momentum-transfer theory is especially useful for discussing ion transport in a semiquantitative way, it is largely neglected in textbooks on kinetic theory. An exception is the book by Present (1958), where both a simplified (Chap. 4) and an advanced (Chap. 8) treatment of diffusion are given. To make contact with the more usual textbook treatments, we give here a brief comparison of momentum-transfer theory with mean-free-path and mean-free-flight theories. Although these latter theories are usually only concerned with the low-field case, where all energies are thermal, we shall be a little more careful and try to indicate where and how the extension to high fields can be made.

Free-path or free-flight arguments go back to Maxwell (1860). Interestingly enough, in the same paper Maxwell also first devised a momentum-transfer theory of diffusion, but was apparently not too happy with it afterwards (because of a mistake, corrected by Clausius). It was independently invented by Stefan (1871, 1872) and later used by Langevin (1905) to give the first quantitatively accurate theory of low-field ion mobility. The theory was then apparently forgotten for many years, until it was invented still another time (apparently independently) by Frankel (1940) and by Present and de Bethune

(1949). One impetus for this revival was the problem of isotope separation in World War II (Present, 1958, p. x).

These theories focus on the mean time between collisions, τ, or the mean distance traveled between collisions, λ. The two are simply related,

$$\lambda = \bar{v}\tau, \tag{5-2-100}$$

where $\bar{v}$ is the mean speed of the particle. The notion of a definite time or distance between collisions means that a collision should be a rather distinct event, so that these theories are somewhat tied to the idea of colliding rigid spheres, or at least to some sort of equivalent rigid-sphere collision. The collision cross section Q is related to τ by the expression

$$\bar{v}_r N Q = \tau^{-1} = \nu, \tag{5-2-101}$$

where $\bar{v}_r$ is the mean relative speed of ions and neutrals, and ν is the collision frequency, as in (5-2-7). The cross section Q is rather ill-defined, except for rigid spheres, but this defect is tolerated because of the simplicity and physical appeal of free-path and free-flight theories.

We first present a free-flight theory of mobility, due largely to Wannier (1952, 1953), which is similar to momentum-transfer theory, and then a more traditional version of the mean-free-path theory of diffusion. As in Section 1-2, we assume that an ion undergoes an acceleration eE/m for a time τ, but recognize that it loses only a fraction of its total momentum and energy on each collision. This fraction must depend on the mass ratio m/M and on the law of force between ions and neutrals. It is straightforward to determine the mass dependence of the momentum loss on collision from the equations of momentum and energy conservation in a single collision. If we then average over all collisions and ignore subtleties about the average of a product and the product of the averages, we obtain (Wannier, 1953, eq. 136)

$$v_d \approx \left(1 + \frac{m}{M}\right)\frac{eE}{m}\tau. \tag{5-2-102}$$

We now substitute for τ from (5-2-101), and approximate $\bar{v}_r$ by the root-mean-square relative speed,

$$\bar{v}_r \approx (\overline{v^2} + \overline{V^2})^{1/2}. \tag{5-2-103}$$

Combining these results, we obtain

$$v_d \approx \left(\frac{1}{m} + \frac{1}{M}\right)\frac{eE}{NQ(\overline{v^2} + \overline{V^2})^{1/2}}. \tag{5-2-104}$$

Except for the indefinite nature of Q, this is exactly the same as the result from momentum-transfer theory, as given by (5-2-7)–(5-2-9). This is not surprising, since both approaches have just used Newton's second law of motion, plus some rough averaging over collisions. The problem of determining $\overline{v^2}$ remains, and this must again be handled by some sort of energy-balance argument, as follows.

The average energy gained by an ion between collisions is $eE\lambda_z$, when λ_z is the component of λ in the field direction. On the average, this energy must be lost by collisions at steady state, so that the energy-balance equation is

$$eE\lambda_z = \tfrac{1}{2}m\overline{v^2} - \tfrac{1}{2}m\overline{v'^2}. \tag{5-2-105}$$

For the component of λ_z we write

$$\lambda_z = \bar{v}_z\tau = v_d\tau, \tag{5-2-106}$$

since only the net velocity v_d down the field contributes to the energy gain. For the collisional loss of energy, we go back to our earlier calculation and from (5-2-17) and (5-2-18) we find

$$\frac{1}{2}m\overline{v^2} - \frac{1}{2}m\overline{v'^2} = \frac{mM}{(m+M)^2}(m\overline{v^2} - M\overline{V^2}). \tag{5-2-107}$$

The final energy-balance equation is thus

$$eEv_d\tau = \frac{mM}{(m+M)^2}(m\overline{v^2} - M\overline{V^2}), \tag{5-2-108}$$

which is the same as the previous result, (5-2-19). Eliminating $eE\tau$ between this equation and (5-2-102) for v_d, we once again obtain the Wannier energy formula,

$$\tfrac{1}{2}m\overline{v^2} = \tfrac{1}{2}M\overline{V^2} + \tfrac{1}{2}(m+M)v_d^2. \tag{5-2-109}$$

In other words, the only real difference between momentum-transfer theory and this version of free-flight theory is the ill-defined nature of Q in the free-flight theory.

The extension to multicomponent mixtures is somewhat simpler in momentum-transfer theory. In the free-flight theory a separate argument is needed to show that it is τ^{-1} that is linear in the mole fractions. Once that is done, the results are identical.

Unlike the foregoing free-flight theory, the usual mean-free-path treatment of diffusion is *not* essentially equivalent to momentum-transfer theory. The treatment of diffusion is, in fact, one of the trickier parts of mean-free-path theory, and it is necessary to be quite careful. To this end, we will not assume a trace concentration of ions at the outset, but will take an arbitrary mixture of two species, 1 and 2, having number densities n_1 and n_2, respectively. This opens up the possibility of having a net flux of particles; to reduce confusion we shall denote the mean *velocities* of the species as $\mathbf{u}_1$ and $\mathbf{u}_2$, and their mean *speeds* by $\bar{v}_1$ and $\bar{v}_2$. We begin by defining the diffusion coefficients. Several reasonable definitions are possible, but we pick the definition in terms of a coordinate system in which there is no net flux of particles. The flux equations for the two species are thus written as

$$\mathbf{J}_1 \equiv n_1 \mathbf{u}_1 = -D_{12} \nabla n_1 + n_1 \mathbf{u}_0, \qquad (5\text{-}2\text{-}110a)$$

$$J_2 \equiv n_2 \mathbf{u}_2 = -D_{21} \nabla n_2 + n_2 \mathbf{u}_0, \qquad (5\text{-}2\text{-}110b)$$

where

$$n\mathbf{u}_0 \equiv n_1 \mathbf{u}_1 + n_2 \mathbf{u}_2 \qquad (5\text{-}2\text{-}111)$$

is the net flux. Any forced flow due to an external field can be included in $\mathbf{u}_0$. These equations tacitly assume a uniform pressure. Strictly speaking, this is an inconsistent assumption, and terms involving the pressure gradient should appear in (5-2-110), but these terms can be made negligibly small by making the apparatus dimensions sufficiently large compared to the mean free path (Mason and Marrero, 1970). With uniform pressure,

$$n_1 + n_2 = n = \text{constant} \quad \text{or} \quad \nabla n_1 = -\nabla n_2, \qquad (5\text{-}2\text{-}112)$$

and addition of (5-2-110a) and (5-2-110b) shows that $D_{12} = D_{21}$. This is one advantage of the coordinate system we have used to define the diffusion coefficients: only one diffusion coefficient occurs, instead of two coefficients related by some more-or-less complicated formula.

The above has nothing to do with mean-free-path theory as such, being only definitions. We now imagine a reference plane perpendicular to the concentration gradients and fixed in the laboratory coordinate system, and calculate the net flux of each species across this plane. The basic idea of mean-free-path theory is that the flux in one direction is equal to the free-molecule flux arising at a distance of about one mean free path from the plane. There is no need to trace back through several mean free paths—a single collision is assumed to give a particle complete amnesia. In an isotropic gas the unidirectional flux of species 1 in any direction is $\frac{1}{4} n_1 \bar{v}_1$. A net flux occurs because the number densities on the two sides of the plane are somewhat different. Expanding the number density as a Taylor series about the reference plane, we can write

$$J_1 = n_1 u_1 = \tfrac{1}{4} \bar{v}_1 (n_1 - \xi \lambda_1 \nabla n_1) - \tfrac{1}{4} \bar{v}_1 (n_1 + \xi \lambda_1 \nabla n_1)$$

$$= -\tfrac{1}{2} \xi \bar{v}_1 \lambda_1 \nabla n_1, \qquad (5\text{-}2\text{-}113a)$$

where ξ is a number of order unity and λ_1 is the mean free path of species 1 in the mixture. Similarly, for species 2,

$$J_2 = n_2 u_2 = -\tfrac{1}{2}\xi \bar{v}_2 \lambda_2 \nabla n_2, \tag{5-2-113b}$$

where λ_2 is the mean free path of species 2 in the mixture. To identify D_{12} we must know u_0, so we add these two equations,

$$nu_0 = J_1 + J_2 = -\tfrac{1}{2}\xi \nabla n_1(\lambda_1 \bar{v}_1 - \lambda_2 \bar{v}_2). \tag{5-2-114}$$

Combining this result with (5-2-113) to form $(u_1 - u_0)$ and $(u_2 - u_0)$, we identify the diffusion coefficient as

$$D_{12} = \tfrac{1}{2}\xi[(n_2/n)\lambda_1 \bar{v}_1 + (n_1/n)\lambda_2 \bar{v}_2]. \tag{5-2-115}$$

(Incidentally, it was difficulty with net drift that led to Maxwell's mistake in his 1860 paper.) This is the final result for conventional mean-free-path theory, and it contains a disastrous flaw—it predicts a strong dependence of D_{12} on the composition of the mixture, whereas the true dependence amounts to only a few percent according to both experiment and accurate theory (Chapman and Cowling, 1970, Sec. 14.31).

Clearly, the flaw must have arisen because we traced the particles back a distance of only one mean free path. We should have traced back through several collisions. The direct calculation of more than one collision history is very cumbersome, but some approximate calculations on the "persistence of velocities" after collision long ago showed the effect to be in the right direction (Jeans, 1925, Secs. 352–362, 416–418). However, it was many years before an accurate calculation was made (Monchick and Mason, 1967), and this showed that the convergence was very poor indeed, although the correct answer was eventually obtained.

The mean-free-path result can nevertheless be fixed up by an indirect argument, as follows (Jeans, 1925, Secs. 419–420; Cowling, 1950, p. 69). The free path λ_1 counts *all* collisions, but it is difficult to see how collisions between two particles of species 1 can interfere with the diffusion of 1 through 2. It is therefore argued that only collisions between 1 and 2 should be included in λ_1. A similar argument applies to λ_2. We therefore drop the cross section Q_{11} from λ_1 and Q_{22} from λ_2, and write

$$\lambda_1 = \bar{v}_1/(n_2 \bar{v}_r Q_{12}), \qquad \lambda_2 = \bar{v}_2/(n_1 \bar{v}_r Q_{12}), \tag{5-2-116}$$

where Q_{12} is the cross section for 1–2 collisions. Thus $n_2 \lambda_1$ and $n_1 \lambda_2$ are constants, and D_{12} does *not* depend on composition,

$$D_{12} = \tfrac{1}{2}\xi(\bar{v}_1^2 + \bar{v}_2^2)/(n\bar{v}_r Q_{12}) \approx \tfrac{1}{2}\xi \bar{v}_r/nQ_{12}, \tag{5-2-117}$$

where we have taken

$$\bar{v}_r^2 \approx \bar{v}_1^2 + \bar{v}_2^2. \qquad (5\text{-}2\text{-}118)$$

This now agrees with the results of momentum-transfer theory, subject to a suitable numerical choice for ξ. At this point we can see why we did not assume a trace concentration of one species at the start. Had we taken $n_1 \to 0$, for instance, we would have lost the last term of (5-2-115) for D_{12}, not realizing that the eventual counting of only unlike collisions would make $\lambda_2 \to \infty$ and $n_1 \lambda_2 = $ constant.

Although the foregoing argument gives a satisfactory result, it has been criticized as ad hoc (Chapman, 1928; Furry, 1948; Present, 1958, Sec. 4-1). There is some justice to this criticism, and a proper calculation including all orders of the persistence of velocities shows that it is not necessary to consider only the unlike collisions (Monchick and Mason, 1967). However, an unfriendly critic can claim that the momentum-transfer theory of diffusion involves a similar ad hoc assumption, only better concealed. That is, the assumption that the gradient of partial pressure is the driving force for diffusion singles out momentum transfer as the controlling mechanism, and collisions between like species do not affect momentum transfer. In other words, the use of momentum transfer is simply a device for counting only unlike collisions. The only safe way out of this sort of controversy is recourse to an accurate theory, so that we can be wise after the event. Fortunately, the criticism does not apply to the calculation of ion mobility, which clearly does involve momentum transfer in ion-neutral collisions.

Even if we accept the argument of counting only unlike collisions in diffusion, momentum-transfer theory still has one decided advantage over mean-free-path theory—it is much easier to extend to multicomponent mixtures of arbitrary composition. The extension of (5-2-110) to multicomponent mixtures is not obvious, and the multicomponent diffusion coefficients that occur turn out to have a complicated dependence on mixture composition. It is simpler to write a momentum-balance equation for each species, equating the partial pressure gradient (the driving force) to the momentum losses by collisions with each of the other species (Williams, 1958; Mason and Evans, 1969; Mason and Marrero, 1970),

$$-\nabla p_1 = n_1 kT[(n_2/nD_{12})(\mathbf{u}_1 - \mathbf{u}_2) + (n_3/nD_{13})(\mathbf{u}_1 - \mathbf{u}_3) + \cdots], \qquad (5\text{-}2\text{-}119)$$

with a similar equation for each of the other species. This result is very accurate, and the D_{ij} that occur are constant and equal to the corresponding binary diffusion coefficients, within a few percent. We shall have no need for this general result in ion transport, however, because the ions are present in trace concentration and the neutral species are distributed uniformly. Moreover, it applies only to the case of zero field strength.

We now consider the extension of the mean-free-path theory to ions in high fields, accepting the counting of only unlike collisions. We pass to the limit of trace ion concentration and revert to our former notation: $n_2 \rightarrow N$, $n_1 \rightarrow n \ll N$, $v_2 \rightarrow V$, $v_1 \rightarrow v$. Rewriting (5-2-117) for D_{12} in terms of τ, we find that we must take $\xi = 2/3$ for the result to obey the low-field Nernst–Townsend–Einstein relation with v_d given by (5-2-102). The result is

$$D = \tfrac{1}{3}\bar{v}_r^2\tau, \qquad (5\text{-}2\text{-}120)$$

which applies only to low fields because we have assumed in the derivation that the system is isotropic. However, we can plausibly extend the result to any field strength by describing the diffusion as an isotropic process superposed on the directed drift velocity v_d. This extension is suggested by the flux equations (5-2-110), with v_d included as part of u_0. We therefore write

$$D_T = (\bar{v}_r^2)_T\tau = (2\bar{\varepsilon}_T)(v_d/eE) = (kT_T/e)K, \qquad (5\text{-}2\text{-}121a)$$

$$D_L = (\bar{v}_r^2)_L\tau = (2\bar{\varepsilon}_L)(v_d/eE) = (kT_L/e)K, \qquad (5\text{-}2\text{-}121b)$$

where we have substituted $\tau = \mu v_d/eE$ from (5-2-102). The factor 1/3 has disappeared because we are dealing with components only. We now have the problem of calculating $\bar{\varepsilon}_T = \tfrac{1}{2}kT_T$ and $\bar{\varepsilon}_L = \tfrac{1}{2}kT_L$, but this is obviously to be handled by an energy-balance argument. The calculation, in fact, will proceed exactly like that in Section 5-2C for the partitioning of ion energy, and there is no need to repeat it. What we see here is that we have lost the correction factor $[1 + (d \ln K/d \ln E)]$ from the expression for D_L. This is of no consequence if τ is a constant, but can otherwise lead to discrepancies of up to a factor of 2. It is hard to see how to restore this factor without some rather devious argumentation.

To sum up, free-flight and free-path theories of mobility and diffusion can be put in close correspondence to one another and to momentum-transfer theory, provided that one is willing to accept the free-path argument for counting only unlike collisions in diffusion. Even so, momentum-transfer theory has three distinct advantages. First, it supplies a well-defined cross section, namely a diffusion or momentum-transfer cross section. Second, it is easier to extend to multicomponent mixtures, both for mobility and for diffusion. Third, it relates D_L to the differential mobility, whereas free-path theory relates D_L to the mobility itself. As we shall see later, it is also more obviously related to accurate versions of kinetic theory based on the Boltzmann equation.

F. Estimation of Cross Sections

Momentum-transfer theory correctly supplies the relevant cross section for mobility and diffusion, namely $Q_D(\varepsilon)$ of (5-2-5), a result which also holds in the accurate theory, as mentioned in connection with (5-2-24). Before going on to develop the latter by attacking the Boltzmann equation, we give some useful rough-and-ready methods for estimating the magnitude and energy dependence of Q_D. The expression (5-2-5) for Q_D is restricted to elastic collisions, but is valid even if quantum mechanics is required to describe the collisions. However, it usually happens that classical mechanics suffices, and it is then convenient to replace the differential cross section $\sigma(\theta)$ by the impact parameter b, and rewrite (5-2-5) as

$$Q_D = 2\pi \int_0^\infty (1 - \cos \theta)b \, db, \qquad (5\text{-}2\text{-}122)$$

where θ is given by the classical-mechanical expression,

$$\theta = \pi - 2b \int_{r_0}^\infty \left[1 - \frac{b^2}{r^2} - \frac{V(r)}{\varepsilon}\right]^{-1/2} \frac{dr}{r^2}. \qquad (5\text{-}2\text{-}123)$$

In this equation r_0 is the distance of closest approach (the outermost zero of the bracketed expression), $V(r)$ is the ion-neutral potential energy, and $\varepsilon = \frac{1}{2}\mu v_r^2$ as before. Although classical theory always fails to describe $\sigma(\theta)$ at small angles, these classical formulas are adequate because the angular weighting factor $(1 - \cos \theta)$ in Q_D suppresses the small-angle contribution.

The accurate evaluation of Q_D for any but trivially simple forms of $V(r)$ requires extensive numerical integration, as discussed in Chapter 7, but often fairly good estimates can be obtained from simple physical arguments, and this will suffice for the present purposes. One accurate result that can be obtained by dimensional arguments alone goes back at least to Lord Rayleigh (1900). For a potential of the form

$$V(r) = \pm \frac{C_n}{r^n}, \qquad (5\text{-}2\text{-}124)$$

where C_n and n are positive constants, the energy dependence of *any* classical cross section is

$$Q \propto (C_n/\varepsilon)^{2/n}. \qquad (5\text{-}2\text{-}125)$$

Of course, nothing is known about the constant of proportionality from this argument. One quick way to verify this result is to note from (5-2-123) that θ is a function of the single dimensionless variable $\varepsilon b^n/C_n$; then (5-2-125) follows because any classical Q is proportional to $b \, db$ in a three-dimensional world.

When the scattering is dominated by an attractive potential of the form $-C_n/r^n$, the magnitude of Q_D can be estimated by an argument given by Wannier (1953) and elaborated somewhat by Dalgarno et al. (1958). At a given energy there is an impact parameter b_0 at which the particles take up an unstable orbiting motion (Hirschfelder et al., 1964, Sec. 8.4). For $b > b_0$ the particles are kept apart by the centrifugal repulsion (i.e., by their angular momentum). The value of b_0 occurs when the maximum in the effective potential is equal to the total collision energy ε; by the effective potential is meant the true potential plus the centrifugal repulsion,

$$V_{\text{eff}} = -\frac{C_n}{r^n} + \frac{\varepsilon b^2}{r^2}. \tag{5-2-126}$$

The diffusion cross section is taken to be equal to the cross section for orbiting or "capture"

$$Q_D(\text{capture}) = \pi b_0^2 = \pi \frac{n}{n-2}\left(\frac{n-2}{2}\right)^{2/n}\left(\frac{C_n}{\varepsilon}\right)^{2/n}. \tag{5-2-127}$$

A somewhat better numerical answer can be obtained from a more careful derivation, as follows. We rewrite (5-2-122) as

$$Q_D = 4\pi \int_0^\infty \sin^2(\theta/2)b\,db. \tag{5-2-128}$$

If orbiting occurs, θ varies rapidly with b out to large b and then slowly tails off. We therefore replace the oscillating function $\sin^2(\theta/2)$ with its mean value of $\frac{1}{2}$ out to some cutoff value, b_c, and with zero for larger b. This can be called a "random-angle" approximation. The precise criterion for the choice of b_c is not too critical; Hahn and Mason (1971) have chosen b_c as that value corresponding to a cutoff angle θ_c such that

$$|\theta_c/2| = 1/\pi. \tag{5-2-129}$$

This gives an implicit equation for b_c via (5-2-123) for θ; for an explicit equation we expand the square root in (5-2-123) and obtain a small-angle approximation (Kennard, 1938, Sec. 70),

$$\theta = \frac{b}{\varepsilon}\int_b^\infty \frac{V(b) - V(r)}{(r^2 - b^2)^{3/2}}r\,dr + \cdots, \tag{5-2-130}$$

with $r_0 = b + \cdots$. This integral can be evaluated for inverse-power potentials,

$$\theta = \pm A_n(C_n/\varepsilon b^n), \tag{5-2-131}$$

where

$$A_n = \pi^{1/2}\Gamma[(n + 1)/2]/\Gamma(n/2). \tag{5-1-132}$$

From this we can solve for b_c and obtain

$$Q_D(\text{random}) = \pi b_c^2 = \pi(\pi A_n/2)^{2/n}(C_n/\varepsilon)^{2/n}. \tag{5-2-133}$$

This formula gives a reasonable result even for repulsive potentials, for which $\sin^2(\theta/2)$ does not oscillate about a mean value of $\frac{1}{2}$, but instead decreases smoothly from a value of 1 at $b = 0$ toward zero for large b.

A comparison of the capture and random-angle approximations with accurate values of Q_D calculated by numerical integration (Higgins and Smith, 1968) is shown in Table 5-2-10. The agreement is not earthshaking, but it is good enough for many purposes, especially when the alternative is double numerical integration. The extension of the random-angle approximation to nonmonotonic potentials is not difficult (Hahn and Mason, 1971).

Many of the values of n in Table 5-2-10 have physical significance: $n = 2$ is the ion-dipole interaction, $n = 3$ is the ion-quadrupole, $n = 4$ is the attractive ion-induced dipole, $n = 6$ is the attractive ion-induced quadrupole and dispersion energy, and so on.

The common feature of both the capture and random-angle approximations is that Q_D is proportional to the square of some characteristic distance that is a function of energy. This observation suggests that we might obtain an easy generalization, suitable for arbitrary potentials, by picking a more general characteristic distance that could be taken as proportional to an effective collision diameter. A simple choice for a repulsive potential is the distance of closest approach in a head-on collision, or the distance at which the potential energy is equal to the initial relative kinetic energy. This obviously underestimates the cross section, since this is the smallest possible separation at a given energy. We therefore pick a larger distance at which the potential energy is some definite fraction of ε; this choice also furnishes a recipe for attractive

Table 5-2-10   Accuracy of Approximate Momentum-Transfer (Diffusion) Cross Sections for the Potentials $V(r) = \pm C_n/r^n$

	Capture (5-2-127)	Random angle (5-2-133)	
n	Attraction	Attraction	Repulsion
2	0.31	0.76	1.55
3	0.71	0.80	1.66
4	0.91	0.87	1.61
6	1.20	1.06	1.50
8	1.35	1.17	1.42
10	1.41	1.22	1.36
25	1.30	1.18	1.18
50	1.18	1.11	1.10
∞	1.00	1.00	1.00

potentials. The result is thus

$$|V(d)| = \gamma\varepsilon, \tag{5-2-134}$$

$$Q_D \approx \pi d^2, \tag{5-2-135}$$

with $\gamma < 1$. For inverse-power potentials we then find

$$\pi d^2 = \pi(C_n/\gamma\varepsilon)^{2/n} = Q_D(\text{approx.}). \tag{5-2-136}$$

A little empirical fiddling indicates that good choices are $\gamma = \frac{2}{3}$ for repulsive potentials and $\gamma = \frac{1}{3}$ for attractive potentials. A comparison of this approximation with the accurate values of Higgins and Smith (1968) is given in Table 5-2-11. The agreement is remarkably good for such a simple approximation.

Short-range repulsion energy can often be represented by an exponential potential,

$$V(r) = V_0 \exp(-r/\rho), \tag{5-2-137}$$

where V_0 and ρ are positive constants. For this potential the effective collision diameter approximation yields

$$\pi d^2 = \pi\rho^2[\ln(V_0/\gamma\varepsilon)]^2 = Q_D(\text{approx.}). \tag{5-2-138}$$

The ratio $\pi d^2/Q_D$ for this potential is a function of energy, unlike that for the inverse-power potential. Some values based on accurate Q_D obtained by

Table 5-2-11   Accuracy of the Effective Collision Diameter Approximation to the Momentum-Transfer (Diffusion) Cross Section for the Potentials $V(r) = \pm C_n/r^n$

	$Q_D(\text{approx.})/Q_D$	
n	Repulsion	Attraction
2	0.943	0.929
3	1.012	0.780
4	1.026	0.784
6	1.030	0.914
8	1.028	1.012
10	1.025	1.061
25	1.014	1.075
50	1.008	1.042
∞	1.000	1.000

Table 5-2-12   Accuracy of the Effective Collision Diameter Approximation to the Momentum-Transfer (Diffusion) Cross Section for the Potential $V(r) = V_0 \exp(-r/\rho)$

ε/V_0	$Q_D(\text{approx.})/Q_D$
0	1.000
10^{-4}	1.021
10^{-3}	1.023
10^{-2}	1.022
10^{-1}	1.006

numerical integration (Monchick, 1959) are given in Table 5-2-12. The agreement is again remarkably good.

Thus reasonable estimates of both the energy dependence and the magnitude of the cross section can be obtained without undue difficulty even for complicated potentials. If the potential is so complicated that (5-2-134) cannot be solved explicitly for d, results can be obtained by treating d parametically; that is, an arbitrary value of d is chosen, the value of Q_D calculated from (5-2-135), and the corresponding value of ε from (5-2-134). Nonmonotonic potentials can be handled by the same method used by Hahn and Mason (1971) for the random-angle approximation.

The effective collision diameter approximation has an importance beyond quick estimates of cross sections. It is the basis for an iterative numerical scheme for the direct inversion of transport coefficient data to find the potential, without the need to assume any particular analytical form for the potential. This scheme is discussed in Chapter 7. The approximation itself seems to have been devised by Hirschfelder and Eliason (1957) as a basis for the rapid estimation of transport properties of gases. A method for the direct inversion of gaseous transport coefficient data to find the potential was given by Gough et al. (1972), as an extension of an earlier method of Dymond (1968). The extension to the inversion of ion mobility data is straightforward (Viehland et al., 1976) and is the basis for much of our present quantitative information on ion-neutral potentials.

Finally, a few words must be given about the estimation of Q_D when an ion moves in its parent gas and resonant charge exchange is possible. Except at very low energies, the mobility of an ion is then dominated by the charge exchange process. The reasons for this are discussed in some detail later; all we need to mention at this point is that Q_D is related to the charge exchange cross section Q_{ex}, and that a sufficiently accurate relation for the present purposes is (Dalgarno, 1958)

$$Q_D \approx 2Q_{ex}. \tag{5-2-139}$$

The value of this relation is that Q_{ex} is often known experimentally from measurements with ion beams. Only a few measurements may be needed, for Q_{ex} can often be accurately represented over a large energy range by an expression of the form

$$Q_{ex}^{1/2} = a_1 - a_2 \ln \varepsilon. \tag{5-2-140}$$

This gives the same energy dependence as (5-2-138) for an exponential potential. The similarity is not accidental—the interaction that drives charge exchange is often exponential in form.

5-3. ACCURATE KINETIC THEORY: SOLUTION OF THE BOLTZMANN EQUATION BY MOMENT METHODS

A. Background

The procedure for constructing an accurate kinetic theory is, in principle, straightforward. It is only necessary to find the distribution function of the ions, $f(\mathbf{v}, \mathbf{r}, t)$, and then any quantity of interest can be found by integration. For instance, the ion density $n(\mathbf{r}, t)$ is

$$n(\mathbf{r}, t) = \int f(\mathbf{v}, \mathbf{r}, t) \, d\mathbf{v}. \tag{5-3-1}$$

This is really only the normalization condition on f; more interesting quantities are, for example, the average ion velocity and energy,

$$\langle \mathbf{v} \rangle = n^{-1} \int \mathbf{v} f \, d\mathbf{v}, \tag{5-3-2}$$

$$\tfrac{1}{2} m \langle v^2 \rangle = n^{-1} \int (\tfrac{1}{2} m v^2) f \, d\mathbf{v}. \tag{5-3-3}$$

In general, for any function of the ion velocity, $\psi(v)$, we define the average quantity as the moment with respect to f,

$$\langle \psi(\mathbf{r}, t) \rangle \equiv n^{-1} \int \psi(\mathbf{v}) f(\mathbf{v}, \mathbf{r}, t) \, d\mathbf{v}. \tag{5-3-4}$$

In this section we use the symbol $\langle \cdots \rangle$ for an average, instead of a bar as in the preceding section. Not only is our definition of an average now more precise, but we will be taking averages of longer expressions.

The equation to be solved in order to find $f(\mathbf{v}, \mathbf{r}, t)$ has been known for over 100 years (Maxwell, 1867; Boltzmann, 1872). It is basically just an equation of continuity for f in the six-dimensional space of $\mathbf{v}$ and $\mathbf{r}$:

$$\frac{\partial f}{\partial t} + \mathbf{v} \cdot \nabla_r f + \mathbf{a} \cdot \nabla_v f \equiv \frac{Df}{Dt}$$

$$= \sum_j \iint [f(\mathbf{v}', \mathbf{r}, t) F_j(\mathbf{V}'_j, \mathbf{r}, t) - f(\mathbf{v}, \mathbf{r}, t) F_j(\mathbf{V}_j, \mathbf{r}, t)]$$

$$\times v_{rj} \sigma_j(\theta, v_{rj}) \, d\Omega_j \, d\mathbf{V}_j, \tag{5-3-5}$$

where

$$\mathbf{a} = e\mathbf{E}/m, \tag{5-3-6}$$

$$v_{rj} = |\mathbf{v} - \mathbf{V}_j|, \tag{5-3-7}$$

$$d\Omega = \sin\theta \, d\theta \, d\phi, \tag{5-3-8}$$

and other quantities are as defined previously. This is the famous Boltzmann equation. The left-hand side describes how f changes by virtue of the independent (collisionless) motion of the ions, and the right-hand side describes how f changes because of binary collisions with the neutral particles (there may be a mixture of neutral species, each species denoted by j). The term $\partial f/\partial t$ simply states that f changes with time at fixed values of $\mathbf{v}$ and $\mathbf{r}$. The term $\mathbf{v} \cdot \nabla_r f$ describes that part of the change due to the free motion of the ions; some ions leave the vicinity of $\mathbf{r}$ and others move in. The term $\mathbf{a} \cdot \nabla_v f$ describes that part of the change due to an external force altering $\mathbf{v}$. The three terms together state that f changes with time as one moves with the stream of phase points. The right-hand side of the Boltzmann equation attributes all of this change to binary collisions. The loss term fF_j describes the loss of ions having velocities in the vicinity of $\mathbf{v}$ by collisions with neutrals of velocity $\mathbf{V}_j$. The gain term $f'F'_j$ describes the gain of ions into the velocity region around $\mathbf{v}$ by collisions of ions of velocity $\mathbf{v}'$ with neutrals of velocity $\mathbf{V}'_j$. Collisions with initial velocities $\mathbf{v}$, $\mathbf{V}_j$ lead to final velocities $\mathbf{v}'$, $\mathbf{V}'_j$, and vice versa (inverse collisions).

There is a substantial literature concerning the range of validity of the Boltzmann equation, and its derivation from more fundamental principles. We will just take for granted that the Boltzmann equation gives an adequate description of the systems we are interested in, and concentrate our efforts on trying to find suitable solutions. As a rough rule, the Boltzmann equation is satisfactory as long as the mean free path is large compared to the collision diameter (binary collisions), but is small compared to any important apparatus dimensions.

As we have written the Boltzmann equation (5-3-5), it still contains some important limitations, of which three are particularly pertinent. First, there is no term for ion-ion collisions (no term involving a product of two f's), because we assume that the ions are present in only trace quantities. Second, conservation of ions means that no ion-neutral chemical reactions are considered. Third, the fact that only positions and velocities appear as variables means that no internal degrees of freedom of ions or neutrals are considered, and hence that all collisions are elastic. We will later drop the last two restrictions, but never the first one. The treatment of ion-ion collisions by the Boltzmann equation involves fundamental difficulties because of the long-range behavior of the Coulomb potential, and is more appropriately considered as belonging to plasma kinetic theory. It should be noted that the neglect of ion-ion collisions means that the motions of different ions in a drift tube are completely independent, so that there is no loss of generality in considering only one ionic species.

A final simplification occurs because the ions are present in trace amounts and because electric fields have little, if any, effect on the motion of neutral species. The velocity distribution function of the neutrals can be taken as an equilibrium Maxwellian distribution at the gas temperature T:

$$F_j = N_j(M_j/2\pi kT)^{3/2} \exp(-M_j V_j^2/2kT), \tag{5-3-9}$$

where we assume (without loss of generality) that the average velocity of the whole system is zero. This makes the Boltzmann equation linear in the unknown ion distribution f, even at high field strengths.

We now turn to the task of trying to find suitable solutions of the Boltzmann equation. A few sobering historical remarks are in order at this point. Maxwell and Boltzmann, although they gave the fundamental formulation which has been the starting point for almost all subsequent workers, never really succeeded in finding general ways to calculate transport coefficients. Maxwell succeeded only for a very special model (the Maxwell model, mentioned in Sec. 5-2A), and Boltzmann failed with the rigid-sphere model (although, in retrospect, he almost succeeded). The solutions came in the second decade of the twentieth century from two independent workers, Sydney Chapman in England and David Enskog in Sweden. (Both were graduate students at the time!) Although they adopted rather different approaches, their results were mathematically equivalent. A historical summary of these developments appears in the first (1939) and second (1952) editions of the well-known treatise by Chapman and Cowling (1970), and at least two interesting personal accounts have been published by Chapman (1967a, b). The Chapman–Enskog theory is now regarded as one of the best-developed theories in statistical physics. Unfortunately, it is of little help for the present problem. The reason is that it is a perturbation theory based on the equilibrium Maxwellian distribution, and is therefore accurate only at low field strengths. At high fields, the ion distribution is far from equilibrium, the perturbation is large instead of small, and the Chapman–Enskog method fails.

Other cases for which high-field solutions are known are also based on some sort of perturbation theory that cannot be extended in any obvious way to the case of arbitrary ion-neutral mass ratios and interactions. For example, electron theory is based on the smallness of the ratio m/M, so that the fractional energy transfer in an ion-neutral collision is small. The electron distribution function will therefore be nearly spherical in velocity space (although it may be far from Maxwellian), since collisions may change the direction of the electron velocity by a large amount, but can affect its magnitude only slightly. The case of $m \ll M$ is often called the Lorentz model, since it was first studied by H. A. Lorentz (1905) as a model for the motion of electrons in metals (Chapman and Cowling, 1970, Sec. 10.5). The opposite extreme of $m \gg M$ can also be solved by a perturbation theory that is based on the smallness of the deflection suffered by a heavy ion in a collision with a light neutral. This case is now usually called the Rayleigh model, in honor of the discussion by Lord Rayleigh (1891) of the motion of a heavy test particle in a gaseous sea of light particles (see also Chapman and Cowling, 1970, Sec. 10.53).

When perturbation theory cannot be used, a standard procedure is to try that most fundamental of all methods—guessing the answer and then refining it. This procedure is usually put into practice with the help of some variational principle or moment method. Unfortunately, no variational principle (or theorem on bounds) is known for the Boltzmann equation except for vanishingly weak electric fields, so we are left with moment methods.

B. Moment Equations

The philosophy here is to give up the attempt to find the complete distribution function and recognize that we would usually be happy just to know the first few moments (or averages) of the distribution—for instance, those corresponding to v_d, T_T, T_L, D_T, and D_L. Indeed, even if we happened to find f in some way, we would probably only use it to calculate a small number of moments and would, in effect, discard almost all of the information contained in f. We therefore reduce the scale of our ambitions and seek a method whereby we try to find the first few moments themselves directly, without the intervention of f. Of course, if we could find *all* the moments, this would be mathematically equivalent to finding f itself, but almost everything of physical interest is contained in just the first few moments. We also do not want to know about all the possible kinds of behavior allowed by the Boltzmann equation. Part of the difficulty in solving the Boltzmann equation arises just because it is so general, and encompasses such a variety of time-dependent phenomena. In drift tubes we are concerned mostly with steady-state phenomena, or at worst with quasi-steady-state situations in which the time variation is slow enough to be regarded as a sort of perturbation. In short, the idea here is that we may have some chance of success in finding useful solutions of the Boltzmann equation if we are not excessively ambitious.

To form moment equations, we multiply the Boltzmann equation (5-3-5) from the left by any function $\psi(\mathbf{v})$ of the ion velocity and integrate over ion velocities,

$$\frac{\partial}{\partial t} \int \psi f \, d\mathbf{v} + \int \psi \mathbf{v} \cdot \mathbf{V}_r f \, d\mathbf{v} + \frac{e}{m} \int \psi \mathbf{E} \cdot \mathbf{V}_v f \, d\mathbf{v}$$

$$= \sum \iiint \psi(f'F_j' - fF_j) v_{rj} \sigma_j \, d\mathbf{\Omega}_j \, d\mathbf{V}_j \, d\mathbf{v}. \qquad (5\text{-}3\text{-}10)$$

We now integrate by parts and use the inverse collision property of the Boltzmann collision operator,

$$v_{rj}\sigma_j(v_{rj})\,d\Omega_j\,d\mathbf{V}_j\,d\mathbf{v} = v'_{rj}\sigma_j(v'_{rj})\,d\Omega'_j\,d\mathbf{V}'_j\,d\mathbf{v}'. \qquad (5\text{-}3\text{-}11)$$

The result is

$$\frac{\partial}{\partial t}(n\langle\psi\rangle) + \mathbf{V}_r\cdot(n\langle\mathbf{v}\psi\rangle) - \frac{ne}{m}\mathbf{E}\cdot\langle\mathbf{V}_v\psi\rangle = -n\sum_j N_j\langle J_j\psi\rangle, \qquad (5\text{-}3\text{-}12)$$

where the linear collision operator J_j is defined by

$$J_j\psi \equiv N_j^{-1}\iint F_j[\psi(\mathbf{v}) - \psi(\mathbf{v}')]v_{rj}\sigma_j\,d\Omega_j\,d\mathbf{V}_j. \qquad (5\text{-}3\text{-}13)$$

The linearity of J_j is a result of the fact that F_j is an equilibrium Maxwellian distribution, in turn a result of the assumption of trace ion concentration. These moment equations are essentially the "equations of change" that were the basis for the treatment of Maxwell (1867), who of course did not start from the Boltzmann equation itself, since Boltzmann did not present it until five years later.

The moment equations still encompass more varieties of behavior than we care about, so we seek reasonable physical assumptions that will allow us to reduce (5-3-12) to something more tractable. The most obvious feature is that we are interested only in nearly steady states, but what do we mean by "nearly"? If we were concerned only with true steady states, we could neglect all time derivatives; perhaps we could neglect some of the time derivatives and keep only a few. Here our knowledge of the physics of ion-neutral collisions can help. We know that ion momentum and energy can be rapidly dissipated locally by collisions, but that ion mass (or number density) cannot. Ion mass can be dissipated only by the ions moving away (unless chemically reactive collisions are allowed), which is a much slower process. Presumably, higher moments of the ion velocity behave similarly. We will therefore neglect time derivatives of all ion velocity moments except $\partial n/\partial t$. It is possible to take a less extreme position and develop a description that includes the time variation of momentum and energy, so that transients and various relaxation effects can be studied, but we postpone this topic until Chapter 8.

We can carry this idea a little further and make an assumption regarding the spatial variation of ion properties. We wish to assume that the gradient of ion density, $\mathbf{V}_r n$, is small in order to be sure that Fick's law of diffusion will be applicable. Since $\mathbf{V}_r n$ can decay to zero only by the motion of ions, whereas gradients of momentum, energy, and higher moments can decay locally by collisions, it is plausible to neglect all spatial gradients of ion properties except $\mathbf{V}_r n$, which is itself assumed to be small. Of course, this assumption implies that we do not impose any gradients externally, by fixed temperature inequalities or large shearing flows, for example, but this is in keeping with our ideas of drift-tube operation. One caution should be kept in mind, however. When $m \ll M$, ion energy can decay only very slowly by elastic collisions, so its time and space variation may not always be small compared to that of number density (Robson, 1976b, 1981). In other words, not all of the present results should be applied uncritically to electrons. Notice that we are neglecting any explicit effects of boundaries, so that we are considering only the so-called "hydrodynamic" regime.

In short, we assume that ion properties other than number density are independent of position in the apparatus, that $\mathbf{V}_r n$ is small, and that all time derivatives are negligible except $\partial n/\partial t$. Since number density is thus to be treated differently from the other moments, it is convenient to incorporate this fact into the moment equation (5-3-12). Setting $\psi = 1$, we obtain the equation of continuity,

$$(\partial n/\partial t) + \mathbf{V}_r \cdot (n\langle \mathbf{v} \rangle) = 0. \tag{5-3-14}$$

Using this result, and dropping various derivatives as discussed above, we reduce the moment equations to

$$(e/m)\mathbf{E} \cdot \langle \mathbf{V}_v \psi \rangle - [\langle \mathbf{v}\psi \rangle - \langle \mathbf{v} \rangle\langle \psi \rangle] \cdot \mathbf{V}_r \ln n = \sum_j N_j \langle J_j \psi \rangle. \tag{5-3-15}$$

This is a whole set of equations, since ψ can be any function of $\mathbf{v}$. A standard procedure for solving such a set of equations is to choose the ψ to be some complete set of functions, usually orthogonal for convenience and ease of manipulation, expand the right-hand side in terms of these functions, and thereby convert the integral equations into an infinite set of algebraic equations for the moments of the ψ's. The hope is that these algebraic equations can be solved for the first few moments by some truncation or iteration procedure.

Adding a formal subscript to the ψ's to denote different members of the complete set, we write the expansion,

$$J_j\psi_p = \sum_q a_{pq}^{(j)}\psi_q. \tag{5-3-16}$$

We suppose that the ψ's are orthogonal with respect to some weight function $g(\mathbf{v})$,

$$\int g\psi_p^\dagger\psi_q \, d\mathbf{v} \equiv (\psi_p, \psi_q) = (\psi_p, \psi_p)\delta_{pq}, \tag{5-3-17}$$

where † indicates complex conjugation and (ψ_p, ψ_p) is the normalization factor. This orthogonality allows us to find the coefficients $a_{pq}^{(j)}$ by multiplying (5-3-16) by $\psi_q^\dagger$ and integrating,

$$a_{pq}^{(j)} = \frac{\int g\psi_q^\dagger(J_j\psi_p) \, d\mathbf{v}}{\int g\psi_q^\dagger\psi_q \, d\mathbf{v}} = \frac{(\psi_q, J_j\psi_p)}{(\psi_q, \psi_q)}. \tag{5-3-18}$$

With the expansion (5-3-16), the moment equations become

$$(e/m)\mathbf{E} \cdot \langle \mathbf{V}_v \psi_p \rangle - [\langle \mathbf{v}\psi_p \rangle - \langle \mathbf{v} \rangle \langle \psi_p \rangle] \cdot \mathbf{V}_r \ln n$$

$$= \sum_j N_j \sum_q a_{pq}^{(j)} \langle \psi_q \rangle = N \sum_q b_{pq} \langle \psi_q \rangle, \qquad (5\text{-}3\text{-}19a)$$

where

$$b_{pq} \equiv \sum_j x_j a_{pq}^{(j)}. \qquad (5\text{-}3\text{-}19b)$$

After a little straightforward manipulation of $\langle \mathbf{V}_v \psi_p \rangle$ and $\langle \mathbf{v}\psi_p \rangle$, this becomes basically a set of algebraic equations for the moments $\langle \psi_p \rangle$, with $\mathbf{V}_r \ln n$ regarded as a small perturbation. The procedure says nothing about how the basis functions ψ_p are to be chosen. This is obviously a crucial matter: a good choice presumably gives rapid convergence to accurate results, whereas a poor choice turns (5-3-19) into an intractable morass. The choice is really largely intuitive; it is usually based on a known solution to some simple related problem ("Look for the hydrogen atom of the problem!"), or on physical arguments of some sort, or even just on a lucky guess. It is therefore worthwhile to examine the question from several points of view.

The foregoing description implies that we first choose the basis functions, whereby the weight factor $g(\mathbf{v})$ is automatically determined. However, we can imagine that we first choose the $g(\mathbf{v})$, and then construct the basis functions by repeated use of the orthogonality relation (5-3-17). We now turn to a description of the use of these two points of view in choosing the basis functions.

C. Choice of Basis Functions

Two principal methods of selecting the basis functions for a moment solution have been used: an eigenfunction procedure and a distribution-function procedure.

In the eigenfunction procedure, the ψ_p used are eigenfunctions of an operator $J_j^{(0)}$ chosen such that the eigenvalue equation,

$$J_j^{(0)} \psi_p = \lambda_p^{(j)} \psi_p, \qquad (5\text{-}3\text{-}20)$$

can be solved, and such that $J_j^{(0)}$ mimics J_j in some sense. The Maxwell model of constant ion-neutral collision frequency is the classic example used in kinetic theory. It is, in fact, the *only* example to date for which the eigenvalue problem has been completely solved. This fact obviously limits the scope of the eigenfunction procedure, but not as badly as might at first be feared. Given the Maxwell-model eigenfunctions, it is always possible to manipulate them in some way—for instance, by varying a parameter in the functions, or by changing the independent variable, say from energy to speed. Usually there is some physical basis for such a change. The fact that the operator is not known for which these manipulated basis functions are the eigenfunctions is of no importance. It is not directly involved in the solution of the moment equations. Once the ψ_p are chosen, the problem comes down to evaluating the coefficients $a_{pq}^{(j)}$ and solving the moment equations (5-3-19). The hope is that the chosen ψ_p will result in only a small number of terms being important in the expansion (5-3-16) of the collision term $J_j\psi_p$, so that the moment equations can be solved accurately by a low-order truncation-iteration procedure. Of course, if the eigenfunctions of the real collision operator J_j are used, only one term is needed in the expansion and it is much easier to solve the moment equations. This is what happens for the Maxwell model, but no other such model is known.

Another way to choose the basis functions is to imagine seeking solutions of the Boltzmann equation by representing the distribution function as a series expansion of the form,

$$f = f^{(0)} \sum_q c_q \phi_q^\dagger, \tag{5-3-21}$$

where $f^{(0)}$ is some zero-order approximation to f, obtained by solving some simple related problem, by making an educated guess based on physical arguments, or even by just inspired guesswork. Once $f^{(0)}$ is chosen, a complete set of orthogonal polynomials ϕ_p is constructed by repeated use of the orthogonality relation,

$$\int f^{(0)} \phi_q^\dagger \phi_p \, d\mathbf{v} = nN_p \delta_{pq}, \tag{5-3-22}$$

where N_p is the normalization factor. The density n appears as a factor only for convenience, because we have normalized the distribution function to the density rather than to unity, and has no special physical significance. We can see that the expansion coefficients c_q are directly related to the moments of ϕ_p with respect to f, by multiplying (5-3-21) by ϕ_p and integrating,

$$c_p = N_p^{-1}\langle \phi_p \rangle. \tag{5-3-23}$$

We now select the functions ϕ_p as the basis functions for forming the moment equations, and use the expansion (5-3-21) for f in the collision operator term,

$$\langle J_j \phi_p \rangle \equiv n^{-1} \int f(J_j\phi_p)\,d\mathbf{v} = n^{-1}\int f^{(0)}\left(\sum_q c_q \phi_q^\dagger\right)(J_j\phi_p)\,d\mathbf{v}$$

$$= n^{-1}\sum_q c_q \int f^{(0)}\phi_q^\dagger(J_j\phi_p)\,d\mathbf{v} = \sum_q \langle \phi_q \rangle f_{pq}^{(j)}, \tag{5-3-24}$$

where

$$f_{pq}^{(j)} \equiv (nN_q)^{-1}\int f^{(0)}\phi_q^\dagger(J_j\phi_p)\,d\mathbf{v}, \tag{5-3-25}$$

and where we have used (5-3-23) for c_q. The moment equations then become

$$(e/m)\mathbf{E}\cdot\langle\mathbf{V}_v\phi_p\rangle - [\langle\mathbf{v}\phi_p\rangle - \langle\mathbf{v}\rangle\langle\phi_p\rangle]\cdot\mathbf{V}_r\ln n = \sum_j N_j \sum_q \langle\phi_q\rangle f_{pq}^{(j)}. \qquad (5\text{-}3\text{-}26)$$

But these are *exactly* of the same form as the moment equations (5-3-19) of the eigenfunction method, with ϕ_p appearing in place of ψ_p. We can make the two sets of moment equations *identical* by choosing the zero-order distribution function $f^{(0)}$ to be equal to the weight function $g(\mathbf{v})$ of the eigenfunction method, since this will make the functions ϕ_p and ψ_p the same.

Thus in the eigenfunction procedure the basis functions are chosen first, which automatically specifies the weight function $g(\mathbf{v})$, whereas in the distribution-function procedure the weight function is chosen first and the basis functions are then automatically determined. The new insight obtained by the distribution-function procedure is that the weight function is a zero-order approximation to the distribution function. Although the two procedures lead to formally identical results, the possibilities for educated guesswork may be rather different.

Kihara (1953) was among the first to suggest that solving the Boltzmann equation could be viewed as an eigenfunction problem, but his applications to ion transport were largely limited to the Maxwell model. His method was adopted and further elaborated in a study of ion mobility in weak electric fields by Mason and Schamp (1958). Expansion of the distribution function itself in generalized orthogonal polynomials seems to have been first formulated in a systematic way by Mintzer (1965), who was primarily concerned with problems in rarefied gas dynamics. As a result, it was some years before his views were recognized as especially relevant to problems of ion transport (Viehland and Mason, 1978).

D.   Summary of Basis Functions Used

We now enumerate some sets of zero-order distributions and basis functions that have been used. Many apparently unconnected versions of kinetic theory can be seen to be variations on the same fundamental theory, but with different basis functions. Many of these sets of functions are associated with special names, which we shall use as convenient labels. There is no reason to believe that this list is exhaustive, and there is certainly scope for ingenious new suggestions. In the following chapter we give more details of the major physical results obtained from some of the more important sets of basis functions.

1.   One-Temperature Theory

Here the weight function, or zero-order ion distribution function, is chosen to be an equilibrium Maxwellian at the gas temperature,

$$f^{(0)} = n(m/2\pi kT)^{3/2}\exp(-w^2), \qquad (5\text{-}3\text{-}27a)$$

$$w^2 \equiv mv^2/2kT. \qquad (5\text{-}3\text{-}27b)$$

If it is assumed that the basis functions are polynomials and that the independent variables are the components of w^2, then the orthogonality condition determines the functions. In a spherical coordinate system they are

$$\psi_{lm}^{(r)} = w^l S_{l+1/2}^{(r)}(w^2) Y_l^m(\theta, \phi), \qquad (5\text{-}3\text{-}28)$$

$$Y_l^m(\theta, \phi) = P_l^{|m|}(\cos \theta) e^{im\phi}, \qquad (5\text{-}3\text{-}29a)$$

$$\cos \theta = v_z/v, \qquad (5\text{-}3\text{-}29b)$$

where $S_{l+1/2}^{(r)}(w^2)$ are Sonine (generalized Laguerre) polynomials, and $Y_l^m(\theta, \phi)$ are the usual spherical harmonics, with the z-axis along the direction of the electric field. There are three indices on these basis functions because there are three independent variables, the three components of w^2. These functions are usually called Burnett functions in honor of David Burnett, who first introduced Sonine polynomials into classical kinetic theory (Burnett, 1935a, b).

There are two reasons why Burnett functions are especially useful in kinetic theory. The first is obvious from the description above: they are the basis functions associated with the equilibrium Maxwellian distribution, and so should be appropriate for describing small departures from equilibrium. The second is not at all obvious: they are the eigenfunctions of the collision operator for the Maxwell model (constant collision frequency), and so should produce rapid convergence whenever the collision frequency is slowly varying.

Sometimes it is advantageous to use a Cartesian coordinate system, and several examples are known in classical kinetic theory where this is the case (Grad, 1949a, b, 1960; Desloge, 1964, 1966, Chap. 34). The basis functions then turn out to be Hermite polynomials,

$$\psi_{pqr} = H_p(w_x) H_q(w_y) H_r(w_z). \qquad (5\text{-}3\text{-}30)$$

These are also eigenfunctions of the Maxwell model collision operator, in Cartesian coordinates.

There is an interesting parallel between the eigenfunctions of the Maxwell model in kinetic theory and the eigenfunctions of the hydrogen atom (Coulomb problem) and the harmonic oscillator in quantum mechanics (Louck and DeVault, 1964).

The foregoing version of one-temperature theory includes as special cases both the classical Chapman–Enskog kinetic theory (Chapman and Cowling, 1970; Hirschfelder et al., 1964; Ferziger and Kaper, 1972) by setting $m = 0, l = 1$, and the Kihara weak-field theory of ion mobility (Kihara, 1953; Mason and Schamp, 1958) by setting $m = 0$. Further applications of this one-temperature theory, including energy partitioning and diffusion, have been made by Hahn and Mason (1973), Kumar and Robson (1973), Robson and Kumar (1973), Robson (1973), and Whealton and Mason (1974).

However, this is not the only possibility for a one-temperature theory, even if $f^{(0)}$ is assumed to be an equilibrium Maxwellian and the basis functions are assumed to be polynomials. It is necessary to assume as well that the independent variable is the energy. Although this is indeed the correct variable for the Maxwell model, and presumably also for similar cases, there may be other cases where a different variable choice is advantageous. One straightforward possibility is to use the speed rather than the energy, that is, to choose w rather than w^2 as the variable in (5-3-28). Shizgal (1979) has shown that this produces much faster convergence for the rigid-sphere Lorentz model (used to discuss electron relaxation problems, among other things). A modest price must be paid for this improvement—the basis functions are not one of the familiar special functions of mathematical physics, and Shizgal had to construct them one by one from the orthogonality condition (Schmidt orthogonalization). However, direct numerical methods can sometimes replace the explicit use of polynomials, while still retaining speed as the variable (Shizgal, 1981b; Shizgal et al., 1981).

2. Two-Temperature Theory

Here $f^{(0)}$ and the basis functions have the same form as in the one-temperature theory, but the "temperature" of the ions is taken to be an adjustable parameter,

$$f^{(0)} = n(m/2\pi k T_b)^{3/2} \exp(-w_b^2), \qquad (5\text{-}3\text{-}31a)$$

$$w_b^2 \equiv mv^2/2kT_b, \qquad (5\text{-}3\text{-}31b)$$

where T_b is the "basis temperature." Physically, we might expect T_b to be closely related to the mean ion energy, say as $\frac{3}{2}kT_b \approx \frac{1}{2}m\langle v^2\rangle$, and in many cases this is so, but there is no necessity for such a requirement. We can choose T_b in various ways for various reasons, such as to improve the rate of convergence for some moment of special interest, or to speed up the mechanics of a truncation-iteration procedure, for example. Some of these choices will appear later. The basis functions are still the Burnett functions of (5-3-28), or the Hermite polynomials of (5-3-30), but with w_b in place of w as the variable.

The change from T to T_b makes an enormous difference in mobility calculations, and makes it possible to carry out computations at arbitrary field strengths. Some difficulties arise with the symmetry of the collision operator in this basis, but they are readily overcome. This representation was devised by Viehland and Mason (1975) on the basis of an eigenfunction approach, and independently by Burnett (1975) by a distribution-function procedure. As already noted, the methods are equivalent (Viehland and Mason, 1978). Although this representation gives good mobility results, the convergence is poor for the energy partitioning and the diffusion coefficients for heavy ions ($m > M$) at high field strengths (Viehland and Mason, 1978).

At high fields, it might seem plausible on physical grounds that a better approximation could be obtained by giving $f^{(0)}$ a displacement v_{dis} in the field direction, that is, by replacing (5-3-31b) with

$$w_b^2 = m(\mathbf{v} - \mathbf{v}_{dis})^2/2kT_b$$
$$= m[v_x^2 + v_y^2 + (v_z - v_{dis})^2]/2kT_b. \qquad (5\text{-}3\text{-}31c)$$

Although this distribution has been used for some first-order calculations on special problems involving electrons (Mozumder, 1980a, b; Paranjape, 1980), it has not yet been applied systematically to accurate calculations. The rationalization seems to be as follows. The displacement of the distribution introduces such complexity into the actual calculations (see Section 5-3E) that it is almost as easy to go one step further and let the basis temperature T_b be different in the longitudinal and transverse directions. This gives the three-temperature theory discussed next. However, many of the mathematical complications have now been worked through in detail and presented in a systematic way (Kumar, 1980a, b), so that this objection is no longer so serious. Moreover, some test calculations with a rigid-sphere cold-gas model indicate that the velocity displacement plays a more important role than the temperature anisotropy, especially for large ion/neutral mass ratios (Kumar et al., 1980, p. 414).

There is also the possibility of choosing speed instead of energy as the variable. This choice has been used in related problems involving the energy relaxation of "hot" atoms (Shizgal, 1981a, b) and "hot" electrons (Shizgal, 1983a, b), but with a discrete ordinate method rather than a moment method.

3. Three-Temperature Theory

Since even elementary arguments suggest that the ion temperature is different in the longitudinal and transverse directions (Section 5-2B), it is plausible to introduce this feature into $f^{(0)}$. We expect this feature to be especially pronounced for heavy ions ($m > M$) at high E/N, a result indicated long ago by some calculations for special cases by Wannier (1953). We also expect the distribution to be strongly peaked in the field direction under these conditions. All these considerations suggest an anisotropic displaced distribution,

$$f^{(0)} = n[m/2\pi kT_b^{(T)}][m/2\pi kT_b^{(L)}]^{1/2}$$
$$\times \exp[-m(v_x^2 + v_y^2)/2kT_b^{(T)} - m(v_z - v_{dis})^2/2kT_b^{(L)}], \qquad (5\text{-}3\text{-}32)$$

where $T_b^{(T)}$ and $T_b^{(L)}$ are the transverse and longitudinal basis temperatures, and v_{dis} is the displacement velocity. These can be treated as adjustable parameters, although physically we might expect to have relations like $T_b^{(T)} \approx T_T$, $T_b^{(L)} \approx T_L$, and $v_{dis} \approx v_d$. Woo et al. (1976) showed that this choice gave substantial agreement with moments calculated by Monte Carlo methods for several different mass ratios and ion-neutral interactions (Skullerud, 1973a).

The above choice of $f^{(0)}$ was used as the basis of moment calculations by Lin et al. (1979c) and Viehland and Lin (1979), who used a Cartesian coordinate system for their basis functions. The latter are then Hermite polynomials,

$$\psi_{pqr} = H_p(w_x)H_q(w_y)H_r(w_z), \tag{5-3-33a}$$

with

$$w_x^2 = mv_x^2/2kT_b^{(T)}, \qquad w_y^2 = mv_y^2/2kT_b^{(T)}, \tag{5-3-33b}$$

$$w_z^2 = m(v_z - v_{\text{dis}})^2/2kT_b^{(L)}. \tag{5-3-33c}$$

This choice tends to obscure the cylindrical symmetry in a drift-tube problem, but was made for the sake of the familiarity of Hermite polynomials, with their well-studied properties. However, Kumar (1980a, b) has since then worked out most of the formal mathematics necessary for the use of a spherical coordinate system, where the symmetry is more obvious. The basis functions are then generalized Burnett functions with a vector argument.

The use of (5-3-32) and (5-3-33) produces good results for the cases where the two-temperature theory fails, namely energy partitioning and diffusion for heavy ions ($m > M$) at high E/N, but naturally at the cost of increased complexity of the calculations (Lin et al., 1979c; Viehland and Lin, 1979).

It might be wondered how good a result would be obtained if $f^{(0)}$ incorporated only temperature anisotropy and not velocity displacement; that is, if we set $v_{\text{dis}} = 0$ in (5-3-32) and (5-3-33). This was, in fact, tried first, but with disappointing results (Viehland and Mason, 1977, unpublished results). Only a modest improvement in accuracy over the two-temperature calculation was obtained, but at the expense of appreciable complications in the computations.

4. Spherical Harmonic Expansion

If we look back at $f^{(0)}$ and the basis functions for the one- and two-temperature theories, we see that the results are equivalent to expansion of the distribution function as follows:

$$f(\mathbf{v}, \mathbf{r}, t) = n(\mathbf{r}, t)(m/2\pi kT_b)^{3/2} \exp(-w_b^2)$$

$$\times \sum_{r=0}^{\infty} \sum_{l=0}^{\infty} \sum_{m=-l}^{l} f_{lm}^{(r)} w_b^l S_{l+1/2}^{(r)}(w_b^2) Y_l^m(\theta, \phi), \tag{5-3-34a}$$

where the coefficients $f_{lm}^{(r)}$ are moments of the distribution function,

$$f_{lm}^{(r)} = \langle \psi_{lm}^{(r)} \rangle / N_{lm}^{(r)}, \tag{5-3-34b}$$

with $\psi_{lm}^{(r)}$ given by (5-3-28) and $N_{lm}^{(r)}$ being a normalization constant. We can, however, imagine using just the expansion in $Y_l^m(\theta, \phi)$ without being so specific about the function of w_b (i.e., of v), and writing

$$f(\mathbf{v}, \mathbf{r}, t) = n(\mathbf{r}, t) \sum_{l=0}^{\infty} \sum_{m=-l}^{l} f_{lm}(v) Y_l^m(\theta, \phi), \tag{5-3-35}$$

where the $f_{lm}(v)$ are unspecified functions of v. This can be a useful procedure in some special cases, of which two are known at present. The first is the case where $m \ll M$, for which only two terms in the expansion are needed. This special feature allows an exact explicit solution to be found, and is responsible for the fact that transport theory for electrons was for many years far more advanced than for ions. The second case has a number of restrictions, all designed so that some ingenuity can be exercised in the choice of functions for expansion of the $f_{lm}(v)$. This ingenuity is largely based on knowledge of the asymptotic behavior of the distribution function in various limits. Since such information is seldom available for real physical systems, and since so many restrictions apply, this case is mainly useful for supplying numerical results against which more flexible methods can be tested. It is, in other words, used as a generator of benchmark test cases. Some details of both special cases follow.

a. Two-Term Approximation (Electrons). The physical basis for this approximation is as follows. In the limit of $m/M \to 0$, an elastic collision merely rotates the relative velocity through some angle, without changing its magnitude. The distribution of ion velocities therefore ought to be isotropic in velocity space. This limit is not physically satisfactory, however, because no energy loss can take place, and hence no steady state can be achieved if an electric field is present. (The fractional energy loss in a collision is of order $2m/M$.) It is necessary to let m/M be nonzero, although it can be considered very small. The first term in the harmonic expansion of (5-3-35), with $l = 0$, $m = 0$, corresponds to the dominant isotropic part of the distribution, $f_{00}(v)$. The energy loss needed for a steady state is taken into account by including one more term in the expansion, with $l = 1$, $m = 0$, which should be adequate if m/M is small enough. Note that $f_{00}(v)$ will usually be far from Maxwellian at high fields, even though the distribution is nearly isotropic.

Substitution of the two-term expansion back into the Boltzmann equation leads, after considerable mathematical manipulation, to two coupled differential equations for $f_{00}(v)$ and $f_{10}(v)$, which can be solved exactly to yield

$$f_{00}(v) = A \exp\left[-\int_0^v \frac{mv\, dv}{kT + \frac{1}{3}M(a/v)^2} \right], \qquad (5\text{-}3\text{-}36a)$$

$$f_{10}(v) = -(a/v)(df_{00}/dv), \qquad (5\text{-}3\text{-}36b)$$

where A is a normalization constant, and $a = eE/m$ is the acceleration. The collision frequency v is essentially the same quantity that arose in momentum-transfer theory,

$$v(v) \equiv 2\pi N v \int_0^\pi (1 - \cos\theta)\sigma(\theta, v)\sin\theta\, d\theta. \qquad (5\text{-}3\text{-}37)$$

Mathematical details are given in Chapman and Cowling (1970, Sec. 19.61). An improved derivation is given by Wannier (1971), with some corrections added by McCormack (1971). A derivation by moment methods, which is more in keeping with the viewpoint of this chapter, is given by Kumar et al. (1980, p. 402).

If more terms in the harmonic expansion are kept, consistency requires that higher powers of m/M also be carried, and the complexity of the problem increases drastically. A consistent solution with four terms (i.e., including $l = 3$, $m = 3$) has been obtained by Cavalleri (1981).

The two-term approximation is excellent for electrons, provided that no inelastic collisions occur. Inelastic collisions spoil the two-term approximation by destroying the quasi-isotropic property of f. An inelastic collision not only rotates the relative velocity, but also may change its magnitude drastically. It is difficult to include inelastic collisions in the two-term approximation in a consistent way, and errors of uncertain magnitude may be introduced (Reid, 1979; Lin et al., 1979b; Pitchford et al., 1981; Haddad et al., 1981; Pitchford and Phelps, 1982; Braglia et al., 1982; Phelps and Pitchford, 1985). Inelastic collisions are usually more serious for electrons than for ions. Even in helium, the inelastic collision losses are equal to the elastic collision losses at about $E/N \approx 10$ Td (Loeb, 1960, p. 295), whereas this field strength is still in the Ohmic regime for most ions.

The history of the two-term approximation (as it is now called) is rather interesting. In a pioneering paper, Pidduck (1916, 1936) essentially solved the problem for rigid-sphere interactions, but his work was apparently overlooked. The problem was revived by Druyvesteyn (1930, 1934), who independently obtained Pidduck's result in the limit of $T = 0$. Davydov (1935) removed the $T = 0$ limit and thus recovered Pidduck's result, but neither Druyvestyn nor Davydov was aware of Pidduck's prior work. The removal of the rigid-sphere restriction was achieved by Morse et al. (1935), but only in the limit of $T = 0$. The general solution, valid for any T and any $v(v)$, was finally achieved by Chapman and Cowling (1970; the result appeared already in the 1939 edition). Subsequent work (Wannier, 1971; McCormack, 1971) has improved the pedigree of the derivation without changing the result.

b. Special Functions. The properties of the spherical harmonics allow some simplifications to be made if we substitute the harmonic expansion (5-3-35) back into the general moment equations (5-3-15), especially if something simple is used for the functions ψ in (5-3-15). It is often convenient to study the power moments of the velocity, $\langle v^s \rangle$, for which we therefore choose $\psi = v^s$, where s is an integer. Even with these simplifications we are still left with an infinite array of coupled equations for all the $\langle v^s \rangle$, which include the recalcitrant collision term $\langle Jv^s \rangle$. In any particular case (i.e., given the ion-neutral mass ratio and interaction), we can always reduce the collision term, and then possibly solve the resulting equations by some brute-force truncation-iteration scheme, given enough computer capacity. This, of course, is not a very appealing prospect.

However, for certain special conditions the set of coupled moment equations can be greatly simplified. This simplified set can then be used as conditions that a distribution function must satisfy. A trial distribution function with undetermined constants is chosen and the constants found by requiring that the first few moment equations be satisfied. More constants and more moment equations can be included to secure improvement. The procedure is thus somewhat like the Rayleigh–Ritz variational method, except that no extremum principle is involved. Success, of course, depends on the degree of shrewdness shown in the choice of trial functions.

This powerful but limited technique was devised by Wannier (1951, 1953) and applied to the case of $m = M$ for the Maxwell model (as a test) and for rigid spheres. Skullerud (1976) extended the procedure, and made a number of numerical calculations for different mass ratios and several inverse-power repulsive potentials, $V(r) = C_n/r^n$. The two essential conditions to be satisfied are, first, that the neutral gas temperature is very low so that its distribution function can be replaced by a delta function, and second, that the differential cross section for ion-neutral scattering can be written as the product of an angle-dependent factor and a velocity-dependent factor,

$$\sigma(\theta, v) = i(\theta)q(v). \tag{5-3-38}$$

An inverse-power potential has this property, and moreover

$$q(v) \propto v^{-4/n}, \tag{5-3-39}$$

which enables the collision term $\langle Jv^s \rangle$ to be reduced to a particularly simple form. The set of moment equations can then be manipulated algebraically to obtain a set of equations involving only moments with respect to the spherical component of the distribution function, and having the form

$$\sum_{j=0}^{l+1} \alpha_j^{(l)} \langle v^{l-1+j(1+\gamma)} \rangle_{00} = \beta^{(l)}, \tag{5-3-40}$$

where

$$\langle v^r \rangle_{00} \equiv \int v^r f_{00}(v) \, d\mathbf{v}, \tag{5-3-41}$$

$$\gamma \equiv 1 - 4/n. \tag{5-3-42}$$

The index l is the same as the one in the spherical harmonic expansion (5-3-35). The ion density does not appear in (5-3-41) because it was removed in the definition of $f_{00}(v)$ in (5-3-35). The quantities $\alpha_j^{(l)}$ and $\beta^{(l)}$ depend on the mass ratio and on a complicated integral of $i(\theta)$ over scattering angle; for a given system they are constants. The set of equations (5-3-40) for the (unknown) moments gives the conditions that the distribution function $f_{00}(v)$ must satisfy.

The function $f_{00}(v)$ is now represented as a linear combination of some shrewdly chosen special functions,

$$f_{00}(v) = \sum_{i=1}^{L} c_i \phi_i(v), \qquad (5\text{-}3\text{-}43)$$

from which the moments appearing in (5-3-40) can be evaluated. Enough equations of (5-3-40) are then taken to determine the coefficients c_i. For instance, if $L = 5$ only four equations from (5-3-40) are needed, since normalization of $f_{00}(v)$ already furnishes one condition. Once $f_{00}(v)$ is approximated with sufficient accuracy, any desired velocity moment $\langle v^r \rangle_{00}$ can be calculated, and then the higher-order moments $\langle v^r \rangle_{l0}$ can be found algebraically by use of the original moment equations. Wannier (1951, 1953) was able to calculate the drift velocity for rigid spheres to five significant figures with only four functions: an exponential integral, a Gaussian, and modified Hankel functions of order 0 and 1. These were chosen to represent the essential behavior of the distribution, which was known from asymptotic analysis and Monte Carlo calculations. He suggested that modified Bessel functions would be a useful basis set. As already mentioned, Skullerud (1976) has obtained a number of accurate results by this method. Skullerud has also tried other basis sets, and found good convergence when the asymptotic behavior of the distribution was correctly represented (Kumar et al., 1980, p. 415).

The method does not work when m/M is either very large or very small (Skullerud, 1976). For m/M small, only a small number of terms in the spherical harmonic expansion are important (as $m/M \to 0$, only two terms survive). Most of the moment equations therefore furnish only uselessly weak constraints on the distribution function. For m/M large, the distribution is strongly skewed in the field direction, and the spherical harmonic expansion itself is not very accurate.

The greatest value of this technique is to supply accurate numbers for testing the results of more general methods.

5. *Bimodal Expansion*

This choice of basis functions does not really pertain to drift tubes, but does furnish an interesting example of how physical insight can be incorporated into the basis functions. The problem concerns the decay toward equilibrium of a swarm of high-energy particles of mass m released in a reservoir of inert buffer gas of mass M and temperature T (the "hot-atom" or "hot-electron" problem). No electric field is necessarily present. It is reasonable to expect that a time-dependent version of the two-temperature theory (Section 5-3D2) should work well, and this indeed is usually the case (Keizer, 1972; Shizgal and Fitzpatrick, 1980; Shizgal, 1980; Knierim et al., 1981). But severe difficulties occur when m and M are nearly equal (within a factor of about 3); this is signaled in the two-temperature theory by convergence difficulties, numerical instabilities, and regions where the approximate distribution function becomes negative. This behavior results from the possibility of transferring the total energy of an energetic particle to the buffer gas in a single head-on collision. It is thus possible to build up a significant number of particles with near zero energy after only a few collisions, giving rise to a bimodal velocity distribution. With such a distribution it is not surprising that an expansion for f based on Sonine polynomials gives poor results (Knierim et al., 1981). With smaller or larger mass ratios, many more collisions are required for a particle to lose its excess energy. In such cases no low-energy peak develops and f may be adequately described by an expansion around a high-temperature Maxwellian, (5-3-31).

The foregoing energy-loss mechanism is not so serious in the drift-tube problem, because the electric field speeds up the resulting slow ions again. However, some vestiges of the difficulty may remain in the form of a singularity in f at $v = 0$ (for a buffer gas with $T = 0$) when $m = M$; this phenomenon was predicted long ago by Wannier (1953).

The accumulation of a group of (formerly) energetic particles at low energy suggests using a bimodal distribution for $f^{(0)}$ that consists of two Maxwellians, one at a high temperature and one at a low temperature, and writing a separate expansion for each one (Knierim et al., 1981),

$$f = h(t)f^{(0)}(T_a) \sum_{k=0}^{\infty} a_k S^{(k)}_{1/2}(w_a)$$

$$+ [1 - h(t)]f^{(0)}(T_b) \sum_{r=0}^{\infty} b_r S^{(r)}_{1/2}(w_b), \qquad (5\text{-}3\text{-}44)$$

where $f^{(0)}(T_{a,b})$ and $w_{a,b}$ are given by (5-3-31). Here $h(t)$ is the fraction of particles allotted to the low-energy component. The problem is isotropic because there is no field, so only the first spherical harmonic is used, corresponding to $m = l = 0$. Although we would expect to have $T_a \approx T$ on physical grounds, there is no need to force this choice. Both T_a and T_b can be chosen to improve convergence, or for any other convenient reason. Bimodal forms have also been used in attempting to describe the velocity distribution in a strong shock wave (Mott-Smith, 1951; Muckenfuss, 1962).

The flexibility introduced by (5-3-44) exacts a mathematical price. In the first place, the functions are no longer all orthogonal, so many of the subsequent mathematical operations become cumbersome. Second, and more serious, the expansion is now overcomplete; extra constraints must be added to avoid a disastrous division by zero when trying to solve the moment equations. The simplest procedure is to drop one term from the high-temperature expansion for each new term included in the low-temperature expansion, but there are many other possibilities. Knierim et al. (1981) obtained very reasonable results in the hot-atom problem by using only the first term of the low-temperature expansion and dropping the $r = 2$ term from the high-temperature expansion.

E. Comments on Computations

It is all very well to introduce lots of flexibility and physical insight through the choice of $f^{(0)}$, but a real price must then be paid in the form of extra labor in the subsequent calculations. This is a very serious matter in practice, and it is worth devoting a few remarks to pointing out which are the easy and which are the hard parts of a moment calculation.

By far the most difficult and time-consuming task is the calculation of the matrix elements—the $a_{pq}^{(j)}$ of (5-3-19) or the $f_{pq}^{(j)}$ of (5-3-26), which are the same. For convenience we repeat the definition here:

$$a_{pq}^{(j)} \equiv f_{pq}^{(j)} \equiv (\psi_q, J_j\psi_p)/(\psi_q, \psi_q) = (nN_q)^{-1} \int f^{(0)}\psi_q^\dagger(J_j\psi_p)\,d\mathbf{v}, \quad (5\text{-}3\text{-}45)$$

where ψ_p and ψ_q are a pair of basis functions, and the collision operator J_j is

$$J_j\psi_p \equiv N_j^{-1} \iint F_j[\psi_p(\mathbf{v}) - \psi_p(\mathbf{v}')]v_{rj}\sigma_j\,d\mathbf{\Omega}_j\,d\mathbf{V}_j. \quad (5\text{-}3\text{-}46)$$

The indices p and q are only symbolic, and may represent several variables, not just one. Inspection of these two expressions shows that each $a_{pq}^{(j)}$ represents a weighted average of the change in some function ψ_p caused by a collision between an ion and a neutral particle of species j. In the language of a function space, a collision transforms the vector ψ_p into the vector $J_j\psi_p$, and $a_{pq}^{(j)}$ is the component of $J_j\psi_p$ in the qth direction, or the projection of $J_j\psi_p$ onto ψ_q. Evidently, the value of $a_{pq}^{(j)}$ must depend on the ion-neutral interaction, on the two masses, and on any parameters occurring in $f^{(0)}$, such as T_b or v_{dis}, as well as on the particular values of p and q.

Once the matrix elements are evaluated, we are left with an infinite set of moment equations, (5-3-19) or (5-3-26), for the various moments $\langle\psi_p\rangle$. Since these are linear algebraic equations, there is no special difficulty in devising some truncation-iteration scheme for finding the particular moment of interest, usually just a few of the lower ones. This can always be done quickly by a computer, on which matrix inversion is usually easy, even if rather large sets of equations have to be handled because the convergence happens to be slow.

There are two stages in the evaluation of the $a_{pq}^{(j)}$, which obviously cannot be completely evaluated until the ion-neutral interaction is specified. An eightfold integration is required—two for $d\mathbf{\Omega}_j$, and three each for $d\mathbf{V}_j$ and $d\mathbf{v}$. By separating out the motion of the center of mass of a colliding ion-neutral pair, plus the direction in space of the relative collision velocity v_{rj}, it is possible to perform six of the integrations without specifying the ion-neutral interaction (other than that it is spherically symmetric). This constitutes the first stage. It is straightforward in principle, but can be quite complicated in practice, especially if $f^{(0)}$ and the ψ's depend on $\mathbf{v}$ in anything but a very simple way. Even in classical (one-temperature) Chapman–Enskog kinetic theory, where $f^{(0)}$ is Maxwellian and the ψ's are low-order Sonine polynomials, the task is nontrivial, and standard treatises often devote a chapter or an appendix to the problem (Chapman and Cowling, 1970, Chap. 9; Ferziger and Kaper, 1972, App. B; the present matrix elements are usually called "bracket integrals" in classical kinetic theory).

The result of this reduction is that each matrix element is expressed as a linear combination of irreducible "collision integrals", which contain the remaining two integrations. (Sometimes, if Cartesian coordinates are used, three integrations will be left in the collision integrals.) The evaluation of these collision integrals constitutes the second stage. One integration covers the details of all the possible ion-neutral collisions, and is customarily expressed as a set of transport cross sections,

$$Q^{(l)}(\varepsilon) \sim 2\pi \int^\pi (1 - \cos^l\theta)\sigma(\theta, \varepsilon) \sin \theta \, d\theta, \qquad (5\text{-}3\text{-}47)$$

where $\varepsilon = \frac{1}{2}\mu v_r^2$ is the relative collision energy. The case $l = 1$ corresponds to the cross section Q_D that appeared in the momentum-transfer theory of Section 5-2A. Sometimes a different set of cross sections involving Legendre polynomials is used (Chapman, 1916, 1917a, b; Enskog, 1917, 1922; Kumar, 1967; Viehland and Mason, 1978; Kumar et al., 1980, p. 370),

$$\sigma^{(l)}(\varepsilon) = 2\pi \int_0^\pi [1 - P_l(\cos \theta)]\sigma(\theta, \varepsilon) \sin \theta \, d\theta. \qquad (5\text{-}3\text{-}48)$$

In any case, these integrals do not depend on the choice of $f^{(0)}$ and ψ; for a given ion-neutral interaction they can be calculated once and for all, and stored or tabulated. Such calculations are discussed in Chapter 7, and a number of useful cross sections are tabulated in Appendix II. The final step in evaluating the collision integrals is the weighted integration of the $Q^{(l)}$ over ε or v_r. The weighting factor depends on the forms chosen for $f^{(0)}$ and ψ, but the integration is usually easily done numerically because the $Q^{(l)}$ are slowly varying functions of ε in most cases.

At the present time, it is the integrations involved in the first stage that usually constitute the bottleneck in moment calculations. The second stage is far from trivial, except for extremely simple forms of ion-neutral interaction, but is well in hand. The difficulty with the first stage is that the reductions get more and more complicated as higher matrix elements are needed (i.e., as p and q get larger), and many more matrix elements are needed for each higher step in any reasonable truncation-iteration scheme for solving the moment equations. Thus the amount of labor involved increases very rapidly, and a moment calculation is usually impractical unless convergence is fairly fast.

There is thus a compromise involved in the choice of $f^{(0)}$. On the one hand, a physically realistic form may produce fast convergence, but make the evaluation of the matrix elements unacceptably complex. On the other hand, a simple form for $f^{(0)}$ may lead to easy calculations for the matrix elements, but make the convergence unacceptably slow. There is still ample scope for the exercise of ingenuity here.

A final point concerning the choice of $f^{(0)}$ involves the physical interpretation of the results. The ideal situation is to pick $f^{(0)}$ so that practically all the physics is apparent in the first approximation, and higher approximations merely serve to refine the numerical results slightly. The physics can then be exhibited by hand, so to speak, and a computer is needed only for refining the numerical accuracy. A poor choice of $f^{(0)}$, however, means that results must be ground out numerically by computer, and physical insight has to be sought in the examination of large quantities of numerical output, a rather dismal prospect.

F.   Relation to Elementary Theories

Now that we have seen the Boltzmann equation, it is possible to obtain further insight into the relations among elementary and accurate kinetic theories. We have already remarked on the parallel through moment equations—the lowest-order moment equations of the two-temperature theory correspond to approximate equations of elementary theories that account for momentum and energy balance. This correspondence is discussed more specifically in Chapter 6. We can now see another aspect of the various interrelations through the Boltzmann equation itself.

Free-flight theories can often be regarded as arising by approximating the collision term of the Boltzmann equation with a simple relaxation term:

$$J_j f \to [f - f^{(0)}]/N_j \tau_j, \qquad (5\text{-}3\text{-}49)$$

where τ_j is the relaxation time and $f^{(0)}$ is the steady-state distribution to which f eventually relaxes by collisions. These theories are sometimes called relaxation theories. It is a great mathematical simplification to replace an integral operator by an algebraic expression. The most elementary free-flight theories arise by making a first-order approximate solution of the resulting differential equation (Chapman and Cowling, 1970, Sec. 6.6).

Free-path theories, on the other hand, can often be regarded as arising by an iterative solution of the Boltzmann equation with the full collision term. One standard way of solving an integral equation is by iteration: a first approximation is inserted into the integrand, and the equation is solved to find a second approximation; this second approximation is then substituted into the integrand, and the equation is solved to find a third approximation; and so on. The resulting series corresponds closely to tracing a collision history back over many free paths (Monchick, 1962; Monchick and Mason, 1967). Some work has been done on the generalization of this kind of procedure, with the aim of finding iterative solutions of the Boltzmann equation (Skullerud and Kuhn, 1983; Ikuta and Murakami, 1987). These approaches are usually called path-integral methods, and represent a generalization of free-path theory, just as moment methods represent a generalization of momentum-transfer theory.

G.   Historical Notes on Moment Methods and Multitemperature Theories

Many of the ideas outlined in this discussion of accurate kinetic theory have been around for a long time, often scattered through apparently unrelated parts of physics. Without in any way attempting a full survey, we give here some brief notes and selected references in the hope of conveying some of the flavor of the subject. An extensive survey has been given by Weinert (1982).

In a sense, moment methods in kinetic theory are older than the Boltzmann equation itself. Maxwell (1867) based his approach on "equations of change," which would now be regarded as moment equations derived from the Boltzmann equation. Chapman (1916, 1917a, b) adopted a similar approach in his successful development of a method for calculating transport coefficients. However, when writing their definitive treatise, Chapman and Cowling (1970, 3rd ed.; the first edition appeared in 1939) decided to adopt the method of Enskog (1917, 1922), who had approached the same problem through the Boltzmann equation and power-series expansions of the distribution function. They also incorporated the important refinement of Burnett (1935a, b), in which power-series expansions are replaced by expansions in orthogonal polynomials. This book was so (deservedly) influential that moment methods dropped for a time into obscurity in kinetic theory, an apparently harmless enough development inasmuch as the two methods are mathematically equivalent. A new moment approach was introduced by Grad (1949b), whose aim was not to calculate transport coefficients, but to obtain a set of differential equations for gases that went beyond the Navier–Stokes equations, and that could therefore take into account boundary effects in rarefied gas dynamics that played no role in Chapman–Enskog theory. He emphasized the moments themselves and used a truncation procedure to obtain a closed set of equations (his 13-moment approximation). He also introduced tensorial Hermite polynomials (Grad, 1949a) as the basis functions for the moment equations.

Gradually more emphasis came to be laid on moment equations as a set of matrix equations, on the transformations from one basis set to another, and on the calculation of the various matrix elements needed (Johnson and Ikenberry, 1958; Johnston, 1960; Ikenberry, 1962; Mintzer, 1965; Coope et al., 1965; Kumar, 1966a, b, 1967, 1970; Ford, 1968; Coope and Snider, 1970; Coope, 1970; Chen et al., 1971; Hunter and Snider, 1974; Aisbett et al., 1974; Weinert, 1979, 1980, 1981). Much of the mathematical machinery had been developed earlier for use in nuclear physics (Talmi, 1952; Smirnov, 1961). The physical motivations were often quite different, ranging over nuclear spectroscopy, rarefied gas dynamics, the extension of Chapman–Enskog theory to polyatomic gases, and magnetic-field effects on neutral-gas transport coefficients. The great economy and power of these methods when applied to the calculation of gas transport coefficients (Kumar, 1967) was largely superfluous, however, because the convergence of the by-now-classic Chapman–Enskog formulas was already extremely good. A pioneering application of these elegant methods to ion transport (Kumar and Robson, 1973; Robson and Kumar, 1973; Robson, 1973) also found their power to be largely wasted. The one-temperature basis functions that were used failed to produce convergent results at even moderate electric field strengths, no matter how high the order of approximation taken. However, the introduction of multitemperature and anisotropic temperature theories has now shown that convergence can be attained at any field strength, although with slower convergence than in the classical Chapman–Enskog theory (Viehland and Mason, 1975, 1978; Lin et al., 1979b, c; Viehland and Lin, 1979). There is thus hope that the power of this assembled mathematical machinery will soon prove useful in ion transport theory (Kumar, 1980a, b; Kumar et al., 1980; Kumar, 1984; Ness and Robson, 1985).

The idea of multiple temperatures in physical and chemical problems must be quite old. Certainly, it appears repeatedly in diverse connections: electron temperatures in gaseous electronics, plasma physics, semiconductors, and superconductors; rotational and vibrational temperatures in sound absorption, shock-wave phenomena, and chemical kinetics; nuclear spin temperatures in nuclear magnetic resonance; as internal variables in general continuum thermodynamics; and others. Aside from electrons, the concept appears in kinetic theory in connection with the relaxation to equilibrium of a mixture of gases of different masses (Grad, 1960; Goldman and Sirovich, 1969; Shizgal and Fitzpatrick, 1974; Goebel et al., 1976, and many papers referred to therein; Weinert, 1978; Fernández de la Mora and Fernández-Feria, 1987). Although collisions soon produce a quasiequilibrium among species of the same mass, the slow interchange of energy between species of different mass on collision gives a much longer time scale for complete equilibrium and justifies the notion of different temperatures for different species. The generalization to the superposition of non-Maxwellian distributions for problems involving large deviations from equilibrium seems eminently reasonable (Weitzsch, 1961; Suchy, 1964; Weinert and Suchy, 1977; Weinert, 1983a, b, 1984). However obvious it may seem in retrospect, the combination of moment methods with multitemperature models for drift-tube theory took time to develop before it broke a bottleneck of some 20 years' standing. The situation as it existed just prior to this combination, consisting of fragments composed of the theories of Chapman and Enskog, of Kihara and of Mason and Schamp, and of Wannier, is described by McDaniel and Mason (1973, Chap. 5).

H.   Final Remark on Moment Methods

Moment methods can be quite powerful, but it should always be kept in mind that the aim is not just to produce numbers, but physical insight as well. If a moment calculation is *only* able to produce accurate numbers, it is worth considering whether it might be better to tackle the problem in a more numerical way, say by a Monte Carlo calculation. This may lead to direct information on the distribution function itself, instead of just some moments of the distribution. A moment calculation of low accuracy in which relations among experimental quantities are apparent may be more valuable than one of high accuracy in which all the relations are obscured. In this connection it is worth quoting some remarks by Wannier (1951). Referring to the general moment method sketched above in Section 5-3D4b, he observed, "It is true that, in principle, the general problem could be solved by the method developed here.... We would be able to produce a number for the drift velocity for a given numerical ratio of the electric field and the temperature, but we would not gain direct information about the functional relationship. This relationship would only reveal itself indirectly after extended numerical computations. It is to be hoped that a more satisfactory way of proceeding can be found."

REFERENCES

Aisbett, J., J. M. Blatt, and A. H. Opie (1974). General calculation of the collision integral for the linearized Boltzmann transport equation, *J. Stat. Phys.* **11**, 441–456.

Amdur, I., J. E. Jordan, K.-R. Chien, W.-M. Fung, R. L. Hance, E. Hulpke, and S. E. Johnson (1972). Scattering of fast potassium ions by helium, neon, and argon, *J. Chem. Phys.* **57**, 2117–2121.

Biondi, M. A., and L. M. Chanin (1961). Blanc's law—Ion mobilities in helium-neon mixtures, *Phys. Rev.* **122**, 843–847.

Blanc, A. (1908). Recherches sur les mobilités des ions dans les gaz, *J. Phys. (Paris)* **7**, 825–839.

Boltzmann, L. (1872). Weitere Studien über das Wärmegleichgewicht unter Gas-molekülen, *Sitzungsber. Akad. Wiss. Wien* **66**, 275–370. An English translation appears in S. G. Brush, *Kinetic Theory, Vol. 2, Irreversible Processes*, Pergamon, Oxford, 1966, pp. 88–175.

Braglia, G. L., L. Romano, and M. Diligenti (1982). Comment on "Comparative calculations of electron-swarm properties in N_2 at moderate E/N values," *Phys. Rev.* **A26**, 3689–3694.

Burnett, D. (1935a). The distribution of velocities in a slightly non-uniform gas, *Proc. London Math. Soc.* **39**, 385–430.

Burnett, D. (1935b). The distribution of molecular velocities and the mean motion in a non-uniform gas, *Proc. London Math. Soc.* **40**, 382–435.

Burnett, D. (1975). Personal communications to E. A. Mason. Unfortunately, these calculations have never been published.

Cavalleri, G. (1981). Solutions with third-order accuracy for the electron distribution function in weakly ionized gases (or in intrinsic semiconductors): Application to the drift velocity, *Aust. J. Phys.* **34**, 361–384.

Chapman, S. (1916). On the law of distribution of molecular velocities, and on the theory of viscosity and thermal conduction, in a non-uniform simple monatomic gas, *Philos. Trans. R. Soc. London* **A216**, 279–348.

Chapman, S. (1917a). On the kinetic theory of a gas: Part II. A composite monatomic gas: Diffusion, viscosity and thermal conduction, *Philos. Trans. R. Soc. London* **A217**, 115–197.

Chapman, S. (1917b). The kinetic theory of simple and composite monatomic gases: Viscosity, thermal conduction, and diffusion, *Proc. R. Soc. London* **A93**, 1–20.

Chapman, S. (1928). On approximate theories of diffusion phenomena, *Philos. Mag.* **5**, 630–636.

Chapman, S. (1967a). The kinetic theory of gases fifty years ago, in *Lectures in Theoretical Physics*, Vol. 9C: *Kinetic Theory*, W. E. Brittin, Ed., Gordon and Breach, New York, pp. 1–13.

Chapman, S. (1967b). Reminiscences, in *Sydney Chapman, Eighty*, S.-I. Akasofu, B. Fogle, and B. Haurwitz, Eds., sponsored by University of Alaska, University of Colorado, and University Corporation for Atmospheric Research, pp. 159–199.

Chapman, S., and T. G. Cowling (1970). *The Mathematical Theory of Non-Uniform Gases*, 3rd ed., Cambridge University Press, London.

Chen, F. M., H. Moraal, and R. F. Snider (1971). On the evaluation of kinetic theory collision integrals: Diamagnetic diatomic molecules, *J. Chem. Phys.* **57**, 542–561.

Coope, J. A. R. (1970). Irreducible Cartesian tensors: 3. Clebsch–Gordan reduction, *J. Math. Phys.* **11**, 1591–1612.

Coope, J. A. R., and R. F. Snider (1970). Irreducible Cartesian tensors: 2. General formulation, *J. Math. Phys.* **11**, 1003–1017.

Coope, J. A. R., R. F. Snider, and F. R. McCourt (1965). Irreducible Cartesian tensors, *J. Chem. Phys.* **43**, 2269–2275.

Cowling, T. G. (1950). *Molecules in Motion*, Hutchinson's University Library, London. Reprinted by Harper, New York, 1960.

Creaser, R. P. (1969). The mobility of ions in gases, thesis, Australian National University, Canberra.

Creaser, R. P. (1974). Measured mobility and derived interaction potentials for K^+ ions in rare gases, *J. Phys.* **B7**, 529–540.

Dalgarno, A. (1958). The mobilities of ions in their parent gases, *Philos. Trans. R. Soc. London* **A250**, 426–439.

Dalgarno, A., M. R. C. McDowell, and A. Williams (1958). The mobilities of ions in unlike gases, *Philos. Trans. R. Soc. London* **A250**, 411–425.

Davydov, B. (1935). Über die Geschwindigkeitsverteilung der sich im elektrischen Felde bewegenden Elektronen, *Phys. Z. Sowjetunion* **8**, 59–70.

Desloge, E. A. (1964). Transport properties of a simple gas, *Am. J. Phys.* **32**, 733–742.

Desloge, E. A. (1966). *Statistical Physics*, Holt, Rinehart and Winston, New York.

Druyvesteyn, M. J. (1930). De invloed der energieverliezen bij elastische botsingen in de theorie der electronendiffusie, *Physica* **10**, 61–70.

Druyvesteyn, M. J. (1934). Bemerkungen zu zwei früheren Arbeiten über die Elektronendiffusion, *Physica* **1**, 1003–1006.

Dymond, J. H. (1968). Repulsive potential-energy curves for the rare-gas atoms, *J. Chem. Phys.* **49**, 3673–3678.

Einstein, A. (1905). Über die von der molekular-kinetischen Theorie der Wärme geforderte Bewegung von in ruhenden Flüssigkeiten suspendierten Teilchen, *Ann. Phys. Leipzig* **17**, 549–560. An English translation appears in *Investigations on the Theory of the Brownian Movement*, R. Fürth, Ed., Dover, New York, 1956, pp. 1–18.

Einstein, A. (1908). Elementare Theorie der Brownschen Bewegung, *Z. Elektrochem.* **14**, 235–239. An English translation appears in *Investigations on the Theory of the Brownian Movement*, R. Fürth, Ed., Dover, New York, 1956, pp. 68–85.

Ellis, H. W., R. Y. Pai, E. W. McDaniel, E. A. Mason, and L. A. Viehland (1976). Transport properties of gaseous ions over a wide energy range, *At. Data Nucl. Data Tables* **17**, 177–210.

Ellis, H. W., E. W. McDaniel, D. L. Albritton, L. A. Viehland, S. L. Lin, and E. A. Mason (1978). Transport properties of gaseous ions over a wide energy range: Part 2, *At. Data Nucl. Data Tables* **22**, 179–217.

Enskog, D. (1917). Kinetische Theorie der Vorgänge in mässig verdünnten Gasen, Dissertation, Uppsala. An English translation appears in S. G. Brush, *Kinetic Theory*, Vol. 3, Pergamon, Oxford, 1972, pp. 125–225.

Enskog, D. (1922). Die numerische Berechnung der Vorgänge in mässig verdünnten Gasen, *Ark. Mat. Astron. Fys.* **16**, No. 16, 1–60.

Fernández de la Mora, J., and R. Fernández-Feria (1987). Kinetic theory of binary gas mixtures with large mass disparity, *Phys. Fluids* **30**, 740–751.

Ferziger, J. H., and H. G. Kaper (1972). *Mathematical Theory of Transport Processes in Gases*, North-Holland, Amsterdam.

Ford, G. W. (1968). Matrix elements of the linearized collision operator, *Phys. Fluids* **11**, 515–521.

Frankel, S. P. (1940). Elementary derivation of thermal diffusion, *Phys. Rev.* **57**, 661.

Furry, W. H. (1948). On the elementary explanation of diffusion phenomena in gases, *Am. J. Phys.* **16**, 63–78.

Goebel, C. J., S. M. Harris, and E. A. Johnson (1976). Two-temperature disparate-mass gas mixtures: A thirteen moment description, *Phys. Fluids* **19**, 627–635.

Goldman, E., and L. Sirovich (1969). Equations for gas mixtures: 2, *Phys. Fluids* **12**, 245–247.

Gough, D. W., G. C. Maitland, and E. B. Smith (1972). The direct determination of intermolecular potential energy functions from gas viscosity measurements, *Mol. Phys.* **24**, 151–161.

Grad, H. (1949a). Note on N-dimensional Hermite polynomials, *Commun. Pure Appl. Math.* **2**, 325–330.

Grad, H. (1949b). On the kinetic theory of rarefied gases, *Commun. Pure Appl. Math.* **2**, 331–407.

Grad, H. (1960). Theory of rarefied gases, in *Rarefied Gas Dynamics*, F. M. Devienne, Ed., Pergamon, Oxford, pp. 100–138.

Haddad, G. N., S. L. Lin, and R. E. Robson (1981). The effects of anisotropic scattering on electron transport, *Aust. J. Phys.* **34**, 243–249.

Hahn, H., and E. A. Mason (1971). Random-phase approximation for transport cross sections, *Chem. Phys. Lett.* **9**, 633–635.

Hahn, H., and E. A. Mason (1972). Field dependence of gaseous-ion mobility: Theoretical tests of approximate formulas, *Phys. Rev.* **A6**, 1573–1577.

Hahn, H., and E. A. Mason (1973). Energy partitioning of gaseous ions in an electric field, *Phys. Rev.* **A7**, 1407–1413.

Higgins, L. D., and F. J. Smith (1968). Collision integrals for high temperature gases, *Mol. Phys.* **14**, 399–400.

Hirschfelder, J. O., and M. A. Eliason (1957). The estimation of the transport properties for electronically excited atoms and molecules, *Ann. N.Y. Acad. Sci.* **67**, 451–461.

Hirschfelder, J. O., C. F. Curtiss, and R. B. Bird (1964). *Molecular Theory of Gases and Liquids*, Wiley, New York.

Hope, S. A., G. Féat, and P. T. Landsberg (1981). Higher-order Einstein relations for nonlinear charge transport, *J. Phys.* **A14**, 2377–2390.

Hoselitz, K. (1941). The mobility of alkali ions in gases: 5. Temperature measurements in the inert gases, *Proc. R. Soc. London* **A177**, 200–204.

Howorka, F., F. C. Fehsenfeld, and D. L. Albritton (1979). H^+ and D^+ ions in He: Observations of a runaway mobility, *J. Phys.* **B12**, 4189–4197.

Hunter, L. W., and R. F. Snider (1974). On the evaluation of kinetic theory collision integrals: 2. Angular momentum coupling schemes, *J. Chem. Phys.* **61**, 1160–1171.

Huxley, L. G. H., and R. W. Crompton (1974). *The Diffusion and Drift of Electrons in Gases*, Wiley, New York.

Iinuma, K., E. A. Mason, and L. A. Viehland (1987). Tests of approximate formulas for the calculation of ion mobility and diffusion in gas mixtures, *Mol. Phys.* **61**, 1131–1150.

Ikenberry, E. (1962). Representation of Grad's Hermite polynomials as sums of products of Sonine polynomials and solid spherical harmonics, *Arch. Ration. Mech. Anal.* **9**, 255–259.

Ikuta, N., and Y. Murakami (1987). Elementary theory of transport phenomena in charged-particle system under electric field, *J. Phys. Soc. Jpn.* **56**, 115–127.

Inouye, H., and S. Kita (1972). Experimental determination of the repulsive potentials between K^+ ions and rare-gas atoms, *J. Chem. Phys.* **56**, 4877–4882.

Jeans, J. H. (1925). *The Dynamical Theory of Gases*, 4th ed., Cambridge University Press, London. Reprinted by Dover, New York, 1954.

Johnson, D. E., and E. Ikenberry (1958). Developments toward a series solution of the Maxwell–Boltzmann equation, *Arch. Ration. Mech. Anal.* **2**, 41–65.

Johnston, T. W. (1960). Cartesian tensor scalar product and spherical harmonic expansions in Boltzmann's equation, *Phys. Rev.* **120**, 1103–1111.

Keizer, J. (1972). On a time dependent theory of hot-atom reactions, *J. Chem. Phys.* **56**, 5958–5962.

Kennard, E. H. (1938). *Kinetic Theory of Gases*, McGraw-Hill, New York.

Kihara, T. (1953). The mathematical theory of electrical discharges in gases: B. Velocity-distribution of positive ions in a static field, *Rev. Mod. Phys.* **25**, 844–852.

Klots, C. E., and D. R. Nelson (1970). The ratio of transverse and longitudinal diffusion coefficients, *Bull Am. Phys. Soc.* **15**, 424.

Knierim, K. D., S. L. Lin, and E. A. Mason (1981). Time-dependent moment theory of hot-atom reactions, *J. Chem. Phys.* **75**, 1159–1165.

Kramers, H. A. (1949). On the behaviour of a gas near a wall, *Nuovo Cimento Suppl.* **6**, 297–304.

Kumar, K. (1966a). Talmi transformation for unequal-mass particles and related formulas, *J. Math. Phys.* **7**, 671–678.

Kumar, K. (1966b). Polynomial expansions in kinetic theory of gases, *Ann. Phys. N.Y.* **37**, 113–141.

Kumar, K. (1967). The Chapman–Enskog solution of the Boltzmann equation: A reformulation in terms of irreducible tensors and matrices, *Aust. J. Phys.* **20**, 204–252.

Kumar, K. (1970). Transport coefficients for a multicomponent mixture of ionized gases, *Aust. J. Phys.* **23**, 505–520.

Kumar, K. (1980a). Matrix elements of the Boltzmann collision operator in a basis determined by an anisotropic Maxwellian weight function including drift, *Aust. J. Phys.* **33**, 449–468.

Kumar, K. (1980b). Relation between polynomials orthogonal with respect to a class of Gaussian weight functions, *Aust. J. Phys.* **33**, 469–479.

Kumar, K. (1984). The physics of swarms and some basic questions of kinetic theory, *Phys. Rep.* **112**, 319–375.

Kumar, K., and R. E. Robson (1973). Mobility and diffusion: 1. Boltzmann equation treatment for charged particles in a neutral gas, *Aust. J. Phys.* **26**, 157–186.

Kumar, K., H. R. Skullerud, and R. E. Robson (1980). Kinetic theory of charged particle swarms in neutral gases, *Aust. J. Phys.* **33**, 343–448.

Langevin, P. (1905). Une formule fondamentale de théorie cinétique, *Ann. Chim. Phys.* **5**, 245–288. A translation is given by E. W. McDaniel, *Collision Phenomena in Ionized Gases*, Wiley, New York, 1964, App. 2, pp. 701–726.

Lin, S. L., I. R. Gatland, and E. A. Mason (1979a). Mobility and diffusion of protons and deuterons in helium—A runaway effect, *J. Phys.* **B12**, 4179–4188.

Lin, S. L., R. E. Robson, and E. A. Mason (1979b). Moment theory of electron drift and diffusion in neutral gases in an electrostatic field, *J. Chem. Phys.* **71**, 3483–3498.

Lin, S. L., L. A. Viehland, and E. A. Mason (1979c). Three-temperature theory of gaseous ion transport, *Chem. Phys.* **37**, 411–424.

Loeb, L. B. (1960). *Basic Processes of Gaseous Electronics*, 2nd ed., University of California Press, Berkeley, Calif.

Lorentz, H. A. (1905). The motion of electrons in metallic bodies: 1, 2, 3, *Proc. Amsterdam Acad.* **7**, 438–453, 585–593, 684–691 (English Transl.) [*K. Akad. Wet. Amsterdam* **13**, 493–508, 565–573, 710–719 (1905)]. See also *The Theory of Electrons*, Dover, New York, 1952, pp. 267–274.

Louck, J. D., and G. P. DeVault (1964). Eigenfunctions of the Boltzmann collision operator, *Phys. Fluids* **7**, 1388–1390.

McCormack, F. J. (1971). Comments on "Derivation of the Davydov distribution from the Boltzmann equation," *Am. J. Phys.* **39**, 1413.

McDaniel, E. W., and E. A. Mason (1973). *The Mobility and Diffusion of Ions in Gases*, Wiley, New York.

Mason, E. A. (1957a). Higher approximations for the transport properties of binary gas mixtures: 1. General formulas, *J. Chem. Phys.* **27**, 75–84.

Mason, E. A. (1957b). Higher approximations for the transport properties of binary gas mixtures: 2. Applications, *J. Chem. Phys.* **27**, 782–790.

Mason, E. A., and R. B. Evans III (1969). Graham's laws: Simple demonstrations of gases in motion: Part 1. Theory, *J. Chem. Educ.* **46**, 358–364.

Mason, E. A., and H. Hahn (1972). Ion drift velocities in gaseous mixtures at arbitrary field strengths, *Phys. Rev.* **A5**, 438–441.

Mason, E. A., and T. R. Marrero (1970). The diffusion of atoms and molecules, *Adv. At. Mol. Phys.* **6**, 155–232.

Mason, E. A., and H. W. Schamp, Jr. (1958). Mobility of gaseous ions in weak electric fields, *Ann. Phys. N.Y.* **4**, 233–270.

Maxwell, J. Clerk (1860). Illustrations of the dynamical theory of gases: Part 2. On the process of diffusion of two or more kinds of moving particles among one another, *Philos. Mag.* **20**, 21–33. Reprinted in *The Scientific Papers of James Clerk Maxwell*, Vol. 1, Dover, New York, 1962, pp. 392–405.

Maxwell, J. Clerk (1867). On the dynamical theory of gases, *Philos. Trans. R. Soc. London* **157**, 49–88. Reprinted in *The Scientific Papers of James Clerk Maxwell*, Vol. 2, Dover, New York, 1962, pp. 26–78; and in S. G. Brush, *Kinetic Theory*, Vol. 2, *Irreversible Processes*, Pergamon, Oxford, 1966, pp. 23–87.

Milloy, H. B., and R. E. Robson (1973). The mobility of potassium ions in gas mixtures, *J. Phys.* **B6**, 1139–1152.

Mintzer, D. (1965). Generalized orthogonal polynomial solutions of the Boltzmann equation, *Phys. Fluids* **8**, 1076–1090.

Monchick, L. (1959). Collision integrals for the exponential repulsive potential, *Phys. Fluids* **2**, 695–700.

Monchick, L. (1962). Equivalence of the Chapman–Enskog and the mean-free-path theory of gases, *Phys. Fluids* **5**, 1393–1398.

Monchick, L., and E. A. Mason (1967). Free-flight theory of gas mixtures, *Phys. Fluids* **10**, 1377–1390.

Morse, P. M., W. P. Allis, and E. S. Lamar (1935). Velocity distributions for elastically colliding electrons, *Phys. Rev.* **48**, 412–419.

Mott-Smith, H. M. (1951). The solution of the Boltzmann equation for a shock wave, *Phys. Rev.* **82**, 885–892.

Mozumder, A. (1980a). Electron thermalization in gases: 1. Helium, *J. Chem. Phys.* **72**, 1657–1664.

Mozumder, A. (1980b). Electron thermalization in gases: 2. Neon, argon, krypton, and xenon, *J. Chem. Phys.* **72**, 6289–6298.

Muckenfuss, C. (1962). Some aspects of shock structure according to the bimodal model, *Phys. Fluids* **5**, 1325–1336.

Nernst, W. (1888). Zur Kinetik der in Losüng befindlichen Körper: 1. Theorie der Diffusion, *Z. Phys. Chem.* **2**, 613–637. An English translation appears in *Cell Membrane Permeability and Transport*, G. R. Kepner, Ed., Dowden, Hutchinson & Ross, Stroudsburg, Pa. 1979, pp. 174–183.

Ness, K. F., and R. E. Robson (1985). Interaction integrals in the kinetic theory of gases, *Transp. Theory Stat. Phys.* **14**, 257–290.

Paranjape, B. V. (1980). Field dependence of mobility in gases, *Phys. Rev.* **A21**, 405–407.

Phelps, A. V., and L. C. Pitchford (1985). Anisotropic scattering of electrons by N_2 and its effect on electron transport, *Phys. Rev.* **A31**, 2932–2949.

Pidduck, F. B. (1916). The kinetic theory of the motion of ions in gases, *Proc. London Math. Soc.* **15**, 89–127.

Pidduck, F. B. (1936). Energy distribution of electrons in a gas, *Q. J. Math.* **7**, 199–201.

Pitchford, L. C., and A. V. Phelps (1982). Comparative calculations of electron-swarm properties in N_2 at moderate E/N values, *Phys. Rev.* **A25**, 540–554.

Pitchford, L. C., S. V. O'Neil, and J. R. Rumble, Jr. (1981). Extended Boltzmann analysis of electron swarm experiments, *Phys. Rev.* **A23**, 294–304.

Present, R. D. (1958). *Kinetic Theory of Gases*, McGraw-Hill, New York.

Present, R. D., and A. J. de Bethune (1949). Separation of a gas mixture flowing through a long tube at low pressure, *Phys. Rev.* **75**, 1050–1057.

Rayleigh, Lord (1891). Dynamical problems in illustration of the theory of gases, *Philos. Mag.* **32**, 424–445. Reprinted in *Scientific Papers*, Vol. 3, Cambridge University Press, London, 1902, pp. 473–490.

Rayleigh, Lord (1900). On the viscosity of argon as affected by temperature, *Proc. R. Soc. London* **66**, 68–74. Reprinted in *Scientific Papers*, Vol. 4, Cambridge University Press, London, 1902, 452–458.

Reid, I. D. (1979). An investigation of the accuracy of numerical solutions of Boltzmann's equation for electron swarms in gases with large inelastic cross sections, *Aust. J. Phys.* **32**, 231–254; *Corrigendum* **35**, 473–474 (1982).

Robson, R. E. (1972). A thermodynamic treatment of anisotropic diffusion in an electric field, *Aust. J. Phys.* **25**, 685–693.

Robson, R. E. (1973). Mobility of ions in gas mixtures, *Aust. J. Phys.* **26**, 203–206.

Robson, R. E. (1976a). On the generalized Einstein relation for gaseous ions in an electrostatic field, *J. Phys.* **B9**, L337–L339.

Robson, R. E. (1976b). Diffusion cooling of electrons in a finite gas, *Phys. Rev.* **A13**, 1536–1542.

Robson, R. E. (1981). Boundary effects in solution of Boltzmann's equation for electron swarms, *Aust. J. Phys.* **34**, 223–241.

Robson, R. E., and K. Kumar (1973). Mobility and diffusion: 2. Dependence on experimental variables and interaction potential for alkali ions in rare gases, *Aust. J. Phys.* **26**, 187–201.

Robson, R. E., and E. A. Mason (1982). Comment on "Field dependence of mobility in gases," *Phys. Rev.* **A25**, 2411–2413.

Shizgal, B. (1979). Eigenvalues of the Lorentz Fokker–Planck equation, *J. Chem. Phys.* **70**, 1948–1951.

Shizgal, B. (1980). Nonequilibrium time dependent theory of hot atom reactions: 2. The hot $^{18}F + H_2$ reaction, *J. Chem. Phys.* **72**, 3156–3162.

Shizgal, B. (1981a). Nonequilibrium time dependent theory of hot atom reactions: 3. Comparison with Estrup–Wolfgang theory, *J. Chem. Phys.* **74**, 1401–1408.

Shizgal, B. (1981b). A Gaussian quadrature procedure for use in the solution of the Boltzmann equation and related problems, *J. Comput. Phys.* **41**, 309–328.

Shizgal, B. (1983a). Electron thermalization in gases, *J. Chem. Phys.* **78**, 5741–5744.

Shizgal, B. (1983b). Energy relaxation of electrons in helium, *Chem. Phys. Lett.* **100**, 41–44.

Shizgal, B., and J. M. Fitzpatrick (1974). Matrix elements of the linear Boltzmann collision operator for systems of two components at different temperatures, *Chem. Phys.* **6**, 54–65.

Shizgal, B., and J. M. Fitzpatrick (1980). Nonequilibrium time dependent theory of hot atom reactions: 1. Model calculations, *J. Chem. Phys.* **72**, 3143–3155.

Shizgal, B., M. J. Lindenfeld, and R. Reeves (1981). Eigenvalues of the Boltzmann collision operator for binary gases: Mass dependence, *Chem. Phys.* **56**, 249–260.

Skullerud, H. R. (1972). Mobility, diffusion and interaction potential for potassium ions in argon, *Tech. Rep. EIP 72-3*, Physics Dept., Norwegian Inst. Tech., Trondheim.

Skullerud, H. R. (1973a). Monte-Carlo investigations of the motion of gaseous ions in electrostatic fields, *J. Phys.* **B6**, 728–742.

Skullerud, H. R. (1973b). Mobility, diffusion and interaction potential for potassium ions in argon, *J. Phys.* **B6**, 918–928.

Skullerud, H. R. (1976). On the relation between the diffusion and mobility of gaseous ions moving in strong electric fields, *J. Phys.* **B9**, 535–546.

Skullerud, H. R., and L. R. Forsth (1979). Perturbation treatment of thermal motions in gaseous ion-transport theory, *J. Phys.* **B12**, 1881–1888.

Skullerud, H. R., and S. Kuhn (1983). On the calculation of ion and electron swarm properties by path-integral methods, *J. Phys.* **D16**, 1225–1234.

Smirnov, Yu. F. (1961). Talmi transformations for particles with different masses, *Nucl. Phys.* **27**, 177–187.

Stefan, J. (1871). Über das Gleichgewicht und die Bewegung, insbesondere die Diffusion von Gasgemengen, *Sitzungsber. Akad. Wiss. Wien* **63**, 63–124.

Stefan, J. (1872). Über die dynamische Theorie der Diffusion der Gase, *Sitzungsber. Akad. Wiss. Wien* **65**, 323–363.

Suchy, K. (1964). Neue Methoden in der kinetische Theorie verdünnter Gase, *Springer Tracts Mod. Phys. (Ergeb. Exakten Naturwiss.)* **35**, 103–294.

Talmi, I. (1952). Nuclear spectroscopy with harmonic oscillator wave functions, *Helv. Phys. Acta* **25**, 185–234.

Townsend, J. S. (1899). The diffusion of ions into gases, *Philos. Trans. R. Soc. London* **A193**, 129–158.

Viehland, L. A., and S. L. Lin (1979). Application of the three-temperature theory of gaseous ion transport, *Chem. Phys.* **43**, 135–144.

Viehland, L. A., and E. A. Mason (1975). Gaseous ion mobility in electric fields of arbitrary strength, *Ann. Phys. N.Y.* **91**, 499–533.

Viehland, L. A., and E. A. Mason (1978). Gaseous ion mobility and diffusion in electric fields of arbitrary strength, *Ann. Phys. N.Y.* **110**, 287–328.

Viehland, L. A., E. A. Mason, and J. H. Whealton (1974). Mean energy distribution of gaseous ions in electrostatic fields, *J. Phys.* **B7**, 2433–2439.

Viehland, L. A., M. M. Harrington, and E. A. Mason (1976). Direct determination of ion-neutral molecule interaction potentials from gaseous ion mobility measurements, *Chem. Phys.* **17**, 433–441.

Waldman, M., and E. A. Mason (1981). Generalized Einstein relations from a three-temperature theory of gaseous ion transport, *Chem. Phys.* **58**, 121–144.

Waldman, M., E. A. Mason, and L. A. Viehland (1982). Influence of resonant charge transfer on ion diffusion and generalized Einstein relations, *Chem. Phys.* **66**, 339–349.

Wannier, G. H. (1951). On the motion of gaseous ions in a strong electric field: 1, *Phys. Rev.* **83**, 281–289.

Wannier, G. H. (1952). Motion of gaseous ions in a strong electric field: 2, *Phys. Rev.* **87**, 795–798.

Wannier, G. H. (1953). Motion of gaseous ions in strong electric fields, *Bell Syst. Tech. J.* **32**, 170–254.

Wannier, G. H. (1971). Derivation of the Davydov distribution from the Boltzmann equation, *Am. J. Phys.* **39**, 281–285.

Wannier, G. H. (1973). On a conjecture about diffusion of gaseous ions, *Aust. J. Phys.* **26**, 897–900.

Weinert, U. (1978). Matrix elements of the linearized collision operator for multi-temperature gas-mixtures, *Z. Naturforsch.* **33a**, 480–492.

Weinert, U. (1979). Matrix representation of the kinetic theory differential operator, *J. Math. Phys.* **20**, 2339–2346.

Weinert, U. (1980). Spherical tensor representation, *Arch. Ration. Mech. Anal.* **74**, 165–196.

Weinert, U. (1981). On the inversion of the linearized collision operator, *Z. Naturforsch.* **36a**, 113–120.

Weinert, U. (1982). Multi-temperature generalized moment method in Boltzmann transport theory, *Phys. Rep.* **91**, 297–399.

Weinert, U. (1983a). Gaussian basis functions in kinetic theory: 1. Incorporation of inelastic processes, *Physica* **121A**, 150–174.

Weinert, U. (1983b). Electrical conductivity for strong fields, *Proc. 16th Int. Conf. Phenomena Ionized Gases*, Düsseldorf, Vol. 1, pp. 62–63.

Weinert, U. (1984). Gaussian basis functions in kinetic theory: 2. Completely transformed kinetic equation, *Physica* **125A**, 497–518.

Weinert, U., and E. A. Mason (1980). Generalized Nernst–Einstein relations for nonlinear transport coefficients, *Phys. Rev.* **A21**, 681–690.

Weinert, U., and K. Suchy (1977). Generalization of the moment method of Maxwell–Grad for multi-temperature gas mixtures and plasmas, *Z. Naturforsch.* **32a**, 390–400.

Weitzsch, F. (1961). Ein neuer Ansatz für die Behandlung gasdynamischer Probleme bei starken Abweichungen vom thermodynamischen Gleichgewicht, *Ann. Phys. Leipzig* **7**, 403–417.

Whealton, J. H., and E. A. Mason (1972). Composition dependence of ion diffusion coefficients in gas mixtures at arbitrary field strengths, *Phys. Rev.* **A6**, 1939–1942.

Whealton, J. H., and E. A. Mason (1974). Transport coefficients of gaseous ions in an electric field, *Ann. Phys. N.Y.* **84**, 8–38.

Whealton, J. H., E. A. Mason, and R. E. Robson (1974). Composition dependence of ion-transport coefficients in gas mixtures, *Phys. Rev.* **A9**, 1017–1020.

Williams, F. A. (1958). Elementary derivation of the multicomponent diffusion equation, *Am. J. Phys.* **26**, 467–469.

Woo, S. B., S. P. Hong, and J. H. Whealton (1976). Semi-empirical joint ion-neutral speed distributions in a weakly ionized gas in electric fields, *J. Phys.* **B9**, 2553–2558.

6

SOME ACCURATE THEORETICAL RESULTS

In Chapter 5 we have given a summary of accurate kinetic theory as represented by moment solutions of the Boltzmann equation. In this chapter we present more details of these solutions and some of the results that follow from them. We first give accounts of the one-, two-, and three-temperature theories, and show how they can be used to improve the generalized Einstein relations developed in Sections 5-2B and 5-2C. We then consider the effect of resonant charge exchange on the accurate calculation of mobilities and diffusion coefficients. Resonant charge exchange affects the magnitude and energy dependence of the ion-neutral cross sections at all field strengths, and the peculiar nature of charge-exchange collisions produces a characteristic distortion at high fields of the ion distribution function, which becomes anisotropic with a long high-velocity tail in the field direction. Finally, we relax the conditions of spherically symmetric ion-neutral interactions and elastic collisions, in order to develop the description of polyatomic systems, both ions and neutrals. A brief review of this material, together with experiments and applications, has been given by McDaniel and Viehland (1984).

This chapter is thus a review of the present state of the accurate theory, including both its successes and its imperfections. Notable among the latter are various convergence difficulties, some of which are unusual and unexpected. Many of these have been "papered over" in a limited practical sense by brute-force computation, but there is still plenty of room for improvement. We hope that this account will encourage efforts toward improvements.

6-1. ONE-TEMPERATURE THEORY

There are three reasons for discussing the one-temperature theory, even though it converges only for weak electric fields. The first is that it provides the connection with the classical Chapman–Enskog kinetic theory of gases, which is perhaps familiar to the reader. The second is that the theory has a structure similar to that of the two-temperature theory, and so forms a useful introduction to a theory that converges at all field strengths (at least for the mobility). The third is that some of the relations obtained are in fact accurate at high fields.

A.   Vanishing Fields: Chapman–Enskog Theory

Here the zero-order ion distribution function is an equilibrium Maxwellian at the gas temperature, as given by (5-3-27). The Burnett basis functions of (5-3-28) are, however, simplified by the conditions of the problem. In the first place, we need consider only systems with cylindrical symmetry, and can therefore set $m = 0$ in the spherical harmonics. Second, because the electric field is vanishingly small we need consider only a first-order distortion of the distribution function in the field direction, and therefore need only the terms with $l = 1$. Third, the Nernst–Townsend–Einstein relation holds exactly for vanishingly small fields, and we do not need to consider diffusion explicitly because $D = (kT/e)K$. We can therefore drop the density gradient term from the moment equations.

The expansion for the distribution function thus takes the following simplified form:

$$f = f^{(0)}\left[1 + w_z \sum_{r=0}^{\infty} c_r S_{3/2}^{(r)}(w^2)\right],\qquad(6\text{-}1\text{-}1)$$

where

$$f^{(0)} = n(m/2\pi kT)^{1/2}\exp(-w^2),$$

$$\mathbf{w} = (m/2kT)^{1/2}\mathbf{v},$$

as before, and the c_r are coefficients. That is, we use only the functions $\psi_{10}^{(r)}$ out of the full set of Burnett functions given by (5-3-28). The moment equations (5-3-19) become, after a little manipulation,

$$(eE/m)(m/2kT)^{1/2}[(\tfrac{3}{2}+r)\langle\psi_{00}^{(r)}\rangle - 2\langle\psi_{20}^{(r-1)}\rangle] = \tfrac{3}{2}N\sum_{s=0}^{\infty} b_{rs}\langle\psi_{10}^{(s)}\rangle,\qquad(6\text{-}1\text{-}2a)$$

where

$$b_{rs} = \sum_{j} x_j a_{rs}^{(j)},\qquad(6\text{-}1\text{-}2b)$$

in which the index j refers to the species of neutral gas if a mixture is considered. The Burnett functions with $l = 0$ and 2 arise from the evaluation of the term $\langle\mathbf{V}_v\psi_{10}^{(r)}\rangle$, through the recursion relation

$$\left(l+\frac{1}{2}\right)\frac{\partial\psi_{lm}^{(r)}}{\partial w_z} = (l+|m|)\left(l+\frac{1}{2}+r\right)\psi_{l-1,m}^{(r)} - (l+1-|m|)\psi_{l+1,m}^{(r-1)},\qquad(6\text{-}1\text{-}3)$$

with $\psi_{lm}^{(-1)} \equiv 0$. To find the mobility this infinite set of equations (6-1-2) must be solved to give $\langle \psi_{10}^{(0)} \rangle$ in terms of the b_{rs}; the mobility then follows directly from

$$v_d = \langle v_z \rangle = (2kT/m)^{1/2} \langle \psi_{10}^{(0)} \rangle. \qquad (6\text{-}1\text{-}4)$$

The equations cannot in general be solved exactly, and some approximation scheme must be adopted. For simplicity we now consider only a single neutral gas. The matrix elements a_{rs} are given by (5-3-18) as

$$a_{rs} = (\psi_{10}^{(s)}, J\psi_{10}^{(r)})/(\psi_{10}^{(s)}, \psi_{10}^{(s)}). \qquad (6\text{-}1\text{-}5)$$

From this expression we can devise a reasonable approximation scheme based on a presumed ordering of the a_{rs} according to magnitude. For the Maxwell model the Burnett functions are eigenfunctions of the collision operator:

$$J\psi^{(r)} = \lambda_r \psi^{(r)},$$

where λ_r is the eigenvalue. Since the $\psi^{(r)}$ are orthogonal, we see from (6-1-5) that $a_{rs} = 0$ for $r \neq s$ for the Maxwell model. For a first approximation for other ion-neutral interactions we therefore set all off-diagonal a_{rs} equal to zero and obtain

$$(eE/m)(m/2kT)^{1/2}[(\tfrac{3}{2} + r)\langle \psi_{00}^{(r)} \rangle_1 - 2\langle \psi_{20}^{(r-1)} \rangle_1] = \tfrac{3}{2}Na_{rr}\langle \psi_{10}^{(r)} \rangle_1, \quad (6\text{-}1\text{-}6)$$

where the notation $\langle \cdots \rangle_1$ means the first approximation. In particular, since $\psi_{00}^{(0)} = 1$ and $\psi_{20}^{(-1)} = 0$, we find

$$Na_{00}\langle \psi_{10}^{(0)} \rangle_1 = (eE/m)(m/2kT)^{1/2}, \qquad (6\text{-}1\text{-}7)$$

or

$$[K]_1 = (e/kT)[D]_1 = e/mNa_{00}, \qquad (6\text{-}1\text{-}8)$$

where $[\cdots]_1$ means the first approximation. This result is exact for the Maxwell model but only approximate for other ion-neutral interactions.

Higher approximations for $\langle \psi_{10}^{(0)} \rangle$ can now be obtained by inserting lower approximations for the other $\langle \psi_{10}^{(s)} \rangle$ in the terms involving the off-diagonal a_{rs} in the summation on the right-hand side of the moment equations (6-1-2a). There is some choice at this point in the particular systematic procedure adopted to generate higher approximations; ultimately all such procedures should converge to the same result, but some might be more effective than others at lower levels of approximation. Two truncation schemes are commonly used in Chapman–Enskog theory, one due to Chapman and Cowling and one due to Kihara. A fairly detailed exposition of these two truncation schemes and comparisons of their numerical accuracy has been given (Mason, 1957a, b), but for K and D the convergence is so good that such details are unimportant for the present purposes. A straightforward iteration of the present moment equations yields (Mason and Schamp, 1958)

$$\frac{K}{[K]_1} = \frac{D}{[D]_1} = 1 + \frac{a_{01}a_{10}}{a_{00}a_{11}} + \frac{a_{02}a_{20}}{a_{00}a_{22}} + \cdots. \tag{6-1-9}$$

This expression is neither the Chapman–Cowling nor the Kihara result, but the differences are unimportant. Numerical testing of the convergence for particular cases is given in Section 6-1C.

All that remains is to evaluate the matrix elements a_{rs}, which is really the hardest part of the whole calculation. As discussed in Section 5-3E, most of the calculations can be carried out without specifying the ion-neutral interaction, so that the a_{rs} are expressed as linear combinations of irreducible collision integrals. In the present case these are conventionally defined as follows:

$$\bar{\Omega}^{(l,s)}(T) \equiv [(s+1)!(kT)^{s+2}]^{-1} \int_0^\infty \bar{Q}^{(l)}(\varepsilon) \exp(-\varepsilon/kT)\varepsilon^{s+1}\, d\varepsilon, \tag{6-1-10}$$

where $\varepsilon = \frac{1}{2}\mu v_r^2$ is the relative energy of the ion-neutral collision. (The indices l, s used here are *not* the same as those in the matrix elements and Burnett functions.) The transport cross sections are

$$\bar{Q}^{(l)}(\varepsilon) \equiv 2\pi \left[1 - \frac{1 + (-1)^l}{2(1 + l)}\right]^{-1} \int_0^\pi (1 - \cos^l \theta)\sigma(\theta, \varepsilon) \sin \theta\, d\theta. \tag{6-1-11}$$

The normalization factors for $\bar{\Omega}^{(l,s)}$ and $\bar{Q}^{(l)}$ here have been chosen so that both quantities are equal to πd^2 for the collision of classical rigid spheres of diameter d. Given the ion-neutral potential, the $\bar{Q}^{(l)}$ and $\bar{\Omega}^{(l,s)}$ can be computed, but numerical integration is usually necessary. The calculations to find the a_{rs} are systematic but tedious and involved; the first few have been collected and tabulated (McDaniel and Mason, 1973, pp. 172–173).

The first approximation to K and D according to (6-1-8) is determined by a_{00}, which is

$$a_{00} = (8/3m)(2\mu kT/\pi)^{1/2}\bar{\Omega}^{(1,1)}, \tag{6-1-12}$$

which leads to

$$[K]_1 = \frac{3e}{16N} \left(\frac{2\pi}{\mu kT}\right)^{1/2} \frac{1}{\bar{\Omega}^{(1,1)}}. \tag{6-1-13}$$

The relation of this result to the momentum-transfer result of (5-2-23) has already been commented on in Section 5-2A. The numerical factors differ, and instead of the thermally averaged cross section $\bar{\Omega}^{(1,1)}$ we had a cross section Q_D evaluated at an average (thermal) energy. We can now see that the correspondence occurs much earlier in the calculation. In particular, the moment equation from which we obtained the expression above for $[K]_1$ is essentially the same as the momentum-balance equation of (5-2-8), which was

$$eE = \mu v_d \nu(\bar{\varepsilon}),$$

where $\nu(\bar{\varepsilon})$ is the collision frequency at the average relative energy $\bar{\varepsilon}$.

To demonstrate this correspondence, we must first be more careful about the difference between $\bar{v}_r^2$ and $\overline{v_r^2}$; when both the ions and the neutrals have a Maxwellian distribution, the mean relative speed is

$$\bar{v}_r = = (8kT/\pi\mu)^{1/2}. \tag{6-1-14}$$

We next define an average collision frequency at the temperature T as

$$\bar{\nu}(T) \equiv \tfrac{4}{3} N \bar{v}_r \bar{\Omega}^{(1,1)}(T), \tag{6-1-15a}$$

in precise analogy to the definition of $\nu(\bar{\varepsilon}) = N\bar{v}_r Q_D$ given by (5-2-7). The numerical factor of $\tfrac{4}{3}$ occurs because the momentum-transfer theory uses the approximation of constant collision frequency (Maxwell model), for which $v_r Q_D$ is a constant and for which (McDaniel and Mason, 1973, p. 196)

$$v_r Q_D = v_r Q^{(1)} = \tfrac{4}{3}\bar{v}_r \bar{\Omega}^{(1,1)}. \tag{6-1-15b}$$

(The factor of $\tfrac{4}{3}$ would be unity for a rigid-sphere model.) With these results we find the matrix element a_{00} in terms of the collision frequency to be

$$Na_{00} = (\mu/m)\bar{\nu}(T), \tag{6-1-16}$$

and the moment equation (6-1-8) in first approximation to be

$$\mu\bar{\nu}(T)\langle v_z\rangle_1 = eE. \tag{6-1-17}$$

This is exactly the momentum-balance equation of the momentum-transfer theory.

Thus the lowest-order moment equation for the drift velocity corresponds to the equation for momentum balance. We do not need an energy-balance equation in this case, because the ion energy is entirely thermal when the electric field is vanishingly small.

To summarize, the foregoing results for K and D are valid only in the limit of vanishingly small fields, binary collision mechanism, elastic collisions, and small deviations from equilibrium. As written, the results are valid in quantum mechanics as well as classical mechanics, provided only that the Boltzmann equation is valid. The difference between quantum and classical mechanics enters only through the expression for the differential cross section $\sigma(\theta, \varepsilon)$, which appears in the formula (6-1-11) for the transport cross sections.

B. Weak Fields: Kihara Theory

The main difference between weak and vanishingly small fields is that we must now consider more than just first-order distortions of the ion distribution function in the field direction, and so must consider Burnett basis functions with general l. This also allows the mean ion energy to differ from $\frac{3}{2}kT$. As a consequence, the Nernst–Townsend–Einstein relation no longer holds exactly, and we must consider diffusion and mobility separately, keeping the density gradient term in the moment equations. The greatest limitation arises from the choice of $f^{(0)}$ as an equilibrium distribution; this generally restricts the results to weak fields, because at high fields the ion distribution function will be far from its equilibrium form, and an expansion about equilibrium is unlikely to be accurate. This problem has also been carried through by the Chapman–Enskog method rather than a moment method (Schruben and Condiff, 1973); the results are of course equivalent.

We first consider the spatially homogeneous case, which allows us to calculate the mobility and mean ion energy (including its partitioning in different directions), and then consider diffusion. For simplicity we consider only a single neutral gas here, and deal separately with mixtures in Section 6-1G. For this case we can set $m = 0$, and the expansion for the distribution function is

$$f = f^{(0)} \sum_{r=0}^{\infty} \sum_{l=0}^{\infty} f_l^{(r)} w^l S_{l+1/2}^{(r)}(w^2) P_l(w_z/w), \qquad (6\text{-}1\text{-}18)$$

where the expansion coefficients $f_l^{(r)}$ are, as usual, moments of the distribution over the basis functions $\psi_{l0}^{(r)}$. The expansion (6-1-1) for the vanishing field is a special case of this expansion, obtained as follows: the $r = 0$, $l = 0$ term gives the unity inside the brackets in (6-1-1), $f_1^{(r)} = c_r$ in (6-1-1), and all other terms are dropped.

For the spatially homogeneous case the moment equations become, on using the recursion relation (6-1-3),

$$(eE/m)(m/2kT)^{1/2}[l(l + \tfrac{1}{2} + r)\langle \psi_{l-1,0}^{(r)} \rangle - (l + 1)\langle \psi_{l+1,0}^{(r-1)} \rangle]$$

$$= (l + \tfrac{1}{2})N \sum_{s=0}^{\infty} a_{rs}(l)\langle \psi_{l0}^{(s)} \rangle, \qquad (6\text{-}1\text{-}19)$$

where

$$a_{rs}(l) \equiv (\psi_{l0}^{(s)}, J\psi_{l0}^{(r)})/(\psi_{l0}^{(s)}, \psi_{l0}^{(s)}) \qquad (6\text{-}1\text{-}20)$$

The a_{rs} for the vanishing field case correspond to $a_{rs}(1)$ here. From this set of equations we can find the mobility and the mean ion energy.

1. Mobility

To find the mobility we focus on the function

$$\psi_{10}^{(0)} = w_z, \qquad (6\text{-}1\text{-}21)$$

and seek some approximation scheme for solving (6-1-19) for $\langle \psi_{10}^{(0)} \rangle$. A number of schemes have been proposed (Kihara, 1953; Mason and Schamp, 1958; Kumar and Robson, 1973; Robson and Kumar, 1973). All are based on the Maxwell model as a first approximation, which yields (6-1-17) again, and all are equivalent to an expansion in powers of $(E/N)^2$, as expected both from the general considerations of Section 5-1B and from the momentum-transfer theory of Section 5-2A, according to the expansion given in (5-2-27). As might be expected from (5-2-27), the ratio E/N always occurs multiplied by the quantity $[e/kT\bar{\Omega}^{(1,1)}]$, so that it is convenient to define a dimensionless field-strength function as

$$\mathscr{E} \equiv \left[\left(\frac{m}{2kT} \right)^{1/2} \frac{e}{m a_{00}(1)} \right] \frac{E}{N} = \frac{3\pi^{1/2}}{16} \left(\frac{m + M}{M} \right)^{1/2} \left[\frac{e}{kT\bar{\Omega}^{(1,1)}} \right] \frac{E}{N}. \quad (6\text{-}1\text{-}22)$$

[This dimensionless $\mathscr{E}$ is analogous to the $\mathscr{E}^*$ of (5-2-34) for rigid spheres; it should not be confused with the same symbol used in Section 5-1C for just E/N.]

The mobility can then be written as the expansion

$$K = [K]_1 (g_0 + g_2 \mathscr{E}^2 + g_4 \mathscr{E}^4 + \cdots), \quad (6\text{-}1\text{-}23)$$

where each g_n is a complicated function of the $a_{rs}(l)$. The first term corresponds to the vanishing field case, and higher approximations extend the results into the weak-field region. Aside from convergence, which is considered in the next section, the main limitation of this expansion is the increasing complexity of the algebra as further terms are added, together with the difficulty of evaluating the complicated new $a_{rs}(l)$ that appear. Since the convergence turns out to be poor, there is little point in giving detailed expressions for the g_n here; they can be found in McDaniel and Mason (1973, Sec. 5-4).

With the benefit of hindsight from the momentum-transfer theory and from the two-temperature theory (Section 6-2), we can see that the above expansion for K is equivalent to a successive refinement of the collisional loss part of the momentum-balance equation. Defining the collision frequency as in (6-1-15a), we find that (6-1-23) is equivalent to a momentum-balance equation of the form

$$\mu \langle v_z \rangle \frac{\bar{v}(T)}{g_0 + g_2 \mathscr{E}^2 + \cdots} = eE. \quad (6\text{-}1\text{-}24)$$

That is, the higher approximations appear to adjust $\bar{v}(T)$ to an effective collision frequency when the field is nonvanishing.

2. Ion Energy

To find the total ion energy we focus on the function

$$\psi_{00}^{(1)} = \tfrac{3}{2} - w^2, \quad (6\text{-}1\text{-}25)$$

and try to solve the moment equations (6-1-19) for $\langle \psi_{00}^{(1)} \rangle$. We pick $l = 0$, and for a first approximation set all the off-diagonal $a_{rs}(l)$ equal to zero, obtaining

$$-(eE/m)(m/2kT)^{1/2}\langle\psi_{10}^{(r-1)}\rangle_1 = \tfrac{1}{2}Na_{rr}(0)\langle\psi_{00}^{(r)}\rangle_1. \qquad (6\text{-}1\text{-}26)$$

With $r = 1$ this becomes

$$-(eE/m)(m/2kT)^{1/2}\langle w_z\rangle_1 = \tfrac{1}{2}Na_{11}(0)(\tfrac{3}{2} - \langle w^2\rangle_1), \qquad (6\text{-}1\text{-}27)$$

and the matrix element $a_{11}(0)$ is (McDaniel and Mason, 1973, p. 172)

$$a_{11}(0) = \frac{16}{3}\left(\frac{2kT}{\pi\mu}\right)^{1/2}\frac{mM}{(m+M)^2}\bar{\Omega}^{(1,1)}. \qquad (6\text{-}1\text{-}28)$$

Using the definition (6-1-15a) for the mean collision frequency $\bar{v}(T)$, and $\bar{v}_r = (8kT/\pi\mu)^{1/2}$, which is consistent with a first approximation, we can rearrange this moment equation to the form

$$eE\langle v_z\rangle_1 = \frac{mM}{(m+M)^2}(m\langle v^2\rangle_1 - 3kT)\bar{v}(T). \qquad (6\text{-}1\text{-}29)$$

This is exactly the energy-balance equation (5-2-19) of the momentum-transfer theory.

If we combine this result with the corresponding first approximation for $\langle w_z\rangle_1$, given by the momentum-balance equation (6-1-17), we obtain the Wannier energy formula again,

$$\tfrac{1}{2}m\langle v^2\rangle_1 = \tfrac{3}{2}kT + \tfrac{1}{2}(m+M)\langle v_z\rangle_1^2. \qquad (6\text{-}1\text{-}30)$$

Even though this appears to be valid only in a first approximation (i.e., only for vanishingly weak fields), the details of the derivation suggest that the range of validity may be much larger. If the effect of higher approximations is mainly to adjust $\bar{v}(T)$ to an effective collision frequency, then (6-1-30) will be valid at all field strengths because the collision frequency is eliminated in obtaining it. Although it is too much to expect higher approximations for the collisional loss of momentum to have exactly the same form as those for the collisional loss of energy, it is not unreasonable to expect them to be rather similar. This argument can be used to justify the success of a correction to the Wannier energy formula, obtained by Viehland et al. (1974) in the following way.

By carrying through the higher approximations, we develop the expression for the mean energy as a series in $(E/N)^2$, assumed small enough to ensure convergence,

$$\tfrac{1}{2}m\langle v^2\rangle = \tfrac{3}{2}kT + \tfrac{1}{2}mv_d^2 + \tfrac{1}{2}Mv_d^2[\beta_0 + \beta_2(E/N)^2 + \cdots]. \qquad (6\text{-}1\text{-}31)$$

This is analogous to the expansion (6-1-23) for the mobility, or to (5-1-5a), which was

$$K(E) = K(0)[1 + \alpha_2(E/N)^2 + \cdots].$$

We then eliminate the explicit power-series field dependence between these two expressions to obtain a formula for the mean energy in terms of K and its field dependence, as expressed by the logarithmic derivative. From the second formula we obtain

$$K' \equiv d \ln K_0/d \ln (E/N) = 2\alpha_2(E/N)^2 + \cdots, \qquad (6\text{-}1\text{-}32)$$

which gives us an expression for $(E/N)^2$ in terms of the experimental quantity K'. Substituting for $(E/N)^2$ back into (6-1-31) and expressing the α's and β's in terms of the matrix elements $a_{rs}(l)$, we find

$$\tfrac{1}{2}m\langle v^2 \rangle = \tfrac{3}{2}kT + \tfrac{1}{2}(m + M)v_d^2(1 + \beta K' + \cdots), \qquad (6\text{-}1\text{-}33)$$

or

$$\tfrac{3}{2}kT_{\text{eff}} \equiv \tfrac{1}{2}\mu\langle v_r^2 \rangle = \tfrac{3}{2}kT + \tfrac{1}{2}Mv_d^2(1 + \beta K' + \cdots), \qquad (6\text{-}1\text{-}34)$$

where to first order in (presumably) small quantities,

$$\beta = \frac{mM(5 - 2A^*)}{5(m^2 + M^2) + 4mMA^*} + \cdots, \qquad (6\text{-}1\text{-}35)$$

and

$$A^* \equiv \bar{\Omega}^{(2,2)}/\bar{\Omega}^{(1,1)}. \qquad (6\text{-}1\text{-}36)$$

The dimensionless ratio A^* is approximately unity and depends weakly on the ion-neutral interaction and the temperarature. It contains some information on the average angular distribution of the ion-neutral scattering; it does not appear in the simple momentum-transfer theory formulas of Section 5-2 because the scattering was there assumed to be isotropic. This correction β is analogous to the semiempirical correction β for T_{eff} given in (5-2-74). The specific expressions do not resemble each other very much, but their general behavior with mass is similar. As shown in the following section, the formulas for $\langle v^2 \rangle$ and T_{eff} above are quite accurate even at high fields.

To find the energy partitioning, we proceed as above, using the function

$$\psi_{20}^{(0)} = \tfrac{1}{2}(3w_z^2 - w^2). \qquad (6\text{-}1\text{-}37)$$

With this choice of $l = 2$, $r = 0$, the moment equations (6-1-19) yield in first approximation the equation

$$5(eE/m)(m/2kT)^{1/2}\langle \psi_{10}^{(0)} \rangle_1 = \tfrac{5}{2}Na_{00}(2)\langle \psi_{20}^{(0)} \rangle_1. \qquad (6\text{-}1\text{-}38)$$

The matrix element is (McDaniel and Mason, 1973, p. 173)

$$a_{00}(2) = \frac{16}{15}\left(\frac{2kT}{\pi\mu}\right)^{1/2}\frac{M}{(m+M)^2}[5m\bar{\Omega}^{(1,1)} + 3M\bar{\Omega}^{(2,2)}]. \qquad (6\text{-}1\text{-}39)$$

On substituting for $\bar{v}(T)$ and $\bar{v}_r$, as before, we can put this moment equation into the form

$$eE\langle v_z\rangle_1 = \frac{1}{10}\frac{mM(5m+3MA^*)}{(m+M)^2}[3\langle v_z^2\rangle_1 - \langle v^2\rangle_1]\bar{v}(T). \qquad (6\text{-}1\text{-}40)$$

This bears a striking resemblance to the second energy-balance equation of the momentum-transfer theory, (5-2-69), but is not exactly the same because of the rather casual averaging over scattering angles that was used there. A little investigation shows that the averaging used corresponds to a model with constant collision frequency and isotropic scattering, for which $A^* = 5/6$ (McDaniel and Mason, 1973, p. 195). With this value of A^*, (6-1-40) transforms into (5-2-69).

We now proceed as in the momentum-transfer theory, combining the momentum-balance equation (6-1-17) with the two energy-balance equations, (6-1-29) and (6-1-40), to obtain first approximations for the ion temperatures,

$$[kT_L]_1 \equiv m[\langle v_z^2\rangle_1 - \langle v_z\rangle_1^2] = kT + \zeta_L M \langle v_z\rangle_1^2, \qquad (6\text{-}1\text{-}41a)$$

$$[kT_T]_1 = m\langle v_x^2\rangle_1 = m\langle v_y^2\rangle_1 = kT + \zeta_T M\langle v_z\rangle_1^2, \qquad (6\text{-}1\text{-}41b)$$

where

$$\zeta_L = \frac{5m - (2m - M)A^*}{5m + 3MA^*}, \qquad (6\text{-}1\text{-}42a)$$

$$\zeta_T = \frac{(m + M)A^*}{5m + 3MA^*}. \qquad (6\text{-}1\text{-}42b)$$

Notice that $\zeta_L + 2\zeta_T = 1$, as it should. These equations become the same as the corresponding equations (5-2-70) of the simple momentum-transfer theory on setting $A^* = 5/6$, and the same as Wannier's equations (5-2-71) for the general Maxwell model on setting $A^* = 5\bar{Q}^{(2)}/4\bar{Q}^{(1)}$. We now argue, as before, that these relations have a larger range of validity than that of a mere first approximation, because the higher approximations mainly affect $\bar{v}(T)$, which has been algebraically eliminated.

Corrections to $[T_L]_1$ and $[T_T]_1$ can be obtained by considering higher approximations and introducing K' in place of $(E/N)^2$ in the expansions (Viehland et al., 1974). The results are

$$kT_L = kT + \zeta_L Mv_d^2(1 + \beta_L K' + \cdots), \qquad (6\text{-}1\text{-}43a)$$

$$kT_T = kT + \zeta_T M v_d^2 (1 + \beta_T K' + \cdots), \tag{6-1-43b}$$

where, to first order in (presumably) small quantities,

$$\beta_L = \frac{m + M}{M}\beta + 2\zeta_T \Delta, \tag{6-1-44a}$$

$$\beta_T = \frac{m + M}{M}\beta - \zeta_L \Delta. \tag{6-1-44b}$$

Here β is the analogous correction for T_{eff}, given by (6-1-35), and Δ is a rather complicated function of the ion-neutral mass ratio and interaction,

$$\Delta = \frac{5}{6}\frac{m}{MA^*\zeta_L}\left[\frac{5m(\kappa - 1) - 3MA^*(1 - \frac{5}{7}\kappa)}{5m + 3MA^*} - \left(2 - \frac{6}{5}A^*\right)\beta\right], \tag{6-1-45a}$$

where

$$\kappa = \frac{98}{5}\left[\frac{(m + M)^2(5m + 3MA^*)}{35m(4m^2 + M^2) + 14M(11m^2 + 3M^2)A^* + 144mM^2F^*}\right]$$
$$\times\left[1 + \frac{90m(3m^2 + M^2 + \frac{8}{5}mMA^*)}{(3m + MA^*)[105(2m^2 - M^2) + 252mMA^* + 240M^2F^*]}\right], \tag{6-1-45b}$$

and $F^* \equiv \bar{\Omega}^{(3,3)}/\bar{\Omega}^{(1,1)}$ is a ratio similar to A^*. Despite its complexity, the quantity κ is never far from unity; it is unity for $m \gg M$, 7/5 for $m \ll M$, and 1.47 for $m = M$, with rigid-sphere cross sections. It thus turns out that Δ depends primarily on m/M and only weakly on the ion-neutral interaction (i.e., on A^* and F^*). It can be fitted by a simple formula to a usable level of accuracy; in terms of the variable $m/(m + M)$ it is a square divided by a linear term,

$$\Delta \approx \frac{3}{2}\left(\frac{m}{m + M}\right)^2\left(1 + \frac{m}{m + M}\right)^{-1} = \frac{1.5(m/M)^2}{(1 + m/M)(1 + 2m/M)}. \tag{6-1-45c}$$

The accuracy of these energy (temperature) formulas is tested in Section 6-1C, but first we consider the diffusion problem.

3. Diffusion

To treat diffusion within the framework of Fick's law, we consider $\nabla_r n$ as a small quantity and systematically neglect higher derivatives, such as $\nabla_r^2 n$. Products and powers, such as $(\nabla_r n)^2$, do not occur anyway because $n \ll N$. That is, we treat diffusion as a small perturbation on the spatially homogeneous case. We imagine that the spatially homogeneous problem has been solved and then ask for the additional ion flux caused by the imposition of a small gradient. It is important to remember that diffusion is *defined* with respect to the spatially homogeneous cases, and that the effect of the imposed gradient on *all* the moments must be considered.

We can add the perturbation directly to the distribution function, or we can start with the moment equations and add perturbations to all the moments. Although the results are equivalent, the first procedure leads to somewhat simpler algebra. We therefore write, using Cartesian coordinates,

$$f = f_0 - \left(f_x \frac{\partial}{\partial x} + f_y \frac{\partial}{\partial y} + f_z \frac{\partial}{\partial z} \right) \ln n, \qquad (6\text{-}1\text{-}46a)$$

with

$$\int f_0 \, d\mathbf{v} = n, \qquad \int f_{x,y,z} \, d\mathbf{v} = 0, \qquad (6\text{-}1\text{-}46b)$$

where f_0 is the distribution function for the spatially homogeneous case. We now need to consider different kinds of moments, corresponding to f_0 and to $f_{x,y,z}$,

$$\langle \psi \rangle^{(0)} \equiv n^{-1} \int \psi f_0 \, d\mathbf{v}, \qquad (6\text{-}1\text{-}47a)$$

$$\langle \psi \rangle^{(x,y,z)} \equiv n^{-1} \int \psi f_{x,y,z} \, d\mathbf{v}. \qquad (6\text{-}1\text{-}47b)$$

The moments $\langle \psi \rangle^{(0)}$ for the spatially homogeneous reference case are the same as those considered in the previous parts of this section. From the expressions above we form $\langle v_x \rangle$ and $\langle v_z \rangle$ for a uniform field in the z-direction, and identify the moments we want as

$$\langle v_x \rangle^{(0)} = \langle v_y \rangle^{(0)} = 0, \qquad D_T = \langle v_x \rangle^{(x)} = \langle v_y \rangle^{(y)}; \qquad (6\text{-}1\text{-}48a)$$

$$\langle v_z \rangle^{(0)} = v_d = KE, \qquad D_L = \langle v_z \rangle^{(z)}. \qquad (6\text{-}1\text{-}48b)$$

The units for $\langle v_x \rangle^{(x)}$, $\langle v_y \rangle^{(y)}$, and $\langle v_z \rangle^{(z)}$ look peculiar because the $f_{x,y,z}$ have an extra dimension of length, owing to the space derivatives included with their definition in (6-1-46).

We now insert (6-1-46) into the moment equations (5-3-19) and separate terms according to the density gradient. This generates three sets of moment equations, one for the spatially homogeneous moments of the reference case,

$$(eE/m)\langle \partial \psi_p / \partial v_z \rangle^{(0)} = N \sum_q b_{pq} \langle \psi_q \rangle^{(0)}, \qquad (6\text{-}1\text{-}49a)$$

one for the spatially inhomogeneous moments perpendicular (x and y) to the field,

$$(eE/m)\langle\partial\psi_p/\partial v_z\rangle^{(x)} + \langle v_x\psi_p\rangle^{(0)} = N\sum_q b_{pq}\langle\psi_q\rangle^{(x)}, \qquad (6\text{-}1\text{-}49b)$$

and one for the spatially inhomogeneous moments parallel (z) to the field,

$$(eE/m)\langle\partial\psi_p/\partial v_z\rangle^{(z)} + [\langle v_z\psi_p\rangle^{(0)} - \langle v_z\rangle^{(0)}\langle\psi_p\rangle^{(0)}] = N\sum_q b_{pq}\langle\psi_q\rangle^{(z)}.$$

$$(6\text{-}1\text{-}49c)$$

For basis functions we choose the Burnett functions of (5-3-28), and focus on the functions that give v_x (or v_y) and v_z,

$$w_x = \tfrac{1}{2}(\psi_{11}^{(0)} + \psi_{1-1}^{(0)}), \qquad (6\text{-}1\text{-}50a)$$

$$w_z = \psi_{10}^{(0)}. \qquad (6\text{-}1\text{-}50b)$$

The derivative terms in (6-1-49) are eliminated through the recursion relation (6-1-3), and the first approximation is obtained by setting all the off-diagonal matrix elements equal to zero.

For simplicity we again consider only a single neutral gas. In first approximation the spatially homogeneous equation (6-1-49a) yields (6-1-17) for the mobility again, as it should. The equations for the spatially inhomogeneous moments yield, after a little algebra,

$$e[D_{T,L}]_1 = [kT_{T,L}]_1[K]_1, \qquad (6\text{-}1\text{-}51)$$

which are just the generalized Einstein relations in lowest order.

Higher approximations for the diffusion coefficients are obtained through some systematic truncation-iteration scheme for solving the moment equations (Kumar and Robson, 1973; Robson and Kumar, 1973; Whealton and Mason, 1974). The result is equivalent to an expansion in powers of $(E/N)^2$, as would be expected. The convergence is, unfortunately, even poorer than the corresponding expansion (6-1-23) for the mobility. However, we can try the same trick that seemed to work for the ion energies, and eliminate the explicit power-series field dependence to obtain formulas for D_T and D_L in terms of K and K'. The results are, to first order (Whealton and Mason, 1974),

$$eD_T/K = kT_T(1 + \cdots), \qquad (6\text{-}1\text{-}52a)$$

$$eD_L/K = kT_L(1 + K' + \cdots), \qquad (6\text{-}1\text{-}52b)$$

$$kT_{T,L} = kT + \zeta_{T,L}Mv_d^2(1 + \cdots), \qquad (6\text{-}1\text{-}52c)$$

with $\zeta_{T,L}$ given by (6-1-42). We can recognize these as the same generalized

Einstein relations that were obtained by momentum-transfer theory in Section 5-2B, but with better formulas for $T_{T,L}$ than were derived in Section 5-2C. This is as far as the one-temperature theory will conveniently carry us with Fickian diffusion. Attempts to improve the GER are not very successful, and this part of the subject rates a separate discussion, given later in Section 6-4. The formulas for $T_{T,L}$ obtained from the one-temperature theory, however, turn out to be rather accurate, as shown in Section 6-1C.

Before we leave the subject of ion diffusion, it is worth noting that the one-temperature theory can also be used to treat non-Fickian diffusion. We can find expressions for some of the higher-order transport coefficients that were discussed in a general way in Section 5-1C. The procedure is straightforward in principle, but rapidly becomes complicated (Whealton and Mason, 1974). The full moment equations (5-3-12), which implicitly include higher derivatives, must be retained, and higher derivatives must be included in the expansion (6-1-46) for f. Explicit results were obtained by Whealton and Mason for only a few of the coefficients, in a first approximation corresponding to the Maxwell model. Even these partial results suggested that some sort of generalized Nernst–Einstein relations might hold among the higher-order transport coefficients, and led to the more general investigation of Weinert and Mason (1980). For the Maxwell model they found several simple relations among these coefficients, which are defined in Section 5-1C, as follows:

$$N^2 q_1^n (e/kT) = D(0)(2d_2^n + d_2^E), \tag{6-1-53a}$$

$$N^2 q_1^E (e/kT) = D(0)d_2^E, \tag{6-1-53b}$$

$$N^2 R(0)(e/kT)^2 = \tfrac{1}{2}N^2(q_1^n + q_1^E)(e/kT) = D(0)(d_2^n + d_2^E). \tag{6-1-53c}$$

The only relation involving these coefficients that is generally true for any ion-neutral interaction is given by (5-1-11b), which is

$$K(0)\alpha_2 - D(0)(d_2^n + d_2^E)(e/kT) + N^2(q_1^n + q_1^E)(e/kT)^2$$
$$- N^2 R(0)(e/kT)^3 = 0.$$

This relation is obtained as the sum of the special relations (6-1-53), with $\alpha_2 = 0$ (true for the Maxwell model). It has been suggested that these special relations might be approximately valid for ion-neutral interactions more general than the Maxwell model, but this conjecture has not been tested, either experimentally or theoretically.

To summarize, the foregoing results are valid for a binary collision mechanism, elastic collisions, and weak fields. The limitation to weak fields is apparent from the form of the expansion in powers of $(E/N)^2$. Although there is no explicit assumption that the deviation from equilibrium is small, the choice of the ion temperature as equal to the gas temperature evidently amounts to an implicit assumption of this sort. Effects of quantum mechanics are automatically included; only the values of the collision integrals are affected, not the form of the kinetic-theory formulas. The Chapman–Enskog vanishing-field result is a special case of these results. In some cases the explicit expansion in $(E/N)^2$ can be algebraically eliminated to obtain relations among experimental quantities. As the convergence tests in the following Section 6-1C show, this trick is often remarkably successful. A proper justification for this success, however, depends on the two-temperature theory described in Section 6-2.

C. Convergence of Approximations

We first consider the mobility, and then the ion temperatures and diffusion coefficients.

1. Mobility

Two sorts of convergence errors appear in (6-1-23) for the mobility: the convergence error of the expansion in powers of $\mathscr{E}^2$ and the convergence errors of the expressions for the expansion coefficients g_n. We first take up the convergence of the expressions for the expansion coefficients and then the convergence of the overall expansion.

Other things being equal, the convergence is usually poorest for $m \ll M$, the Lorentz model, and we confine our comparisons to this case. This has the added advantage that the exact answer can be found independently by the two-term approximation (Section 5-3D4a). The other two soluble cases, the Maxwell model and the Rayleigh model ($m \gg M$), are not very interesting for this comparison because the one-temperature theory is correct in its first approximation for these models.

The leading coefficient, g_0, gives the vanishing-field Chapman–Enskog result. Writing out the first two terms of (6-1-9) in detail, we obtain

$$\frac{g_0}{[g_0]_1} = \frac{K(0)}{[K(0)]_1} = \frac{D(0)}{[D(0)]_1} = 1 + \frac{a_{01}(1)a_{10}(1)}{a_{00}(1)a_{11}(1)} + \cdots$$

$$= 1 + \frac{M^2(6C^* - 5)^2}{30m^2 + M^2(25 - 12B^*) + 16mMA^*} + \cdots \qquad (6\text{-}1\text{-}54a)$$

$$= 1 + \frac{M^2(6C^* - 5)^2}{30m^2 + 10M^2 + 16mMA^*} + \cdots, \qquad (6\text{-}1\text{-}54b)$$

where

$$A^* \equiv \bar{\Omega}^{(2,2)}/\bar{\Omega}^{(1,1)}, \qquad (6\text{-}1\text{-}55a)$$

$$B^* \equiv [5\bar{\Omega}^{(1,2)} - 4\bar{\Omega}^{(1,3)}]/\bar{\Omega}^{(1,1)}, \qquad (6\text{-}1\text{-}55b)$$

$$C^* \equiv \bar{\Omega}^{(1,2)}/\bar{\Omega}^{(1,1)}. \qquad (6\text{-}1\text{-}55c)$$

The difference between (6-1-54a) and (6-1-54b) comes from replacing $a_{11}(1)$ by its Maxwell-model eigenvalue, which is equivalent to setting $B^* = 5/4$. The difference is minor, and we use (6-1-54b) in what follows.

The general expression for the next coefficient, g_2, is rather complicated (McDaniel and Mason, 1973, p. 177), but for $m \ll M$ it simplifies to

$$\frac{g_2}{[g_2]_1} = 1 + \tfrac{1}{10}[4(6C^* - 5) - 3(5 - 4B^*)] + \cdots. \qquad (6\text{-}1\text{-}56)$$

The coefficient g_4 has only been evaluated in first approximation, and its convergence error can be estimated only by analogy with the convergence errors in g_0 and g_2.

Some numerical results are given in Table 6-1-1 for g_0 and g_2 for inverse-power potentials, $V(r) = \pm C_n/r^n$. In these cases the exact value of g_0 is also known. For g_0 it is apparent that the first approximation may not always be sufficiently accurate but that the second approximation will usually be satisfactory. The pattern of convergence for g_2 appears to be rather similar to that for g_0, but perhaps slightly poorer.

It is worth examining the convergence for g_0 and g_2 for an ion-neutral potential that is physically more realistic than a single inverse power. A useful potential model includes the long-range attractive polarization energy and represents the short-range repulsion energy as an inverse power. It is convenient to write such an $(n, 4)$ potential in the form

$$V(r) = \frac{n\varepsilon_0}{n - 4}\left[\frac{4}{n}\left(\frac{r_m}{r}\right)^n - \left(\frac{r_m}{r}\right)^4\right], \qquad (6\text{-}1\text{-}57)$$

Table 6-1-1   Convergence of the Expansion Coefficients g_0 and g_2 of (6-1-23) for K in Powers of $(E/N)^2$: Results for the Potential $V(r) = \pm C_n/r^n$ for the Unfavorable Case of $m \ll M$ (Lorentz Model)

n	$g_0/[g_0]_1$ Exact	$[g_0]_2/[g_0]_1$ (6-1-54b)	$[g_2]_2/[g_2]_1$ (6-1-56)
2	1.132	1.100	0.700
3	1.014	1.011	0.922
4	1.000	1.000	1.000
6	1.014	1.011	1.056
8	1.031	1.025	1.075
10	1.045	1.036	1.084
25	1.091	1.071	1.097
50	1.110	1.085	1.099
∞	1.132	1.100	1.100

where ε_0 is the depth of the potential minimum and r_m is the position of the minimum. The needed collision integrals for this model have been calculated by numerical integration and tabulated (Viehland et al., 1975). Some numerical results are given in Table 6-1-2 as a function of temperature for $n = 8, 12, \infty$. At low temperatures the first approximation is exact, since the potential (6-1-57) behaves like the r^{-4} Maxwell model at low energies. At high temperatures the behavior approaches that of an r^{-n} repulsive potential. From the results shown we may again conclude that if the first approximation is not sufficiently accurate, the second approximation usually will be.

The expansion in powers of $\mathscr{E}^2$ can be tested by comparison with the exact results for the rigid-sphere Lorentz model, calculated by numerical integration from the two-term expansion discussed in Section 5-3D4a (Hahn and Mason, 1972). The results, valid at all values of T and E/N, are shown in Fig. 6-1-1 in terms of the dimensionless field strength and mobility defined in (5-2-34)–(5-2-36). The results are disappointing, even though this is a very unfavorable case. It is not so surprising that the expansion diverges, since $\mathscr{E}$ is unbounded as $E/N \to \infty$ or $T \to 0$, and a power series in such a variable must inevitably diverge. What is disappointing is that the radius of convergence appears to be very small indeed.

The question naturally arises whether the useful range of the expansion could be extended by including more terms. Whealton and Mason (1974) have estimated an upper bound for the radius of convergence by conjecturing that some patterns in the behavior of the first few $a_{rs}(l)$ and g_n are general, at least for repulsive ion-neutral potentials. The result turns out to be simply that the series diverges unless

$$|g_2|\mathscr{E}^2 < 1, \qquad (6\text{-}1\text{-}58a)$$

Table 6-1-2 Convergence of the Expansion Coefficients g_0 and g_2 of (6-1-23) for K in Powers of $(E/N)^2$: Results for $(n, 4)$ Potentials for the Unfavorable Case of $m \ll M$ (Lorentz Model)

$\dfrac{kT}{\varepsilon_0}$	$[g_0]_2/[g_0]_1$			$[g_2]_2/[g_2]_1$		
	$n = 8$	$n = 12$	$n = \infty$	$n = 8$	$n = 12$	$n = \infty$
0	1.000	1.000	1.000	1.000	1.000	1.000
0.5	1.044	1.020	1.005	0.822	0.881	1.013
1	1.017	1.005	1.012	0.899	0.840	1.058
2	1.000	1.004	1.043	1.005	0.926	1.088
4	1.013	1.024	1.072	1.058	1.015	1.098
9	1.024	1.039	1.088	1.075	1.068	1.100
∞	1.025	1.044	1.100	1.075	1.089	1.100

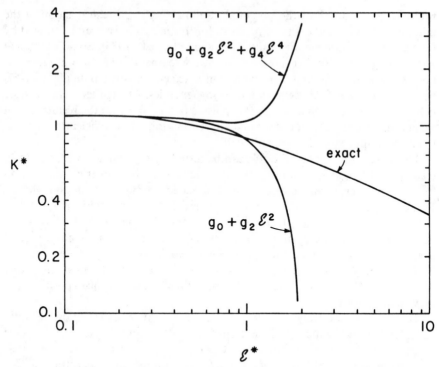

Figure 6-1-1. Mobility vs. field strength for the rigid-sphere Lorentz model ($m \ll M$). The one-temperature theory expansion of (6-1-23) is compared with accurate results from numerical integration (Hahn and Mason, 1972). This figure should be compared with Fig. 5-2-3, showing similar results for the momentum-transfer theory. The dimensionless mobility K^* and field strength $\mathscr{E}^*$ are defined by (5-2-34)–(5-2-36).

or in terms of the expansion of (5-1-5a),

$$|\alpha_2|(E/N)^2 < 1. \qquad (6\text{-}1\text{-}58b)$$

In terms of the masses and the collision integral ratios of (6-1-55), this is approximately

$$\mathscr{E}^2 < \frac{1}{10} \frac{m}{M} \frac{(5m + 3MA^*)(30m^2 + 10M^2 + 16mMA^*)}{(m + M)^2(3m + MA^*)}. \qquad (6\text{-}1\text{-}59)$$

For the example shown in Fig. 6-1-1 this formula gives the bound on the radius of convergence as $\mathscr{E}^* < 1.7$, which seems consistent with the curves shown. Robson and Kumar (1973) have made numerical calculations with a truncation procedure that gives K as the ratio of two polynomials in $\mathscr{E}^2$, and have gone to higher orders of approximation than that given here by just three terms of the series expansion. Their results for rigid spheres appear to break down at a value of $\mathscr{E}$ close to that given by (6-1-59).

We conclude that the one-temperature theory is not very useful as far as expansions in $(E/N)^2$ are concerned.

2. Ion Temperatures and Diffusion

The situation is more promising when we eliminate the explicit expansion in $(E/N)^2$ in favor of K', which is bounded, and obtain relations among experimental quantities. In particular, we have (6-1-33) for the mean ion energy and (6-1-43) for T_T and T_L, which relate these quantities to m, M, v_d, K', and A^*. We also have the generalized Einstein relations of (6-1-52), but these have already been tested in connection with the momentum-transfer theory (Tables 5-2-3 and 5-2-4). The one-temperature theory has merely provided an alternate derivation of these relations.

Some numerical comparisons with accurate results (Skullerud, 1976) for ion energy and temperatures are given in Tables 6-1-3, 6-1-4, and 6-1-5. These should be compared with the corresponding semiempirical results in Tables 5-2-7, 5-2-8, and 5-2-9. The accuracy is remarkably good. Except for some of the rigid-sphere results, the discrepancies are mostly down to the 5% level or better. The present correction factors, β and $\beta_{T,L}$, vanish for $m/M \to 0$, so we must go to

Table 6-1-3 Accuracy of the Kinetic-Theory Formula (6-1-33) for the Mean Ion Energy in a Cold Gas[a]

	$m\langle v^2\rangle[(m + M)v_d^2(1 + \beta K')]^{-1}$			
$\dfrac{m}{M}$	$n = 4$	$n = 8$	$n = 12$	$n = \infty$
0	1	1.0202	1.0324	1.0606
0.1	1	1.02	1.03	1.058
0.2	1	1.016	1.027	1.056
0.5	1	1.006	1.010	1.028
0.8	1	0.999	1.001	1.009
1.0	1	0.997	0.997	1.002
1.5	1	0.996	0.995	0.997
2.0	1	0.996	0.995	0.997
3.0	1	0.998	0.997	0.997
4.0	1	0.998	0.998	0.998
∞	1	1.0000	1.0000	1.0000
$\langle|\mathrm{dev}|\rangle$	0.0%	0.7%	1.1%	2.0%

[a]The value of β is given by (6-1-35). These results should be compared with the semiempirical results of Table 5-2-9.

Table 6-1-4 Accuracy of the Kinetic-Theory Formula (6-1-43) for T_T in a Cold Gas[a]

$\dfrac{m}{M}$	$kT_T[\zeta_T M v_d^2(1 + \beta_T K')]^{-1}$			
	$n = 4$	$n = 8$	$n = 12$	$n = \infty$
0	1	1.0202	1.0324	1.0606
0.1	1	1.01	1.02	1.067
0.2	1	1.018	1.030	1.073
0.5	1	1.017	1.028	1.073
0.8	1	1.015	1.022	1.067
1.0	1	1.009	1.016	1.052
1.5	1	1.002	1.004	1.030
2.0	1	0.999	1.002	1.012
3.0	1	0.995	0.994	0.998
4.0	1	0.992	0.981	1.002
∞	1	0.950	0.933	0.901
$\langle|\text{dev}|\rangle$	0.0%	1.4%	2.2%	4.9%

[a]The value of ζ_T is given by (6-1-42) and the value of β_T by (6-1-44) and (6-1-45). These results should be compared with the semiempirical results of Table 5-2-7.

Table 6-1-5 Accuracy of the Kinetic-Theory Formula (6-1-43) for T_L in a Cold Gas[a]

$\dfrac{m}{M}$	$kT_L[\zeta_L M v_d^2(1 + \beta_L K')]^{-1}$			
	$n = 4$	$n = 8$	$n = 12$	$n = \infty$
0	1	1.0202	1.0324	1.0606
0.1	1	1.03	1.05	1.064
0.2	1	1.022	1.036	1.057
0.5	1	0.997	1.002	1.008
0.8	1	0.977	0.975	0.964
1.0	1	0.973	0.965	0.954
1.5	1	0.975	0.963	0.951
2.0	1	0.978	0.969	0.959
3.0	1	0.983	0.975	0.974
4.0	1	0.989	0.987	0.967
∞	1	1.022	1.021	1.010
$\langle|\text{dev}|\rangle$	0.0%	2.0%	2.8%	3.9%

[a]The value of ζ_L is given by (6-1-42) and the value of β_L by (6-1-44) and (6-1-45). These results should be compared with the semiempirical results of Table 5-2-8.

a higher order of approximation to improve the light-ion case, but this refinement must await the description of the two-temperature theory.

These results represent one of the more successful aspects of the one-temperature theory. The most successful aspect—the temperature dependence of the zero-field mobility—is described next.

D. Temperature Dependence of Mobility

The good convergence for the coefficient g_0 shown in Tables 6-1-1 and 6-1-2 means that the one-temperature theory gives an accurate description of zero-field mobility (and diffusion). This is no surprise, because in this limit the theory is equivalent to the classical Chapman–Enskog kinetic theory. As we can see from (6-1-13), the temperature dependence of the mobility is given by the temperature dependence of the quantity $T^{1/2}\bar{\Omega}^{(1,1)}$, except for the small contribution from the higher-order corrections contained in g_0. Since the entire dependence of the mobility on the ion-neutral interaction is contained in $\bar{\Omega}^{(1,1)}$, the temperature dependence of the mobility is closely connected with the behavior of the ion-neutral potential. Somewhat surprisingly, the connection is sensitive enough that mobility measurements are an excellent source of information on the potential, despite the layers of integration intervening—from $V(r)$ to $\sigma(\theta, \varepsilon)$ to $\bar{Q}^{(1)}(\varepsilon)$ to $\bar{\Omega}^{(1,1)}(T)$. Moreover, the field dependence of the mobility is essentially equivalent to the temperature dependence, with $T_{\rm eff}$ replacing T. This feature appears already in the momentum-transfer theory of Section 5-2A, and is made more precisely quantitative by the two-temperature theory of Section 6-2.

Although ion-neutral interactions occur in considerable variety, one common feature stands out. Unless the neutral molecule possesses an appreciable permanent dipole or quadrupole moment, the longest-ranged component of the potential is the r^{-4} polarization energy, which arises from the interaction of the ion charge with the dipole it induces in the neutral molecule,

$$V_{\rm pol} = -\frac{e^2 \alpha_d}{2r^4}, \qquad (6\text{-}1\text{-}60)$$

where α_d is the dipole polarizability of the neutral. This is the potential that dominates at very low energies, which means in the limit $E/N \to 0$, $T \to 0$. For this potential $\bar{\Omega}^{(1,1)}$ is proportional to $(e^2 \alpha_d / kT)^{1/2}$, as can be shown by dimensional analysis (Section 5-2F), and all mobilities therefore approach a common finite polarization limit as the temperature is lowered. The constant of proportionality must be found by numerical integration; the first evaluation was carried out by Langevin (1905) and improved by Hassé (1926), and the most accurate recent value has been reported by Heiche and Mason (1970). The resulting polarization limit of the mobility at standard gas density is

$$K_{\rm pol} \equiv K_0(E/N \to 0,\ T \to 0) = \frac{13.853}{(\alpha_d \mu)^{1/2}}\ {\rm cm}^2/\text{V-s}, \qquad (6\text{-}1\text{-}61)$$

where α_d is in $Å^3$ and μ is the reduced mass in atomic mass units (daltons or g/mol). In this case $g_0 = 1$ and the first Chapman–Enskog approximation is exact. For ion-dipole (r^{-2}) and ion-quadrupole (r^{-3}) potentials the mobility should continue to decrease as T approaches zero. But these potentials are orientation dependent and are zero for the lowest rotational state of the molecule, so that the mobility should ultimately pass through a minimum at some very low temperature and rise back toward the polarization limit (Arthurs and Dalgarno, 1960). However, even the translational motion becomes quantum-mechanical at low enough energies and the classical-mechanical formula (6-1-61) must ultimately fail (see Section 6-1E).

At closer distances, ion-neutral potentials show a variety of distinctive features, each of which affects the mobility in a characteristic manner as T is increased from 0 K. If the short-range interaction is repulsive, the mobility rises from the polarization limit with increasing temperature, as if the repulsion were partially cancelling the polarization attraction as the collisions became more energetic. The mobility then reaches a broad maximum, and finally slowly decreases as the repulsion eventually dominates the collisions. If the short-range interaction is strongly attractive, as in the case of a chemical valence force, the mobility may decrease with increasing temperature, as if the valence attraction were augmenting the polarization. For a combination of short-range attraction and repulsion, the mobility may first show a decrease from the polarization limit, then an increase to a maximum, followed by a final decrease with increasing temperature.

Some of these features are shown for two real systems in Fig. 6-1-2. The system Li^+ in He is a case of short-range repulsion, because the Li^+ ion has the same closed-shell electronic structure as a He atom. The maximum energy of attraction, due mostly to the polarization energy before it is swamped by the repulsion, is only about 0.074 eV or 850 K. The system H^+ (or D^+) in He, on the other hand, is a case of short-range valence attraction. The HeH^+ molecular ion has the electronic structure of a heteronuclear hydrogen molecule, and is a stable species. The maximum energy of attraction, or bond strength, is about 2.04 eV or 23,700 K. The initial rise and subsequent bending over to approach a maximum is shown by Li^+ in He, whereas an initial decrease is shown by H^+ in He. The results for both systems are normalized in Fig. 6-1-2 by dividing by K_{pol}, so that both curves start from unity ($K_{pol} = 19.23$ cm^2/V-s for Li^+ in He, and 34.16 cm^2/V-s for H^+ in He). By the time the temperature has risen to 900 K, the mobilities for the two systems (relative to K_{pol}) differ by nearly a factor of 2.

Several other features that have previously been mentioned are also illustrated in Fig. 6-1-2, and it is worth a short digression to point them out. First, the $\mu^{-1/2}$ scaling for K_0 is illustrated for the system H^+ in He. This scaling rule is predicted by both the momentum-transfer and the one-temperature theories. Some of the results shown were really obtained from measurements on D^+ in He, but the normalization by K_{pol} removes the $\mu^{-1/2}$ mass dependence. It can be seen that the results for both H^+ and D^+ in He fall together within the

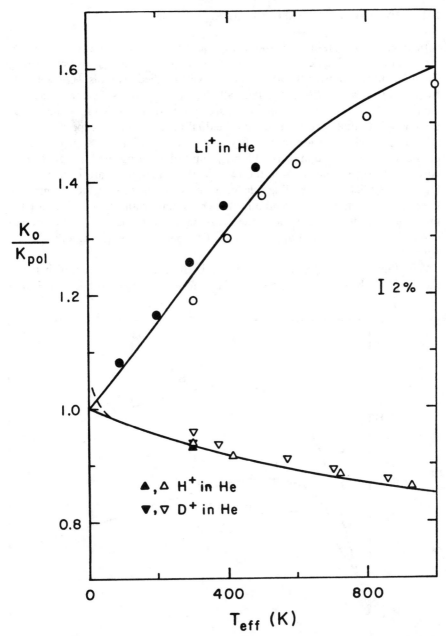

Figure 6-1-2. Differing effects of short-range repulsion (Li⁺ in He) and valence attraction (H⁺ in He) on the temperature dependence of the mobility. The mobilities are normalized by K_{pol}, the $T = 0$ limit. The curves are theoretical calculations, independent of any mobility data (Hariharan and Staemmler, 1976; Kolos, 1976; Kolos and Peek, 1976; Gatland et al., 1977; Dickinson and Lee, 1978; Lin et al., 1979a). The points correspond to experimental measurements reported by Hoselitz (●, 1941), Ellis et al. (○, 1976a), Orient (▲, 1971; ▼, 1972), and Howorka et al. (△, ▽, 1979). Other features shown are mass dependence, field dependence, and quantum effects; see the text for a discussion.

experimental uncertainty of about 2%. Second, the scaling together of gas temperature and electric field strength into a single variable, T_{eff}, as predicted by the momentum-transfer theory, is shown. All the filled symbols were obtained by varying T at $E/N \rightarrow 0$, whereas all the open symbols were obtained by varying E/N at $T = 300$ K. When both are plotted together as a function of T_{eff}, there is substantial agreement within the dual uncertainties of experimentation (about 2%) and of the use of only first approximations. These approximations consist of taking $\beta = 0$ in (6-1-34) for T_{eff}, and of ignoring similar higher-order corrections for K_0 that require the two- or the three-temperature theory. The failure to predict the T_{eff} scaling rule for K_0 is one of the biggest shortcomings of the one-temperature theory, and one of the most important successes of both the momentum-transfer theory and the two-temperature theory. In the present cases, the use of only first approximations probably introduces errors of a few percent.

The third feature shown in Fig. 6-1-2 is the accurate connection between mobility and the ion-neutral potential. The curves shown for both Li$^+$ in He and

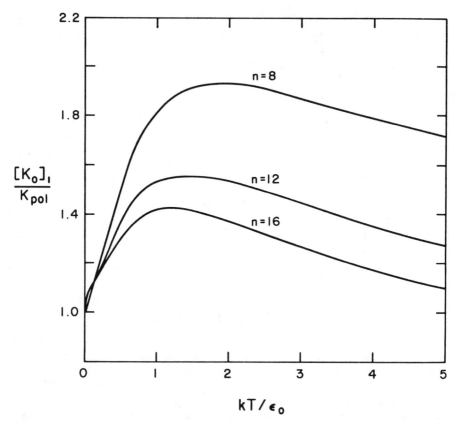

Figure 6-1-3. Effect of short-range repulsion on the temperature dependence of mobility, as illustrated for a series of $(n, 4)$ potentials (Viehland et al., 1975).

H^+ in He are completely independent of any mobility measurements—they were calculated via $\bar{\Omega}^{(1,1)}(T_{\text{eff}})$ from potential energy curves obtained by ab initio quantum-mechanical calculations. The potential curve for Li^+ in He was obtained by Hariharan and Staemmler (1976) with a moderately accurate wave function, and that for H^+ in He by Kolos (1976) and Kolos and Peek (1976) with an extremely accurate wave function. The kinetic-theory calculations, which included higher-order corrections from the two-temperature theory, were carried out by Gatland et al. (1977) for Li^+ in He and by Dickinson and Lee (1978) and Lin et al. (1979a) for H^+ in He.

A fourth feature shown in Fig. 6-1-2 concerns the accuracy of using classical mechanics to describe the motions of the ions and neutrals, but we defer discussion of this point to Section 6-1E.

Returning now from the digression over Fig. 6-1-2, we explore in more detail how the ion-neutral potential affects the mobility. This is conveniently done by means of potential models, in which some feature can be varied in a systematic way. To show the effect of adding a short-range repulsion to the polarization potential, we use the $(n, 4)$ potential model of (6-1-57). Variation of n changes the steepness of the repulsion and the relative width of the potential well, and the effect on the mobility is shown in Fig. 6-1-3 as $[K_0]_1/K_{\text{pol}}$ vs. reduced temperature kT/ε_0 for $n = 8, 12, 16$. The numerical results are taken from the tabulations of Viehland et al. (1975). The height of the maximum is seen to depend strongly on the parameter n, and the temperature at which the maximum occurs to depend strongly on the potential well depth and only weakly on n.

The effect of both short-range repulsion and short-range attraction can be investigated by adding some r^{-6} attraction energy to the $(n, 4)$ potential to give an $(n, 6, 4)$ potential,

$$V(r) = \frac{n\varepsilon_0}{n(3 + \gamma) - 12(1 + \gamma)} \left[\frac{12}{n}(1 + \gamma)\left(\frac{r_m}{r}\right)^n - 4\gamma\left(\frac{r_m}{r}\right)^6 - 3(1 - \gamma)\left(\frac{r_m}{r}\right)^4 \right],$$

$$(6\text{-}1\text{-}62)$$

where ε_0, r_m, and n have the same meaning as for the $(n, 4)$ potential, and γ is a dimensionless fourth parameter that measures the relative strength of the r^{-6} and r^{-4} attraction energies. In Fig. 6-1-4 curves of $[K_0]_1/K_{\text{pol}}$ vs. kT/ε_0 are shown for $n = 12$ with $\gamma = 0, 0.4, 0.8$, taken from the tabulations of Viehland et al. (1975). The curve for $\gamma = 0$ is the same as the curve for $n = 12$ in Fig. 6-1-3, for comparison purposes. The addition of r^{-6} attraction energy is seen to have the effect of decreasing the height of the maximum relative to the polarization limit, and in that sense simulates the effect of increasing the steepness of the repulsion. A minimum can even appear in the mobility curve for large enough values of γ. The addition of sufficient r^{-6} attraction energy can cause the mobility to fall lower than the polarization limit at all temperatures.

The foregoing models assumed that the centers of repulsion, attraction,

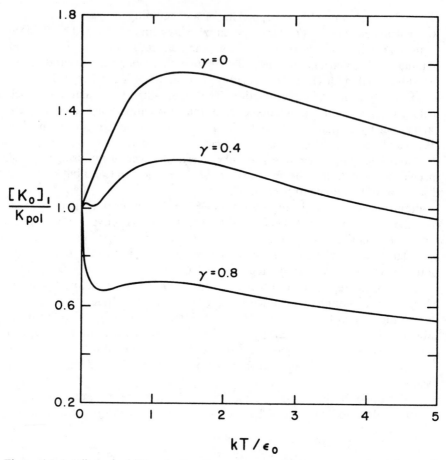

Figure 6-1-4. Effect of additional attraction energy on the temperature dependence of mobility, as illustrated for a series of (12, 6, 4) potentials (Viehland et al., 1975).

polarization, etc., were all located at the centers of mass of the ion and neutral. Although this is a very reasonable assumption for small ions and neutral molecules, it may well be inadequate for a large molecular ion, whose center of charge may not be located near its center of mass, or inadequate even for a large polyatomic neutral. A model that attempts to account for such features in a simple way adds a rigid spherical core to one of the point-center models. One for which numerical calculations are available is the (12, 4) core model (Mason et al., 1972),

$$V(r) = \frac{\varepsilon_0}{2} \left[\left(\frac{r_m - a}{r - a} \right)^{12} - 3 \left(\frac{r_m - a}{r - a} \right)^{4} \right], \qquad (6\text{-}1\text{-}63)$$

where ε_0 and r_m have their usual meanings, and a is the effective core diameter. Additional calculations for more general $(n, 6, 4)$ core models have been carried out by Iinuma et al. (1982), but no numerical tables have been published. Figure 6-1-5 shows $[K_0]_1/K_{pol}$ vs. kT/ε_0 for several values of the reduced core diameter

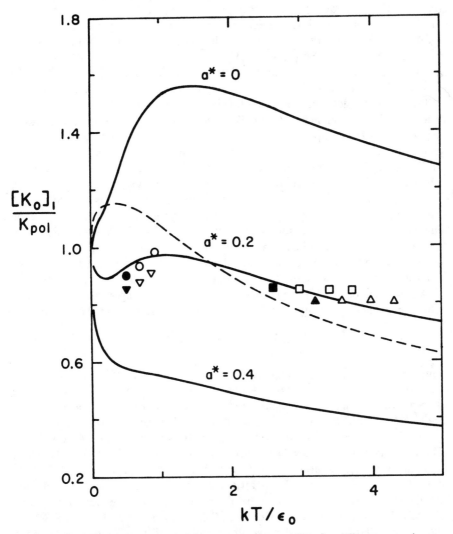

Figure 6-1-5. Effect of added repulsive core on the mobility for different core sizes, as illustrated for a series of (12, 4) core potentials. The added core suppresses the maximum and eventually reduces the mobility everywhere below its polarization limit. This is not true for the $(\infty, 4)$ model of a rigid sphere with an added polarization potential, shown as the dashed curve. The temperature dependence is inferred from the field dependence (open symbols). The points correspond to the following ions in SF_6 drift gas (Patterson, 1970a): ●, ○ SF_5^-; ▲, △ SF_6^-; ■, □ $(SF_6)SF_6^-$; ▲, △ $(SF_6)_2SF_6^-$.

$a^* \equiv a/r_m$ for the (12, 4) model. The effect of the added rigid core is marked. A value of $a^* = 0.2$ is sufficient to reduce the mobility everywhere below the polarization limit. The effect, however, is *not* the same as the simple addition of a rigid core to a polarization potential, which leads to the $(\infty, 4)$ model of (6-1-57). The curve for the $(\infty, 4)$ model is shown dashed in Fig. 6-1-5 (Hassé, 1926). The reason is that the core model corresponds roughly to the charge being located on the surface of the sphere, whereas the $(\infty, 4)$ model corresponds to the charge located at the center of the sphere. Also shown in Fig. 6-1-5 are some measured mobilities for SF_5^-, SF_6^-, $(SF_6)SF_6^-$, and $(SF_6)_2SF_6^-$ in SF_6 drift gas (Patterson, 1970a). It is apparent that either the core model or the $(\infty, 4)$ model can account for the magnitudes of the mobilities of the two large clustered ions, but the $(\infty, 4)$ model cannot account for the low mobilities of SF_5^- and SF_6^-. In Fig. 6-1-5 the dependence of K_0 on E/N has been used to infer the temperature dependence of K_0 (open symbols). Notice that the mobilities of the large clustered ions are essentially independent of temperature, whereas the mobilities of SF_5^- and SF_6^- increase with increasing temperature. This latter behavior is consistent with the core model, but not the $(\infty, 4)$ model.

Comparison of Figs. 6-1-4 and 6-1-5 shows that the effects of adding r^{-6} energy and adding a core are rather similar, although on close inspection the curves are found to have somewhat different shapes. The development of a minimum and the accompanying lowering of the maximum thus appear to depend on the feature of the potential showing *more* attraction than V_{pol} as the particles approach each other from infinity. Both models show this feature.

A new effect occurs for the special case of an ion moving in its parent gas, in which an electron is easily transferred during a collision, as in the example

$$He^+ + He = He + He^+. \qquad (6\text{-}1\text{-}64)$$

This resonant charge exchange converts a collision having a center-of-mass deflection angle of θ into one of $\pi - \theta$; a large number of glancing collisions are thereby transformed into apparent almost head-on collisions. As a result, $\bar{Q}^{(1)}$ and $\bar{\Omega}^{(1,1)}$ are greatly increased, except at very low energies or temperatures, where the polarization interaction still dominates (Heiche and Mason, 1970). The mobility is thereby decreased below the polarization limit, and decreases with increasing temperature. The effect is illustrated in Fig. 6-1-6 for He^+ in He (for which $K_{pol} = 21.63$ cm^2/V-s). Both calculated and experimental results are shown. Here we see again some of the other features that were illustrated in Fig. 6-1-2 for Li^+ in He and H^+ in He. First, there is the scaling of T and E/N into the single variable T_{eff} (filled circles for variation of T at $E/N \to 0$, and open circles for variation of E/N at $T = 300$ K). Second, there is the accurate connection between mobility and the ion-neutral potential. The curves are calculated from potentials for He^+ in He that are independent of any mobility measurements (Dickinson, 1968a; Sinha et al., 1979). Third, there are the quantum effects at low temperatures, discussed in Section 6-1E.

The transfer of an ionic fragment during collision has an effect similar to that of resonant charge exchange, for much the same reason. For instance, He_2^+ ions

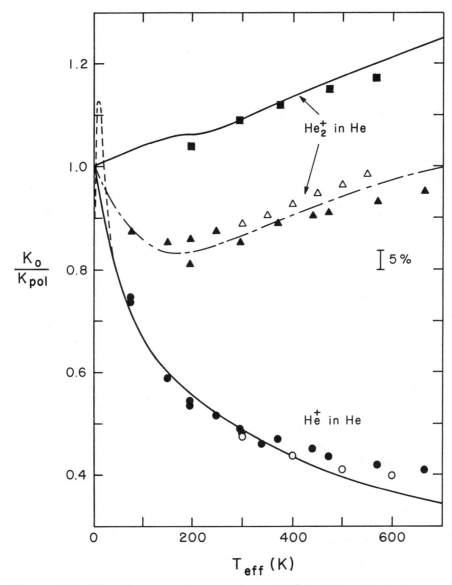

Figure 6-1-6. Effect of resonant charge exchange (He^+ in He) and ionic fragment exchange (He_2^+ in He) on the temperature dependence of the mobility. Both exchange processes substantially reduce the mobility and alter its expected temperature dependence. The solid curve for He_2^+ in He is calculated from a potential with neglect of ion exchange (Geltman, 1953; Mason and Schamp, 1958), and the associated points (■ Orient, 1967) refer to an excited state $(He_2^+)^*$ for which ion exchange is nonresonant. The dash-dot curve for ground-state He_2^+ merely connects the measured points with the polarization limit. The curves for He^+ in He represent theoretical calculations including charge exchange, independent of any mobility data (Dickinson, 1968a; Sinha et al., 1979). The dashed portion shows low-temperature quantum effects. The points correspond to the following measurements: ●, ▲ Orient (1967), Patterson (1970b), Helm (1976a, 1977); ○, △ Ellis et al. (1976a). The scaling of T and E/N into the single variable T_{eff} is also shown.

moving through He gas can exchange a He^+ ion,

$$He_2^+ + He = He + He_2^+. \tag{6-1-65}$$

The process is nearly resonant (and hence highly probable) if the He_2^+ ion is in its ground electronic state. Experimental results for this system are shown also in Fig. 6-1-6, together with a curve calculated with neglect of the exchange reaction (Geltman, 1953; Mason and Schamp, 1958). The polarization limit for this system is $K_{pol} = 18.76 \text{ cm}^2/\text{V-s}$. Here again we see the scaling of T and E/N into the single variable T_{eff}, with filled triangles for measurements as a function of T at $E/N \to 0$ and open triangles for measurements as a function of E/N at $T = 300 \text{ K}$.

The foregoing examples show how the mobility depends on the ion-neutral interaction in a rather definite way.

E.   Quantum Effects

All the preceding kinetic-theory formulas of this section are valid in both classical and quantum mechanics. The entire difference between classical and quantal results resides in the differential cross section $\sigma(\theta, \varepsilon)$ that appears in the expression (6-1-11) for the transport cross sections $\bar{Q}^{(l)}(\varepsilon)$. All quantum effects originate in $\sigma(\theta, \varepsilon)$, as do all effects of the ion-neutral interaction. Special phenomena associated with electron-neutral collisions also originate in $\sigma(\theta, \varepsilon)$.

Given the interaction, we could first calculate $\sigma(\theta, \varepsilon)$ and then obtain the $\bar{Q}^{(l)}(\varepsilon)$ and $\bar{\Omega}^{(l,s)}(T)$ by integration. This is an inefficient and inaccurate procedure from a computational standpoint. In a classical calculation it is better to convert to the impact parameter formulation,

$$\sigma(\theta) \sin \theta \, d\theta = b \, db, \tag{6-1-66}$$

calculate $\theta(b)$ from Newton's second law of motion according to (5-2-123), and then obtain the $\bar{Q}^{(l)}$ by integration over b. Similarly, in a quantal calculation it is better to perform a phase-shift analysis and then obtain the $\bar{Q}^{(l)}$ directly from the phase shifts without the intermediate use of $\sigma(\theta)$.

In this section we first show how a phase-shift calculation proceeds and then indicate the typical quantum effects that can occur. More detailed discussion is reserved for Chapter 7.

In a phase-shift calculation a plane wave incident on a scattering center is first decomposed into component partial waves, each partial wave corresponding to a different value of the angular-momentum quantum number l. (This quantum number has nothing to do with the index l of the transport cross sections.) Each partial wave is then followed through the scattering region by integration of the corresponding radial wave equation. Since energy and angular momentum are conserved, the only effect of the scattering potential is to shift the phase of the partial wave by an amount δ_l. The partial waves are then recombined to form the scattered wave, from which the differential cross section is identified. This is a standard calculation described in almost all textbooks on quantum mechanics. In particular, the scattering amplitude $f_s(\theta)$ is given in terms of the phase shifts as

$$f_s(\theta) = \frac{1}{2i\kappa} \sum_{l=0}^{\infty} (2l + 1)(e^{2i\delta_l} - 1)P_l(\cos \theta), \qquad (6\text{-}1\text{-}67a)$$

where $\kappa = \mu v/\hbar$ is the wave number of relative motion, and the differential cross section is

$$\sigma(\theta) = |f_s(\theta)|^2. \qquad (6\text{-}1\text{-}67b)$$

On substituting (6-1-67) back into the expression (6-1-11) for the transport cross sections, we find that the integration over θ can be carried out explicitly. After considerable trigonometric transformation the expressions for the first four transport cross sections can be put in the following neat forms:

$$\bar{Q}^{(1)} = \frac{4\pi}{\kappa^2} \sum_{l=0}^{\infty} (l + 1) \sin^2(\delta_l - \delta_{l+1}), \qquad (6\text{-}1\text{-}68a)$$

$$\bar{Q}^{(2)} = \frac{4\pi}{\kappa^2} \left(\frac{3}{2}\right) \sum_{l=0}^{\infty} \frac{(l + 1)(l + 2)}{2l + 3} \sin^2(\delta_l - \delta_{l+2}), \qquad (6\text{-}1\text{-}68b)$$

$$\bar{Q}^{(3)} = \frac{4\pi}{\kappa^2} \sum_{l=0}^{\infty} \left(\frac{l + 1}{2l + 5}\right)\left[\frac{(l + 2)(l + 3)}{2l + 3} \sin^2(\delta_l - \delta_{l+3})\right.$$
$$\left. + \frac{3(l^2 + 2l - 1)}{2l - 1} \sin^2(\delta_l - \delta_{l+1})\right], \qquad (6\text{-}1\text{-}68c)$$

$$\bar{Q}^{(4)} = \frac{4\pi}{\kappa^2} \left(\frac{5}{4}\right) \sum_{l=0}^{\infty} \frac{(l + 1)(l + 2)}{(2l + 3)(2l + 7)} \left[\frac{(l + 3)(l + 4)}{2l + 5} \sin^2(\delta_l - \delta_{l+4})\right.$$
$$\left. + \frac{2(2l^2 + 6l - 3)}{2l - 1} \sin^2(\delta_l - \delta_{l+2})\right]. \qquad (6\text{-}1\text{-}68d)$$

The factors $\frac{3}{2}$ and $\frac{5}{4}$ come from the normalization factor in the definition of the transport cross sections and have no special significance. The general formulation of all the transport cross sections in terms of phase shifts is rather complicated and has been given by Wood (1971).

Before indicating specific quantum effects, we should see how the foregoing expressions reduce to the corresponding classical ones. Two conditions are sufficient to obtain the classical limit in the case of transport cross sections:

1. The phase shifts can be calculated by the semiclassical Jeffreys–Wentzel–Kramers–Brillouin (JWKB) approximation, including the Langer modification.

2. The quantum number l can be taken as a continuous variable, so that differences of phase shifts can be replaced by differentials and sums over l replaced by integrals.

From condition 1 we make the identification between l and b,

$$l + \tfrac{1}{2} = \kappa b, \tag{6-1-69}$$

and the phase shift is given by the JWKB formula

$$\delta_l = \delta(b) = \kappa \int_{r_0}^{\infty} \left[1 - \frac{b^2}{r^2} - \frac{V(r)}{\varepsilon} \right]^{1/2} dr - \kappa \int_{b}^{\infty} \left(1 - \frac{b^2}{r^2} \right)^{1/2} dr. \tag{6-1-70}$$

Equivalent expressions for δ_l, which are sometimes more convenient for numerical computation, have been given by F. T. Smith (1965). Differentiation of (6-1-70) and comparison with the expression (5-2-123) for the deflection angle establishes the semiclassical connection between δ_l and θ,

$$\theta(b) = 2 \frac{d\delta_l}{dl} = \frac{2}{\kappa} \frac{d\delta(b)}{db}. \tag{6-1-71}$$

Phase-shift differences are thus directly related to the deflection angle through condition 2:

$$\delta_{l+n} - \delta_l = (n/2)\theta. \tag{6-1-72}$$

Substitution of (6-1-72) into the expressions (6-1-68) for the transport cross sections and replacement of the sums by integrals yields

$$\bar{Q}^{(1)} \approx \frac{4\pi}{\kappa^2} \int_{0}^{\infty} (l + 1) \sin^2(\theta/2) \, dl \to 2\pi \int_{0}^{\infty} (1 - \cos \theta) b \, db, \tag{6-1-73}$$

with similar expressions for the other $\bar{Q}^{(l)}$. Thus the two conditions above lead *exactly* to the classical limit, and $\hbar$ drops out identically. This result is special for transport cross sections and is due to the occurrence of phase-shift differences. Other types of cross sections usually depend on the phase shifts themselves rather than just on their differences, and the two conditions then lead to semiclassical results, in which $\hbar$ still appears and which still exhibit many quantum effects. Further approximations must then be made to obtain classical limits (if, indeed, these limits exist).

Quantum effects on transport cross sections can usually be ascribed to either of two causes:

1. Inaccuracy of the JWKB approximation
2. The discrete nature of l

It is the latter that produces the most striking effects in the form of maxima and minima in the momentum-transfer cross section, caused by resonances in the phase shifts (i.e., rapid changes in δ_l through multiples of $\pi/2$). These resonances are associated with the phenomenon of classical orbiting that occurs at a maximum in the effective potential (the true potential plus the centrifugal repulsion potential); orbiting was discussed in Section 5-2F in connection with the estimation of cross sections. Failure of the JWKB approximation usually modifies quantitative features but does not produce qualitative changes. The orbiting resonances, however, are often washed out by the averaging of the cross sections to form the collision integrals; only for light particles do the effects persist through to the mobility. The net result is that both of the causes above may contribute comparably to produce nonspectacular quantum deviations.

The foregoing effects are illustrated in Fig. 6-1-7, which shows the momentum-transfer cross section as a function of energy calculated by Munn et al. (1964) for a (12, 6) potential,

$$V(r) = 4\varepsilon_0 \left[\left(\frac{\sigma}{r}\right)^{12} - \left(\frac{\sigma}{r}\right)^6 \right], \qquad (6\text{-}1\text{-}74)$$

where σ is the value of r such that $V(\sigma) = 0$. This potential is not quantitatively

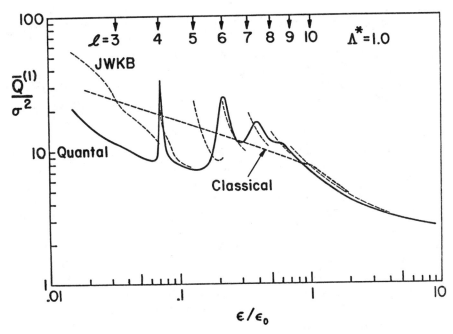

Figure 6-1-7. Quantal, semiclassical JWKB, and classical momentum-transfer cross sections for a (12, 6) potential, showing orbiting resonances and other quantum effects discussed in the text.

suitable as a model for ion-neutral interactions because it lacks the asymptotic r^{-4} form, but it shows the same general behavior. A measure of the quantum-mechanical behavior of the system is the ratio $\lambda/\sigma = 2\pi/\kappa\sigma$, where λ is the de Broglie wavelength; for potentials like (6-1-74) it is common practice to use the de Boer parameter Λ^*,

$$\Lambda^* \equiv \frac{h}{\sigma(2\mu\varepsilon_0)^{1/2}}, \qquad (6\text{-}1\text{-}75)$$

which is the value of λ/σ for collisions of energy $\varepsilon = \varepsilon_0$. The results in Fig. 6-1-7 are for $\Lambda^* = 1.0$. The correct quantal curve shows the orbiting resonances clearly; they occur at those energies for which one of the phase shifts shows a rapid change, as indicated by the corresponding value of l at the top of the figure. The energy at which an orbiting resonance occurs for a particular l value is easily found by calculating the height of the maximum in the effective potential, with the centrifugal potential set equal to $l(l + 1)\hbar^2/2\mu r^2 \approx (l + \frac{1}{2})^2\hbar^2/2\mu r^2$. The curve marked JWKB in the figure was calculated by properly summing (not integrating) the phase shifts calculated by the JWKB approximation. It also shows structure due to orbiting resonances but differs from the exact quantal curve, especially at low energies, because of the inaccuracies of the JWKB approximation for Λ^* as large as 1.0. The classical curve shows no trace of the orbiting resonances and roughly represents an average drawn through the resonances; this classical average representation becomes more accurate as Λ^* is decreased. Although classical orbiting is possible only for $\varepsilon/\varepsilon_0 \leqslant 0.8$ with the (12, 6) model, close inspection shows a small undulation on the classical curve at $\varepsilon/\varepsilon_0 \approx 1$, caused by rainbow scattering, which in this sense is a sort of classical resonance behavior. There is no sharp distinction between orbiting and rainbow scattering from a quantum-mechanical viewpoint because of tunneling near the top of the centrifugal barrier. These effects are discussed in more detail in Chapter 7.

Quantum effects on mobilities are shown in Fig. 6-1-2 for H^+ in He and in Fig. 6-1-6 for He^+ in He. Dickinson and Lee (1978) carried out a quantum-mechanical calculation of the collisions of H^+ with He, and from the results computed the mobility of H^+ in He down to 10 K. Their mobility, shown as a dashed curve in Fig. 6-1-2, coincides with the classical calculation of Lin et al. (1979a) down to 50 K; deviations can be seen below that temperature. Similar results occur with He^+ in He. The quantal calculations of Dickinson (1968a) agree with the classical calculations of Sinha et al. (1979) down to about 50 K. In this case there is a remnant of the orbiting resonances around 20 K that survived the averaging. Most other systems, being heavier than both H^+ in He and He^+ in He, would not deviate from classical behavior until temperatures even lower than 50 K were reached.

The quantal mobility does not approach the classical polarization limit at $T = 0$ K because at low energies only a few phase shifts contribute to the cross section. It can be shown that, for potentials falling off faster than r^{-3}, the phase shifts for $l \neq 0$ vanish as a power of κ but δ_0 is linear in κ (Landau and Lifshitz, 1977, Sec. 132). The scattering therefore becomes dominated by δ_0 at low energies (so-called s-wave scattering), and

$$\tan \delta_0 = -\kappa a, \tag{6-1-76a}$$

where a is called the scattering length. Moreover, in the limit of zero energy,

$$\delta_0(0) = n_0 \pi, \tag{6-1-76b}$$

where n_0 is the number of bound states in the potential well (Levinson's theorem). Thus at low energies,

$$\delta_0 = n_0 \pi + \kappa a + \cdots, \tag{6-1-76c}$$

and all the transport cross sections take on the same limiting form,

$$\bar{Q}^{(l)} \to \frac{4\pi}{\kappa^2} \sin^2 \delta_0 = 4\pi a^2, \tag{6-1-77}$$

as for rigid spheres. If there is a very low-lying resonance, the cross section may be increased enormously (Landau and Lifshitz, 1977, Sec. 133); the low-energy form for the cross sections is then

$$\bar{Q}^{(l)} \to \frac{4\pi a^2}{(1 - \frac{1}{2} a r_0 \kappa^2)^2 + \kappa^2 a^2}, \tag{6-1-78}$$

where r_0 is called the effective range. However, even (6-1-78) approaches $4\pi a^2$ at zero energy. The mobility at very low temperatures must therefore ultimately diverge as $T^{-1/2}$, but it happens at a temperature so low that it is experimentally unimportant. Even for the light system of H^+ in He, calculations with a model potential show no trace of such behavior at 1 K (Dickinson, 1968b).

F. Inelastic Collisions

All the foregoing results apply strictly only to monatomic ions and neutrals; a proper account of polyatomic ions and neutrals must allow for inelastic collisions involving rotational and vibrational degrees of freedom. In the case of electrons it may be necessary to allow for electronic excitation of the neutrals as well. This requires starting over again at the very beginning with a reformulation of the Boltzmann equation. We postpone the details of this development to Section 6-6, and give here just an outline of the results for the zero-field case, which were worked out much earlier by the Chapman–Enskog method. The zero-field results supply a useful limiting case against which the more general results can be viewed.

We can anticipate two results without any calculations at all. First, the simple Nernst–Townsend–Einstein relation will continue to hold rigorously, because it is more general than the Boltzmann equation or any extension thereof (Section 5-1A). Second, since diffusion involves mass transfer and mass is conserved in an inelastic collision as well as in an elastic one, we should not expect inelastic collisions to modify diffusion and mobility very much. It is an empirical fact that there is nothing remarkable about the diffusion coefficients of molecular gases as compared with atomic gases (Marrero and Mason, 1972). Both of these conclusions can be expected to hold only at zero field strengths, however. We already know that energy transfer is important at high fields in determining T_{eff}, T_L, and T_T (Sections 5-2A and 5-2C). Inelastic collisions would be expected to modify energy transfer, and so influence the high-field mobility and diffusion. This expectation is supported by the knowledge that the thermal conductivities of molecular gases, which involve energy transport, are indeed quite different from those of atomic gases (Mason and Monchick, 1962; Monchick et al., 1965).

The simplest extension of the Boltzmann equation is the Wang Chang–Uhlenbeck–de Boer (WUB) equation, which treats each internal state of ions and neutrals as a separate chemical species, and considers inelastic collisions as special chemical reactions (Wang Chang et al., 1964). Thus there is a Boltzmann equation for each internal state, just as there is for each species in a multicomponent mixture. Moreover, the differential cross section for elastic scattering $\sigma(\theta, \varepsilon)$ is replaced by a differential cross section $\sigma(\alpha\beta, \alpha'\beta'; \theta, \phi, \varepsilon)$ that describes collisions between two particles initially in internal states α and β which emerge from the collision in final states α' and β' at the angle θ, ϕ, the initial translational energy being ε. A more general extension of the Boltzmann equation is the Waldmann–Snider (WS) equation, which allows for quantum-mechanical interference effects that can occur when the internal states are degenerate (Waldmann, 1965, 1968, 1973). These effects produce only small changes in the transport coefficients, which can be detected experimentally by the application of an external magnetic field (Beenakker and McCourt, 1970).

The expansion (6-1-1) for the distribution function must also be extended to contain terms involving internal degrees of freedom as well as translational velocities. These added terms are of two types. One involves functions of only the internal energy; the other involves tensor functions constructed from the various angular momenta associated with a molecular ion, such as molecular rotation, electron spin, and nuclear spin. The latter are somewhat loosely called "spin polarization" terms, and describe a small alignment of the ion angular momentum caused by the imposed gradient. These spin polarization terms have only a small effect on the numerical value of the diffusion coefficient, as judged by calculations for the model of loaded spheres (Sandler and Dahler, 1967; Sandler and Mason, 1967), and the dominant effect comes from just the internal energy.

Detailed formal calculations for diffusion coefficients, based on the WUB equation and without the spin polarization terms, have been carried through by Monchick et al. (1963, 1966, 1968) and by Alievskiĭ and Zhdanov (1969). The external appearances of the expressions for D and K, including higher approximations, remain the same as for elastic collisions, but the explicit expressions for the transport cross sections and collision integrals are more complicated. The first few are as follows:

$$\bar{\Omega}^{(l,s)}(T) = 2[(s+1)!Z_i Z_j]^{-1} \sum_{\alpha\beta\alpha'\beta'} \exp[-(\varepsilon_i^\alpha + \varepsilon_j^\beta)/kT]$$

$$\times \int_0^\infty \bar{Q}^{(l)}(\alpha\beta, \alpha'\beta'; \varepsilon)\gamma^{2s+3}e^{-\gamma^2}d\gamma \qquad \text{for } l,s \leqslant 2, \qquad (6\text{-}1\text{-}79)$$

$$\gamma^2 \bar{Q}^{(1)}(\alpha\beta, \alpha'\beta'; \varepsilon) = \int_0^{2\pi} d\phi \int_0^\pi (\gamma^2 - \gamma\gamma' \cos\theta)$$

$$\times \sigma(\alpha\beta, \alpha'\beta'; \theta, \phi, \varepsilon) \sin\theta \, d\theta, \qquad (6\text{-}1\text{-}80a)$$

$$\gamma^4 \bar{Q}^{(2)}(\alpha\beta, \alpha'\beta'; \varepsilon) = \frac{3}{2}\int_0^{2\pi} d\phi \int_0^\pi [\gamma^2(\gamma^2 - \gamma'^2 \cos^2\theta) - \tfrac{1}{6}(\gamma^2 - \gamma'^2)^2]$$

$$\times \sigma(\alpha\beta, \alpha'\beta'; \theta, \phi, \varepsilon) \sin\theta \, d\theta, \qquad (6\text{-}1\text{-}80b)$$

where

$$\gamma^2 \equiv \varepsilon/kT, \qquad \gamma'^2 \equiv \varepsilon'/kT, \qquad (6\text{-}1\text{-}81a)$$

$$\varepsilon - \varepsilon' = \varepsilon_i^{\alpha'} + \varepsilon_j^{\beta'} - \varepsilon_i^\alpha - \varepsilon_j^\beta, \qquad (6\text{-}1\text{-}81b)$$

in which primes denote quantities after collision, ε_i^α and $\varepsilon_i^{\alpha'}$ are internal energies of the ion before and after collision, and ε_j^β and $\varepsilon_j^{\beta'}$ are internal energies of the neutral before and after collision. The quantities Z_i and Z_j in the normalization factor are the internal partition functions,

$$Z_i = \sum_\alpha \exp(-\varepsilon_i^\alpha/kT), \qquad Z_j = \sum_\beta \exp(-\varepsilon_j^\beta/kT). \qquad (6\text{-}1\text{-}82)$$

Computing power is now sufficiently great that some calculations of inelastic transport cross sections and collision integrals have begun to appear for model nonspherical potentials (Evans and Watts, 1976; Evans, 1977; Parker and Pack, 1978; Maitland et al., 1981; Nyeland et al., 1984; and other papers referred to therein). The results generally conform to the expectations discussed above. Ratios of collision integrals, such as A^* and B^* of (6-1-55), may differ from those for spherical potentials by about 10%, but nothing dramatic appears. The following argument suggests that inelastic effects on mobility and diffusion are generally small. Inelastic collisions enter $\bar{\Omega}^{(1,1)}$ mainly through the term $\gamma\gamma' \cos\theta$; to a first approximation $\gamma' \approx \gamma$ and there is no effect from inelastic collisions. For a second approximation γ' can be written as γ plus some terms in

$\Delta\varepsilon = \varepsilon - \varepsilon'$; inelastic correction terms are then of the form of integrals of $\gamma(\Delta\varepsilon)\cos\theta$ over angles and energies. Such terms would vanish for isotropic scattering, and even for anisotropic scattering would be expected to be small unless some special correlation existed between θ and $\Delta\varepsilon$. However, even if these extra terms are ignored, so that the formulas have the same appearance as those for elastic collisions only, the differential cross sections $\sigma(\alpha\beta, \alpha'\alpha'; \theta, \phi, \varepsilon)$ still include all the inelastic scattering. Thus the use of an effective spherical potential can be justified, but the final collision integrals should be regarded as made up of elastic plus inelastic excitation and deexcitation processes. This argument probably does not apply at very low temperatures, however.

As far as mobilities are concerned, the most important results of all these elaborate formal kinetic-theory calculations for zero field strength are that the Nernst–Townsend–Einstein relation remains valid, that the external forms of the transport equations and transport coefficients remain unchanged, and that numerical values of mobilities are only moderately affected by inelastic collisions.

G. Mixtures

The treatment of mixtures in the one-temperature theory is formally very simple. The moment equations remain exactly the same in form, but the matrix elements $a_{rs}(l)$ are replaced by matrix elements $b_{rs}(l)$ that are linear combinations of the $a_{rs}(l)$ for the different neutral species, weighted by their mole fractions,

$$b_{rs}(l) = \sum_j x_j a_{rs}^{(j)}(l), \qquad (6\text{-}1\text{-}83)$$

where the sum runs over all the neutral species in the mixture. However, in first approximation the one-temperature theory has very little to add to the results of the momentum-transfer theory for mixtures, except a more elegant derivation based on the Boltzmann equation. The reason is that, in effect, the one-temperature theory only repeats the momentum- and energy-balance equations of the momentum-transfer theory with modified expressions (6-1-15) for the average collision frequencies. Moreover, the expressions for K_{mix} and $\langle D_{T,L}\rangle_{mix}$ in Section 5-2D were carefully arranged to eliminate the collision frequencies and obtain expressions involving only experimental quantities, so that even these differences are suppressed.

Nevertheless, the one-temperature theory has two important contributions to make to the theory of ion mobility and diffusion in gas mixtures. The first is improved expressions for the ion temperatures in a mixture, which are needed in the computation of the diffusion coefficients. The second is the calculation of the deviations from Blanc's law that can occur even at zero field strength, which are of importance in connection with the question of the chemical identity of an ion.

It is useful to repeat the mixture formulas here, together with the improved expressions for the ion temperatures. Only one minor change will be later introduced by the more accurate two-temperature theory. The mobility of an ion in a mixture is given in terms of its mobilities in the single component gases and their field derivatives as

$$\frac{1}{K_{\text{mix}}} = \sum_j \frac{x_j}{K_j}(1 + \Delta_j), \tag{6-1-84a}$$

where

$$\Delta_j = \frac{1}{2}\left(\frac{K'_j}{1 + K'_j}\right)(1 - \delta_j), \tag{6-1-84b}$$

$$\delta_j^{-1} = (m + M_j)K_j^2 \left(\sum_i \frac{x_i}{K_i}\right)\left[\sum_i \frac{x_i}{(m + M_i)K_i}\right], \tag{6-1-84c}$$

$$K'_j \equiv \frac{d \ln K_{0j}}{d \ln(E/N)}. \tag{6-1-85}$$

These are identical to the formulas (5-2-91) and (5-2-92) of the momentum-transfer theory. The diffusion coefficients in a mixture are given by

$$\frac{\langle T_{T,L}\rangle_{\text{mix}}}{\langle D_{T,L}\rangle_{\text{mix}}}(1 + \gamma_{T,L}K'_{\text{mix}}) = \sum_j x_j \frac{(T_{T,L})_j}{(D_{T,L})_j}(1 + \Delta_j)(1 + \gamma_{T,L}K'_j) \tag{6-1-86a}$$

where

$$\gamma_T = 0, \qquad \gamma_L = 1, \tag{6-1-86b}$$

$$K'_{\text{mix}} = \left(\sum_j \frac{x_j K'_j}{K_j}\right)\left(\sum_j \frac{x_j}{K_j}\right)^{-1}. \tag{6-1-86c}$$

These are identical to (5-2-97) and (5-2-98) of the momentum-transfer theory. It is assumed that the diffusion coefficients in the single gases, $(D_{T,L})_j$, are known. If not, they can be calculated from the generalized Einstein relations of (5-2-56), namely

$$(eD_{T,L})_j/(kT_{T,L})_j = K_j(1 + \gamma_{T,L}K'_j), \tag{6-1-87}$$

or from one of the more accurate GER given later in Section 6-4.

The expressions for the single-gas ion temperatures are given by (6-1-42)–(6-1-45), which we repeat here in order to have all the formulas collected together:

$$(kT_{T,L})_j = kT + \zeta_{T,L}^{(j)} M_j v_{dj}^2 (1 + \beta_{T,L}^{(j)} K_j') \qquad (6\text{-}1\text{-}88)$$

$$\zeta_T^{(j)} = \frac{(m + M_j)A_j^*}{5m + 3M_j A_j^*}, \qquad (6\text{-}1\text{-}89a)$$

$$\zeta_L^{(j)} = \frac{5m - (2m - M_j)A_j^*}{5m + 3M_j A_j^*}, \qquad (6\text{-}1\text{-}89b)$$

$$\beta_T^{(j)} = \frac{5m(m + M_j)(5 - 2A_j^*)}{5(m^2 + M_j^2) + 4mM_j A_j^*} - \frac{1.5m^2 \zeta_L^{(j)}}{(m + M_j)(2m + M_j)}, \qquad (6\text{-}1\text{-}90a)$$

$$\beta_L^{(j)} = \frac{5m(m + M_j)(5 - 2A_j^*)}{5(m^2 + M_j^2) + 4mM_j A_j^*} + \frac{3.0m^2 \zeta_T^{(j)}}{(m + M_j)(2m + M_j)}. \qquad (6\text{-}1\text{-}90b)$$

These give somewhat more accurate results (Tables 6-1-4 and 6-1-5) than the corresponding semiempirical formulas of Skullerud. The formulas for the ion temperature in a gas mixture are calculated from the one-temperature theory to be (Whealton and Mason, 1974),

$$\langle kT_{T,L}\rangle_{\text{mix}} = kT + \langle \zeta_{T,L}\rangle_{\text{mix}} \langle M\rangle_{\text{mix}} \langle v_d\rangle_{\text{mix}}^2 (1 + \langle \beta_{T,L}\rangle_{\text{mix}} K_{\text{mix}}'),$$

$$(6\text{-}1\text{-}91)$$

$$\langle M\rangle_{\text{mix}} \langle \zeta_T\rangle_{\text{mix}} = \frac{(m + \langle M\rangle_{\text{mix}})\langle MA^*\rangle_{\text{mix}}}{5m + 3\langle MA^*\rangle_{\text{mix}}}, \qquad (6\text{-}1\text{-}92a)$$

$$\langle M\rangle_{\text{mix}} \langle \zeta_L\rangle_{\text{mix}} = \frac{5m\langle M\rangle_{\text{mix}} - (2m - \langle M\rangle_{\text{mix}})\langle MA^*\rangle_{\text{mix}}}{5m + 3\langle MA^*\rangle_{\text{mix}}}, \qquad (6\text{-}1\text{-}92b)$$

where

$$\langle M\rangle_{\text{mix}} \equiv \sum_j \omega_j M_j \Big/ \sum_j \omega_j, \qquad (6\text{-}1\text{-}93a)$$

$$\langle MA^*\rangle_{\text{mix}} \equiv \sum_j \omega_j M_j A_j^* \Big/ \sum_j \omega_j, \qquad (6\text{-}1\text{-}93b)$$

$$\omega_j = x_j(1 + \Delta_j)/(m + M_j)K_j, \qquad (6\text{-}1\text{-}94)$$

with Δ_j given in (6-1-84). These are precise results from the one-temperature theory, but the following expressions for $\langle \beta_{T,L}\rangle_{\text{mix}}$ are only semiempirical conjectures:

$$\langle \beta_T\rangle_{\text{mix}} \approx \frac{5m(m + \langle M\rangle_{\text{mix}})(5 - 2\langle A^*\rangle_{\text{mix}})}{5(m^2 + \langle M\rangle_{\text{mix}}^2) + 4m\langle MA^*\rangle_{\text{mix}}} - \frac{1.5m^2 \langle \zeta_L\rangle_{\text{mix}}}{(m + \langle M\rangle_{\text{mix}})(2m + \langle M\rangle_{\text{mix}})},$$

$$(6\text{-}1\text{-}95a)$$

$$\langle \beta_L \rangle_{\text{mix}} \approx \frac{5m(m + \langle M \rangle_{\text{mix}})(5 - 2\langle A^* \rangle_{\text{mix}})}{5(m^2 + \langle M \rangle^2_{\text{mix}}) + 4m\langle MA^* \rangle_{\text{mix}}} + \frac{3.0m^2\langle \zeta_T \rangle_{\text{mix}}}{(m + \langle M \rangle_{\text{mix}})(2m + \langle M \rangle_{\text{mix}})},$$

$$(6\text{-}1\text{-}95b)$$

where

$$\langle A^* \rangle_{\text{mix}} \equiv \langle MA^* \rangle_{\text{mix}} / \langle M \rangle_{\text{mix}}. \qquad (6\text{-}1\text{-}95c)$$

This essentially completes the kinetic-theory treatment of mixtures, insofar as it can be put into a relatively simple form. The only change introduced by the two-temperature theory is to make the collision-integral ratio A^* a function of T_{eff} rather than of T. It should be kept in mind that the foregoing formulas are all based on Taylor expansions, and are to be regarded with caution if any of the K'_j happens to be large. If greater accuracy is needed, it is probably best to fall back on the full implicit higher-order expressions supplied by the two- and three-temperature theories.

The second contribution of the one-temperature theory concerns deviations from Blanc's law at zero field strength, which occur in the higher kinetic-theory approximations. The reason for worrying enough about these deviations to bother working through the tedious calculations involved is that experimental deviations from Blanc's law are often taken as evidence that an ion changes its chemical identity as the gas composition is varied (Loeb, 1960, pp. 129ff.; Biondi and Chanin, 1961). It is therefore important to know how much deviation can be attributed to purely kinetic-theory effects in which the ion remains unchanged.

A general variational proof can be given that the deviations from Blanc's law at zero field strength are always positive, and suggests that they are of the same order of magnitude as the higher-order corrections to the mobilities of the single component gases (Holstein, 1955; Biondi and Chanin, 1961). Explicit expressions for the deviations have been worked out by the Chapman–Enskog method (Sandler and Mason, 1968), and by the one-temperature theory (Whealton and Mason, 1974). The special case of electrons has been considered in detail by Petrović (1986).

The simple Nernst–Townsend–Einstein relation remains valid for multicomponent mixtures at zero field strength to all orders of kinetic-theory approximation.

The following expressions, which are valid when only elastic collisions occur, refer to zero field strength and include only the leading term of the deviation:

$$\frac{1}{K_{\text{mix}}} = \sum_j \frac{x_j}{K_j} + \Delta(\text{Blanc}), \qquad (6\text{-}1\text{-}96)$$

$$\Delta(\text{Blanc}) = \frac{\frac{1}{4}\sum_i \sum_{j>i}(x_i q_i/K_i)(x_j q_j/K_j)G^2_{ij}}{\sum_i x_i q_i/K_i}, \qquad (6\text{-}1\text{-}97a)$$

where

$$q_i \equiv \frac{15m^2 + 5M_i^2 + 8mM_iA_i^*}{2(m + M_i)^2}, \tag{6-1-97b}$$

$$G_{ij} \equiv \frac{M_i}{m + M_i}\frac{6C_i^* - 5}{q_i} - \frac{M_j}{m + M_j}\frac{6C_j^* - 5}{q_j}, \tag{6-1-97c}$$

and the C^* and A^* are the dimensionless ratios of collision integrals already defined in (6-1-55). The summation in the numerator of (6-1-97a) runs over all pairs of neutral gases in the mixture, and the summation in the denominator over all single neutral gases. Since the q_i are always positive, it is clear from the form of (6-1-97a) that Δ(Blanc) is always positive, in accord with Holstein's theorem. We can immediately note several cases for which Blanc's law is exact. The first is the Maxwell model, for which $(6C^* - 5) = 0$. The second is the Rayleigh model, for which $m \gg M_i, M_j$. These are just the cases for which we have already noted that the first approximation for the mobility of an ion in a single gas is exact; thus it is no surprise that Blanc's law is also exact.

For a binary mixture the maximum deviation from Blanc's law is

$$\text{max } \Delta(\text{Blanc}) = \frac{1}{4}\left(\frac{q_1}{K_1}\right)\frac{q_2}{K_2}\left[\left(\frac{q_1}{K_1}\right)^{1/2} + \left(\frac{q_2}{K_2}\right)^{1/2}\right]^{-2} G_{12}^2, \tag{6-1-98a}$$

and occurs at a composition

$$(x_1)_{\text{max}} = 1 - (x_2)_{\text{max}} = \left(\frac{q_2}{K_2}\right)^{1/2}\left[\left(\frac{q_1}{K_1}\right)^{1/2} + \left(\frac{q_2}{K_2}\right)^{1/2}\right]^{-1}. \tag{6-1-98b}$$

The major sensitivity to temperature and to the ion-neutral potentials arises from the factors $(6C^* - 5)$, and the major percentage variation of Δ(Blanc) occurs through G_{12}. To demonstrate this we rearrange (6-1-98a) to the form

$$\frac{\text{max } \Delta(\text{Blanc})}{1/K_{\text{mix}}(\text{Blanc})} = \frac{\frac{1}{4}q_1q_2G_{12}^2}{q_1 + q_2 + (K_1 + K_2)(q_1q_2/K_1K_2)^{1/2}}$$

$$\leqslant \frac{\frac{1}{4}q_1q_2G_{12}^2}{(q_1^{1/2} + q_2^{1/2})^2}. \tag{6-1-99}$$

The last step follows because $\frac{1}{2}(K_1 + K_2) \geqslant (K_1K_2)^{1/2}$. Since the q's lie between 2.5 and 7.5, the contribution of the factors other than G_{12} lies between about 5/32 and 15/32. It is possible to concoct fairly large deviations from Blanc's law by suitable choices of mass ratios and ion-neutral potentials; for instance, $(6C^* - 5)$ is $+1$ for a rigid-sphere potential and -1 for an r^{-2} potential. With these values and with $m \ll M_1, M_2$ we find a maximum deviation of about 10%. An even greater deviation can be obtained by use of a screened Coulomb potential in place of the r^{-2} potential, which leads to a maximum deviation of more than 50%; this corresponds physically to the mobility of a foreign ion through a partially ionized gas. Most other combinations of masses and potentials lead to much smaller deviations, usually of the order of a few percent.

Values of $(6C^* - 5)$ are shown as a function of temperature in Fig. 6-1-8 for some $(n, 4)$ potentials and in Fig. 6-1-9 for some $(12, 6, 4)$ potentials.

An advantage of the full one-temperature theory over the Chapman–Enskog theory is that it allows the sensitive factors $(6C^* - 5)$ to be replaced by experimental quantities, namely the initial field dependence of the mobilities in

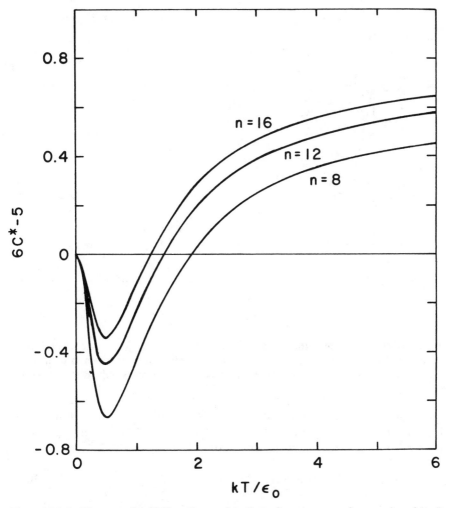

Figure 6-1-8. The quantity $(6C^* - 5)$ as a function of temperature for a series of $(n, 4)$ potentials. This quantity is used to describe higher-order kinetic-theory approximations, deviations from Blanc's law, and thermal diffusion. Data from Viehland et al. (1975).

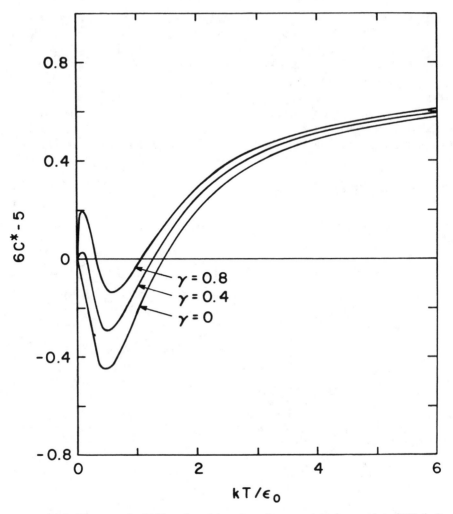

Figure 6-1-9. The quantity $(6C^* - 5)$ as a function of temperature for a series of $(12, 6, 4)$ potentials. This quantity is used to describe higher-order kinetic-theory approximations, deviations from Blanc's law, and thermal diffusion. Data from Viehland et al. (1975).

the single component gases. The reason is that both kinds of quantities arise from a dominant matrix element $b_{01}(1)$. If we express the initial field dependence in terms of the expansion coefficient α_2 of (5-1-5a),

$$K(E) = K(0)[1 + \alpha_2(E/N)^2 + \cdots], \qquad (6\text{-}1\text{-}100a)$$

or

$$\alpha_2 = \lim_{E/N \to 0} \frac{d \ln K(E)}{d(E/N)^2}, \qquad (6\text{-}1\text{-}100b)$$

then the deviations from Blanc's law can be written as

$$G_{ij} = \frac{4kT}{5N^2}\left[\frac{5m + 3M_iA_i^*}{(m + M_i)(3m + M_iA_i^*)}\frac{\alpha_{2i}}{K_i^2} - \frac{5m + 3M_jA_j^*}{(m + M_j)(3m + M_jA_j^*)}\frac{\alpha_{2j}}{K_j^2}\right],$$

$$(6\text{-}1\text{-}101)$$

with the rest of (6-1-97) remaining the same. This expresses the deviations from Blanc's law in terms of experimental quantities except for the ratios A^*, which usually do not depend sensitively on the ion-neutral potential.

In summary, kinetic-theory deviations from Blanc's law at zero field strength are usually small, although it is possible to imagine systems with rather large deviations. In any case, the deviations are given explicitly by (6-1-97) and/or (6-1-101).

H. Effect of Temperature Gradients: Thermal Diffusion

In all the discussion so far it has been assumed that the temperature is uniform. Temperature gradients can, however, cause diffusion in mixtures; this phenomenon, known as thermal diffusion, requires the addition of another term to the flux equation (5-1-1) for the ions, which becomes

$$\mathbf{J} = nK\mathbf{E} - D\nabla n - nD\alpha_T\nabla\ln T, \qquad (6\text{-}1\text{-}102)$$

where α_T is a dimensionless quantity known as the thermal diffusion factor. At this stage we can only discuss the low-field behavior of α_T, and must postpone consideration of its field dependence until the two-temperature theory is developed in Section 6-2. Even so, the dependence of α_T on the ion-neutral mass ratio and potential, and on the temperature, is fairly complicated. We shall not go into much detail, since our main concern here is only to assess the importance of thermal diffusion as a source of error in mobility and diffusion measurements. The reader interested in further information on thermal diffusion can find it in the standard monographs by Chapman and Cowling (1970), Hirschfelder et al. (1964), and Ferziger and Kaper (1972). Specialized reviews on thermal diffusion are also available (Grew and Ibbs, 1952, 1962; Mason et al., 1966).

For an estimate of the effect of thermal diffusion on drift-tube measurements we can compare the magnitude of the thermal diffusive flux with the magnitudes of the ordinary diffusive flux and the forced flux terms in (6-1-102). The ratio of thermal flux to diffusive flux is

$$\frac{\text{thermal flux}}{\text{diffusive flux}} = \frac{nD\alpha_T\nabla\ln T}{D\nabla n} = \alpha_T\frac{\nabla\ln T}{\nabla\ln n}. \qquad (6\text{-}1\text{-}103)$$

For ordinary neutral gases α_T is of magnitude unity, so the thermal flux will be negligible if $\nabla\ln T$ is much smaller than $\nabla\ln n$. The ratio of thermal flux to forced flux is

$$\frac{\text{thermal flux}}{\text{forced flux}} = \frac{nD\alpha_T \mathbf{V} \ln T}{nK\mathbf{E}} = 8.62 \times 10^{-5}\alpha_T \frac{\mathbf{V}T(\text{K/cm})}{\mathbf{E}(\text{V/cm})}, \quad (6\text{-}1\text{-}104)$$

for singly charged ions, where we have used the Nernst–Townsend–Einstein relation. If we take a rather large value of 100 K/cm for $\mathbf{V}T$ and a rather small value of 1 V/cm for $\mathbf{E}$, we find that the ratio of thermal flux to forced flux is only of order 10^{-2}, a rather small contribution.

Thus thermal diffusion is probably not an important source of error in mobility and diffusion measurements unless α_T is very large. When we examine the kinetic-theory expression for α_T, we shall see that it is possible for α_T to be large if the ions are large, even in the low-field limit.

The Chapman–Enskog first approximation for α_T for ions in a neutral gas can be written as

$$[\alpha_T]_1 = (6C^* - 5)(S_\alpha/Q_\alpha), \quad (6\text{-}1\text{-}105a)$$

where

$$S_\alpha = \frac{15}{2}\left[\frac{m(m - M)}{(m + M)^2}\right] + \frac{4mMA^*}{(m + M)^2} - \frac{5}{3}\left(\frac{M^2}{m + M}\right)\frac{ND}{\eta}, \quad (6\text{-}1\text{-}105b)$$

$$Q_\alpha = \frac{4}{3}qM\frac{ND}{\eta}, \quad (6\text{-}1\text{-}105c)$$

in which C^* and A^* are as already defined, η is the viscosity of the pure neutral gas, and q is as defined in (6-1-97b). This expression is strictly valid only for elastic collisions, but can be used for rough estimates for polyatomic systems (Monchick et al., 1966, 1968).

The sign convention used in writing the flux equation (6-1-102) is that the ions move "down" the temperature gradient towards the lower temperature region if α_T is positive, and towards the higher temperature region if α_T is negative. The sign of α_T depends on $(6C^* - 5)$ and on the mass ratio. If $(6C^* - 5)$ is positive, then α_T is positive for $m \gg M$ but negative for $m \ll M$, and vice versa. The limiting form for $m \ll M$ is

$$[\alpha_T]_1 = -\tfrac{1}{2}(6C^* - 5), \quad (6\text{-}1\text{-}106a)$$

and for $m \gg M$ is

$$[\alpha_T]_1 = \tfrac{3}{4}(6C^* - 5)\frac{\eta}{ND}. \quad (6\text{-}1\text{-}106b)$$

However, $(6C^* - 5)$ depends sensitively on the ion-neutral potential and can be positive, negative, or zero. For an r^{-n} potential it is equal to $(1-4/n)$, and so is positive for $n > 4$, zero for $n = 4$, and negative for $n < 4$. Thus thermal diffusion always vanishes in the polarization limit. For more complicated potentials like the $(n, 4)$ and $(12, 6, 4)$ models $(6C^* - 5)$ can have a fairly complicated temperature dependence, as shown in Figs. 6-1-8 and 6-1-9. The curves in these figures can be used for making quick estimates of α_T.

If values of $(6C^* - 5)$ are not readily available but the mobility is known as a function of temperature or field strength, an estimate can be obtained as follows. By direct differentiation of the definition (6-1-10) of the $\bar{\Omega}^{(l,s)}$ we obtain the exact recursion relation

$$C^* \equiv \frac{\bar{\Omega}^{(1,2)}}{\bar{\Omega}^{(1,1)}} = 1 + \frac{1}{3} \frac{d \ln \bar{\Omega}^{(1,1)}}{d \ln T}. \qquad (6\text{-}1\text{-}107a)$$

Combining this with the first approximation for the mobility, we obtain the approximate formula

$$6C^* - 5 \approx -2 \frac{d \ln K_0}{d \ln T_{\text{eff}}}, \qquad (6\text{-}1\text{-}107b)$$

in which we have borrowed the scaling result from the momentum-transfer theory (or the two-temperature theory) that condenses T and E/N into the single variable T_{eff}.

Examination of (6-1-106a) shows that α_T can never be very large for light ions, but (6-1-106b) reveals that α_T can be large for heavy ions and indeed enormous for ions as large as charged aerosol or dust particles. The reason is that D becomes very small for large particles, whereas η does not change. An order-of-magnitude estimate can be readily obtained by remembering that $\bar{\Omega}^{(1,1)} \approx \pi d^2$, where d is an equivalent rigid-sphere diameter. So for ions of micron size moving through neutrals of angstrom size,

$$\alpha_T \sim (10^{-4} \text{ cm}/10^{-8} \text{ cm})^2 = 10^8. \qquad (6\text{-}1\text{-}108)$$

Temperature gradients can therefore be important for large ions.

6-2. TWO-TEMPERATURE THEORY

The structure of the two-temperature theory is quite similar to that of the one-temperature theory, and we can therefore dispense with most of the formal details in this section, which would only be repetitious. The results, however, can be quite different, especially the convergence for strong electric fields. It might be wondered why the two-temperature theory was not developed much sooner than it actually was, since an apparently minor parameter change is all that is involved—the substitution of an ion basis temperature T_b as a parameter in place of the gas temperature T. The reason is that the collision operator J is not symmetric on the inner product space defined by the Burnett functions unless $T_b = T$. The symmetry of J was used in an apparently essential way in the derivation of the moment equations by Kihara (1953) in his development of the one-temperature theory, and there was a substantial delay before it was realized that this symmetry is not really needed (Viehland and Mason, 1975). With the wisdom of hindsight we have avoided this symmetry difficulty in the derivation of the general moment equations in Section 5-3B, and suppressed the problem entirely. Such is progress.

It might also reasonably be wondered why the replacement of T by T_b produces such dramatic improvement in the calculation of the mobility as a function of E/N. The two-temperature theory gives reasonably good results at all E/N, whereas the one-temperature theory gives the miserable convergence illustrated in Fig. 6-1-1. The reason is that the expansion variable is not really E/N, but is the dimensionless quantity $\mathscr{E}$ of (6-1-22), in which E/N is divided by $[kT\bar{\Omega}^{(1,1)}(T)]$. This divisor is constant at fixed T, so that the variable $\mathscr{E}$ increases without limit as E/N increases, a behavior that is almost guaranteed to produce bad convergence. In the two-temperature theory, however, the divisor becomes $\{(kT_b)^{1/2}[(kT_{\text{eff}})^{1/2}\bar{\Omega}^{(1,1)}(T_{\text{eff}})]\}$, which cancels the unbounded increase of E/N and causes $\mathscr{E}$ to remain finite, provided that T_b is chosen properly. The proper choice is to take kT_b approximately proportional to the mean ion energy. The Wannier energy formula then shows that $kT_b \sim v_d^2 = K^2(E/N)^2$ at high fields, and $\mathscr{E}$ therefore remains finite and the convergence is reasonable. It is this finiteness of $\mathscr{E}$ that most distinguishes the two-temperature theory from the one-temperature theory. It is not necessary that $\frac{3}{2}kT_b$ be precisely equal to the mean ion energy; some calculations with a rigid-sphere cold-gas model indicate that rough equality within a factor of about 2 is sufficient for convergence (Kumar et al., 1980, pp. 411–412).

A. Moment Equations and Truncation-Iteration Scheme

The two-temperature expansion for the ion distribution function is

$$f = f^{(0)} \sum_{r=0}^{\infty} \sum_{l=0}^{\infty} \sum_{m=-l}^{l} f_{lm}^{(r)} w_b^l S_{l+1/2}^{(r)}(w_b^2) Y_l^m(\theta, \phi), \qquad (6\text{-}2\text{-}1)$$

where

$$f^{(0)} = n(m/2\pi kT_b)^{3/2} \exp(-w_b^2), \qquad (6\text{-}2\text{-}2a)$$

$$w_b^2 = mv^2/2kT_b, \qquad (6\text{-}2\text{-}2b)$$

and the expansion coefficients $f_{lm}^{(r)}$ are moments of the distribution over the basis functions $\psi_{lm}^{(r)}$. This is the same as the general expansion of the one-temperature theory, except that T_b replaces T. The moment equations have the same form as the equations for the one-temperature theory, with T replaced by T_b. This replacement changes the character of the dimensionless field-strength function $\mathscr{E}$, as already mentioned, and also changes the explicit expressions for the matrix elements $a_{rs}(l)$, which are still defined as

$$a_{rs}(l) \equiv (\psi_{lm}^{(s)}, J\psi_{lm}^{(r)})/(\psi_{lm}^{(s)}, \psi_{lm}^{(s)}). \tag{6-2-3}$$

The $a_{rs}(l)$ are independent of the index m because J is a scalar operator, but $a_{sr}(l) \neq a_{rs}(l)$ unless $T_b = T$.

To emphasize the role played by $\mathscr{E}$, we exhibit the general moment equations for the case of a spatially homogeneous distribution of ions in a single neutral gas,

$$\mathscr{E}[l(l + \tfrac{1}{2} + r)\langle\psi_{l-1,0}^{(r)}\rangle^{(0)} - (l + 1)\langle\psi_{l+1,0}^{(r-1)}\rangle^{(0)}]$$

$$= (l + \tfrac{1}{2}) \sum_{s=0}^{\infty} \gamma_{rs}(l)\langle\psi_{l0}^{(s)}\rangle^{(0)}, \tag{6-2-4}$$

where

$$\mathscr{E} \equiv \left[\left(\frac{m}{2kT_b}\right)^{1/2} \frac{e}{ma_{00}(1)}\right]\frac{E}{N}, \tag{6-2-5}$$

$$\gamma_{rs}(l) \equiv a_{rs}(l)/a_{00}(1), \tag{6-2-6}$$

and it is understood that any $\psi_{lm}^{(r)}$ with any negative indices is zero. We do not bother writing down the equations for the spatially inhomogeneous moments because the two-temperature theory does not give a good account of diffusion when $m > M$ (Viehland and Mason, 1978). The best that can be done is to produce a minor improvement in the generalized Einstein relations already obtained from the one-temperature theory. A reasonable treatment of diffusion for all values of m/M is given by the three-temperature theory described in Section 6-3.

The calculations for the matrix elements are tedious and complicated, but the $a_{rs}(l)$ can finally be expressed as linear combinations of irreducible collision integrals, $\bar{\Omega}^{(l,s)}(T_{\text{eff}})$. These collision integrals are defined by (6-1-10) of the one-temperature theory, but the temperature parameter T is now replaced by T_{eff}, defined as

$$T_{\text{eff}} \equiv (mT + MT_b)/(m + M). \tag{6-2-7}$$

The first few $a_{rs}(l)$ are given in Table 6-2-1. These suffice for many low-order calculations, in which most of the useful physical content of the theory appears, largely through the close connection with the momentum-transfer theory of Section 5-2. However, the numerical accuracy is usually poorer than people have come to expect from contact with the classical Chapman–Enskog version of kinetic theory, which here corresponds to vanishing fields. For accurate

Table 6-2-1 Matrix Elements for the Two-Temperature Theory, in units of
$$\frac{M}{m+M}\left(\frac{2kT_{\text{eff}}}{\pi\mu}\right)^{1/2}\bar{\Omega}^{(1,1)}(T_{\text{eff}})$$

The following condensed dimensionless notation is used:

$$e \equiv M/(m+M)$$
$$d \equiv MT_b/(mT + MT_b) = e(T_b/T_{\text{eff}})$$
$$f \equiv \mu(T_b - T)/(mT + MT_b) = d(1-e)(T_b - T)/T_b$$
$$A^* \equiv \bar{\Omega}^{(2,2)}/\bar{\Omega}^{(1,1)}$$
$$B^* \equiv [5\bar{\Omega}^{(1,2)} - 4\bar{\Omega}^{(1,3)}]/\bar{\Omega}^{(1,1)}$$
$$C^* \equiv \bar{\Omega}^{(1,2)}/\bar{\Omega}^{(1,1)}$$
$$D^* \equiv \bar{\Omega}^{(1,4)}/\bar{\Omega}^{(1,1)}$$
$$E^* \equiv \bar{\Omega}^{(2,3)}/\bar{\Omega}^{(2,2)}$$
$$F^* \equiv \bar{\Omega}^{(3,3)}/\bar{\Omega}^{(1,1)}$$
$$G^* \equiv \bar{\Omega}^{(2,4)}/\bar{\Omega}^{(2,2)}$$

For rigid spheres, A^*, B^*, C^*, D^*, E^*, F^*, and G^* are all unity. For the Maxwell model, $B^* = \frac{5}{4}$, $C^* = \frac{5}{6}$, $D^* = \frac{21}{32}$, $E^* = \frac{7}{8}$, and $G^* = \frac{63}{80}$; A^* and F^* are constants whose numerical values depend on details of the angular scattering pattern.

$$a_{0s}(0) = 0$$
$$a_{10}(0) = -8(f/d)$$
$$a_{11}(0) = (8/3)[2(1-e) + f(6C^* - 5)]$$
$$a_{12}(0) = -(8d/15)[4(1-d)(6C^* - 5) - 3f(4B^* - 5)]$$
$$a_{20}(0) = -(4/d^2)[10ef^2 - f(e^2 + f^2)(6C^* - 5) - 4f^2 eA^*]$$
$$\begin{aligned}a_{21}(0) = -(4/3d)[&10f(1 - 4e + 4e^2) + 2\{d^2(1-d) - fd(2 - 3d + 11f) \\ &+ f^2(4 + 7f)\}(6C^* - 5) - 3f(e^2 + f^2)(4B^* - 5) + 16f(1-e)eA^* \\ &+ 4f^2 d(8E^* - 7)eA^*]\end{aligned}$$
$$\begin{aligned}a_{22}(0) = (4/15)[&40(1-e)(1 - 2e + 2e^2) + f(44 - 128d + 80f - d^2 \\ &+ 90df - 170f^2)(6C^* - 5) + 3\{5f(d^2 + 2e^2 + 2f^2) \\ &- 4(1-d)(e^2 + 3f^2)\}(4B^* - 5) + 15f(e^2 + f^2)(32D^* - 21) \\ &+ 32(1-e)^2 eA^* + 8f\{4(1-d) - 5f\}(8E^* - 7)eA^* \\ &+ 4f^2(80G^* - 63)eA^*]\end{aligned}$$
$$a_{00}(1) = 8/3$$
$$a_{01}(1) = -(8d/15)(6C^* - 5)$$
$$a_{02}(1) = -(8d^2/105)[4(6C^* - 5) + 3(4B^* - 5)]$$
$$a_{10}(1) = -(4/3d)[10f(1 - 2e) + (e^2 + 3f^2)(6C^* - 5) + 8feA^*]$$
$$\begin{aligned}a_{11}(1) = (4/15)[&10(3 - 6e + 4e^2) + \{5(e^2 + 3f^2) - 5d^2 + 22f(1-d)\}(6C^* - 5) \\ &- 3(e^2 + 3f^2)(4B^* - 5) + 16(1-e)eA^* + 8f(8E^* - 7)eA^*]\end{aligned}$$
$$a_{00}(2) = (16/15)[5(1-e) + f(6C^* - 5) + 3eA^*]$$
$$a_{01}(2) = (16d/105)[\{7(d-1) - 3f\}(6C^* - 5) + 3f(4B^* - 5) - 3(8E^* - 7)eA^*]$$

Note: A misprint in the published expression (Viehland and Mason, 1975) for $a_{02}(1)$ has been corrected here.

numerical work, more matrix elements are needed than appear in Table 6-2-1, and the original literature should be consulted (Viehland and Mason, 1975, 1978; Weinert, 1978, 1982; Lindenfeld and Shizgal, 1979; Kumar, 1980; Kumar et al., 1980; Viehland et al., 1981; and other references given in these papers).

1. First Approximations

To devise a truncation-iteration procedure we again refer to the Maxwell model for some clue as to the likely magnitudes of the $a_{rs}(l)$. Unlike the case of the one-temperature theory, not all the off-diagonal $a_{rs}(l)$ vanish for the Maxwell model when $T_b \neq T$, but only those for which $s > r$, so a little caution is now in order. To see what is involved, and to make contact with the momentum-transfer theory, we consider the equations for momentum and energy balance, which we obtain from the moment equations (6-2-4) by setting (l, r) equal to $(1, 0)$, $(0, 1)$, and $(2, 0)$, respectively,

$$\mathscr{E} = \langle \psi_{10}^{(0)} \rangle^{(0)} + \sum_{s=1}^{\infty} \gamma_{0s}(1) \langle \psi_{10}^{(0)} \rangle^{(0)}, \qquad (6\text{-}2\text{-}8a)$$

$$-\mathscr{E}\langle \psi_{10}^{(0)} \rangle^{(0)} = \tfrac{1}{2}\gamma_{10}(0) + \tfrac{1}{2}\gamma_{11}(0)\langle \psi_{00}^{(1)} \rangle^{(0)} + \tfrac{1}{2}\sum_{s=2}^{\infty} \gamma_{1s}(0)\langle \psi_{00}^{(s)} \rangle^{(0)}, \qquad (6\text{-}2\text{-}8b)$$

$$5\mathscr{E}\langle \psi_{10}^{(0)} \rangle^{(0)} = \tfrac{5}{2}\gamma_{00}(2)\langle \psi_{20}^{(0)} \rangle^{(0)} + \tfrac{5}{2}\sum_{s=1}^{\infty} \gamma_{0s}(2)\langle \psi_{20}^{(s)} \rangle^{(0)}. \qquad (6\text{-}2\text{-}8c)$$

Here we have kept all the presumably small terms in the summations and exhibited the presumably large terms explicitly. The troublesome term is $\gamma_{11}(0)\langle \psi_{00}^{(1)} \rangle^{(0)}$ in (6-2-8b); it can be eliminated by taking

$$\langle \psi_{00}^{(1)} \rangle^{(0)} = 0, \qquad (6\text{-}2\text{-}9a)$$

which is equivalent to setting

$$\tfrac{3}{2}kT_b = \tfrac{1}{2}m\langle v^2 \rangle. \qquad (6\text{-}2\text{-}9b)$$

The first approximation then consists of taking Maxwell-model values for the $a_{rs}(l)$ and assigning T_b a value corresponding to the mean ion energy, according to (6-2-9). To complete the identification with momentum-transfer theory we define an average collision frequency at T_{eff} as

$$\bar{v}(T_{\text{eff}}) \equiv \frac{4}{3} N \left(\frac{8kT_{\text{eff}}}{\pi\mu} \right)^{1/2} \bar{\Omega}^{(1,1)}(T_{\text{eff}}), \qquad (6\text{-}2\text{-}10)$$

which corresponds to the definition (6-1-15a) used in the one-temperature theory, but with T_{eff} in place of T.

The three moment equations in first approximation then become

$$eE = \mu\bar{v}(T_{\text{eff}})\langle v_z\rangle_1,\tag{6-2-11a}$$

$$eE\langle v_z\rangle_1 = \frac{mM}{(m+M)^2}(m\langle v^2\rangle_1 - 3kT)\bar{v}(T_{\text{eff}}),\tag{6-2-11b}$$

$$eE\langle v_z\rangle_1 = \frac{1}{10}\frac{mM(5m+3MA^*)}{(m+M)^2}(3\langle v_z^2\rangle_1 - \langle v^2\rangle_1)\bar{v}(T_{\text{eff}}).\tag{6-2-11c}$$

The first equation is the momentum-balance equation, corresponding to (5-2-8) of the momentum-transfer theory. The second equation is the overall energy-balance equation, corresponding to (5-2-19) of the momentum-transfer theory. The third equation is the energy-partitioning equation, corresponding to (5-2-69) of the momentum-transfer theory on setting $A^* = \frac{5}{6}$ for Maxwell-model isotropic scattering. These equations have exactly the same appearance as the corresponding equations of the one-temperature theory, namely (6-1-17), (6-1-29), and (6-1-40), respectively, except that $\bar{v}(T_{\text{eff}})$ replaces $\bar{v}(T)$—a crucial difference. A less important difference is that the collision-integral ratio A^* is evaluated at T_{eff} instead of at T. Notice that the right-hand side of the energy-balance equation (6-2-11b) arises from the off-diagonal matrix element $\gamma_{10}(0)$, and not from the term with $\gamma_{11}(0)$, as it did in the one-temperature theory.

In first approximation, the mobility is found from (6-2-11a) to be

$$[K]_1 = \frac{3e}{16N}\left(\frac{2\pi}{\mu kT_{\text{eff}}}\right)^{1/2}\frac{1}{\bar{\Omega}^{(1,1)}(T_{\text{eff}})},\tag{6-2-12}$$

with T_{eff} given in terms of T and T_b by (6-2-7) and T_b given by (6-2-9). This formula for $[K]_1$ combines the virtues of the momentum-transfer theory and the one-temperature theory, while avoiding their main faults. The numerical factors are correct, and a thermally averaged cross section at temperature T_{eff} appears, corresponding to the random energy of the ions (thermal plus random field energy). This identification of T_{eff} follows from combining (6-2-11a) and (6-2-11b) to eliminate $\bar{v}(T_{\text{eff}})$; the Wannier energy formula is then obtained again,

$$\tfrac{1}{2}m\langle v^2\rangle_1 = \tfrac{3}{2}kT + \tfrac{1}{2}(m+M)\langle v_z\rangle_1^2 = \tfrac{3}{2}[kT_b]_1,\tag{6-2-13a}$$

or

$$\tfrac{1}{2}\mu\langle v_r^2\rangle_1 = \tfrac{3}{2}kT + \tfrac{1}{2}M\langle v_z\rangle_1^2 = \tfrac{3}{2}[kT_{\text{eff}}]_1.\tag{6-2-13b}$$

Of course, we must still investigate the accuracy of these formulas and how to improve them by developing convergent higher approximations.

We can obtain formulas for the longitudinal and transverse ion temperatures by combining all three equations (6-2-11) and eliminating $\bar{v}(T_{\text{eff}})$. The expressions are exactly the same as those from the one-temperature theory, equations (6-1-41) and (6-1-42), except that A^* is evaluated at T_{eff} instead of at T.

For the diffusion coefficients, the best general result we can obtain is just the generalized Einstein relations again. There is no point in repeating all these expressions here.

2. Higher Approximations

Higher approximations are obtained by selecting some systematic procedure for including more moments and more moment equations, keeping a closed set of equations at every stage. That is, the infinite set of moment equations is truncated to produce a finite set that can be solved algebraically or numerically, probably necessarily on a computer in high-order approximations. Many plausible truncation schemes are possible, and a number have indeed been used, mostly chosen on the rather pragmatic basis of the first thing that seems to converge reasonably well. Of course, all reasonable truncation schemes should eventually converge to the same result, but some may be more efficient than others. There is still considerable scope for investigation here. For example, one useful scheme retains all the moments $\langle \psi_{10}^{(r)} \rangle$ with $l + r \leqslant n$ in the nth approximation, and sets all the others equal to zero (Viehland and Mason, 1978). This is not very efficient when $m \ll M$ (e.g., electrons) because only $l = 0, 1$ are required in this case, as pointed out in Section 5-3D4a, although quite large values of r may be needed for accuracy. It is then better to truncate independently in l and r, picking values of $l_{\max}$ and $r_{\max}$, and retaining only those moments for which both $l \leqslant l_{\max}$ and $r \leqslant r_{\max}$ (Lin et al., 1979b). It is not even necessary to set higher moments equal to zero to obtain closure of the moment equations. Instead, higher moments can be expressed as linear combinations of the lower moments, the particular combinations being chosen to represent some special feature of the distribution function that is believed to be important. For instance, anisotropy of the distribution function can be taken into account by using a linear combination that is characteristic of some simple model of extreme anisotropy (Baraff, 1964). General procedures for the economical management of the computations involved in higher approximations have been outlined (Kumar, 1980; Kumar et al., 1980), but much numerical work remains to be done to verify or modify these in the light of experience.

A choice of T_b must be made before any truncation-iteration scheme can be started, and must be made again at each stage of the scheme in order to continue. The choice of T_b turns out to be rather subtle, and can be more important than the choice of the particular truncation-iteration scheme itself. It would appear reasonable to choose T_b to be equal to the ion temperature (mean ion energy) not only in the first approximation of (6-2-9), but also at each stage in the higher approximations. Although this choice can usually be made to work, there are three important reasons against it.

1. If T_b changes at each stage of the calculation, then T_{eff} also changes according to (6-2-7), and all the $\bar{\Omega}^{(l,s)}(T_{\text{eff}})$ and $a_{rs}(l)$ must be recalculated at each stage. This iteration becomes very laborious, especially when it is realized that the hardest part of the whole computation is the generation of the needed $a_{rs}(l)$.

2. If diffusion is also to be treated, which is feasible in the two-temperature theory when $m < M$, then the ion temperature must have a dependence on the ion density gradient, as is apparent even in the elementary discussion of Section 5-2B. This dependence greatly increases the algebraic complexity in higher approximations, although it is necessary in order to get sensible results in a first approximation (Section 5-2B).

3. Other aspects of the distribution may be more important than its moment with respect to v^2, which gives the ion temperature, and T_b can be used as a parameter to make $f^{(0)}$ give a better representation of some important feature of the distribution, such as its tail. For example, with resonant charge exchange an ion loses most of its velocity after a collision and regains it from the electric field starting nearly from rest. The distribution function thus has its maximum near the origin and has a long tail extending in the field direction (Skullerud, 1969). This behavior can be partially represented by adjustment of T_b, with a substantial improvement in convergence (Lin and Mason, 1979).

The following treatment of T_b was originally designed for use when the tail of the distribution was particularly important, either because of resonant charge exchange (Lin and Mason, 1979) or because of high-energy thresholds for inelastic processes (Lin et al., 1979b). Nevertheless, many of its features are probably generally useful. The most important feature is to keep T_b fixed throughout each stage of the truncation procedure. The rationale for doing this can be understood by mentally interchanging the roles of T_b and $\mathscr{E}$ (or E/N). In experimental work, E/N is the independently chosen variable and the ion temperature is a dependent variable. In calculations, it is better to fix T_b as the independent variable and then calculate $\mathscr{E}$ as a dependent variable, so that the $a_{rs}(l)$ need to be computed only once. There is no special necessity for T_b to be the ion temperature, and the best method for calculating $\mathscr{E}$ for a given T_b may depend on the particular system under consideration. One method that has proved satisfactory in a number of cases is to keep the equations for overall momentum and energy balance nearly invariant in their first-approximation form as the truncation proceeds. This can be understood by reference to the moment equations (6-2-8). The momentum balance is given by (6-2-8a), whose right-hand side is usually automatically dominated by just the first term because the $\gamma_{0s}(1)$ are small for $s \geqslant 1$. Thus the equation does not change much as more terms are included in the truncation. This is not true for the overall energy balance equation (6-2-8b) because the second term on the right-hand side, $\gamma_{11}(0)\langle\psi_{00}^{(1)}\rangle^{(0)}$, is not automatically small. We therefore *demand* that the right-hand side of (6-2-8b) be dominated by just the first term, $\gamma_{10}(0)$. In other words, we imagine that T_b has been selected so that the sum of all the terms on the right-hand side of (6-2-8b), except for $\gamma_{10}(0)$, amounts to zero; then the field strength is given by

$$\mathscr{E} = -\gamma_{10}(0)/2\langle\psi_{10}^{(0)}\rangle^{(0)}. \tag{6-2-14}$$

Since $\langle \psi_{10}^{(0)} \rangle^{(0)}$ is not known exactly, but is only successively approximated by solving the truncated moment equations, a starting value of $\mathscr{E}$ is obtained by use of (6-2-8a) as well,

$$[\mathscr{E}]_1 = [-\tfrac{1}{2}\gamma_{10}(0)]^{1/2}. \tag{6-2-15}$$

In many cases, (6-2-15) alone is sufficient to give reasonably fast convergence of the truncated moment equations (Lin and Mason, 1979), but better convergence can usually be achieved by adjusting $\mathscr{E}$ according to (6-2-14) after each higher-order solution is found for $\langle \psi_{10}^{(0)} \rangle^{(0)}$ (Lin et al., 1979b).

To summarize, T is initially given and then values of T_b are chosen. For each fixed value of T_b, $\mathscr{E}$ and the necessary moments are found by solving the truncated moment equations. After converged results are obtained for one value of T_b, the process is repeated for another value of T_b, thereby generating the desired moments as a function of E/N for a given T. The choice of the truncation scheme and the choice of the equation to determine $\mathscr{E}$ may be varied according to the physics of the particular problem, in order to improve the speed of convergence.

Even though good numerical accuracy may require high-order approximations, it is handy to have fairly simple formulas from which the accuracy of the first approximations can at least be quickly estimated. The following formulas have been obtained for this purpose from the two-temperature theory (Viehland and Mason, 1978), using the experimental quantity $K' \equiv d \ln K_0 / d \ln (E/N)$ as an expansion parameter.

a. *Mobility.* The expression for the mobility can be written as

$$K = [K]_1(1 + \alpha_0 K' + \cdots), \tag{6-2-16a}$$

where $[K]_1$ is given by (6-2-12) and

$$\alpha_0 = \frac{m(m + M)}{5(3m^2 + M^2) + 8mMA^*} \left[\frac{10(m + M)}{5m + 3MA^*} - \frac{5(m - M) + 4MA^*}{m + M} \right] \tag{6-2-16b}$$

This formula is obtained from the second-approximation moment equations by taking the quantity $(6C^* - 5)$ in the matrix elements as the expansion parameter, and using the relations

$$6C^* - 5 = 1 + 2 \frac{d \ln \bar{\Omega}^{(1,1)}}{d \ln T_{\text{eff}}}, \tag{6-2-17a}$$

$$\approx -2 \frac{d \ln K_0}{d \ln T_{\text{eff}}}, \tag{6-2-17b}$$

where (6-2-17a) follows exactly from differentiation of the expression for $\bar{\Omega}^{(1,1)}$ and (6-2-17b) follows from differentiation of the formula (6-2-12) for $[K]_1$.

Differentiation of the expression (6-2-13b) for $[T_{\text{eff}}]_1$ then yields

$$\frac{d \ln K_0}{d \ln T_{\text{eff}}} \approx \frac{1}{2}\left(\frac{T_{\text{eff}}}{T_{\text{eff}} - T}\right)\frac{K'}{1 + K'}. \tag{6-2-18}$$

In addition, the Maxwell-model value of $B^* = 5/4$ is used and only terms first order in K' are kept in obtaining (6-2-16).

b. Ion Temperatures and Diffusion. Here we simply recover the expressions (6-1-33) and (6-1-43) of the one-temperature theory for the ion temperatures, and the generalized Einstein relations for the diffusion coefficients. The only practical difference is that A^* is now a function of T_{eff} instead of T. However, there is an important esthetic difference in that the justification for the temperature formulas is now substantially improved, and no longer depends on dubious manipulations of divergent series in $(E/N)^2$. This improvement in pedigree now allows us to obtain an improvement in the corrections to the ion temperatures for the important limiting case of $m/M \to 0$ (e.g., electrons), for which the first-order corrections vanish. The result is

$$\tfrac{1}{2}m\langle v^2\rangle = \tfrac{3}{2}kT + \tfrac{1}{2}Mv_d^2\left[1 + \frac{(6C^* - 5)^2}{25 - 12B^*}\frac{T_{\text{eff}} - T}{T_{\text{eff}}} + \cdots\right], \tag{6-2-19a}$$

$$= \tfrac{3}{2}kT + \tfrac{1}{2}Mv_d^2\left[1 + \frac{1}{25 - 12B^*}\frac{T_{\text{eff}}}{T_{\text{eff}} - T}\left(\frac{K'}{1 + K'}\right)^2 + \cdots\right]. \tag{6-2-19b}$$

The corrections for T_L and T_T are exactly the same,

$$kT_L = kT_T = kT + \tfrac{1}{3}Mv_d^2\left[1 + \frac{(6C^* - 5)^2}{25 - 12B^*}\frac{T_{\text{eff}} - T}{T_{\text{eff}}} + \cdots\right]. \tag{6-2-20}$$

B. Convergence of Approximations

Here we consider mainly the mobility, because the two-temperature theory does not converge well for the diffusion coefficients and ion temperatures when $m > M$. However, this defect can be partially remedied through the use of the generalized Einstein relations and the improved relations for the ion temperatures, as we shall illustrate.

1. *Choice of Basis Temperature*

The first matter to check is the choice of T_b. We pick models for which difficulties are expected, and first try to model resonant charge exchange in an extreme way. At high collision energies the charge-exchange cross section varies only slowly with energy (Section 5-2F), and can therefore reasonably be approximated by a constant. Since ion-atom collisions without charge exchange are usually forward-peaked, we make the extreme assumption that half the collisions are strictly forward ($\theta = 0$) and that the other half involve resonant charge exchange ($\theta = \pi$). This assumption corresponds to a model with a geometric cross section of Q_0, in which the ion is undeflected in half the "collisions" and comes to rest in the other half. The momentum-transfer cross section is equal to Q_0 and the charge-exchange cross section to $Q_0/2$. The gas temperature does not play any significant role, and can be taken to be zero for convenience. Because there is no way to transfer kinetic energy to the perpendicular direction, the ion motion is one-dimensional. The ion distribution function in this case is obviously strongly distorted from a Maxwellian form and is highly anisotropic. This extreme model can be solved exactly because of its simplicity (Skullerud, 1969), and is therefore a good test case. Results are shown in Table 6-2-2 for two choices of T_b: T_b equal to the ion temperature, and T_b fixed and the energy-balance equation kept invariant (Lin and Mason, 1979). Both choices give the same result in first approximation, but the first choice fails to proceed beyond the fourth-order approximation because of its particular iteration procedure in which higher-order approximations depend critically on the cruder lower-order solutions. The second choice does not suffer from this fault and produces reasonable results. Only the odd-order approximations are shown because the special features of this extreme model make the even-order approximations peculiar. In Table 6-2-2, the nth approximation means that only the moments $\langle \psi_{r0}^{(l)} \rangle$ with $l + r \leqslant n$ are kept. The distortion of the distribution function is reflected in the deviation of T_b from the ion temperature—$T_b = 1.15 T_{ion}$.

We next examine the choice of T_b for the Lorentz model ($m \ll M$) with rigid-sphere interactions and $T = 0$. In this case the ion distribution function remains nearly isotropic but is strongly distorted from a Maxwellian form. We therefore truncate the moments independently in the indices l and r, using the accurate two-term approximation for l ($l_{max} = 1$, Section 5-3D4a), but carrying r to high values. This is possible because the matrix elements are especially easy to

Table 6-2-2 Convergence of Drift Velocity and Ion Energies for an Extreme Charge-Exchange Model, with Two Choices of Ion Basis Temperature (Lin and Mason, 1979)

	Approximate/Accurate[a]					
	v_d		$\langle v^2 \rangle$		$\langle v_z^2 - v_d^2 \rangle$	
Order of Approximation	$T_b = T_{ion}$	T_b fixed	$T_b = T_{ion}$	T_b fixed	$T_b = T_{ion}$	T_b fixed
1	0.9509	0.9509	1.151	1.151	−0.501	−0.501
3	1.0284	1.0002	1.205	1.084	1.275	0.855
5	—	1.0001	—	1.071	—	0.933

[a]Accurate results from Skullerud (1969).

calculate when $m/M \to 0$. Convergence is fairly fast for v_d, $\langle v^2 \rangle$, and D_T with both choices of T_b—both choices yield results that agree to better than 1% with the accurate calculations of Skullerud (1974, 1976) after $r_{max} = 8$, and to better than 0.1% after $r_{max} = 18$—but the convergence for D_L is comparatively slow (Lin et al., 1979b). Results for D_L are shown in Fig. 6-2-1, where the percentage

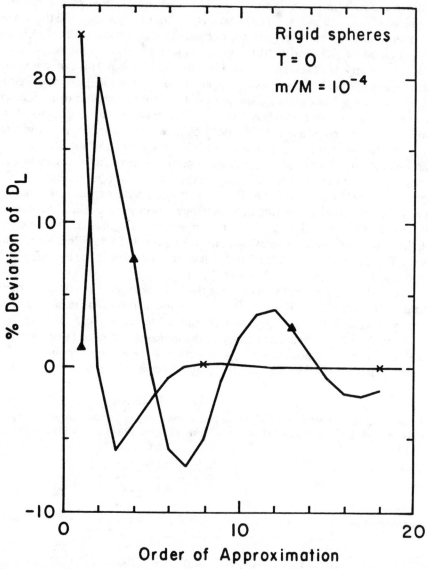

Figure 6-2-1. Rate of convergence of D_L for the two different choices of the ion basis temperature T_b described in the text (Lin et al., 1979b). Convergence is faster for v_d, $\langle v^2 \rangle$, and D_T. The symbols only identify the curves: ▲ $T_b = T_{ion}$; × T_b fixed.

deviations from Skullerud's accurate result are plotted as a function of the order of approximation (the value of r_{max}). Clearly, the results from both schemes converge to the same correct value, but in quite different ways. The choice of $T_b = T_{\text{ion}}$ gives a better result in the first approximation because it incorporates the density gradient into the basis functions through T_b, but the advantage is lost in higher-order approximations. The choice of T_b as a fixed parameter with an invariant energy-balance equation greatly speeds up the convergence and damps out the oscillations.

Incidentally, an oscillatory approach to the converged result seems to be characteristic of many moment solutions at high field strengths. This behavior cannot occur at vanishing field strengths, where a variational principle applies (Hirschfelder et al., 1964, p. 474; Ferziger and Kaper, 1972, Sect. 5.5). One criterion for selecting an effective truncation-iteration scheme would be the damping of such oscillations.

We conclude from the foregoing extreme examples that it is better *not* to require T_b to represent the true ion temperature. Nevertheless, in many cases the choice of $T_b = T_{\text{ion}}$ does give reasonable results (Viehland and Mason, 1975, 1978).

2. *Mobility*

Having settled on a choice for T_b, we now turn to testing the convergence of the mobility as a function of field strength. We again choose the rigid-sphere Lorentz model, which served as a test case for the momentum-transfer theory (Fig. 5-2-3) and the one-temperature theory (Fig. 6-1-1). For this model the variables T and E/N can be scaled together exactly through the use of a dimensionless mobility K^* and a dimensionless field strength $\mathscr{E}^*$ defined by (5-2-34)–(5-2-36), and the results compared with the accurate numerical calculations of Hahn and Mason (1972). The first approximation has the same form as the momentum-transfer expression of (5-2-37) but has a different numerical factor,

$$v_d^*(1 + \tfrac{2}{3}v_d^{*2})^{1/2} = \mathscr{E}^*. \tag{6-2-21}$$

Higher approximations are much more complicated and we do not write them down explicitly. The first two approximations are compared with the accurate results in Fig. 6-2-2. There is no divergence, as in the one-temperature theory, and the numerical agreement is better than that of the momentum-transfer theory. The two-temperature theory thus overcomes the main drawbacks of these other theories. However, the convergence is not all that might be desired; we see in Fig. 6-2-2 that the second approximation is somewhat better than the first, but still differs from the accurate results by a noticeable amount. It is reassuring that nothing startling happens as the field strength is varied. This occurs because $\mathscr{E}$ (or $\mathscr{E}^*$) remains finite, as mentioned at the beginning of this section.

The first approximation (6-2-21) is in fact obtained for all ion/neutral mass

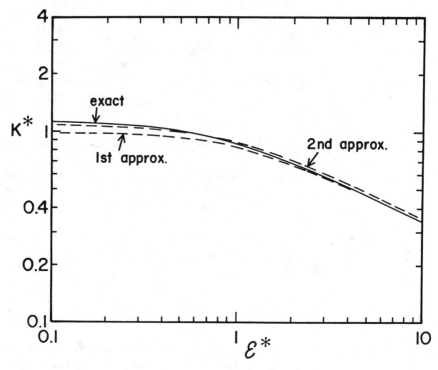

Figure 6-2-2. Mobility vs. field strength for the rigid-sphere Lorentz model ($m \ll M$). The first two approximations of the two-temperature theory are compared with accurate results from numerical integration (Hahn and Mason, 1972). This figure should be compared with similar results in Fig. 5-2-3 (momentum-transfer theory) and Fig. 6-1-1 (one-temperature theory). The dimensionless mobility K^* and field strength $\mathscr{E}^*$ are defined by (5-2-34)–(5-2-36).

ratios, not just for $m \ll M$, and thus indicates that there is no dependence of the reduced mobility on the mass ratio. Accurate calculations show that this is not true, either at vanishing fields (Mason, 1957a, b) or at high fields (Skullerud, 1976), and that the deviations can amount to about 10%. Curiously, the deviations at vanishing fields are large for $m \ll M$, but at high fields are large for $m \gg M$. Convergence at vanishing fields was already examined in connection with the one-temperature theory in Section 6-1C1, where it was concluded that the second approximation would usually be satisfactory. We therefore now examine just the high-field (cold-gas) limit as a function of mass ratio, for the extreme case of rigid-sphere interactions. The results are shown in Fig. 6-2-3. The second approximation (6-2-16) does improve matters for $m \gg M$, but clearly is not uniformly accurate. Matters are somewhat better for softer interactions (Viehland and Mason, 1978).

The foregoing results shown in Figs. 6-2-2 and 6-2-3 merely illustrate a much larger body of numerical calculations carried out to test the convergence of the

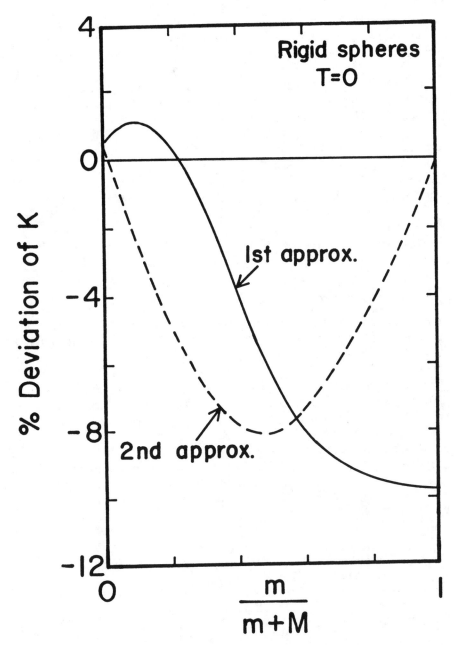

Figure 6-2-3. Accuracy of the two-temperature theory for the high-field (cold-gas) mobility as a function of ion/neutral mass ratio. The solid curve is the first approximation of (6-2-12) and the dashed curve is the second approximation of (6-2-16). Percentage deviations are from the accurate results of Skullerud (1976).

two-temperature theory (Viehland and Mason, 1975, 1978; Lin and Mason, 1979; Lin et al., 1979a, b; Kumar et al., 1980). From these various calculations we can conclude that the first approximation usually gives a reasonable account of the mobility and contains most of the interesting physics, but can easily be quantitatively in error by about 10%. This feature, incidently, accounts for the considerable success of the simple momentum-transfer theory described in Section 5-2. However, good numerical accuracy requires higher approximations, and the second approximation formula can only be relied upon to give an estimate of the likely inaccuracy of the first approximation, and not to produce accurate results itself. To achieve accuracies within 1% it is usually necessary to be prepared to go to at least the third approximation, and sometimes as high as the fifth or sixth approximation. Even this is just a rough rule of thumb. Electrons, for example, may require even higher approximations, although this is feasible because the matrix elements are easy to calculate in this case. As always, it is also necessary to be alert for an occasional pathological case, such as ion "runaway" (Lin et al., 1979a; see also Chapter 8).

3. Ion Temperatures and Diffusion

As far as ion temperatures are concerned, the two-temperature theory mainly gives a better foundation to the formulas for the ion temperatures first obtained from the one-temperature theory, with a slightly different interpretation of the quantity A^*. These formulas are (6-1-33) for the total ion energy and (6-1-43) for T_L and T_T, together with the equations that define the various factors in these formulas. As was illustrated in Tables 6-1-3, 6-1-4 and 6-1-5, these formulas are remarkably accurate. The only new result from the two-temperature theory is a correction to the simple Wannier formulas for the special limiting case of $m/M \to 0$, given by (6-2-19) and (6-2-20). This correction is compared with the accurate results of Skullerud (1976) for some inverse-power potentials in Table 6-2-3. The agreement is quite good, a result of some interest for electrons.

As far as ion diffusion coefficients are concerned, the two-temperature theory

Table 6-2-3   Accuracy of the Relations (6-2-19) and (6-2-20) for Ion Temperatures in the Limiting Case of $m/M \to 0$ (e.g., Electrons)

	$kT_{L,T}/(\frac{1}{3}Mv_d^2)$			
	$n = 4$	$n = 8$	$n = 12$	$n = \infty$
Approximate	1	1.0222	1.0377	1.0769
Accurate[a]	1	1.0202	1.0324	1.0606

[a]Accurate values are from Skullerud (1976) for potentials $V(r) = C_n/r^n$, for the case of a cold gas ($T = 0$).

converges well only for $m < M$. When m/M lies between 1 and 4, convergence is obtained only at low or intermediate values of E/N, and when m/M is greater than 4, the two-temperature theory is essentially useless for diffusion (Viehland and Mason, 1978). Nevertheless, this difficulty can be circumvented to some degree by the use of the generalized Einstein relations (GER) and the improved relations for the ion temperatures. There are two drawbacks to this procedure, one theoretical and one practical. The theoretical drawback is that the two-temperature theory itself does not supply a satisfactory derivation of the GER. They must either be accepted on the basis of physical arguments like those of Section 5-2B, or must be derived more elaborately from a three-temperature theory, as is done in Section 6-4. The practical drawback is that the two-temperature theory does not furnish any way of improving the accuracy of these relations. The sort of accuracy attainable can be estimated by an examination of Tables 5-2-3 and 5-2-4 for the GER and of Tables 6-1-4 and 6-1-5 for the ion temperatures. From these tables we can see that errors in the GER of about 5% can be expected for D_T, and about 10 to 15% for D_L, and that errors in the ion temperatures can be expected at about the 5% level. These are somewhat pessimistic estimates because they are based in part on the extreme case of rigid-sphere interactions, and real systems usually give better results. Nevertheless, it is prudent to expect overall errors of up to about 10% in D_T and 20% in D_L. Such errors may well be acceptable, given the experimental difficulties involved in measuring ion diffusion coefficients accurately.

Here we give just one illustration of the use of the foregoing procedure for a real system, and postpone more extensive discussion to Section 6-4. Of the three systems that were illustrated in Fig. 5-2-4, which was a test of the GER with semiempirical formulas for the ion temperatures, we pick the system with the largest value of m/M, namely K^+ in He. We also take this opportunity to show how to present diffusion results in a convenient way that removes their main quadratic dependence on E/N and reveals structure that is usually concealed in the more customary plot of $\log D_{L,T}$ vs. $\log(E/N)$ (see Fig. 5-2-4). Since the approximate variation of $D_{L,T}$ as $(E/N)^2$ arises from the ion temperatures, it would seem reasonable to remove it by dividing $D_{L,T}$ by

$$[kT_{L,T}]_1 = kT + \zeta_{L,T}M(KE)^2. \qquad (6\text{-}2\text{-}22)$$

A difficulty arises if we wish to compare calculated and measured values of $D_{L,T}$ in this way. Which value of K, calculated or measured, should be used in (6-2-22)? Moreover, what should be used for $\zeta_{L,T}$? To avoid these ambiguities, it is convenient to define an ion temperature involving the unambiguous polarization limit K_{pol},

$$kT_{pol} \equiv kT + \tfrac{1}{3}M[N_0 K_{pol}(E/N)]^2, \qquad (6\text{-}2\text{-}23)$$

where N_0 is the standard gas density of about 2.69×10^{19} cm^{-3} and K_{pol} is given by (6-1-61). The only quantity (besides mass) needed to find K_{pol} is the dipole polarizability of the neutral. The quantity N_0 appears because K_{pol} is defined as the mobility at standard density. Division of $D_{L,T}$ by (kT_{pol}/e) yields a quantity similar to the mobility, and it is convenient to normalize this by dividing by K_{pol}. We therefore define reduced diffusion coefficients, $\tilde{D}_{L,T}$, as

$$\tilde{D}_{L,T} \equiv eND_{L,T}/N_0 K_{pol} kT_{pol}, \qquad (6\text{-}2\text{-}24)$$

which are dimensionless and independent of gas density.

Values of $\tilde{D}_L$ and $\tilde{D}_T$ calculated from the GER as a function of E/N are shown in Fig. 6-2-4 for K^+ in He. For this system it makes little difference whether the semiempirical formulas (5-2-73) or the better theoretical formulas (6-1-43) are

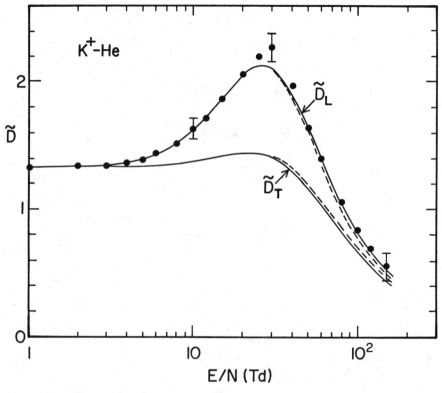

Figure 6-2-4. Reduced longitudinal and transverse diffusion coefficients for K^+ in He. The curves are from the generalized Einstein relations, $D_L = (kT_L/e)K(1 + K')$ and $D_T = (kT_T/e)K$. The solid curves use the theoretical formulas (6-1-43) for the ion temperatures, and the dashed curves use the semiempirical formulas (5-2-73); the differences are trivial in this case. The circles are experimental measurements of D_L from the survey of Ellis et al. (1978); representative error bars are shown. This reduced representation of the diffusion coefficients according to (6-2-24) should be compared with that of Fig. 5-2-4.

used for $T_{L,T}$. Agreement with the experimental values of $\tilde{D}_L$ is quite good—only one point is off by more than 5%—indicating that the estimate of 20% error given above is probably conservative for many real systems. The two-temperature theory could be used directly to calculate $D_{L,T}$ for a system like K^+ in He only at low E/N, even if the interaction potential $V(r)$ were known, because poor convergence would be obtained above about 75 Td for a value of m/M nearly equal to 10. Nevertheless, good results could be obtained indirectly by calculating K and then using the GER.

The representation of diffusion coefficients as $\tilde{D}_{L,T}$ in Fig. 6-2-4 should be compared with the representation as $\log D_{L,T}$ in Fig. 5-2-4.

In summary, the two-temperature theory converges satisfactorily for mobility. It does not work well for diffusion unless $m < M$, and for $m > M$ it is better to use the generalized Einstein relations.

C.   Temperature and Field Dependence of Mobility and Diffusion

Perhaps the most important result of the two-temperature theory is the scaling rule that, in first approximation, condenses the two variables T and E/N into the single variable T_{eff}. Although the same rule was already obtained from the momentum-transfer theory (Section 5-2A), the two-temperature theory makes it more precise and furnishes procedures for calculating corrections. It is worth repeating here the formulas for K and T_{eff},

$$K = \frac{3e}{16N}\left(\frac{2\pi}{\mu k T_{\text{eff}}}\right)^{1/2}\frac{1+\alpha}{\bar{\Omega}^{(1,1)}(T_{\text{eff}})}, \qquad (6\text{-}2\text{-}25)$$

$$\tfrac{3}{2}kT_{\text{eff}} = \tfrac{3}{2}kT + \tfrac{1}{2}Mv_d^2(1 + \beta K' + \cdots), \qquad (6\text{-}2\text{-}26)$$

in which we have used α to represent all the higher corrections to the mobility. For vanishing fields, $1 + \alpha$ represents the coefficient g_0 given by (6-1-54); for stronger fields, α represents the correction terms $\alpha_0 K' + \cdots$ of (6-2-16). The quantity β is given by (6-1-35), but with A^* a function of T_{eff} instead of T. A number of tests of the scaling rule for mobility have already been exhibited: Fig. 5-2-1 for K^+ ions in He, Ne, Ar; Fig. 5-2-2 for electrons in He, Ne, Ar; Fig. 6-1-2 for Li^+, H^+, and D^+ in He; Fig. 6-1-5 for complex ions in SF_6; and Fig. 6-1-6 for He^+ and He_2^+ in He. The agreement was almost always within experimental uncertainty, even though the scaling rule is strictly correct only in the first approximation, where the corrections represented by α and β are ignored. These results suggest that the scaling rule may have wider validity than just the first approximation. This could occur in several ways besides the corrections being small, and so is not very surprising (Viehland and Mason, 1978). For example, β might be small and α, although not small, might be primarily a function of T_{eff} rather than of T and E/N separately. This is known to occur for very heavy ions ($m \gg M$) at high fields, and for fairly light ions ($m < M$) at low fields. Another possibility is that neither α nor β is small but that their effects tend to compensate, as when $\alpha > 0$ and $\beta < 0$. This is known to occur for alkali ions in noble gases at moderate to high field strengths.

Since the scaling rule shows that variation of E/N is essentially equivalent to a variation of temperature, there is nothing further to be added to the discussion of the temperature dependence of mobility given in Section 6-1D for the one-temperature theory. The temperature axis only needs to be read as T_{eff} to include the field dependence of the two-temperature theory. This interpretation was in fact already used in Figs. 6-1-2, 6-1-5, and 6-6-6.

D.   Quantum Effects

All the results of the two-temperature theory are valid in quantum mechanics as well as in classical mechanics. All quantum effects occur through the cross sections, not through the kinetic theory itself. Since exactly the same cross sections occur in the two-temperature theory as in the one-temperature theory, nothing needs to be added to the discussion of Section 6-1E as far as cross sections are concerned. The collision integrals $\bar{\Omega}^{(l,s)}$ of the two-temperature theory are identical to those of the one-temperature theory, with T replaced by T_{eff} as defined by (6-2-7) in terms of T, T_b, m, and M. This is an exact definition of T_{eff}; approximations enter in solving the moment equations to find T_{eff} in terms of v_d or E/N, as in (6-2-26). Thus virtually nothing needs to be added to the discussion of Section 6-1E for quantum effects in the one-temperature theory. The only real difference is that quantum effects on the $\bar{\Omega}^{(l,s)}$ may affect the convergence of the approximations differently in the one- and two-temperature theories.

E.   Inelastic Collisions

The effects of inelastic collisions go much deeper when E/N is not vanishingly small. It is necessary to recognize that the internal degrees of freedom of the ions may require a different basis temperature than do the translational degrees of freedom, and to include this fact in the expansion for the ion distribution function. This changes the discussion sufficiently that we give the whole problem a separate treatment in Section 6-6.

F.   Mixtures

The treatment of mixtures in the two-temperature theory is formally almost the same as in the one-temperature theory, which was discussed in Section 6-1G. The matrix elements $a_{rs}(l)$ for the single gas are replaced by linear combinations, $b_{rs}(l)$, weighted by the mole fractions of the neutral species in the mixture, as in (6-1-83). The moment equations otherwise remain the same. The results can be put into a simple form only in the first approximation; the relevant formulas have already been given in (6-1-84)–(6-1-95) for the one-temperature theory. The two-temperature theory uses the same formulas in this approximation, but makes the collision-integral ratio A^* a function of T_{eff} instead of T. The same cautions given in Section 6-1G regarding accuracy if any of the K_j' is large apply here as well.

G. Effect of Temperature Gradients: Thermal Diffusion

An externally imposed temperature gradient in the neutral gas can affect the ion motion, as was discussed in Section 6-1H for the low-field case. A strong electric field does not change the overall form of the ion flux equation (6-1-102), but does affect the thermal diffusion factor α_T. (A gradient in T produces an equal gradient in T_{eff}, and this is absorbed in α_T.) The expression for α_T is changed in two ways. First, as might be expected, the low-field expression of Section 6-1H is modified so that T_{eff} appears in place of T in the ion-neutral collision integrals. Second, and more important, an additional term appears that is anisotropic and strongly field dependent. The following results have been obtained by Weinert (1980) for the two-temperature theory; the explicit formulas are only first approximations, but these should suffice for estimating possible errors in swarm experiments:

$$\alpha_T = (\alpha_T)_{\text{iso}} + (\alpha_T)_{T,L}, \tag{6-2-27}$$

where $(\alpha_T)_{\text{iso}}$ is the isotropic portion, which is only a modified version of the low-field result, and $(\alpha_T)_{T,L}$ is the anisotropic field-dependent portion. The expression for $(\alpha_T)_{\text{iso}}$ is given by the low-field formula (6-1-105), with the collision-integral ratios A^* and C^* functions of T_{eff} instead of T. The effect on A^* is minor, but is more influential on C^*, which appears in the temperature-sensitive factor $(6C^* - 5)$. Since T_{eff} is large at high field strengths, the factor $(6C^* - 5)$ will tend to be large and positive, as shown in Figs. 6-1-8 and 6-8-9 for several potential models. The quantity ND/η that appears in (6-1-105) is only a convenient representation for the ratio of two collision integrals, one of which becomes a function of T_{eff} and the other of which remains a function of T. Thus ND/η must be replaced as follows:

$$\frac{ND}{\eta} \rightarrow \left(\frac{T}{T_{\text{eff}}}\right)^{1/2} \frac{ND_{\text{iso}}}{\eta}, \tag{6-2-28a}$$

where η remains the viscosity of the neutral gas, and

$$D_{\text{iso}} \equiv (kT_{\text{eff}}/e)K(T_{\text{eff}}). \tag{6-2-28b}$$

The net result is that the quantity ND/η increases with increasing field strength.

The anisotropic field-dependent term $(\alpha_T)_{T,L}$ is always negative, and in first approximation has a simple dependence on field strength (Weinert, 1980),

$$(\alpha_T)_{T,L} = -\frac{(E/N)^2}{Z_{T,L} + (E/N)^2},\qquad (6\text{-}2\text{-}29)$$

where

$$Z_{T,L} = \frac{3kT}{(m+M)(NK)^2}\left[1 + (-1)^{m'}(2 - m'^2)\frac{5m}{5m + 3MA^*}\right]^{-1},\qquad (6\text{-}2\text{-}30a)$$

$$m' = \begin{cases} \pm 1 & \text{for } Z_T \\ 0 & \text{for } Z_L. \end{cases}\qquad (6\text{-}2\text{-}30b)$$

The magnitude of $(\alpha_T)_{T,L}$ varies smoothly from 0 to 1 as E/N is increased, and will therefore tend to overshadow $(\alpha_T)_{\text{iso}}$ unless the ions are large. The fact that $(\alpha_T)_{T,L}$ is negative means that the effect of strong electric fields is to cause the ions to move toward regions of higher gas temperature. Thus $(\alpha_T)_{\text{iso}}$ and $(\alpha_T)_{T,L}$ will usually have opposite signs for large ions in strong fields, which should be a help to experimentalists.

6-3.   THREE-TEMPERATURE THEORY

The three-temperature theory is a natural extension of the two-temperature theory, in that a more flexible choice is made for $f^{(0)}$ in order to improve the convergence. The two-temperature theory gives poor results for the diffusion coefficients of heavy ions ($m > M$) at high E/N, and it is natural to ascribe this failure to anisotropy and skewness of the distribution function, and to try to build these features into $f^{(0)}$ from the beginning. The general line of attack is the same as for the two-temperature theory, but the actual computations are substantially more complicated. As was discussed in Section 5-3E, the complication is mainly caused by the increased difficulty of evaluating the matrix elements involved. It is almost always necessary to resort to a computer from the outset. This is no great practical disadvantage, but unfortunately the simple physical interpretations that came out of the two-temperature theory in first approximation are now mostly lost. The only place where the three-temperature theory has contributed both improved insight and improved numerical results is in the derivation of generalized Einstein relations, which are discussed separately in Section 6-4. Most of the present section is devoted only to a sketch of the details of the three-temperature theory, and to some brief numerical illustrations of its improved convergence properties.

A. Moment Equations and Truncation-Iteration Scheme

Since we want to treat diffusion in the three-temperature theory, we follow the general procedure already outlined in Section 6-1B3 on the one-temperature theory, and first expand f with the ion density gradient as a small perturbation,

$$f = f_0 - \left(f_x \frac{\partial}{\partial x} + f_y \frac{\partial}{\partial y} + f_z \frac{\partial}{\partial z} \right) \ln n, \qquad (6\text{-}3\text{-}1)$$

which is the same as (6-1-46). The moment equations are formed in the same way as before, but the choice of basis functions for forming the moments is now different. The basis functions are chosen to be orthogonal over a zero-order distribution function that is an anisotropic displaced distribution. As explained in Section 5-3C, this is equivalent to expanding the distribution function itself in terms of these basis functions. The three-temperature expansion for the components f_0, f_x, f_y, and f_z of the distribution function is

$$f_{0,x,y,z} = f^{(0)} \sum_{p=0}^{\infty} \sum_{q=0}^{\infty} \sum_{r=0}^{\infty} f_{pqr}^{(0,x,y,z)} H_p(w_x) H_q(w_y) H_r(w_z), \qquad (6\text{-}3\text{-}2)$$

where

$$f^{(0)} = n[m/2\pi k T_b^{(T)}][m/2\pi k T_b^{(L)}]^{1/2} \exp(-w_x^2 - w_y^2 - w_z^2), \qquad (6\text{-}3\text{-}3a)$$

$$w_x^2 = mv_x^2/2kT_b^{(T)}, \qquad w_y^2 = mv_y^2/2kT_b^{(T)}, \qquad (6\text{-}3\text{-}3b)$$

$$w_z^2 = m(v_z - v_{\text{dis}})^2/2kT_b^{(L)}, \qquad (6\text{-}3\text{-}3c)$$

and the basis functions H_p, H_q, H_r are Hermite polynomials. The expansion coefficients $f_{pqr}^{(0,x,y,z)}$ are moments of the components of the distribution function over the basis functions. The three temperatures from which the theory takes its name are the gas temperature T and the two ion basis temperatures $T_b^{(T)}$ and $T_b^{(L)}$. The parameter v_{dis} is the "displacement" of $f^{(0)}$ in the drift direction. Physically, we expect v_{dis} to be approximately equal to the drift velocity, but this is not a requirement. These basis functions are orthogonal with respect to $f^{(0)}$ as a weight function,

$$\int f^{(0)} \psi_{pqr} \psi_{stu} \, d\mathbf{v} \equiv (\psi_{pqr}, \psi_{stu}) = (\psi_{pqr}, \psi_{pqr}) \delta_{ps} \delta_{qt} \delta_{ru}, \qquad (6\text{-}3\text{-}4a)$$

where

$$\psi_{pqr} \equiv H_p(w_x) H_q(w_y) H_r(w_z), \qquad (6\text{-}3\text{-}4b)$$

and (ψ_{pqr}, ψ_{pqr}) is the normalization factor.

The moment equations are formed as usual, but the matrix elements that appear are of course different because of the different basis functions. In particular, the matrix elements now have more indices,

$$a(pqr; stu) \equiv (\psi_{stu}, J\psi_{pqr})/(\psi_{stu}, \psi_{stu}). \tag{6-3-5}$$

The three sets of moment equations for the spatially homogeneous moments, the spatially inhomogeneous moments perpendicular (x and y) to the field, and the spatially inhomogeneous moments parallel (z) to the field, corresponding to (6-1-49), now become, respectively,

$$(eE/m)\langle \partial\psi_{pqr}/\partial v_z \rangle^{(0)} = N \sum_{s=0}^{\infty} \sum_{t=0}^{\infty} \sum_{u=0}^{\infty} a(pqr; stu)\langle \psi_{stu} \rangle^{(0)}, \tag{6-3-6a}$$

$$(eE/m)\langle \partial\psi_{pqr}/\partial v_z \rangle^{(x)} + \langle v_x \psi_{pqr} \rangle^{(0)}$$

$$= N \sum_{s=0}^{\infty} \sum_{t=0}^{\infty} \sum_{u=0}^{\infty} a(pqr; stu)\langle \psi_{stu} \rangle^{(x)}, \tag{6-3-6b}$$

$$(eE/m)\langle \partial\psi_{pqr}/\partial v_d \rangle^{(z)} + [\langle v_z \psi_{pqr} \rangle^{(0)} - \langle v_z \rangle^{(0)}\langle \psi_{pqr} \rangle^{(0)}]$$

$$= N \sum_{s=0}^{\infty} \sum_{t=0}^{\infty} \sum_{u=0}^{\infty} a(pqr; stu)\langle \psi_{stu} \rangle^{(z)}. \tag{6-3-6c}$$

These results are for ions in a single neutral gas. For mixtures, the only change is to replace the matrix elements $a(pqr; stu)$ by linear combinations weighted by the mole fractions of the neutral species in the mixture,

$$b(pqr; stu) \equiv \sum_j x_j a_j(pqr; stu), \tag{6-3-7}$$

where the index j labels the neutral species, just as in the one- and two-temperature theories.

There are some constraints on the matrix elements and moments because of normalization and the cylindrical symmetry in a drift tube. The normalization of f requires that

$$\langle \psi_{000} \rangle^{(0)} = 1, \qquad \langle \psi_{000} \rangle^{(x)} = \langle \psi_{000} \rangle^{(y)} = \langle \psi_{000} \rangle^{(z)} = 0, \tag{6-3-8}$$

and cylindrical symmetry implies that

$$\langle \psi_{pqr} \rangle^{(0,x,y,z)} = \langle \psi_{qpr} \rangle^{(0,x,y,z)}, \tag{6-3-9a}$$

$$a(pqr; stu) = a(qpr; tsu). \tag{6-3-9b}$$

In addition, the fact that an ion cannot be directly accelerated in the x or y

direction by an electric field in the z direction leads to the conditions

$$\langle \psi_{pqr} \rangle = 0 \qquad \text{if } p \text{ or } q \text{ is odd,} \qquad (6\text{-}3\text{-}10a)$$

$$a(pqr; stu) = 0 \qquad \text{if } (p + s) \text{ or } (q + t) \text{ is odd.} \qquad (6\text{-}3\text{-}10b)$$

It is also understood that $a \langle \psi_{pqr} \rangle$ with any negative indices vanishes.

There are also some additional, more obscure, relations known among the matrix elements (Lin et al., 1979c), such as $a(00r; 220) = 2a(00r; 400)$.

Carrying out the differentiations on the left-hand side of (6-3-6), we obtain the final moment equations,

$$2r\mathscr{E}\langle \psi_{pq(r-1)} \rangle^{(0)} = \sum_{stu} \gamma(pqr; stu)\langle \psi_{stu} \rangle^{(0)}, \qquad (6\text{-}3\text{-}11)$$

$$2r\mathscr{E}\langle \psi_{pq(r-1)} \rangle^{(x,z)} + h_{pqr}^{(x,z)} = \sum_{stu} \gamma(pqr; stu)\langle \psi_{stu} \rangle^{(x,z)}, \qquad (6\text{-}3\text{-}12)$$

where

$$\mathscr{E} \equiv \left[\frac{m}{2kT_b^{(L)}} \right]^{1/2} \frac{e}{ma(001; 000)} \frac{E}{N}, \qquad (6\text{-}3\text{-}13)$$

$$\gamma(pqr; stu) \equiv a(pqr; stu)/a(001; 000), \qquad (6\text{-}3\text{-}14)$$

and

$$h_{pqr}^{(x)} \equiv \left[\frac{kT_b^{(T)}}{2m} \right]^{1/2} \frac{\langle \psi_{(p+1)qr} \rangle^{(0)} + 2p\langle \psi_{(p-1)qr} \rangle^{(0)}}{Na(001; 000)}, \qquad (6\text{-}3\text{-}15a)$$

$$h_{pqr}^{(z)} = \left[\frac{kT^{(L)}}{2m} \right]^{1/2}$$

$$\times \frac{\langle \psi_{pq(r+1)} \rangle^{(0)} + 2r\langle \psi_{pq(r-1)} \rangle^{(0)} - \langle \psi_{001} \rangle^{(0)}\langle \psi_{pqr} \rangle^{(0)}}{Na(001; 000)}. \qquad (6\text{-}3\text{-}15b)$$

These equations are to be solved for the drift velocity,

$$v_d = v_{\text{dis}} + [kT_b^{(L)}/2m]^{1/2}\langle \psi_{001} \rangle^{(0)}, \qquad (6\text{-}3\text{-}16)$$

and the diffusion coefficients,

$$D_T = [kT_b^{(T)}/2m]^{1/2}\langle \psi_{100} \rangle^{(x)}, \qquad (6\text{-}3\text{-}17a)$$

$$D_L = [kT_b^{(L)}/2m]^{1/2}\langle \psi_{001} \rangle^{(z)}. \qquad (6\text{-}3\text{-}17b)$$

The solution of these moment equations requires some procedure for choosing the parameters $T_b^{(T)}$, $T_b^{(L)}$, and v_{dis}, and some systematic truncation-iteration scheme for generating successive approximations to the unobtainable true solutions of the infinite set of equations. But first the matrix elements must be generated.

As would be expected, the calculations of the matrix elements are even more tedious and complicated for the three-temperature theory than for the two-temperature theory. The reductions to linear combinations of irreducible collision integrals have been described by Lin et al. (1979c); the results are sufficiently complicated that anyone intending to make serious quantitative calculations would be well advised to program the results for a computer. The three-temperature irreducible collision integrals are defined as

$$[p, q]^{(l)} \equiv 2\pi^{-1/2} \int_0^\infty d\gamma_r \gamma_r^{p+1} \exp(-\gamma_r^2)$$

$$\times \int_{-\infty}^\infty d\gamma_z \gamma_z^q \exp(-\gamma_z^2) v_r Q^{(l)}(\varepsilon), \qquad (6\text{-}3\text{-}18)$$

where the transport cross sections are

$$Q^{(l)}(\varepsilon) \equiv 2\pi \int_0^\pi (1 - \cos^l\theta)\sigma(\theta, \varepsilon) \sin\theta \, d\theta, \qquad (6\text{-}3\text{-}19)$$

which are just unnormalized versions (a minor annoyance) of the $\bar{Q}^{(l)}(\varepsilon)$ defined by (6-1-11). The variables are

$$\varepsilon = \tfrac{1}{2}\mu v_r^2 = \gamma_r^2 k T_{\text{eff}}^{(T)} + (\gamma_z + \tilde{v}_{dis})^2 k T_{\text{eff}}^{(L)}, \qquad (6\text{-}3\text{-}20)$$

where

$$\tilde{v}_{dis} \equiv [\mu v_{dis}^2 / 2k T_{\text{eff}}^{(L)}]^{1/2}, \qquad (6\text{-}3\text{-}21a)$$

$$T_{\text{eff}}^{(T,L)} \equiv [mT + M T_b^{(T,L)}]/(m + M), \qquad (6\text{-}3\text{-}21b)$$

and γ_x, γ_y, and γ_z are just the (reduced) x, y, and z components of the relative velocity $\mathbf{v}_r$, with $\gamma_r^2 = \gamma_x^2 + \gamma_y^2$. The $[p, q]^{(l)}$ are unnormalized three-dimensional analogues of the irreducible collision integrals $\bar{\Omega}^{(l,s)}(T_{\text{eff}})$ of the two-temperature theory, with indices that correspond according to $l \to l$ and $(p + q) \to 2s$. But the $[p, q]^{(l)}$ are now functions of two effective temperatures, $T_{\text{eff}}^{(T)}$ and $T_{\text{eff}}^{(L)}$, and of the displacement velocity v_{dis}. It is this dependence of $[p, q]^{(l)}$ on v_{dis} that makes it so difficult to obtain any physical insight from the three-temperature theory, even in a low order of approximation. As shown by (6-3-16), the drift velocity v_d (and hence the mobility) depend on v_{dis} very directly. Thus solving the moment equations to get v_d in terms of $[p, q]^{(l)}$, even in low order, gives only an implicit equation for v_d, because v_d is also buried in the $[p, q]^{(l)}$ through v_{dis}.

In short, the three-temperature theory is more trouble to use than is the two-temperature theory, but the troubles are only computational ones. This is the penalty to be paid for choosing $f^{(0)}$ elaborate enough to give a rapidly convergent theory. Complications result even in first approximation, as will be apparent in the following outline of a reasonable truncation-iteration scheme.

Guided by the experience of the two-temperature theory, we want to pick the basis temperatures as independent variables and keep them fixed throughout the successive stages of approximation, in order to avoid having to recompute all the matrix elements at each stage. But now there are two basis temperatures, not one, which ought to bear some reasonable relation to one another. There is also a third parameter, v_{dis}, which should also bear some reasonable relation to the basis temperatures. That is, we should not choose $T_b^{(T)}$, $T_b^{(L)}$, and v_{dis} completely independently at the beginning, but should have them related to one another in a physically reasonable way. The procedure currently in use is to relate the three parameters to each other through the lowest-order moment equations, which correspond to equations for overall momentum and energy balance, and then to retain these initial values in all subsequent approximations (Viehland and Lin, 1979; Viehland, 1982). Even this requires numerical iteration to get the values of $T_b^{(T)}$, $T_b^{(L)}$, and v_{dis}, as seen below.

The general truncation-iteration scheme is based, as usual, on the properties of the Maxwell model, for which it can be shown that the matrix elements $\gamma(pqr; stu)$ vanish if $(s + t + u) > (p + q + r)$. Truncation is performed for the nth approximation by setting all $\gamma(pqr; stu)\langle \psi_{stu} \rangle^{(0)}$ equal to zero in (6-3-11) for $(s + t + u) \geq (n + 2)$, and all $\gamma(pqr; stu)\langle \psi_{stu} \rangle^{(x,y,z)}$ equal to zero in (6-3-12) for $(s + t + u) \geq (n + 1)$ (Lin et al., 1979c; Viehland and Lin, 1979). Values of the parameters are chosen to satisfy the physically reasonable conditions

$$v_{\text{dis}} = v_d = \langle v_z \rangle^{(0)}, \tag{6-3-22a}$$

$$\tfrac{1}{2}kT_b^{(T)} = \tfrac{1}{2}kT_T = \tfrac{1}{2}m\langle v_x^2 \rangle^{(0)} = \tfrac{1}{2}m\langle v_y^2 \rangle^{(0)}, \tag{6-3-22b}$$

$$\tfrac{1}{2}kT_b^{(L)} = \tfrac{1}{2}kT_L = \tfrac{1}{2}m\langle (v_z - v_{\text{dis}})^2 \rangle^{(0)}, \tag{6-3-22c}$$

which is equivalent to

$$\langle \psi_{001} \rangle^{(0)} = \langle \psi_{002} \rangle^{(0)} = \langle \psi_{200} \rangle^{(0)} = 0. \tag{6-3-23}$$

The lowest-order moment equations from (6-3-11) are then

$$2\mathscr{E} = 1, \tag{6-3-24a}$$

$$a(002; 000) = 0, \tag{6-3-24b}$$

$$a(200; 000) = 0. \tag{6-3-24c}$$

But (6-3-22) and (6-3-24) form only an implicit set of equations for $T_b^{(T)}$, $T_b^{(L)}$ and v_{dis}, because the collision integrals $[p; q]^{(l)}$ that appear in the matrix elements of

(6-3-24) themselves depend on the parameters $T_b^{(T)}$, $T_b^{(L)}$, and v_{dis}. The solutions must be found by numerical iteration if one wants the parameters to satisfy (6-3-22) accurately. The usual practice is first to pick v_{dis}, which can remain fixed for the remainder of the cycle, next to pick starting values of $T_b^{(T)}$ and $T_b^{(L)}$ from the Maxwell-model relations,

$$kT_b^{(T)} = kT + \frac{m + M}{5m + 3M} Mv_{dis}^2, \qquad (6\text{-}3\text{-}25a)$$

$$kT_b^{(L)} = kT + \frac{3m + M}{5m + 3M} Mv_{dis}^2, \qquad (6\text{-}3\text{-}25b)$$

and then to iterate until $T_b^{(T)}$ and $T_b^{(L)}$ satisfy (6-3-22) and (6-3-24). After this initial iteration the values of v_{dis}, $T_b^{(T)}$, and $T_b^{(L)}$ are kept fixed for all the higher approximations, and values of E/N, K, T_T, T_L, D_T, D_L, etc. are calculated as dependent variables up to some predetermined order of accuracy or approximation. Of course, in the higher approximations the relations $v_{dis} = v_d$, $T_b^{(T)} = T_T$, and $T_b^{(L)} = T_L$ do not hold, and v_{dis}, $T_b^{(T)}$, and $T_b^{(L)}$ are to be regarded simply as computational parameters. A new value of v_{dis} is chosen and the whole procedure is repeated until the entire desired range of E/N is covered.

There is evidence that the initial iteration, to make $T_b^{(T)}$ and $T_b^{(L)}$ conform to (6-3-22) in first approximation, may be overdoing it. It may be sufficient to use fixed values of the basis temperatures from (6-3-25) and avoid the initial iteration entirely. It may in fact be most important to pick just $T_b = 2T_b^{(T)} + T_b^{(L)}$ properly, and not worry too much about the individual values of $T_b^{(T)}$ and $T_b^{(L)}$ (Kumar et al., 1980, pp. 413–414). However, the iteration is very quick on a computer, and is probably worthwhile (Viehland, 1985, personal communication).

B. Convergence of Approximations

A large number of calculations have by now been carried out on various systems that test the convergence properties of the three-temperature theory (Viehland and Lin, 1979; Viehland, 1982, 1983, 1984a). The sort of results obtainable can be illustrated with the rigid-sphere cold-gas model, as shown in Table 6-3-1. Results for softer potentials are usually even better. The following remarks roughly summarize the situation.

First, oscillations of the successive approximations about the correct value are noticeable in almost all cases. In this regard the three-temperature theory is similar to the two-temperature theory (see Fig. 6-2-1). In any new case encountered, the only reasonable indication of convergence is the stability of the calculated results for all three transport coefficients in successive approximations.

Second, convergence seems best for K, somewhat less good for T_T and D_T, and worst for T_L and D_L. As a rough rule of thumb, the fifth approximation is

Table 6-3-1   Convergence of Drift Velocity and Diffusion
Coefficients Calculated from the Three-Temperature
Theory for a Rigid-Sphere Cold-Gas Model, from Viehland
and Lin (1979)

$\dfrac{m}{M}$	Order of Approximation	Approximate/Accurate		
		v_d or K	D_T	D_L
0.1	1	1.009	0.977	1.856
	2	1.021	0.959	1.027
	3	1.004	1.002	1.080
	4	1.007	0.998	1.002
	5	1.001	1.001	0.984
0.5	1	1.003	1.162	2.130
	2	0.994	1.093	1.028
	3	0.999	0.967	0.983
	4	1.000	0.996	1.084
	5	0.999	1.010	1.036
1.0	1	0.988	1.159	1.908
	2	0.992	1.031	0.967
	3	0.996	0.962	1.006
	4	0.998	1.004	1.087
	5	0.998	1.008	1.007
4.0	1	0.996	1.070	1.294
	2	1.000	0.975	0.973
	3	1.000	0.971	1.010
	4	1.000	0.978	1.006
	5	1.000	0.975	1.001

usually accurate within about 0.5% for K, 2% for D_T, and 5% for D_L, which is better than most experimental results. However, the convergence does depend on the nature of the ion-neutral interaction and caution is advisable, as discussed further below.

Third, computer programs for the systematic numerical calculation of the cross sections, the matrix elements, and the transport coefficients by the three-temperature theory are available (Viehland, 1982, 1984a). These general programs become rather awkward and time consuming beyond the fifth approximation. Some special cases, such as electrons, need to have a truncation-iteration procedure tailored to their own particular characteristics.

Finally, there is one frequently occurring situation for which convergence is annoyingly poor. This occurs in the range of E/N for which K_0 or the reduced diffusion coefficients $\tilde{D}_{T,L}$ are increasing with increasing E/N. For example, this range occurs roughly from 8 to 20 Td in Fig. 6-2-4 for $\tilde{D}_L$ of K^+ in He. This behavior is not an artifact of the three-temperature theory, but is rather general and deserves some discussion at this point. Basically, it occurs because some peculiarity of the ion-neutral interaction allows a feature to develop in the ion velocity distribution function that is not reasonably represented by the zero-order approximation, $f^{(0)}$.

To see in a simple physical way how this happens, we use momentum- and energy-balance arguments (as in Section 5-2A), and ignore various fine points such as the distinction between $\bar{v}_r^2$ and $\overline{v_r^2}$. Equating the average momentum gained from the field to the average momentum lost by collisions, we have

$$eE = \mu v_d \bar{v}. \tag{6-3-26}$$

We next write the mean collision frequency $\bar{v}$ in terms of the momentum-transfer cross section,

$$\bar{v} \approx N\bar{v}_r Q^{(1)} \approx N(2\bar{\varepsilon}/\mu)^{1/2} Q^{(1)}, \tag{6-3-27}$$

where $Q^{(1)}$ is given by (6-3-19), and then use an energy-balance argument to evaluate the mean relative collision energy $\bar{\varepsilon}$,

$$\bar{\varepsilon} \equiv \tfrac{1}{2}\mu \overline{v_r^2} \approx \tfrac{3}{2}kT + \tfrac{1}{2}Mv_d^2 \approx \tfrac{1}{2}Mv_d^2. \tag{6-3-28}$$

Substituting these expressions back into (6-3-26) and rearranging, we obtain

$$\bar{\varepsilon} Q^{(1)} \approx [(e/2)(M/\mu)^{1/2}](E/N). \tag{6-3-29}$$

We can use this expression as the basis for a simple graphical interpretation, noting that the left-hand side represents the average momentum loss per collision and the right-hand side represents the average momentum gain from the field.

We plot $\varepsilon Q^{(1)}$ vs. ε for a particular ion-neutral interaction, and then draw a horizontal line corresponding to some imposed electric field as specified by the right-hand side of (6-3-29). The point of intersection gives the mean energy (or the value of T_{eff}) that the ions will have at this field strength. There will of course be some spread of ion energies around this mean, and the nature of the spread will be determined by the shape of the $\varepsilon Q^{(1)}$ curve in the vicinity of the intersection. Such a plot is shown in Fig. 6-3-1 for the (8, 4) potential of (6-1-57), in reduced units. As long as the curve of $\varepsilon Q^{(1)}$ vs. ε is fairly steep, the distribution of ion energies around the mean will be fairly sharp, and a Maxwellian form for $f^{(0)}$ should give reasonable results. This is indicated in Fig. 6-3-1 for a field E_1 for which the attractive part of the potential dominates, and for a field E_3 for which the repulsive part of the potential dominates. But when $\varepsilon Q^{(1)}$ vs. ε has an undulation or a flat region, the spread of ion energies for field strengths falling in this region is large, and a Maxwellian may be a poor form for $f^{(0)}$. This region is

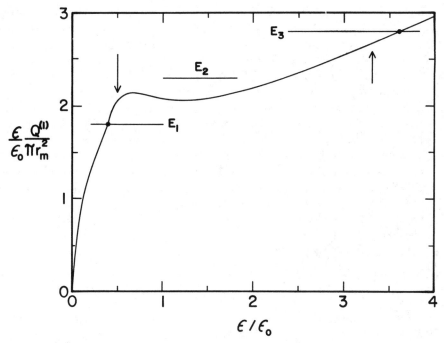

Figure 6-3-1. Average momentum loss per collision as a function of relative collision energy for an (8, 4) potential. The horizontal lines represent average momentum gains for various field strengths, and the intersections give the average relative energy at steady state. When an intersection occurs in the undulatory region between the arrows, the ion distribution is distorted and convergence becomes poorer for the two-temperature and three-temperature theories.

shown in the figure for fields, such as E_2, which fall between the arrows. These arrows mark approximately the region where K_0 increases with increasing E/N. (The first arrow marks where $\bar{v}$ starts to decrease rapidly, and the second arrow marks the minimum in $\bar{v}$.) The range between the arrows in the figure represents the energies where the ion-neutral interaction is shifting from domination by attraction to domination by repulsion.

As one might conclude from Fig. 6-3-1, even a moderately flat region in $\varepsilon Q^{(1)}$ vs. ε may be sufficient to cause convergence troubles. Such a region can result from a "soft" ion-neutral interaction, for example. Skullerud (1984) has made a mathematical investigation of such cases, and concludes that their true ion velocity distribution f falls off asymptotically more slowly than an exponential Maxwellian. As a result, the condition for the convergence of an expansion of f about a Maxwellian $f^{(0)}$ is not met, and a moment calculation with such an $f^{(0)}$ converges poorly or even fails altogether.

There are several cures for the poor convergence in the region where K_0 vs. E/N rises, none very pleasant. The simplest is to make a brute-force moment calculation to as high an order of approximation as seems feasible, and then to try to estimate a converged value through the oscillations of the approximations. There are various mathematical tricks for making such estimates (Shanks, 1955; Baker, 1975). Further smoothing by interpolation across the difficult region of E/N may also help. Another approach is to suppose that the situation shown in Fig. 6-3-1 leads to something close to a bimodal energy distribution in the region marked off between the arrows. But rather than go to the trouble of using a bimodal expansion as in Section 5-3D5, one tries to approximate the bimodal distribution by a single Maxwellian centered around an intermediate energy. That is, one tries to make a judicious choice of the basis temperature T_b, or the temperatures $T_b^{(T)}$ and $T_b^{(L)}$, in order to improve the convergence. Schemes for making such a choice have been worked out for both the two-temperature (Viehland, 1983) and three-temperature (Viehland, 1984a) theories. They improve the convergence somewhat, but not as much as one would like.

The last, and least pleasant, cure for poor convergence is to make T_b a function of v instead of a constant, and write

$$f^{(0)} \propto \exp(-w_b^2) \qquad (6\text{-}3\text{-}30a)$$

with

$$w_b^2 = \int_0^v \frac{mv'\,dv'}{kT_b(v')}, \qquad (6\text{-}3\text{-}30b)$$

or analogous expressions involving $T_b^{(T)}$ and $T_b^{(L)}$. Of course, we could represent f itself in this way by a sufficiently complicated choice of $T_b(v')$, so that to represent just the zero-order $f^{(0)}$ we can be satisfied with something fairly simple. A rough argument based on momentum and energy balance, as in Section 5-2A, suggests that a reasonable choice might have the form

$$kT_b(v') = kT + \frac{eE}{NQ^{(1)}(v')}, \qquad (6\text{-}3\text{-}31)$$

in which we have ignored possible constant factors involving m and M. Skullerud (1984) has shown that this choice produces improved convergence in a number of test cases, although at the expense of some increased computational complexity. However, he found some cases with a very slow asymptotic decay of f for which convergence always seemed to fail, and in such pathological cases it seems likely that all moment methods will fail regardless of the choice of $f^{(0)}$.

The moral of all the foregoing convergence studies is that the three-temperature theory will usually produce satisfactory results for most real systems. But convergence should never be blindly taken for granted—Murphy's laws can be relied upon to produce an unexpected pathological case from time to time.

C. Temperature and Field Dependence of Mobility and Diffusion

The three-temperature theory allows accurate calculations of mobility and diffusion coefficients to be made as a function of T and E/N regardless of the value of m/M, but the simple physical interpretation and the scaling rules of the two-temperature theory are lost. The three-temperature theory gives numbers, but not much physical insight except for the generalized Einstein relations (Section 6-4).

D. Quantum Effects

Quantum effects enter only through the cross sections $Q^{(l)}(\varepsilon)$. Whether these are subsequently manipulated by a one-, two-, or three-temperature theory has nothing to do with quantum effects in the kinetic theory itself. If the cross sections are quantum-mechanically correct, the kinetic theory will also be quantum-mechanically correct.

E. Inelastic Collisions

The same remarks apply as for the two-temperature theory. The problem is treated separately in Section 6-6.

F. Mixtures

Here again the three-temperature theory can give numbers, but not much insight. The moment equations for a mixture have the same form as for a single gas, but the matrix elements $a_j(pqr; stu)$ for the single gas are replaced by the linear combinations $b(pqr; stu)$ of (6-3-7). Simple but useful approximate formulas are those already given in Sections 6-1G and 6-2F.

6-4. GENERALIZED EINSTEIN RELATIONS

The generalized Einstein relations (GER) that connect D_T and D_L to K and its field dependence are so successful that many attempts have been made to improve the pedigree of their derivation, and even to calculate correction terms to improve accuracy. The underlying physical reasons for the success of the GER have already been emphasized in the momentum-transfer calculation given in Section 5-2B. The most crucial feature is that D_T and D_L are related to the *differential* mobility,

$$D_{T,L} = \frac{kT_{T,L}}{e}\left(\frac{dv_d}{dE}\right)_{T,L},\qquad (6\text{-}4\text{-}1)$$

which is (5-2-56). Written out for the components, this is

$$D_T = (kT_T/e)K, \tag{6-4-2a}$$

$$D_L = (kT_L/e)K(1 + K'), \tag{6-4-2b}$$

where, as usual, $K' \equiv d \ln K_0/d \ln (E/N)$. In turn, the differential mobility arises because the driving force for diffusion, $\nabla \cdot \mathbf{p}_i$, is given the same status as the external force for mobility, which is the electric field strength $\mathbf{E}$. The equivalence, in some sense, of gradients to true external forces is a basic notion in statistical physics that goes back at least to Einstein's classic work on Brownian motion (and precursors of the notion appeared even earlier).

These essential underlying physical reasons appeared most clearly in the work of Robson (1972) and of Whealton et al. (1974). The discussion in Section 5-2B is largely based on their ideas. A more formal mathematical version was given in a short conference paper by Kumar (1977). The more elaborate moment theories discussed so far in this chapter tend to obscure this simple piece of physics, and so their immediate contributions to justifying and improving the GER have been disappointing. Their main direct contribution to the GER has been the development of reasonably simple formulas for the calculation of the ion temperatures, T_T and T_L. However, once these points are recognized, it is not too difficult to incorporate them into the moment theories. This is best done in the context of the three-temperature theory, inasmuch as the two-temperature theory does not converge for diffusion when $m \gg M$, and has been carried out by Waldman and Mason (1981), whose presentation we follow here.

We first treat the case where all ion-neutral collisions are elastic, which has been worked out in some detail, and give a useful parameterization scheme for the results. We then consider the modifications to be introduced when resonant charge exchange can occur. Finally, we report the fragmentary results available when inelastic collisions become important.

A.   Elastic Collisions

The crucial idea is to introduce the differential mobility as a tensorial quantity, and to show how to use the Boltzmann equation to solve directly for the differential mobility rather than the mobility. We consider two kinds of perturbations to the usual spatially homogeneous reference case. The first kind is the perturbation caused by a small gradient of ion density with $\mathbf{E}$ held fixed, and the second kind is the perturbation caused by adding infinitesimal perturbing electric fields δE_x, δE_y, and δE_z to the already existing field $\mathbf{E}$ (assumed as usual to point in the z-direction). We need keep only perturbations linear in the density gradient, because we do not care about deviations from Fick's law of diffusion in this problem. However, we must keep more than just terms linear in the δE if we wish to go beyond the standard GER of (6-4-2). The reason is that linear perturbation terms just rotate $\mathbf{E}$ through a small angle while increasing its magnitude by δE_z; the result is the first-order GER, as was explained in more detail in Section 5-2B. We therefore write

$$\langle \psi_{pqr} \rangle = \langle \psi_{pqr} \rangle^{(0)} - \langle \psi_{pqr} \rangle_D^{(z)} \frac{\partial \ln n}{\partial z} - \langle \psi_{pqr} \rangle_D^{(x)} \frac{\partial \ln n}{\partial x} + \cdots \qquad (6\text{-}4\text{-}3)$$

for the first perturbation, and

$$\langle \psi_{pqr} \rangle = \langle \psi_{pqr} \rangle^{(0)} + \langle \psi_{pqr} \rangle_E^{(z)} \delta E_z + \langle \psi_{pqr} \rangle_E^{(x)} \delta E_x + \langle \psi_{pqr} \rangle_E^{(zz)} (\delta E_z)^2$$

$$+ \langle \psi_{pqr} \rangle_E^{(xx)} (\delta E_x)^2 + \langle \psi_{pqr} \rangle_E^{(xz)} \delta E_x \, \delta E_z + \cdots \qquad (6\text{-}4\text{-}4)$$

for the second perturbation. In both these expansions we have omitted terms involving the y-direction because of the symmetry between x and y.

These expansions are substituted back into the general moment equations obtained from the Boltzmann equation, and the different kinds of perturbing terms are separated off to produce sets of moment equations for each of the different kinds of moments. The equations for the spatially homogeneous moments $\langle \psi_{pqr} \rangle^{(0)}$ are the same as those already given by (6-3-11) in the discussion of the three-temperature theory. Similarly, the equations for the density-perturbed moments $\langle \psi_{pqr} \rangle_D^{(x,z)}$ are the same as (6-3-12), with D_T and D_L given by (6-3-17). The same set of moment equations for the $\langle \psi_{pqr} \rangle_D^{(x,z)}$ is obtained whether the density-gradient perturbation is applied to f, as in (6-3-1), or to $\langle \psi_{pqr} \rangle$, as in (6-4-3) above. The equations for the field-perturbed moments $\langle \psi_{pqr} \rangle_E^{(\alpha)}$ are new, and are found by letting

$$\mathbf{E} = (E + \delta E_z)\hat{z} + (\delta E_x)\hat{x} \qquad (6\text{-}4\text{-}5)$$

in the general moment equation, where $\hat{z}$ and $\hat{x}$ are unit vectors in the z- and x-directions, and then separating terms to obtain

$$2r\mathscr{E} \langle \psi_{pq(r-1)} \rangle_E^{(\alpha)} + 2(\mathscr{E}/E) g_{pqr}^{(\alpha)}$$

$$= \sum_{stu} \gamma(pqr; stu) \langle \psi_{stu} \rangle_E^{(\alpha)}, \qquad \alpha = z, x, zz, xx, xz, \qquad (6\text{-}4\text{-}6)$$

where $\mathscr{E}$ and $\gamma(pqr; stu)$ are defined by (6-3-13) and (6-3-14), and

$$g_{pqr}^{(z)} \equiv r \langle \psi_{pq(r-1)} \rangle^{(0)}, \qquad (6\text{-}4\text{-}7a)$$

$$g_{pqr}^{(x)} \equiv p[T_b^{(L)}/T_b^{(T)}]^{1/2} \langle \psi_{(p-1)qr} \rangle^{(0)}, \qquad (6\text{-}4\text{-}7b)$$

$$g_{pqr}^{(zz)} \equiv r \langle \psi_{pq(r-1)} \rangle_E^{(z)}, \qquad (6\text{-}4\text{-}7c)$$

$$g_{pqr}^{(xx)} \equiv p[T_b^{(L)}/T_b^{(T)}]^{1/2} \langle \psi_{(p-1)qr} \rangle_E^{(x)}, \qquad (6\text{-}4\text{-}7d)$$

$$g_{pqr}^{(xz)} \equiv r \langle \psi_{pq(r-1)} \rangle_E^{(x)} + p[T_b^{(L)}/T_b^{(T)}]^{1/2} \langle \psi_{(p-1)qr} \rangle_E^{(z)}. \qquad (6\text{-}4\text{-}7e)$$

All these sets of moment equations are to be solved subject to the usual symmetry constraints on the moments (Waldman and Mason, 1981), which we need not write down here.

The idea of differential mobility enters through (6-4-4). We note that δE_x and δE_y will perturb the drift velocity, and expand as in (6-4-4),

$$\langle \mathbf{v} \rangle = [\langle v_z \rangle^{(0)} + K_L^z \, \delta E_z + K_L^{zz}(\delta E_z)^2 + K_L^{xx}(\delta E_x)^2 + \cdots]\hat{z}$$

$$+ [K_T^x \, \delta E_x + K_T^{xz} \, \delta E_x \, \delta E_z + \cdots]\hat{x}, \qquad (6\text{-}4\text{-}8)$$

in which we have omitted terms involving δE_y, as before. Here $\langle v_z \rangle^{(0)} = v_d$, and K_L^z and K_T^x are longitudinal and transverse differential mobilities, as used in the momentum-transfer calculation of Section 5-2B. The quantities K_L^{zz}, K_L^{xx}, and K_T^{xz} represent nonlinear higher-order differential mobilities which will eventually generate correction terms to the standard GER. The relations among the differential mobilities and the $\langle \psi_{pqr} \rangle_E^{(\alpha)}$ follow from the definition (6-3-4b) of the ψ_{pqr}, which gives

$$\langle \psi_{001} \rangle = 2[m/2kT_b^{(L)}]^{1/2}\langle v_z - v_{\text{dis}} \rangle, \qquad (6\text{-}4\text{-}9a)$$

$$\langle \psi_{100} \rangle = 2[m/2kT_b^{(T)}]^{1/2}\langle v_x \rangle. \qquad (6\text{-}4\text{-}9b)$$

Comparing the expansions in (6-4-4) and (6-4-8), we obtain

$$v_d = v_{\text{dis}} + [kT_b^{(L)}/2m]^{1/2}\langle \psi_{001} \rangle^{(0)}, \qquad (6\text{-}4\text{-}10a)$$

which is the same as (6-3-16), and

$$K_L^{(\alpha)} = [kT_b^{(L)}/2m]^{1/2}\langle \psi_{001} \rangle_E^{(\alpha)}, \qquad \alpha = z, \, zz, \, xx, \qquad (6\text{-}4\text{-}10b)$$

$$K_T^{(\alpha)} = [kT_b^{(T)}/2m]^{1/2}\langle \psi_{100} \rangle_E^{(\alpha)}, \qquad \alpha = x, \, xz. \qquad (6\text{-}4\text{-}10c)$$

The various $K_L^{(\alpha)}$ and $K_T^{(\alpha)}$ can be related to scalar field derivatives of the standard mobility K_0, by comparison of (6-4-8) with a Taylor expansion for the mobility. Details have been given by Waldman and Mason (1981); the results are

$$K_L^z = K(1 + K'), \qquad K_T^x = K, \qquad (6\text{-}4\text{-}11a)$$

$$K_L^{zz} = (K/2E)[K'(1 + K') + K''], \qquad K_L^{xx} = K_T^{xz} = (K/2E)K', \qquad (6\text{-}4\text{-}11b)$$

where

$$K' \equiv d \ln K_0/d \ln (E/N), \qquad (6\text{-}4\text{-}12a)$$

$$K'' \equiv d^2(\ln K_0)/d[\ln (E/N)]^2. \qquad (6\text{-}4\text{-}12b)$$

The problem is now reduced to solving the sets of moment equations for the $\langle \psi_{pqr} \rangle_D^{(\alpha)}$ and $\langle \psi_{pqr} \rangle_E^{(\alpha)}$; the GER then follow by substitution into the relations

$$\frac{D_L}{K_L^z} = \frac{\langle \psi_{001} \rangle_D^{(z)}}{\langle \psi_{001} \rangle_E^{(z)}}, \tag{6-4-13a}$$

$$\frac{D_T}{K_T^x} = \frac{\langle \psi_{100} \rangle_D^{(x)}}{\langle \psi_{100} \rangle_E^{(x)}}. \tag{6-4-13b}$$

The hope is that the results will be simpler and easier to use than those obtained by brute-force solutions of the moment equations for D_L, D_T, and K. But of course we cannot solve the moment equations exactly, we can only obtain approximations according to some truncation scheme. A slightly modified version of the truncation scheme for the three-temperature theory (Section 6-3A) is used. First, the parameters are required to meet the following physical conditions: $v_{dis} = v_d = \langle v_z \rangle^{(0)}$, $T_b^{(T)} = T_T$, and $T_b^{(L)} = T_L$. Second, in the nth approximation for the moment equations, all $\gamma(pqr; stu) \langle \psi_{stu} \rangle^{(0)}$ with $(s + t + u) \geqslant (n + 2)$ or with $(s + t + u) \geqslant (p + q + r + n)$ are set equal to zero, and all $\gamma(pqr; stu) \langle \psi_{stu} \rangle_{D,E}^{(\alpha)}$ with $(s + t + u) \geqslant (n + 1)$ or with $(s + t + u) \geqslant (p + q + r + n)$ are set equal to zero. The first approximation $(n = 1)$ is very simple, and yields

$$D_L/K_L^z = kT_L/e, \tag{6-4-14a}$$

$$D_T/K_T^x = kT_T/e, \tag{6-4-14b}$$

which reduce to the usual GER of (6-4-2) on substituting for K_L^z and K_T^x. This is gratifying but hardly surprising, since the whole procedure with differential mobilities was designed with this end in view. However, it does serve the important function of giving the usual GER a firm basis in the Boltzmann equation, and thus a solid pedigree.

To find corrections to the foregoing first-order GER, we must consider the second approximation for the moment equations. These are substantially more complicated, and involve many more moments and matrix elements. Somewhat surprisingly, most of these can be made to drop out by suitably clever algebra (Waldman and Mason, 1981), and the corrections involve only two different moments in addition to the differential mobilities, as follows:

$$\frac{D_T}{K} = \frac{kT_T}{e} \left[1 + \frac{1}{4} \left(\frac{2kT_L}{mv_d^2} \right)^{1/2} \langle \psi_{201} \rangle^{(0)} \frac{K'}{1 + K'} \right], \tag{6-4-15a}$$

$$\frac{D_L}{K} = \frac{kT_L}{e} \left[1 + K' + \frac{1}{8} \left(\frac{2kT_L}{mv_d^2} \right)^{1/2} \frac{T_T}{T_L} \langle \psi_{201} \rangle^{(0)} K' \right.$$

$$\left. + \frac{1}{16} \left(\frac{2kT_L}{mv_d^2} \right)^{1/2} \langle \psi_{003} \rangle^{(0)} \frac{K'(1 + K') + K''}{1 + K'} \right]. \tag{6-4-15b}$$

It is not clear whether this happy simplification is simply fortuitous, or whether it indicates some deeper physical point. It does not seem to happen when inelastic collisions are taken into account (see Section 6-6).

At this point it is worth making contact with a momentum-transfer calculation by Robson (1976), in which the energy-balance equation is modified by the addition of a term corresponding to an ionic heat flux driven by the ion density gradient. This leads to a correction to the GER for D_L, but not for D_T. Since momentum-transfer theory is equivalent to a moment theory having a very limited set of moment equations, we would expect that Robson's result should in some sense correspond to the results in (6-4-15) above. Although a detailed comparison has never been published, inspection shows that Robson's correction corresponds in general form to the correction term in D_L that involves $\langle \psi_{003} \rangle^{(0)}$, with K'' neglected. The terms involving $\langle \psi_{201} \rangle^{(0)}$ do not appear. A more detailed analysis of higher-order momentum-transfer theory in terms of the moment theory above would be interesting for the possible physical insight obtained, especially with a view towards incorporating effects of inelastic collisions.

Two difficulties remain with the GER of (6-4-15), one old and one new. The old difficulty is that the ion temperatures T_T and T_L are not directly observable, and some way must be found to calculate them. The new difficulty is that two new unobservable quantities now appear in (6-4-15), namely $\langle \psi_{201} \rangle^{(0)}$ and $\langle \psi_{003} \rangle^{(0)}$, and these must also be calculated. It might be thought that these moments could be estimated with sufficient accuracy from the Maxwell model, but the results do not turn out to be satisfactory.

Consider first the ion temperatures. In first order, the three-temperature moment theory gives

$$kT_{T,L} = kT + \zeta_{T,L} M v_d^2, \qquad (6\text{-}4\text{-}16)$$

where

$$\zeta_T = \frac{(m + M)\tilde{A}}{4m + 3M\tilde{A}}, \qquad (6\text{-}4\text{-}17a)$$

$$\zeta_L = \frac{4m - (2m - M)\tilde{A}}{4m + 3M\tilde{A}}, \qquad (6\text{-}4\text{-}17b)$$

and

$$\tilde{A} \equiv [0,\, 0]^{(2)}/[0,0]^{(1)}, \qquad (6\text{-}4\text{-}18)$$

in which the irreducible collision integrals $[p, q]^{(l)}$ are defined in (6-3-18). These formulas are quite analogous to the corresponding formulas (6-1-42) of the one- and two-temperature theories, which involve A^* instead of $\tilde{A}$. Indeed, they are identical in the special case of the Maxwell model, for which $\tilde{A} = \frac{4}{5}A^*$, but in general they differ. For instance, for inverse-power potentials, $V(r) = C_n/r^n$, the relation is

$$\frac{A^*}{\bar{A}} = \frac{3}{2} - \frac{1}{n}. \qquad (6\text{-}4\text{-}19)$$

The formulas (6-4-17) are identical to the Wannier formulas (5-2-71) used by Skullerud (1976) in his semiempirical parameterization scheme for the GER, with $\bar{A} = Q^{(2)}/Q^{(1)}$. Unfortunately, these formulas give poor results for the ion temperatures, especially for T_L (the results are similar to those shown in Table 5-2-6, for example), and so are useless for the GER. A reasonable step at this point would be to use the second approximation to find corrections for the ion temperatures of the form

$$kT_{T,L} = kT + \zeta_{T,L} M v_d^2 (1 + \beta_{T,L} K'), \qquad (6\text{-}4\text{-}20)$$

which give fairly good results in the analogous case of the two-temperature theory (see, e.g., Tables 6-1-4 and 6-1-5). However, this involves rather cumbersome calculations, and Waldman and Mason (1981) chose to develop a consistent closed scheme for the simultaneous numerical calculation of T_T, T_L, $\langle \psi_{201} \rangle^{(0)}$, and $\langle \psi_{003} \rangle^{(0)}$. The formulas (6-4-20) are, however, useful in a parameterization scheme for the GER, which is described in Section 6-4B.

The procedure adopted was to obtain a closed set of nine equations for nine unknowns by approximating certain ratios of higher-order irreducible collision integrals and neglecting some of the less important higher-order matrix elements. In particular, it was assumed that

$$[p, q]^{(l)}/[p, q]^{(1)} \approx [0, 0]^{(l)}/[0, 0]^{(1)} \qquad (6\text{-}4\text{-}21)$$

for all (p, q). These relations are exact for inverse-power potentials. In addition, matrix elements $a(100; stu)$ and $a(001; stu)$ with $(s + t + u) > 2$ were set to zero. The nine unknowns were T_T, T_L, $\langle \psi_{201} \rangle^{(0)}$, $\langle \psi_{003} \rangle^{(0)}$ plus the five matrix elements $a(100; 100)$, $a(100; 101)$, $a(001; 001)$, $a(001; 002)$, and $a(001; 200)$. The equations were nonlinear and had to be solved on a computer, but the overall process was still a much faster way of finding D_T and D_L than solving the original matrix equations to some high order of approximation (faster by about a factor of 30 in CPU time for the third approximation to D_T and D_L).

Numerical experimentation with various models indicates that the GER of (6-4-15), plus the self-consistent scheme for calculating T_T, T_L, $\langle \psi_{201} \rangle^{(0)}$, and $\langle \psi_{003} \rangle^{(0)}$, gives values of D_T and D_L accurate to a few percent in most cases. However, we do not give details here because a simple parameterization of the formulas gives results almost as accurate with even less effort. This parameterization scheme is discussed in the following section, but first a few comments are in order.

The first comment concerns mixtures. To extend the foregoing results to mixtures, it is only necessary to replace the matrix elements $a(pqr; stu)$ for a single gas by the linear combinations $b(pqr; stu)$ of (6-3-7), just as in the regular three-temperature theory (Section 6-3F). No applications of these GER to mixtures have been made, however.

The second comment concerns the connection between mobility and diffusion as given approximately by GER, as compared with an accurate numerical connection provided by a full high-order moment calculation. The latter connection could be developed as follows. First, mobility measurements as a function of field strength would be inverted to find the ion-atom potential, as described in detail in Chapter 7. Then the moment theory would be used to calculate diffusion coefficients from the potential. Aside from the massive computation involved, the main feature of this procedure is that it requires mobility data over a very large range of field strength. In contrast, the GER provide a procedure for calculating the diffusion coefficients at a *particular* field strength from mobility measurements at the *same* field, and in the immediate neighborhood of that field strength to evaluate derivatives of the mobility. In addition, the computations involved are far easier. However, the GER provide only approximate, rather than exact (in principle), relations.

Finally, it should be emphasized that the entire foregoing discussion is based on elastic collisions only. Although some of the low-order GER remain approximately valid when inelastic collisions occur, as discussed later, a complete restudy of the theory is needed in order to proceed to the level of the more refined GER considered in this section.

B.   GER Parameterization

The numerical calculations of Waldman and Mason (1981) have shown that the second derivative K'' can safely be dropped from the correction term for D_L in (6-4-15b). When this is done, the expressions for D_T and D_L can be written in a more compact form as follows:

$$\frac{D_T}{K} = \frac{kT_T}{e}\left(1 + \Delta_T \frac{K'}{1 + K'}\right), \tag{6-4-22a}$$

$$\frac{D_L}{K} = \frac{kT_L}{e}[1 + (1 + \Delta_L)K']. \tag{6-4-22b}$$

These numerical calculations also supply values of Δ_T and Δ_L to use as a basis for a parameterization scheme. The search for parameterized relations for Δ_T and Δ_L is simplified by the following considerations. Since the corrections are appreciable only at high fields, the search can reasonably be restricted to high-field, cold-gas cases. This means that only repulsive ion-atom interactions need to be considered, since these dominate at high fields except for special cases, such as those involving strong valence interactions (e.g., H^+ in He) or resonant charge exchange (e.g., He^+ in He). Numerical calculations of Δ_T and Δ_L in the cold-gas case for r^{-8}, r^{-12}, and rigid-sphere potentials show that the results depend primarily on the mass ratio and are rather insensitive to the potential (except in the limit $m/M \to 0$, where an analytical rather than a numerical result can be obtained). It is therefore possible to prepare a table of numerical values of Δ_T and Δ_L as a function of mass ratio; the values recommended by Waldman and Mason (1981) are given in Table 6-4-1.

As already mentioned, Robson (1976) used momentum-transfer theory to obtain similar expressions for D_T and D_L, but with $\Delta_T = 0$. Robson estimated Δ_L by using Maxwell-model formulas with a rigid-sphere interaction; unfortunately, this estimate is higher than the results in Table 6-4-1 by about a factor of 2.

There remains the perpetual problem of the estimation of the ion temperatures T_T and T_L. All the moment theories yield an expression of the form of (6-4-20), which allows T_T and T_L to be calculated from measured values of T, v_d, and K' provided that the quantities $\zeta_{T,L}$ and $\beta_{T,L}$ are known. Since we are parameterizing a three-temperature theory, we should use (6-4-17) for $\zeta_{T,L}$. It is possible to derive expressions for $\beta_{T,L}$ in the three-temperature theory, analogous to (6-1-44) and (6-1-45) of the one- and two-temperature theories, but the results are not quite accurate enough to be satisfactory for use in an improved GER. They are somewhat poorer than the numerical results given in Tables 6-1-4 and 6-1-5 for the one- and two-temperature theories, for example. In such a situation it is probably better tactics not to attempt a still more elaborate theoretical approximation, but instead to follow Skullerud (1976) and set up a parameterization scheme for $\beta_{T,L}$. The considerations that guide such a scheme are much like those cited above for $\Delta_{T,L}$, and it turns out the results depend primarily on the mass ratio and only weakly on the potential. The resulting set of numerical values recommended by Skullerud are listed in Table 6-4-1 along with the values of $\Delta_{T,L}$.

Table 6-4-1   Recommended Values of Parameters to be Used for the Calculation of D_T and D_L by Generalized Einstein Relations

$\dfrac{m}{m+M}$	Δ_T	Δ_L	β_T	β_L
0	$\frac{2}{3}K'(2+K')^{-1}$	$\frac{1}{2}K'(2+K')^{-1}$	0	0
0.1	-0.050	0.028	0	0.06
0.2	0.051	0.128	0	0.22
0.3	0.124	0.183	0	0.39
0.4	0.161	0.187	0	0.57
0.5	0.164	0.169	0	0.72
0.6	0.142	0.140	0	0.82
0.7	0.109	0.101	0	0.90
0.8	0.070	0.060	0	0.96
0.9	0.003	0.025	0	0.99
1.0	0	0	0	1.00

Thus the calculation procedure for T_T and T_L is essentially that proposed by Skullerud and described in Section 5-2C, but with appropriate values of the quantity $\tilde{A}$ in place of a rough value for $Q^{(2)}/Q^{(1)}$. That is, (6-4-17) is used for $\zeta_{T,L}$, and Table 6-4-1 is used for $\beta_{T,L}$. It remains to find $\tilde{A}$, a dimensionless ratio of three-temperature collision integrals defined by (6-4-18). Given the cross sections $Q^{(1)}$ and $Q^{(2)}$ for the ion-atom interaction, it is a straightforward process of numerical integration to evaluate $\tilde{A}$, but this is very inconvenient. Because $\tilde{A}$ is a ratio, it tends to be insensitive to details, so we approximate the anisotropic averages of the three-temperature $[p, q]^{(l)}$ by isotropic averages that correspond to the two-temperature collision integrals $\bar{\Omega}^{(l,s)}$, which have been extensively tabulated (Viehland, et al., 1975). This yields

$$\tilde{A} \equiv [0, 0]^{(2)}/[0, 0]^{(1)} \approx \tfrac{2}{3}\bar{\Omega}^{(2,2)}/\bar{\Omega}^{(1,2)}, \qquad (6\text{-}4\text{-}23)$$

which can be put into the following convenient form by the use of the recursion relations given in (6-1-107) and (6-2-18):

$$\tilde{A} \approx \frac{4}{5} A^* \left[1 - \frac{1}{5}\left(\frac{T_{\text{eff}}}{T_{\text{eff}} - T} \right) \frac{K'}{1 + K'} \right]^{-1}. \qquad (6\text{-}4\text{-}24)$$

This result is exact for inverse-power potentials if $T_{\text{eff}} \gg T$, and yields (6-4-19). For alkali ions in noble gases Waldman and Mason have given numerical values of $\tilde{A}$ as a function of a dimensionless effective temperature $\tilde{T}_{\text{eff}}$, defined as the ratio of $(kT + \tfrac{1}{3}Mv_d^2)$ to its value at the mobility maximum. These values are summarized in Table 6-4-2.

The accuracy of these parameterized formulas for T_T and T_L is tested for inverse-power potentials in Tables 6-4-3 and 6-4-4, using the accurate results of Skullerud (1976). The agreement is quite good, and is slightly better than the corresponding results for the one- and two-temperature theories given in Tables 6-1-4 and 6-1-5. This is only to be expected, inasmuch as the $\beta_{T,L}$ were adjusted for a good fit to $T_{T,L}$.

The mean ion energy can be calculated in a similar way,

$$\tfrac{1}{2}m\langle v^2 \rangle = \tfrac{3}{2}kT + \tfrac{1}{2}(m + M)v_d^2(1 + \beta K'). \qquad (6\text{-}4\text{-}25)$$

Consistency requires that $\zeta_L + 2\zeta_T = 1$ and that

$$\beta = \frac{M}{m + M}(\zeta_L \beta_L + 2\zeta_T \beta_T), \qquad (6\text{-}4\text{-}26)$$

so that β is specified by the numerical values of β_T and β_L given in Table 6-4-1. The accuracy for inverse-power potentials is illustrated in Table 6-4-5. It is somewhat better than that of the one- and two-temperature theories shown in Table 6-1-3.

**Table 6-4-2   Recommended Values of the
Parameter $\tilde{A}$ for Alkali Ion-Noble Gas Systems as
a Function of $\tilde{T}_{\text{eff}}$, the Ratio of $(kT + \frac{1}{3}Mv_d^2)$ to Its
Value at the Mobility Maximum (Waldman and
Mason, 1981)**

$\tilde{T}_{\text{eff}}$	$\tilde{A}$
0	0.696
0.08	0.767
0.1	0.773
0.2	0.812
0.4	0.886
0.6	0.916
0.8	0.927
1.0	0.930
1.5	0.916
2.0	0.896
4.0	0.862
6.0	0.848
8.0	0.842
10.0	0.840
∞	0.806

Table 6-4-3   Accuracy of the Kinetic-Theory Formula (6-4-20) for T_T in a Cold Gas[a]

$\dfrac{m}{M}$	$kT_T[\zeta_T Mv_d^2(1 + \beta_T K')]^{-1}$			
	$n = 4$	$n = 8$	$n = 12$	$n = \infty$
0	1	1.0202	1.0324	1.0606
0.1	1	1.01	1.02	1.067
0.2	1	1.010	1.021	1.068
0.5	1	0.998	1.005	1.044
0.8	1	0.988	0.991	1.023
1.0	1	0.982	0.983	1.004
1.5	1	0.976	0.973	0.984
2.0	1	0.978	0.978	0.975
3.0	1	0.984	0.984	0.982
4.0	1	0.991	0.983	1.004
∞	1	1.0000	1.0000	1.0000
$\langle|\text{dev}|\rangle$	0.0%	1.3%	1.7%	3.0%

[a]The value of ζ_T is given by (6-4-17a) and the value of β_T is from
Table 6-4-1. These results should be compared with those of the
one- and two-temperature theories in Table 6-1-4.

Table 6-4-4   Accuracy of the Kinetic-Theory Formula (6-4-20) for T_L in a Cold Gas[a]

$\dfrac{m}{M}$	$kT_L[\zeta_L M v_d^2(1 + \beta_L K')]^{-1}$			
	$n = 4$	$n = 8$	$n = 12$	$n = \infty$
0	1	1.0202	1.0324	1.0606
0.1	1	1.00	1.02	1.007
0.2	1	0.999	1.002	0.997
0.5	1	0.999	0.997	0.981
0.8	1	0.999	0.995	0.969
1.0	1	1.001	0.993	0.970
1.5	1	0.999	0.988	0.965
2.0	1	1.001	0.994	0.976
3.0	1	1.002	0.997	0.994
4.0	1	1.000	1.002	0.980
∞	1	1.0000	1.0000	1.0000
$\langle|\text{dev}|\rangle$	0.0%	0.3%	0.8%	2.1%

[a]The value of ζ_L is given by (6-4-17b) and the value of β_L is from Table 6-4-1. These results should be compared with those of the one- and two-temperature theories in Table 6-1-5.

Table 6-4-5   Accuracy of the Kinetic-Theory Formula (6-4-25) for the Mean Ion Energy in a Cold Gas[a]

$\dfrac{m}{M}$	$m\langle v^2\rangle[(m + M)v_d^2(1 + \beta K')]^{-1}$			
	$n = 4$	$n = 8$	$n = 12$	$n = \infty$
0	1	1.0202	1.0324	1.0606
0.1	1	1.01	1.02	1.038
0.2	1	1.004	1.011	1.032
0.5	1	0.998	1.000	1.011
0.8	1	0.995	0.996	0.998
1.0	1	0.995	0.994	0.994
1.5	1	0.994	0.993	0.991
2.0	1	0.995	0.994	0.993
3.0	1	0.998	0.998	0.997
4.0	1	0.999	0.999	0.999
∞	1	1.0000	1.0000	1.0000
$\langle|\text{dev}|\rangle$	0.0%	0.5%	0.8%	1.5%

[a]The value of β is given in terms of β_T and β_L by (6-4-26). These results should be compared with those of the one- and two-temperature theories in Table 6-1-3.

Table 6-4-6 Accuracy of the Improved Generalized Einstein Relation (6-4-22a) for D_T in a Cold Gas[a]

$\dfrac{m}{M}$	Approximate/Accurate			
	$n = 4$	$n = 8$	$n = 12$	$n = \infty$
0	1	1.002	1.004	1.007
0.1	1	1.02	0.97	0.950
0.2	1	1.01	0.965	0.914
0.5	1	0.974	0.906	0.864
0.8	1	0.968	0.937	0.857
1.0	1	0.973	0.946	0.875
1.5	1	0.980	0.967	0.910
2.0	1	0.990	0.974	0.939
3.0	1	0.987	0.973	0.953
4.0	1	0.987	0.984	0.946
∞	1	1.000	1.000	1.000
$\langle|\mathrm{dev}|\rangle$	0.0%	1.6%	3.5%	7.3%

[a] The value of T_T is from (6-4-20) and the parameters Δ_T and β_T from Table 6-4-1.

Table 6-4-7 Accuracy of the Improved Generalized Einstein Relation (6-4-22b) for D_L in a Cold Gas[a]

$\dfrac{m}{M}$	Approximate/Accurate			
	$n = 4$	$n = 8$	$n = 12$	$n = \infty$
0	1	0.994	0.962	0.979
0.1	1	1.03	0.98	0.961
0.2	1	1.01	0.982	0.959
0.5	1	1.005	1.010	0.996
0.8	1	1.000	1.000	1.016
1.0	1	1.000	0.997	1.013
1.5	1	1.001	1.003	1.012
2.0	1	0.997	0.998	1.005
3.0	1	0.998	1.000	0.993
4.0	1	0.999	0.998	1.006
∞	1	1.000	1.000	1.000
$\langle|\mathrm{dev}|\rangle$	0.0%	0.5%	0.9%	1.5%

[a] The value of T_L is from (6-4-20) and the parameters Δ_L and β_L from Table 6-4-1.

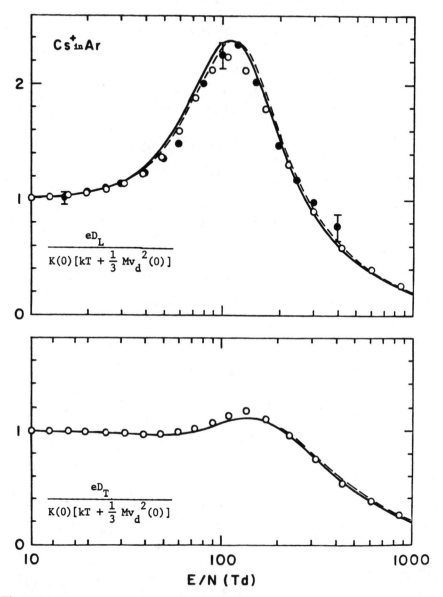

Figure 6-4-1. Reduced diffusion coefficients for Cs^+ in Ar ($m/M = 3.3$) as a function of E/N at 300 K. The solid curves represent the generalized Einstein relations from a full Waldman–Mason calculation, and the dashed curves represent the parameterized results from (6-4-20) and (6-4-22). The filled circles are experimental data from the survey of Ellis et al. (1978), with representative error bars shown. The open circles are computer-generated data based on an ion-atom potential from the inversion of mobility data by Viehland and Lin (1979).

We next test the overall accuracy of the improved GER for inverse-power potentials in Tables 6-4-6 and 6-4-7. That is, D_T and D_L are calculated from (6-4-22), with T_T and T_L calculated from (6-4-20), using the parameters from Table 6-4-1. From the results in these tables we can estimate that both D_T and D_L can be calculated from the improved GER within about 5 to 10% error. This is a modest improvement over the simple GER with ion temperatures from the two-temperature theory, for which errors of up to 10% in D_T and 20% in D_L were estimated in Section 6-2B3.

Finally, we illustrate the accuracy of the improved GER for the real system Cs^+ in Ar in Fig. 6-4-1. The experimental data consist of direct measurements of D_L (Ellis et al., 1978) and of computer-generated values of D_L and D_T based on an ion-atom potential obtained by the inversion of mobility data (Viehland and Lin, 1979). The solid curves are the results of a full Waldman–Mason calculation, using fixed values of $\tilde{A} = 0.85$ and $\tilde{F} = 1.20$. The dashed curves are the results of the parameterized calculation, using $\tilde{A} = 0.85$. The differences are trivial, and the agreement with the data is quite good. The diffusion coefficients have been reduced with $K(0)$ and $v_d(0)$, the mobility and drift velocity at zero field strength and temperature $T = 300$ K, in order to remove their approximate quadratic dependence on E/N. Results for other alkali ion-noble gas systems are quite similar (Waldman and Mason, 1981).

We conclude that the parameterized version of the improved GER enables D_T and D_L to be calculated from mobility data within the accuracy to which they are usually measured experimentally (5 to 10%). In cases where the parameterization may be questioned, the full Waldman–Mason calculation should be used. For the highest possible accuracy, it would be necessary to resort to a full three-temperature moment calculation, and proceed through the ion-atom potential obtained by inversion of the mobility data. Accurate diffusion measurements can, in fact, go beyond the GER and give some important information on the potential, as shown by the work of Skullerud et al. (1986) on Li^+ in He.

C.   Effect of Resonant Charge Exchange

It has already been pointed out in Section 6-1D that resonant charge exchange strongly affects ion-atom scattering by converting the center-of-mass deflection angle from θ to $\pi - \theta$. As a result, $Q^{(1)}$ is greatly increased and the temperature (and field) dependence of the mobility is drastically altered (Fig. 6-1-6). At high electric fields another effect also arises. An ion loses most of its energy after a charge-exchange collision, and regains it from the electric field starting nearly from rest. The ion velocity distribution function thus has its maximum near the origin and has a long tail extending in the field direction. Moreover, the ion kinetic energies parallel and perpendicular to the field differ substantially because energy transfer perpendicular to the field by collisions is difficult, and as a consequence the distribution function is strongly anisotropic (Skullerud, 1969).

It is therefore relevant to be concerned whether the GER are significantly altered by either or both of these effects of resonant charge exchange.

The problem has been studied by Waldman et al. (1982), who used as a guide some extreme models of charge exchange for which accurate results were known. The first question is whether a conventional moment solution still converges adequately in the presence of resonant charge exchange. This question had been examined earlier for the two-temperature theory by Lin and Mason (1979), who found that convergence for mobility and ion temperatures could still be obtained if care was taken with the truncation scheme; they did not examine ion diffusion, however. A similar result for the three-temperature theory, including ion diffusion, was found by Waldman et al. (1982); they also found that the improved GER were still adequate. However, the parameterized version of the GER needed refinement, largely because the parameter $\tilde{A}$ approaches zero for extreme resonant charge exchange. This exposes the fact that the parameters $\beta_{T,L}$ and $\Delta_{T,L}$ do not depend solely on the mass ratio, which is fixed at unity for resonant charge exchange, but depend somewhat on the ion-atom interaction. An adequate parameterization can be obtained by letting $\beta_{T,L}$ depend on $\tilde{A}$, and $\Delta_{T,L}$ depend on $\tilde{A}$ and K'. Other than that, the recipe for the parameterized GER is the same with resonant charge exchange as without. In particular, D_T and D_L are still obtained from (6-4-22), and T_T and T_L are still obtained from (6-4-20), where ζ_T and ζ_L are given by (6-4-17) with $m = M$. However, the parameters $\beta_{T,L}$ that appear in (6-4-20) are now to be taken from Table 6-4-8 rather than from Table 6-4-1. The values in Table 6-4-8 were selected empirically to reproduce known accurate results for some extreme models of charge exchange.

The correction terms $\Delta_{T,L}$ were similarly adjusted empirically to fit known results. The procedure is to first evaluate the moments $\langle \psi_{201} \rangle^{(0)}$, and $\langle \psi_{003} \rangle^{(0)}$ according to the Maxwell model to find expressions for $\Delta_{T,L}^{\text{Max}}$, and then to apply an empirical correction term involving K', as follows:

$$\Delta_{T,L} = \Delta_{T,L}^{\text{Max}}(1 + K'), \qquad (6\text{-}4\text{-}27)$$

where

$$\Delta_T^{\text{Max}} = \frac{1}{5}\frac{Mv_d^2}{kT + \zeta_T Mv_d^2}\left(\frac{2}{4 + 3\tilde{A}}\right)\left(\frac{24 - 5\tilde{A}^2}{2 + \tilde{A}} - \frac{96}{3 + 9\tilde{A} + 5\tilde{F}}\right), \qquad (6\text{-}4\text{-}28a)$$

$$\Delta_L^{\text{Max}} = \frac{Mv_d^2}{kT + \zeta_L Mv_d^2}\left(\frac{1}{2 + \tilde{A}}\right)\frac{4 - \tilde{A} - \tilde{A}^2}{4 + 3\tilde{A}}. \qquad (6\text{-}4\text{-}28b)$$

Notice that these formulas contain, in addition to $\tilde{A}$, a similar ratio $\tilde{F}$, defined as

$$\tilde{F} \equiv [0,0]^{(3)}/[0,0]^{(1)}. \qquad (6\text{-}4\text{-}29)$$

To complete the parameterization scheme we thus need to estimate $\tilde{A}$ and $\tilde{F}$.

Table 6-4-8   Recommended Values of the Parameters $\beta_{T,L}$ as a Function of $\tilde{A}$, to Be Used for the Calculation of T_T and T_L with Resonant Charge Exchange

$\tilde{A}$	β_T	β_L
0	0	0.877
0.2	0	0.848
0.4	0	0.816
0.6	0	0.779
0.8	0	0.736
1.0	0	0.684

In the absence of charge exchange it was possible to take $\tilde{A}$ as a function of a single reduced temperature, $\tilde{T}_{\text{eff}}$ (Table 6-4-2), or even as a constant, and a similar procedure would also work for $\tilde{F}$. But this procedure breaks down in the presence of resonant charge exchange for the following reason. A deflection angle of θ is converted by resonant charge exchange into an apparent deflection of $\pi - \theta$, so that glancing collisions ($\theta \approx 0$) are converted into apparently nearly head-on collisions ($\theta \approx \pi$). The effect on the collision integrals $[p, q]^{(l)}$ and the cross sections $Q^{(l)}$, defined in (6-3-18) and (6-3-19), respectively, is quite different for odd l and even l, owing to the weight factor $(1 - \cos^l\theta)$ in the integrands. If θ is converted to $\pi - \theta$, then $\cos^l\theta$ is converted to

$$\cos^l\theta \to \cos^l(\pi - \theta) = (-1)^l \cos\theta, \tag{6-4-30}$$

and charge exchange as such has no effect if l is even, but a marked effect if l is odd. In more physical language, cross sections with even l give zero weight to both glancing and head-on collisions, so that conversion of one to the other by charge exchange is unimportant, whereas cross sections with odd l give zero weight to glancing collisions but maximum weight to head-on collisions. Thus the important parameter $\tilde{A}$, being a ratio of an $l = 2$ collision integral to an $l = 1$ collision integral, is affected strongly by resonant charge exchange, but the less important $\tilde{F}$, being a ratio involving $l = 3$ and $l = 1$, is affected very little. For $\tilde{F}$ it is thus adequate to take an isotropic average and approximate it by a ratio of two-temperature collision integrals,

$$\tilde{F} \approx \bar{\Omega}^{(3,3)}/\bar{\Omega}^{(1,3)}, \tag{6-4-31}$$

as was done for $\tilde{A}$ in the absence of charge exchange in (6-4-23). The numerical range of $\tilde{F}$ turns out to be very small: it is 1.000 for rigid spheres, is 1.154 for an r^{-4} polarization potential, and has a maximum value of about 1.30 for alkali ion–noble gas systems. Waldman et al. (1982) recommend a fixed value of $\tilde{F} = 1.15$ as satisfactory for use in the parameterized GER.

The estimation of $\tilde{A}$ is, however, more difficult, and requires some additional information. From (6-4-23) and (6-4-24) we see that we need the collision integrals $\bar{\Omega}^{(2,2)}$ and $\bar{\Omega}^{(1,2)}$ as a function of T_{eff}. The latter is related by the recursion relation (6-1-107a) to $\bar{\Omega}^{(1,1)}$, which can be estimated directly from the measured mobility by the first-approximation formula (6-2-12), but $\bar{\Omega}^{(2,2)}$ cannot be estimated solely from mobility data for the following reasons. Resonant charge exchange always involves *pairs* of potential energy curves of *gerade-ungerade* (g, u) symmetry. As is shown in Section 6-5 in detail, the cross sections for odd l are related to the charge-exchange cross section, which depends primarily on just the energy *difference* between the (g, u) potential pairs. The cross sections for even l, however, must be found by calculating a cross section for every individual potential energy curve, and then averaging according to the statistical weights of the curves. In other words, mobility data with resonant charge exchange give information only on the *splitting* between (g, u) potentials, and this is insufficient information for the estimation of $\bar{\Omega}^{(2,2)}$. We must, therefore, have some estimate of the potential curves involved in the scattering, and not just on their differences. Given these potential curves, it is straightforward to find $\bar{\Omega}^{(2,2)}$, at least in principle. It is especially straightforward if the potential curves can be reasonably represented by standard potential models for which the $\bar{\Omega}^{(l,s)}$ have been tabulated numerically.

The accuracy of the foregoing parameterized GER with resonant charge exchange have been tested for some simple models and for He^+ in He (Waldman et al., 1982). The most extreme model is the same as the one discussed in Section 6-2B1, in which half the collisions are strictly forward ($\theta = 0$) and the other half involve resonant charge exchange ($\theta = \pi$). The transport cross sections with odd l are all equal to the geometric cross section, Q_0, and those with even l are all equal to zero. Thus $\tilde{A} = 0$ for this extreme model, which can be solved analytically because the ion motion is essentially one-dimensional (Smirnov, 1967; Fahr and Müller, 1967; Skullerud, 1969). Although this model does not resemble any real physical situation, it can be made more realistic by mixing it with a simple rigid-sphere model without charge exchange, to simulate the three-dimensional scattering of a real case. The transport cross sections for this mixed model are

$$Q^{(l)} = \begin{cases} Q_0 & \text{for odd } l, & (6\text{-}4\text{-}32a) \\ (1 - \delta)Q_0 & \text{for even } l, & (6\text{-}4\text{-}32b) \end{cases}$$

where δ is the fractional contribution of charge exchange. The choice $\delta = 1$ corresponds to extreme one-dimensional charge exchange, and the choice $\delta = 0$ corresponds to isotropic scattering without charge exchange. The relevant parameters for this model are

$$\tilde{A} = \tfrac{2}{3}(1 - \delta), \qquad \tilde{F} = 1, \qquad K' = -\tfrac{1}{2}. \qquad (6\text{-}4\text{-}33)$$

The parameterized GER can then be applied to this model and the results compared to results obtained by other methods.

The comparison is shown in Table 6-4-9. The exact results for $\delta = 1$ were obtained by Skullerud (1969), as were the accurate results for $\delta = 0$ (Skullerud, 1976). Monte Carlo calculations of $T_{T,L}$ for $\delta = 0.25$ and 0.70 were carried out by Lin and Mason (1979). Three-temperature calculations were carried to the fifth approximation by Waldman et al. (1982). The results seem quite reasonable in view of the simplicity of the parameterized GER.

The only real system for which at least a partial test of the parameterized GER is possible is He$^+$ in He. Scattering in this system is described by the two lowest potential curves of He$_2{}^+$, $^2\Sigma_u$ and $^2\Sigma_g$, which were obtained by Sinha et al. (1979) from a variety of experimental and theoretical sources. They used these potentials to calculate the transport cross sections, which they then used in a two-temperature kinetic theory to calculate K_0, T_T, and T_L at two gas temperatures over a range of electric field strengths. Their calculated mobilities agreed very well, within 1.5%, with the experimental measurements of Helm (1976a, 1977). Unfortunately, they did not carry out a full calculation of the diffusion coefficients, but estimated them by means of a two-temperature generalized Einstein relation, so that only their temperatures can be compared with those from the parameterized GER. Waldman et al. (1982) used their potential curves to calculate $\tilde{A}$, took $\tilde{F} = 1.15$, obtained $\beta_{T,L}$ according to Table

Table 6-4-9 Comparison of the Parameterized Generalized Einstein Relations for Resonant Charge-Exchange Models with Other Results[a]

	kT_T/Mv_d^2	kT_L/Mv_d^2	eD_T/Mv_d^2K	eD_L/Mv_d^2K
$\tilde{A} = 0\,(\delta = 1)$				
GER	0	0.562	0	0.211
Exact	0	0.571	0	0.215
$\tilde{A} = 0.2\,(\delta = 0.7)$				
GER	0.087	0.476	0.072	0.184
Three-temp.	0.085	0.503	0.094	0.198
Monte Carlo	0.076 ± 0.002	0.485 ± 0.010		
$\tilde{A} = 0.5\,(\delta = 0.25)$				
GER	0.182	0.382	0.165	0.156
Three-temp.	0.178	0.393	0.166	0.162
Monte Carlo	0.173 ± 0.004	0.389 ± 0.009		
$\tilde{A} = 2/3\,(\delta = 0)$				
GER	0.222	0.343	0.208	0.144
Three-temp.	0.224	0.348	0.216	0.148
Accurate	0.223	0.345	0.212	0.146

[a]See text for details.

Table 6-4-10 Comparison of Ion Temperatures for He$^+$ in He as a Function of Field Strength at 294 K

E/N (Td)	$\tilde{A}$	kT_T(meV) Approx.[a]	Accurate	kT_L(meV) Approx.[a]	Accurate
48.1	0.375	33.1	33.4	56.3	56.4
55.5	0.370	34.9	35.3	63.0	63.4
62.7	0.366	36.7	37.2	69.7	70.3
69.7	0.364	38.6	39.1	76.8	77.3
76.6	0.362	40.4	41.0	83.8	84.2
83.3	0.361	42.3	43.1	91.3	91.0

[a]The approximate values are from the GER as modified for resonant charge exchange by Waldman et al. (1982), and the accurate values are from the kinetic-theory calculations of Sinha et al. (1979).

6-4-8, found K' from their calculated mobilities, and then calculated $T_{T,L}$ according to (6-4-20). The results are shown in Table 6-4-10. The agreement is better than 2% for T_T and better than 1% for T_L. Notice that T_T is much less than T_L.

From the foregoing results we conclude that the parameterized GER can be successful even when resonant charge exchange occurs, provided that reasonable estimates of the interaction potentials are available for the estimation of $\tilde{A}$. Without such information on the potentials, the GER can be used only in a crude way, by making a guess for the value of $\tilde{A}$.

D. Effect of Inelastic Collisions

Only fragmentary results are available for cases where inelastic collisions occur. The simple GER without the $\Delta_{T,L}$ corrections do fairly well in correlating experimental measurements of D_L for atomic ions in simple molecular gases, indicating that inelastic collisions do not have an important effect on diffusion for these systems (McDaniel and Moseley, 1971; Volz et al., 1971; Thomson et al., 1973; James et al., 1975; Pai et al., 1976; Ellis et al., 1976b; Eisele et al., 1977; Thackston et al., 1983). The situation is quite different for electrons in molecular gases, however, where the phenomenon of negative differential conductivity (or mobility) can occur, in which there is a region of E for which

$$\left(\frac{dv_d}{dE}\right)_L < 0. \qquad (6\text{-}4\text{-}34)$$

This phenomenon requires inelastic collisions for its existence. In the simple GER of (6-4-1) and (6-4-2), such a result implies the physically absurd conclusion that $D_L < 0$, leading to a violation of the second law of thermodynamics. The correction factor Δ_L in the improved GER must therefore assume a fundamental role if negative differential conductivity occurs, and cannot be just a minor correction term. However, the corresponding quantity Δ_T is unlikely to play such a role, inasmuch as the transverse differential conductivity, $(dv_d/dE)_T$, represents to first order only a small rotation of the electric field vector and not a change in its magnitude.

A survey of negative differential conductivity and a semiquantitative discussion has been given by Petrović et al. (1984), in terms of simple models for elastic and inelastic cross sections and the "two-term approximation" discussed in Section 5-3D4a. This is sufficient to establish the main physical features needed for the phenomenon to occur. Inelastic collisions are always absolutely necessary (but of course not sufficient), and the phenomenon is enhanced if the elastic momentum-transfer cross section increases with relative energy while the inelastic cross section decreases, although other possibilities exist.

Somewhat surprisingly, the inclusion of just that part of Δ_L which appears in the momentum-transfer calculation of Robson (1976) seems sufficient to give a fair account of D_L in the region of negative differential conductivity. This part presumably corresponds to the term involving the moment $\langle \psi_{003} \rangle^{(0)}$ appearing in (6-4-15). This success was first noticed by Lin et al. (1979b) in some calculations on electrons in methane, and recently a more thorough treatment of the GER and negative differential conductivity based on momentum-transfer theory has been given by Robson (1984). The Robson expression for Δ_L is

$$\Delta_L = \frac{Q}{2kT_L v_d}, \tag{6-4-35}$$

where Q is the heat flux per ion (really just a higher moment of the ion velocity distribution), evaluated for the spatially homogeneous reference state. Lin et al. merely expressed the moments in (6-4-35) in terms of their counterparts in the two-temperature theory to obtain

$$\Delta_L = \frac{(\frac{5}{2}T_b - \frac{3}{2}T_{ion} - T_L)\langle \psi_{10}^{(0)} \rangle^{(0)} - T_b \langle \psi_{10}^{(1)} \rangle^{(0)}}{2T_L \langle \psi_{10}^{(0)} \rangle^{(0)}}, \tag{6-4-36}$$

where

$$T_{ion} = m\langle v^2 \rangle/3k = T_b[1 - \tfrac{2}{3}\langle \psi_{00}^{(1)} \rangle^{(0)}]. \tag{6-4-37}$$

They then used the two-temperature moments they had already calculated for e^- in CH_4 in this formula, with the results shown in Fig. 6-4-2. Also shown in the figure are the accurate values of D_L obtained in the same moment calculation, a curve representing experimental results, a single point from a Monte Carlo calculation that was run as an independent check, and several points calculated by Kleban and Davis (1978) from an iterative numerical solution of the Boltzmann equation. Calculations of Kleban and Davis at higher values of E/N

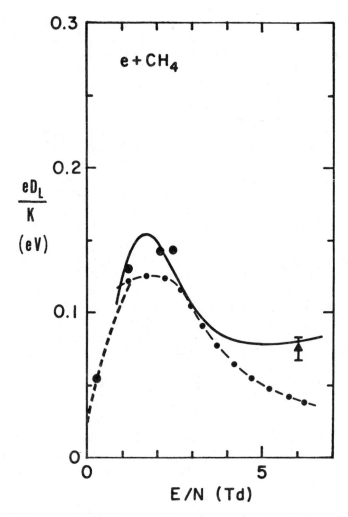

Figure 6-4-2. Longitudinal diffusion coefficient of electrons in methane as a function of E/N in the region of negative differential conductivity. The dot-dash curve represents the GER with Δ_L from (6-4-36). The dashed curve represents experimental results, and the solid curve represents the moment calculations of Lin et al. (1979b). The circles represent the calculations of Kleban and Davis (1978), and the triangle represents the Monte Carlo calculation of Lin et al. (1979b).

have been omitted as probably suffering from numerical convergence errors (Braglia, 1981). The qualitative success of (6-4-36) is surprising, inasmuch as Δ_L single-handedly has to rescue the whole effect from physical absurdity. This success, plus some similar results by Robson (1984), gives hope that the GER might be extendable to include inelastic collisions.

In summary, a simple momentum-transfer calculation is the basis for the best theoretical discussion to date of the role of inelastic collisions in the GER

(Robson, 1976, 1984; Lin et al., 1979b). Unfortunately, the results show only qualitative accuracy, but they are good enough to encourage hope that a more quantitative treatment could be developed. The method used by Waldman and Mason (1981) for elastic collisions, discussed in Section 6-4A, might be extended to include inelastic collisions, but this would require starting over from scratch using the theory for polyatomic species (Viehland et al., 1981), which is discussed in Section 6-6. In addition, the momentum-transfer theory suffers from the usual lack of any independent way of determining T_T and T_L short of a full moment calculation. This lack might be remedied through the results of Viehland et al. (1981) for polyatomic species, but much work clearly remains to be done.

6-5. RESONANT CHARGE AND ION EXCHANGE

When ions move in their parent gas, an ion and a neutral can interchange roles by the resonant transfer of an electron. This resonant charge exchange is usually so probable that it dominates all other elastic scattering processes, except at very low energies where the long-range ion-induced dipole scattering eventually dominates. The result is that the collision dynamics is very different, and the temperature and field dependences of the mobilities and diffusion coefficients are greatly changed from what they would have been in the absence of charge exchange. These effects arise mainly through the influence of charge exchange on the momentum-transfer cross section $Q^{(1)}$, and have been briefly illustrated in Section 6-1D and Fig. 6-1-6. They do not affect the kinetic-theory part of the discussion, they only change the cross sections and collision integrals.

At high fields another effect arises, in which the ion velocity distribution function becomes strongly distorted from a Maxwellian form and highly anisotropic. This affects the kinetic theory itself, in particular the convergence of the successive moment approximations, and has already been mentioned in this connection in Sections 6-2B1 and 6-4C.

The transfer of an ionic fragment during an ion-neutral collision can also have the effect of interchanging the roles of the colliding partners, and has been briefly illustrated in Section 6-1D and Fig. 6-1-6. Unfortunately, the theoretical treatment of ion transfer is much more complicated than that of electron transfer, and no useful results yet exist.

In this section we first treat the convergence problem at high fields, since it has essentially been solved in a simple way. We then discuss resonant electron transfer in some detail. This discussion is concerned entirely with the description and calculation of the cross sections. Finally, we give some comments on ion transfer, largely phenomenological.

A. Convergence at High Fields

The key to satisfactory convergence at high fields in the presence of resonant charge exchange lies in a careful choice of basis temperature. By choosing T_b so that the energy-balance moment equation always retained its first-approximation form, Lin and Mason (1979) found that satisfactory convergence could be obtained for the mobility and ion temperatures with the two-temperature theory. A similar result for the three-temperature theory, with $T_b^{(T)}$, $T_b^{(L)}$, and v_{dis} chosen to force the energy- and momentum-balance moment equations to retain their first-approximation forms, was found to hold for the mobility, ion temperatures, and diffusion coefficients in the presence of resonant charge exchange by Waldman et al. (1982). However, forcing this constraint in *all* approximations seems not to be crucial in three-temperature calculations (Viehland and Hesche, 1986).

In other words, resonant charge exchange forces one to be careful with the truncation-iteration scheme at high field strengths, but is otherwise manageable by moment methods.

B. Electron Transfer

First we give a simple but substantially correct description of ion mobility and diffusion with resonant charge exchange, in which the results appear in a physically transparent form. Then we discuss the special nature of the ion-neutral interaction when resonant electron transfer is possible; this is a necessary preliminary for the full quantum-mechanical treatment that follows. Finally, we give some methods for obtaining quick estimates of the magnitudes and energy dependences of the relevant cross sections.

1. Semiclassical Description

The basis of this description was first given by Holstein (1952), and elaborated by Mason et al. (1959). The nuclei are assumed to follow classical trajectories, which are not affected by an electron exchange that takes place with a probability P_{ex}. The only effect of exchange is to convert the apparent classical deflection angle of the ion from θ to $\pi - \theta$, so that a glancing collision is converted into an apparent head-on collision. In this picture nothing can be said about P_{ex} except that it may depend on the energy and the deflection angle (or impact parameter); actual values would have to come from experiment or from some quantum-mechanical calculation.

If the probability of conversion of θ to $\pi - \theta$ is P_{ex}, the probability of θ being unaffected is $(1 - P_{\text{ex}})$ and the transport cross sections of (6-3-19) become

$$Q^{(l)}(\varepsilon) = 2\pi \int_0^\pi (1 - P_{\text{ex}})(1 - \cos^l\theta)\sigma(\theta, \varepsilon) \sin\theta \, d\theta$$

$$+ 2\pi \int_0^\pi P_{\text{ex}}[1 - \cos^l(\pi - \theta)]\sigma(\theta, \varepsilon) \sin\theta \, d\theta. \qquad (6\text{-}5\text{-}1)$$

This expression immediately shows the distinct difference of character for even and odd l, which was mentioned in Section 6-4C, since

$$\cos^l(\pi - \theta) = (-1)^l \cos \theta. \qquad (6\text{-}5\text{-}2)$$

Thus for even l the value of P_{ex} cancels out in (6-5-1) and the value of $Q^{(l)}$ is unchanged. But for odd l we obtain, after replacement of the differential cross section $\sigma(\theta, \varepsilon)$ by its equivalent in terms of the impact parameter b, and a little rearrangement,

$$Q^{(l)}(\varepsilon) = 4\pi \int_0^\infty P_{ex}b\, db + 2\pi \int_0^\infty (1 - 2P_{ex})(1 - \cos^l\theta)b\, db. \qquad (6\text{-}5\text{-}3)$$

If exchange is dominant, the second integral in (6-5-3) will be small because $P_{ex} \approx \frac{1}{2}$ for small impact parameters (close encounters) and $\cos \theta \approx 1$ for large impact parameters (distant encounters). If exchange is minor, the first integral will be small and the usual expression for $Q^{(l)}$ without exchange will be recovered. This is what happens at low energies, where the long-range ion-induced dipole interaction keeps θ large out to large enough b for P_{ex} to become small.

The total cross section for charge exchange may be defined as

$$Q_{ex}(\varepsilon) = 2\pi \int_0^\infty P_{ex}b\, db, \qquad (6\text{-}5\text{-}4)$$

a quantity that is experimentally measurable (say, in a beam experiment). Thus when exchange is dominant, (6-5-3) becomes approximately

$$Q^{(l)} \approx 2Q_{ex}, \qquad l \text{ odd}, \qquad (6\text{-}5\text{-}5)$$

a result we have already quoted without proof in Section 5-2F. Since $l = 1$ gives the momentum-transfer or diffusion cross section, it is apparent that the charge-exchange cross section controls ion mobility and diffusion.

The foregoing treatment is a bit oversimplified because the phenomenon of resonant charge exchange is always associated with pairs of ion-neutral potentials, one of each pair corresponding to a total wavefunction that is symmetric under exchange of nuclei and the other to the antisymmetric wavefunction. The deflection angle θ will thus depend not only on ε and b, but also on which potential of the pair happens to control the trajectory in a particular collision. It is easy to show that the transport cross sections in such cases are simple averages of the cross sections for scattering by the individual potentials (Mason et al., 1959). For a single pair of potentials, for example,

$$Q^{(l)} = \tfrac{1}{2}[Q_s^{(l)} + Q_a^{(l)}], \qquad (6\text{-}5\text{-}6)$$

where s refers to the symmetric potential and a to the antisymmetric one. The semiclassical description then proceeds as above to yield

$$Q^{(l)}(l \text{ odd}) = 2Q_{ex} + \pi \int_0^\infty (1 - 2P_{ex})[1 - \tfrac{1}{2}(\cos^l\theta_s + \cos^l\theta_a)]b \, db,$$

(6-5-7a)

$$Q^{(l)}(l \text{ even}) = 2\pi \int_0^\infty [1 - \tfrac{1}{2}(\cos^l\theta_s + \cos^l\theta_a)]b \, db. \qquad (6\text{-}5\text{-}7b)$$

Thus for odd l we still obtain $Q^{(l)} \approx 2Q_{ex}$ even when the multiple potentials are taken into account.

2. *Ion-Neutral Potential*

The potential energy curve for the interaction between an ion A^+ and its parent neutral atom A can be considered to be the energy of the molecule $A^+ + A$ as a function of internuclear separation; that is, $V(r)$ is the eigenvalue for the molecular wavefunction. But if Ψ_A is the wavefunction for the configuration $A^+ + A$, there is another wavefunction Ψ_B corresponding to the configuration $A + A^+$. If the two nuclei are identical, the correct total wavefunctions must be either symmetric or antisymmetric with respect to nuclear interchange:

$$\Psi_s = 2^{-1/2}(\Psi_A + \Psi_B), \qquad (6\text{-}5\text{-}8a)$$

$$\Psi_a = 2^{-1/2}(\Psi_A - \Psi_B). \qquad (6\text{-}5\text{-}8b)$$

These two wavefunctions give rise to two different potential curves, $V_s(r)$ and $V_a(r)$. In briefer language, the exact resonance splits the degenerate potentials corresponding to Ψ_A and Ψ_B into a pair of potentials. Both must be taken into account in a full quantum-mechanical calculation of the collisions between A^+ and A.

As an example, the two curves for $He^+ + He$ are shown in Fig. 6-5-1, based on the results chosen by Sinha et al. (1979) for their mobility calculations. The weak minimum at large r in the $V_a(r)$ curve, due to the polarization potential, does not show on the scale of this figure. For comparison, the potentials for the related but nonresonant systems $H^+ + He$ and $Li^+ + He$ are shown, as obtained from quantum-mechanical calculations (Hariharan and Staemmler, 1976; Kolos, 1976; Kolos and Peek, 1976). The profound effect of exact resonance is apparent.

The foregoing two-state description holds only if the wavefunction Ψ_A is itself nondegenerate. It thus applies only to cases in which both ion and atom are in S states, giving rise to Σ_g and Σ_u molecular states. Examples are H^+ in H, H^- in H, He^+ in He, Li^+ in Li, Na^+ in Na, and so on, but if either one is in a P state or higher, more molecular states are possible and more potential energy curves result. For instance, in the cases of Ne^+ in Ne, Ar^+ in Ar, Kr^+ in Kr, and Xe^+ in

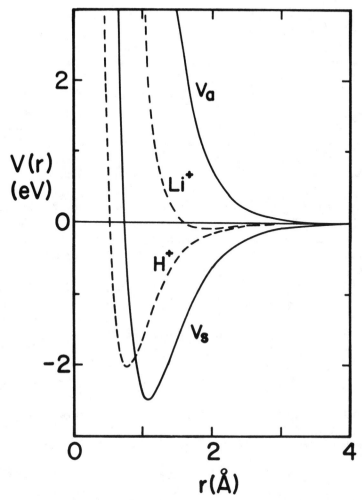

Figure 6-5-1. Interaction of He^+ in He, showing the effect of resonance (solid curves). The dashed curves show the related but nonresonant interactions of H^+ and Li^+ ions with He for comparison.

Xe, the ion is in a ^2P state and the atom in a ^1S state. Four molecular states then arise: $^2\Sigma_g$, $^2\Sigma_u$, $^2\Pi_g$, and $^2\Pi_u$, with four associated potential curves. The determination of the molecular states arising from a given pair of atomic states is a straightforward exercise in the coupling of electronic angular momentum and is summarized by the Wigner–Witmer rules (Herzberg, 1950, Sec. VI, 1).

Since Σ states have zero electronic angular momentum along the internuclear axis and Π, Δ, etc. states have nonzero angular momentum, the Σ states have only half the statistical weight of other states of the same multiplicity, for which the angular momentum vector can point in either direction along the axis.

If spin-orbit coupling of electronic angular momentum is taken into account, the description and determination of the atomic and molecular states becomes more complicated, although the kinetic-theory treatment remains the same. All the noble-gas ions except He^+ show a spin-orbit splitting into a $^2P_{3/2}$ ground state and a $^2P_{1/2}$ metastable excited state. The energy difference ranges from about 0.1 eV for Ne^+ to about 1.3 eV for Xe^+. The molecular states resulting from the interaction of an ion with its parent neutral atom are then indexed by the component of the *total* electronic angular momentum along the internuclear axis, including spin, rather than by the component of just the orbital angular momentum. The quantum number Λ ($\Lambda = 0$ for Σ states, $\Lambda = 1$ for Π states, etc.) is replaced by Ω (Herzberg, 1950, Sec. VI, 1). There are still four molecular states resulting from the interaction of the $^2P_{3/2}$ ground-state ion with its $^1S_0$ parent atom: a pair with $\Omega = \frac{1}{2}$ that corresponds to the original pair of $^2\Sigma$ states, and a pair with $\Omega = \frac{3}{2}$ that corresponds to the original pair of $^2\Pi$ states. The calculation of the cross sections from the potential energy curves proceeds in exactly the same way in either case, and the relative statistical weights of the two pairs are still in the ratio 2:1. In addition, the interaction of the metastable $^2P_{1/2}$ ion with the $^1S_0$ parent atom produces a pair of molecular states with $\Omega = \frac{1}{2}$, for which the calculation of cross sections also proceeds in the same way. The reason for mentioning this matter is not theoretical but experimental, for the two ions can be observed separately in the same experiment if their mobilities are sufficiently different. Helm (1975, 1976b) found that the mobilities of the $^2P_{3/2}$ ions of krypton and of xenon in their parent gases were about 5% lower than those of the $^2P_{1/2}$ ions, and this observation was consistent with the theoretical calculations of Q_{ex} by Sinha and Bardsley (1976). Helm and Elford (1977a) found that the effect was of opposite sign for Ne^+ ions, and decreased with increasing ion energy. This was in qualitative agreement with the theoretical calculations of Cohen and Schneider (1975), although the absolute values of the experimental and theoretical mobilities differed by about 10%. No difference in the mobilities of the $^2P_{3/2}$ and $^2P_{1/2}$ states of Ar^+ in argon was found, however (Helm and Elford, 1977b). Koizumi et al. (1987) found four distinct mobilities for Kr^{2+} in Kr, presumably associated with the four electronic states of Kr^{2+}, namely $^3P_2$, $^3P_1$, $^3P_0$, and $^1D_2$.

Thus spin-orbit effects do not require any changes in the kinetic-theory calculations, once the potential energy curves are given. If, however, one is interested in deducing information about the potential curves from mobility measurements, then it is important to be aware of the spin-orbit effects.

3.   Quantum-Mechanical Description

Since resonant charge exchange involves at least two states, there should be wave-interference effects in the cross sections from scattering by the two states analogous to Newton's rings in optics. There might also be resonances due to the existence of virtual bound states (H_2^+ and He_2^+ are stable molecular ions, for instance), analogous to the orbiting resonances discussed in Section 6-1E.

These effects, as well as the effects of nuclear spin and Bose–Einstein (BE) or Fermi–Dirac (FD) statistics, should all appear in a proper quantum-mechanical description. In addition, we should expect to find the physical simplicities of the semiclassical description contained in the full quantum-mechanical description, preferably in a transparent way.

If we assume that the total Hamiltonian for the system of electrons and nuclei contains only electrostatic terms and that the Born–Oppenheimer separation of nuclear and electronic motion is valid, a general result can be proved fairly straightforwardly (Smith, 1967; Heiche, 1967). This result states that there is no coupling between molecular states of different electronic orbital angular momentum, so that the cross sections are sums of terms, each of which depends only on one g-u pair of molecular states, multiplied by a suitable statistical weight factor. In the case of Ne^+ in Ne, for example, there will be one cross section for the $^2\Sigma_{g,u}$ pair and another for the $^2\Pi_{g,u}$ pair, and the complete cross section will be

$$Q^{(l)} = \tfrac{1}{3}Q_\Sigma^{(l)} + \tfrac{2}{3}Q_\Pi^{(l)}. \qquad (6\text{-}5\text{-}9)$$

We therefore lose no generality by limiting the following discussion to a single g-u pair. A similar result holds when spin-orbit coupling is taken into account.

Since there is no way in principle to tell whether a scattered ion has undergone exchange, the correct scattering amplitude must be a linear combination of $f(\theta)$ and $f(\pi - \theta)$, the sign depending on whether the nuclei are bosons or fermions. For the nuclear-symmetric potential the correct combination is

$$f^s\binom{\text{BE}}{\text{FD}} = 2^{-1/2}[f^s(\theta) \pm f^s(\pi - \theta)], \qquad (6\text{-}5\text{-}10a)$$

where BE and the $+$ sign go together. For the nuclear-antisymmetric potential the result is

$$f^a\binom{\text{BE}}{\text{FD}} = 2^{-1/2}[f^a(\theta) \mp f^a(\pi - \theta)], \qquad (6\text{-}5\text{-}10b)$$

where BE and the $-$ sign go together. In both cases we are neglecting nuclear spin degeneracy for the moment. The total scattering amplitude for the pair of potentials is the sum of the two,

$$f\binom{\text{BE}}{\text{FD}} = 2^{-1/2}\left[f^s\binom{\text{BE}}{\text{FD}} + f^a\binom{\text{BE}}{\text{FD}}\right], \qquad (6\text{-}5\text{-}11a)$$

which can be arranged into the form

$$f\binom{\text{BE}}{\text{FD}} = \tfrac{1}{2}[f^s(\theta) + f^a(\theta)] \pm \tfrac{1}{2}[f^s(\pi - \theta) - f^a(\pi - \theta)]. \quad (6\text{-}5\text{-}11b)$$

If the scattering is strongly peaked around 0 and π, so that interference between the two peaks is negligible, it is reasonable to attribute the forward peak to direct scattering and the backward peak to exchange scattering, and then to write a differential cross section for charge exchange as the square of the second part of the last equation,

$$\sigma_{\text{ex}} = \tfrac{1}{4}|f^s(\pi - \theta) - f^a(\pi - \theta)|^2. \tag{6-5-12}$$

To obtain the expression for the total exchange cross section, we first need the expression for the scattering amplitudes in terms of the phase shifts, which is essentially given by (6-1-67a),

$$f^{s,a}(\theta) = \frac{1}{2i\kappa} \sum_{l=0}^{\infty} (2l + 1)[\exp(2i\delta_l^{s,a}) - 1]P_l(\cos \theta), \tag{6-5-13}$$

where, as usual, $\kappa = \mu v/\hbar$, δ_l^s is the lth partial-wave phase shift for scattering by $V_s(r)$, and δ_l^a is the lth partial-wave phase shift for scattering by $V_a(r)$. On substituting (6-5-13) into (6-5-12) and integrating over all angles to obtain Q_{ex}, we can manipulate the result into the following form (Massey and Smith, 1933):

$$Q_{\text{ex}} = \frac{\pi}{\kappa^2} \sum_{l=0}^{\infty} (2l + 1) \sin^2(\delta_l^s - \delta_l^a). \tag{6-5-14}$$

The expressions for the transport cross sections are a little more complicated, but do not involve any assumption about the separation of a forward and a backward scattering peak. Because of the odd parity of the Legendre polynomials,

$$P_l[\cos(\pi - \theta)] = (-1)^l P_l(\cos \theta), \tag{6-5-15}$$

the expression (6-5-11b) looks like a similar expression for scattering without change exchange, but the even and odd phases must be calculated alternately for the two different potential curves (Massey and Mohr, 1934). The momentum-transfer cross sections for the hypothetical spinless particles under discussion then become

$$Q_0^{(1)}(\text{BE}) = \frac{4\pi}{\kappa^2} \sum_{l=0}^{\infty} (l + 1) \sin^2(\beta_l - \beta_{l+1}), \tag{6-5-16}$$

$$Q_0^{(1)}(\text{FD}) = \frac{4\pi}{\kappa^2} \sum_{l=0}^{\infty} (l + 1) \sin^2(\gamma_l - \gamma_{l+1}), \tag{6-5-17}$$

where

$$\begin{aligned} \beta_l = \delta_l^a \quad &\text{and} \quad \gamma_l = \delta_l^s \qquad \text{for } l \text{ even,} \\ \beta_l = \delta_l^s \quad &\text{and} \quad \gamma_l = \delta_l^a \qquad \text{for } l \text{ odd.} \end{aligned} \tag{6-5-18}$$

The subscript zero on $Q^{(1)}$ emphasizes that the nuclei are imagined to be spinless for the moment. The expressions for $Q^{(2)}$ are

$$Q_0^{(2)}(\text{BE}) = \frac{4\pi}{\kappa^2} \sum_{l=0}^{\infty} \frac{(l+1)(l+2)}{2l+3} \sin^2(\beta_l - \beta_{l+2}), \tag{6-5-19}$$

$$Q_0^{(2)}(\text{FD}) = \frac{4\pi}{\kappa^2} \sum_{l=0}^{\infty} \frac{(l+1)(l+2)}{2l+3} \sin^2(\gamma_l - \gamma_{l+2}). \tag{6-5-20}$$

The characters of $Q^{(1)}$ and of $Q^{(2)}$ can be seen to be quite different. The sum of $Q^{(2)}$ is made up of two types of terms, one depending only on $V_s(r)$ through $(\delta_l^s - \delta_{l+2}^s)$, and the other only on $V_a(r)$ through $(\delta_l^a - \delta_{l+2}^a)$. Nothing resembling interference occurs. The sum for $Q^{(1)}$, however, involves $V_s(r)$ and $V_a(r)$ in the same terms, through $(\delta_l^s - \delta_{l+1}^s)$ and $(\delta_l^a - \delta_{l+1}^a)$, so that interference effects are important. The physical significance of this will become more apparent when we make the connection with the semiclassical results.

Similar expressions that hold for $Q^{(3)}$ and $Q^{(4)}$ are obtained from the expressions (6-1-68c) and (6-1-68d) for no exchange by replacement of δ_l with β_l for BE and γ_l for FD. The difference in character of the transport cross sections depends on whether the index of the cross section is odd or even. (Notice that in this section we are using the cross sections without the normalization factors $\frac{3}{2}$, $\frac{5}{4}$, etc. for even indices; the cross sections that include these factors are denoted by a bar.)

Only minor modification is needed to include the effects of nuclear spin. A nucleus of spin s has $(2s+1)$ states. A system containing two such identical nuclei therefore has $(2s+1)^2$ degenerate states, of which $(s+1)(2s+1)$ are symmetric and $s(2s+1)$ are antisymmetric. The transport cross sections therefore become

$$Q^{(l)}\binom{\text{BE}}{\text{FD}} = \frac{s+1}{2s+1} Q_0^{(l)}\binom{\text{BE}}{\text{FD}} + \frac{s}{2s+1} Q_0^{(l)}\binom{\text{FD}}{\text{BE}}. \tag{6-5-21}$$

This expression is physically transparent, but it gives the impression that the final results depend strongly on the nuclear spin, which is not usually true. A better way to exhibit the effects of spin and statistics is to rewrite (6-5-21) as

$$Q^{(l)}\binom{\text{BE}}{\text{FD}} = \tfrac{1}{2}[Q_0^{(l)}(\text{BE}) + Q_0^{(l)}(\text{FD})] \pm \frac{1}{2(2s+1)}[Q_0^{(l)}(\text{BE}) - Q_0^{(l)}(\text{FD})], \tag{6-5-22}$$

where the upper BE notation goes with the upper $(+)$ sign, as usual. The second term is usually much smaller than the first.

All of the foregoing expressions hold only for Σ, Δ, etc. (Λ even) states of the ion-atom pair. The expressions for Π, Φ, etc. (Λ odd) states are obtained by interchange of β_l and γ_l or, equivalently, of the labels BE and FD (Smith, 1967; Heiche, 1967).

The foregoing formulas for the various cross sections are not in a form that readily suggests relations among them, but the semiclassical description of Section 6-5B1 indicates how to rearrange the exact quantum-mechanical results to exhibit such relations (Heiche and Mason, 1970). Consider first the even-index transport cross sections $Q^{(2)}$, $Q^{(4)}$, etc. According to the semiclassical description, these cross sections are unaffected by exchange, so we expect them to be simple averages of the corresponding cross sections for $V_s(r)$ and $V_a(r)$ individually. Moreover, recalling the semiclassical relations $\kappa b = l + \frac{1}{2}$ and $\theta = 2(d\delta_l/dl)$ given in (6-1-69) and (6-1-71), respectively, we define "angles" as

$$\theta_{ln}^{s,a} \equiv (2/n)(\delta_{l+n}^{s,a} - \delta_l^{s,a}), \tag{6-5-23}$$

just as in (6-1-72). Then from (6-5-22) we see that we can regroup terms in the summations over phase shifts and obtain

$$Q^{(2)} = \tfrac{1}{2}[Q_s^{(2)} + Q_a^{(2)}] + \frac{(-1)^{2s+\Lambda}}{2s+1} \frac{2\pi}{\kappa^2} \sum_{l=0}^{\infty} (-1)^l C_l^{(2)}, \tag{6-5-24}$$

where $\Lambda = 0, 1, 2, \ldots$ for $\Sigma, \Pi, \Delta, \ldots$ states, respectively. The statistics is now taken into account automatically by the $(-1)^{2s}$ factor, since particles with integral s obey BE statistics and those with half-integral s obey FD statistics. (The use of s for nuclear spin should not be confused with the subscript s for "symmetric.") The $Q_{s,a}^{(2)}$ in the first term are given by

$$Q_{s,a}^{(2)} = \frac{4\pi}{\kappa^2} \sum_{l=0}^{\infty} \frac{(l+1)(l+2)}{2l+3} \sin^2\theta_{12}^{s,a}. \tag{6-5-25}$$

Thus we obtain the sort of average expected from the semiclassical description. The spin-dependent second term of (6-5-24) has no simple semiclassical analogue, since it involves spin and since the factor $(-1)^l$ causes the individual terms to oscillate in sign. The expression for $C_l^{(2)}$, however, can be written in a form having a semiclassical appearance:

$$C_l^{(2)} = \frac{(l+1)(l+2)}{2l+1} (\sin^2\theta_{12}^a - \sin^2\theta_{12}^s). \tag{6-5-26}$$

Analogous expressions hold for the other even-index cross sections, and have been given by Viehland and Hesche (1986).

A simple result also holds for P_{ex} and Q_{ex}. Comparing the semiclassical (6-5-4) with the quantum-mechanical (6-5-14), we see that we can define

$$\sin^2(\delta_l^s - \delta_l^a) \equiv \sin^2\zeta_l \equiv P_{ex}, \tag{6-5-27}$$

with Q_{ex} given by

$$Q_{ex} = \frac{2\pi}{\kappa^2} \sum_{l=0}^{\infty} (l + \tfrac{1}{2})P_{ex}. \tag{6-5-28}$$

The results are less obvious for the odd-index transport cross sections, and require some trigonometric manipulation. Substituting for $\theta_{l1}^{s,a}$ and ζ_l, we obtain from (6-5-22) the result

$$Q^{(1)} = \frac{2\pi}{\kappa^2} \sum_{l=0}^{\infty} (l+1)[\sin^2(\zeta_l - \tfrac{1}{2}\theta_{l1}^s) + \sin^2(\zeta_l + \tfrac{1}{2}\theta_{l1}^a)]$$

$$+ \frac{(-1)^{2s+\Lambda}}{2s+1} \frac{2\pi}{\kappa^2} \sum_{l=0}^{\infty} (-1)^l (l+1)[\sin^2(\zeta_l - \tfrac{1}{2}\theta_{l1}^s) - \sin^2(\zeta_l + \tfrac{1}{2}\theta_{l1}^a)]. \qquad (6\text{-}5\text{-}29)$$

With the help of the trigonometric identity,

$$\sin^2(\zeta \pm \tfrac{1}{2}\theta) = \sin^2\zeta + \tfrac{1}{4}(1 - 2\sin^2\zeta)(1 - \cos\theta) \pm \sin^2\zeta \sin\theta, \qquad (6\text{-}5\text{-}30)$$

we can rewrite this expression in the form

$$Q^{(1)} = \frac{4\pi}{\kappa^2} \sum_{l=0}^{\infty} (l+\tfrac{1}{2}) \sin^2\zeta_l$$

$$+ \frac{2\pi}{\kappa^2} \sum_{l=0}^{\infty} (l+\tfrac{1}{2})(1 - 2\sin^2\zeta_l)[1 - \tfrac{1}{2}(\cos\theta_{l1}^s + \cos\theta_{l1}^a)]$$

$$+ \frac{\pi}{\kappa^2} \sum_{l=0}^{\infty} [1 - \tfrac{1}{2}(1 - 2\sin^2\zeta_l)(\cos\theta_{l1}^s + \cos\theta_{l1}^a)]$$

$$- \frac{\pi}{\kappa^2} \sum_{l=0}^{\infty} (l+1)(\sin 2\zeta_l)(\sin\theta_{l1}^s - \sin\theta_{l1}^a)$$

$$- \frac{(-1)^{2s+\Lambda}}{2s+1} \frac{2\pi}{\kappa^2} \sum_{l=0}^{\infty} (-1)^l C_l^{(1)}, \qquad (6\text{-}5\text{-}31)$$

where

$$C_l^{(1)} = \tfrac{1}{2}(l+1)[(1 - 2\sin^2\zeta_l)(\cos\theta_{l1}^s - \cos\theta_{l1}^a)$$

$$+ (\sin 2\zeta_l)(\sin\theta_{l1}^s + \sin\theta_{l1}^a)]. \qquad (6\text{-}5\text{-}32)$$

If we interpret $\sin^2\zeta_l$ as P_{ex} and replace summations by integrations according to the semiclassical relation $\kappa b = (l + \tfrac{1}{2})$, the first two summations in $Q^{(1)}$ give the semiclassical result of (6-5-7a). The third summation is a small piece left over from the first two summations, owing to the difference between $(l+1)$ and $(l + \tfrac{1}{2})$. Since it lacks the factor $(l + \tfrac{1}{2})$, it is probably negligible except at very low energies at which only a few phase shifts contribute. The fourth summation has no obvious physical interpretation, but it has a resemblance to Landau–Zener transition effects coming from the term (Landau and Lifshitz, 1977, p. 349; Mott and Massey, 1965, pp. 352–353),

$$\sin 2\zeta_l = 2[\sin^2\zeta_l(1 - \sin^2\zeta_l)]^{1/2} \to 2[P_{ex}(1 - P_{ex})]^{1/2}. \qquad (6\text{-}5\text{-}33)$$

The last summation, involving $C_l^{(1)}$, involves spin and has no semiclassical analog. Its effect tends to wash out because of the oscillations caused by the $(-1)^l$ factor.

Similar results hold for $Q^{(3)}$, but this cross section appears only in kinetic-theory correction terms and is thus much less important than $Q^{(1)}$. The general expressions for all the odd-index cross sections have been given by Viehland and Hesche (1986).

The behavior of the quantum-mechanical cross sections is illustrated for He$^+$ in He in Fig. 6-5-2. The effect of charge exchange is shown by comparison of $\bar{Q}^{(1)}$ calculated correctly according to (6-5-16) (since $s = 0$ for ^{4}He) with values of $\bar{Q}^{(1)}$ calculated separately for $V_s(r)$ and $V_a(r)$ without charge exchange. The scale factors in the figure are r_e and D_e, the position and depth of the minimum in $V_s(r)$; the potential actually used in the calculations is not as accurate as that shown in Fig. 6-5-1, but it is similar enough to exhibit the features of interest here (Heiche and Mason, 1970). The most striking feature at first sight is the oscillatory nature of the curves. This is due to the presence of orbiting resonances, which correspond semiclassically to the possibility of unstable classical orbits for integral values of orbital angular momentum, as shown in Fig. 6-1-7. The resonances can be seen to coincide approximately with the integral values of l marked at the top of the figure, until for large l they become

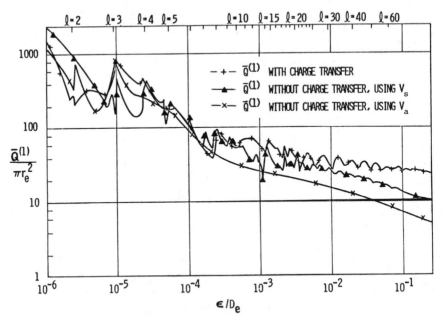

Figure 6-5-2. Increase of the momentum-transfer cross section $\bar{Q}^{(1)}$ by charge exchange, illustrated by calculations for He$^+$ in He (Heiche and Mason, 1970). The symbols merely identify the curves and are not the calculated points. The structures in the curves are orbiting resonances, due mostly to the deep well of the $V_s(r)$ potential.

too close together and tend to wash out. The resonances disappear for the $V_a(r)$ cross section above a relative energy of about $\varepsilon/D_e = 10^{-5}$ but for the other curves persist to much higher energies because of the deep well of the $V_s(r)$ potential.

Disregarding the resonance structure, we see that up to an energy slightly greater than 10^{-4} all three curves are about the same, on the average, because the scattering is dominated by the polarization tail of the potential, which is the same for all three. At higher energies charge exchange begins to dominate, and $\bar{Q}^{(1)}$ with charge exchange becomes more than twice as big as *either* cross section without charge exchange.

The different characteristics of Q_{ex}, $\bar{Q}^{(1)}$, $\bar{Q}^{(2)}$, and $\bar{Q}^{(3)}$ are illustrated in Fig. 6-5-3 for He^+ in He (Heiche and Mason, 1970). A number of features stand out. First, the curves for $\bar{Q}^{(1)}$ and $\bar{Q}^{(3)}$ are similar but that for $\bar{Q}^{(2)}$ is quite different. In particular, $\bar{Q}^{(2)}$ is slightly greater than $\bar{Q}^{(1)}$ and $\bar{Q}^{(3)}$ in the polarization region, which is the normal situation when charge exchange does not occur, but is much smaller and has a markedly different energy dependence in the charge-exchange region. The reason is that charge exchange affects only the odd-index transport cross sections. Second, the resonances are more numerous in Q_{ex} than in $\bar{Q}^{(1)}$ or $\bar{Q}^{(3)}$. This phenomenon has been attributed to the fact that only resonances involving odd l appear in $\bar{Q}^{(1)}$, since $s = 0$ for ^{4}He (Dickinson, 1968a). Third, the approximation $\bar{Q}^{(1)} \approx 2Q_{ex}$ is remarkably good in an average sense, even into

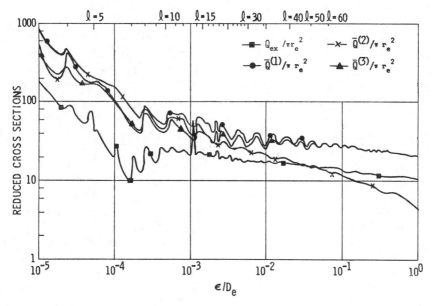

Figure 6-5-3. Charge-exchange cross section Q_{ex} and transport cross sections $\bar{Q}^{(1)}$, $\bar{Q}^{(2)}$, and $\bar{Q}^{(3)}$ calculated for He^+ in He (Heiche and Mason, 1970). The symbols merely identify the curves. The different energy dependence of $\bar{Q}^{(2)}$ is apparent.

the region in which the polarization potential dominates the scattering. The agreement in the polarization region happens to be fortuitous in this case, but it is a general phenomenon in the charge-exchange region.

Similar results hold for heavier systems, such as Cs^+ in Cs, but the oscillatory behavior of the cross sections is much less pronounced (Heiche and Mason, 1970). This is partly due to the fact that heavier systems have shorter de Broglie wavelengths, and partly due to the fact that they also often have nonzero nuclear spins. This allows the resonances for the BE and FD contributions to combine, and they tend to cancel one another out.

The influence of charge exchange on the temperature and field dependence of mobility and diffusion are considered next.

4.   Temperature and Field Dependence of Mobility and Diffusion

The major effect of resonant charge exchange is to change the nature of the temperature and field dependence of mobility and diffusion through the pronounced effect on the cross sections. The effects on the ion velocity distribution function are still satisfactorily described by moment methods (Section 6-5A), so that scaling rules for isotropic masses and the scaling of T and E/N into T_{eff} are still valid. The latter scaling has already been illustrated in Fig. 6-1-6 for He^+ in He.

However, the most prominent features of the cross sections, namely orbiting resonances such as those illustrated in Fig. 6-5-3, are mostly washed out by the integration over energy to form the collision integrals, leaving at most a few gradual undulations. What remains is only the general magnitude and energy dependence of the cross sections. Since charge exchange effectively converts a glancing collision into a head-on collision, the cross section is very large and the mobility correspondingly low. The net result is that the mobility as a function of effective temperature drops from its polarization limit to a much lower value controlled by charge exchange.

This behavior is clearly seen in Fig. 6-1-6 for He^+ in He. The only remnant of the many orbiting resonances is the undulation around 20 K in the dashed curve, which was calculated quantum mechanically by Dickinson (1968a). The solid curve calculated classically by Sinha et al. (1979) does not show this feature. The two curves essentially coincide above about 50 K.

The effect of resonant charge exchange on ion diffusion is somewhat more complicated, and is most easily understood in terms of the generalized Einstein relations (GER) given in (6-4-22). Taking D_T as an example, we write

$$D_T = K \frac{kT_T}{e} \left(1 + \Delta_T \frac{K'}{1 + K'} \right), \tag{6-5-34}$$

where

$$kT_T = kT + \zeta_T M v_d^2 (1 + \beta_T K'), \tag{6-5-35a}$$

$$\zeta_T = \frac{(m + M)\tilde{A}}{4m + 3M\tilde{A}}, \qquad (6\text{-}5\text{-}35b)$$

$$\beta_T \approx 0. \qquad (6\text{-}5\text{-}35c)$$

The main effect on D_T (or D_L) is through K, to which it is proportional, and the behavior of K has already been considered. But another appreciable effect occurs through the ion temperature T_T (or T_L), and a smaller effect, which we ignore in this discussion, is contained in the correction term involving Δ_T (or Δ_L).

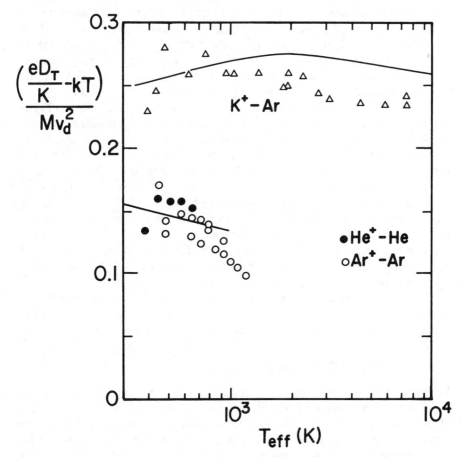

Figure 6-5-4. Effect of resonant charge exchange on ion temperature, illustrated for He^+ in He and Ar^+ in Ar, with the nonresonant system K^+ in Ar shown for comparison. This is a residual plot that is approximately equivalent to $\zeta_T = (m + M)\tilde{A}/(4m + 3M\tilde{A})$ as a function of T_{eff}. The measurements for He^+ in He are taken from Märk and Märk (1984), for Ar^+ in Ar from Sejkora et al. (1984), and for K^+ in Ar from Skullerud (1972, 1973). The curve for K^+ in Ar is based on the values of $\tilde{A}$ in Table 6-4-2, and for He^+ in He on the values of $\tilde{A}$ in Table 6-4-10.

The effect of charge exchange on the ion temperatures is particularly interesting, because in principle it gives different information on the ion-neutral potential than does the mobility. This would be especially valuable when charge exchange occurs because at least two potentials are always involved, and a single experimental quantity (mobility) is insufficient to determine two potentials. Specifically, the extra information enters through the dimensionless ratio $\tilde{A}$, which involves the two cross sections $Q^{(1)}$ and $Q^{(2)}$. Without charge exchange these two cross sections are not greatly different, and the numerical value of $\tilde{A}$ (or of the corresponding two-temperature ratio A^*) is insensitive to details of the potential. But $Q^{(1)}$ and $Q^{(2)}$ differ dramatically in the presence of charge exchange, as was illustrated in Fig. 6-5-3. In qualitative terms, $Q^{(1)}$ measures the splitting between $V_s(r)$ and $V_a(r)$, whereas $Q^{(2)}$ measures the average of the scattering produced by $V_s(r)$ and $V_a(r)$ separately. (This remark is made more quahtitative in the following discussion on the estimation of cross sections.) There is thus some hope of determining both $V_s(r)$ and $V_a(r)$ if both K and D_T (or D_L) are measured as a function of temperature and/or field strength.

The foregoing remarks are illustrated for He^+ in He and Ar^+ in Ar in Fig. 6-5-4, with the nonresonant system K^+ in Ar shown for comparison. The quantity $[(eD_T/K) - kT]/Mv_d^2$ is plotted as a function of $T_{eff} \approx T + Mv_d^2/3k$; according to (6-5-34) and (6-5-35), this is approximately the same as ζ_T vs. T_{eff}. It is clear from this figure that charge exchange has a pronounced effect on the measurements. It is also clear that there is appreciable experimental scatter; this is not so surprising because this is a residual plot that magnifies scatter, but it means that any information on the potentials contained in the measurements is subject to large errors. In addition, the fact that the results for He^+ in He and for Ar^+ in Ar are virtually the same is more bad news for the determination of potentials, because the system He^+-He involves two potentials $(^2\Sigma_{g,u})$ but Ar^+-Ar involves four potentials $(^2\Sigma_{g,u}$ and $^2\Pi_{g,u})$.

In summary, resonant charge exchange normally causes the mobility to decrease with increasing temperature or field strength, and causes an extra decrease in the diffusion coefficients through its influence on ion temperatures.

5. *Estimation of Cross Sections*

In Section 5-2F we discussed how quick estimates of the magnitude and energy dependence of momentum-transfer cross sections could be made. In Section 6-1E we also discussed how quantum-mechanical cross sections went over to their semiclassical and classical limits. Here we give the corresponding discussion for resonant charge exchange, but extra results are obtained because of the importance of two new cross sections, Q_{ex} and $Q^{(2)}$.

We consider three commonly used approximations: the semiclassical, impact-parameter, and random-phase approximations, which are all closely related to approximations discussed for the case of no charge exchange.

In the semiclassical approximations the sums over l are replaced by integrations over dl, small bits left over from the difference between $(l + 1)$ and

$(l + \frac{1}{2})$ are ignored, and the phase shifts are calculated by the JWKB approximation of (6-1-70). The integration over dl destroys the effects of nuclear spin and the difference between Σ, Π, etc., states; that is, the terms involving $C^{(1)}$, $C^{(2)}$, etc., integrate to zero because of sign oscillation caused by the factor $(-1)^l$. The transport cross sections with even index thereby reduce to the completely classical form, just as in the case of no charge exchange. However, the cross sections with odd index do not reduce to a completely classical form; in particular, (6-5-31) does not reduce to (6-5-7a), for the fourth summation in (6-5-31) involving $[(\sin 2\zeta)(\sin \theta^s - \sin \theta^a)]$ survives the passage to the semiclassical limit. The semiclassical approximation is expected to be accurate except for light particles at low energies.

In the impact-parameter approximation the additional step is made of expanding the JWKB expression for the phase shift to obtain (Massey and Mohr, 1934)

$$\delta_l^{s,a} \approx - \frac{\kappa}{2\varepsilon} \int_b^\infty \frac{V_{s,a}(r)}{(r^2 - b^2)^{1/2}} r \, dr, \qquad (6\text{-}5\text{-}36)$$

where $\kappa b = l + \frac{1}{2}$. This expression is closely related to the Kennard small-angle approximation (5-2-130) for the classical deflection angle; they are connected by the semiclassical formula $\theta = 2(d\delta_l/dl)$. The approximation corresponds physically to replacing the actual collision trajectory with a straight line and is thus expected to be accurate only at high energies or large impact parameters. It is often usable, however, at low energies if the large phase shifts corresponding to small impact parameters happen to contribute to the cross section in an essentially random manner. The advantage of this approximation over the semiclassical approximation is that (6-5-36) is much easier to evaluate than the full JWKB expression and can often be integrated in terms of known functions.

The random-phase approximation is a further simplification of the impact-parameter approximation, and is the quantum-mechanical analog of the classical "random-angle" approximation of Section 5-2F. It approximates the integrations over dl (or db), which are all of the form

$$\int_0^\infty F(b) \sin^2 g(b) \, b \, db. \qquad (6\text{-}5\text{-}37)$$

If $g(b)$ is large and varies rapidly with b, then $\sin^2 g(b)$ oscillates rapidly between 0 and 1 and is replaced by its mean value of $\frac{1}{2}$ out to some value b^*, after which it is taken as zero. With the Firsov (1951) choice of b^*,

$$g(b^*) = 1/\pi, \qquad (6\text{-}5\text{-}38)$$

(6-5-37) is replaced by

$$\frac{1}{2} \int_0^{b^*} F(b) b \, db, \qquad (6\text{-}5\text{-}39)$$

and the integration can then usually be done easily. For the transport cross sections, which involve differences of phase shifts with different l values, the further approximation

$$\delta_{l+n} = \delta_l + n \frac{d\delta_l}{dl} + \cdots \qquad (6\text{-}5\text{-}40)$$

is made, which is consistent with taking l as a continuous variable in the integrations.

Comparisons of these approximations with accurate quantum-mechanical calculations for He^+ in He and Cs^+ in Cs have been carried out by Heiche and Mason (1970). Their results can be summarized briefly as follows.

The integration over dl in the semiclassical approximation washes out the structure due to the orbiting resonances for all the cross sections, including Q_{ex}, $Q^{(1)}$, and $Q^{(2)}$, and essentially puts a smooth average curve through the orbiting resonances. The advantage of the integration is that in practice the integrand needs to be evaluated at many fewer values of l than are needed in the summation. When the resonances are small, as for the heavy system Cs^+ in Cs, the semiclassical approximation is almost identical to the full quantum-mechanical results. From these results we can conclude that the semiclassical approximation will almost always give an accurate result for the transport coefficients, with the possible exception of very light systems at very low effective temperatures where a few orbiting resonances may survive as weak undulations (as in Fig. 6-1-6 for He^+ in He).

In the impact-parameter approximation the expressions for the phase shifts can usually be integrated analytically, but the integrations for the cross sections must still be carried out numerically. The agreement with the accurate transport cross sections is still very good, apart from the expected loss of the orbiting resonances, although not quite as good as the full semiclassical approximation. However, a conspicuous failure occurs for Q_{ex} in the low-energy region where the polarization interaction is dominant over the resonance splitting into V_s and V_a potentials (Fig. 6-5-1). This failure is to be expected because Q_{ex} depends only on the difference $(\delta_l^s - \delta_l^a)$, which in the impact-parameter approximation of (6-5-36) depends on $(V_s - V_a)$, so that the polarization potential cancels out.

The random-phase approximation permits the integrations for the cross sections to be carried out analytically. The calculations of Heiche and Mason show that the much simpler random-phase approximation gives results remarkably close to the impact-parameter approximation in most cases.

Finally, detailed examination of the calculations shows that the approximation,

$$Q^{(1)} \approx 2Q_{ex}, \qquad (6\text{-}5\text{-}41)$$

is very good indeed at energies above the region where polarization scattering dominates. Moreover, the location of the transition between dominant polarization and dominant charge exchange is easily located at the intersection of $Q_{pol}^{(1)}(\varepsilon)$ and the random-phase approximation for $Q_{ex}(\varepsilon)$. Nothing peculiar appears to happen in this transition region. These observations can be used as the basis for a simple approximation for the mobility over the entire temperature range corresponding to elastic scattering, as follows.

The polarization potential is

$$V_{pol}(r) = -\frac{e^2\alpha_d}{2r^4}, \tag{6-5-42}$$

where α_d is the dipole polarizability of the neutral. This leads to

$$Q_{pol}^{(1)} = 4.9175(e^2\alpha_d/\varepsilon)^{1/2}, \tag{6-5-43}$$

which in turn leads to

$$K_{pol} = 13.853/(\alpha_d\mu)^{1/2} \text{ cm}^2/\text{V-s}, \tag{6-5-44}$$

where α_d is in Å^3 and μ is the reduced mass in atomic mass units, as already discussed in Section 6-1D. These classical results are essentially exact. We do not have exact numerical results for Q_{ex}, even in the random-phase approximation, because there is no universal expression for the exchange splitting as there is for V_{pol}. However, since this splitting is an exchange effect, it can often be represented over a large range by a simple exponential,

$$V_a(r) - V_s(r) \approx A \exp(-r/a), \tag{6-5-45}$$

where A and a are positive constants. From (6-5-36) the phase-shift differences can then be calculated to be

$$\delta^s - \delta^a \approx (\kappa Ab/2\varepsilon)K_1(b/a), \tag{6-5-46}$$

where $K_1(x)$ is the modified Bessel function of the second kind. Using the random-phase approximation with the asymptotic expression for $K_1(x)$,

$$K_1(x) \sim (\pi/2x)^{1/2}e^{-x}, \tag{6-5-47}$$

we obtain

$$Q_{ex} = \tfrac{1}{2}\pi b^{*2}, \tag{6-5-48}$$

$$(\kappa Ab^*/2\varepsilon)(\pi a/2b^*)^{1/2} \exp(-b^*/a) = 1/\pi. \tag{6-5-49}$$

With $b^* \sim Q_{ex}^{1/2}$ from (6-5-48), we see that (6-5-49) predicts that $Q_{ex}^{1/2}$ vs. $\ln(\varepsilon/Q_{ex}^{1/4})$ should be a straight line, as indeed turns out to be the case in practice. But an even simpler result can be obtained by assuming that the major dependence on b^* in (6-5-49) comes from the exponential term; the energy dependence of Q_{ex} is then given by (Dalgarno, 1958)

$$Q_{ex}^{1/2} = a_1 - a_2 \ln \varepsilon, \tag{6-5-50}$$

where a_1 is a weak (logarithmic) function of b^*. In practice, Q_{ex} can usually be fitted over a wide range of energy by an expression like (6-5-50) with constant a_1 and a_2.

The formula (6-5-50) was previously given in Section 5-2F, where it was noted that the energy dependence was the same as that of $Q^{(1)}$ without charge exchange for an exponential repulsion potential. We can now see that the basic reason for the similar energy dependence is the exponential form of the interaction, and that it will follow from almost any approximation of a dimensional character, whether it be a random-angle, random-phase, or effective collision diameter approximation.

To summarize, (6-5-43) for $Q_{pol}^{(1)}$ and (6-5-50) for Q_{ex} can be combined to give a simple result for the mobility over the entire range of effective temperatures. If a_1 and a_2 are found from experimental results, than $Q^{(1)} \approx 2Q_{ex}$ at high energies and $Q^{(1)} = Q_{pol}^{(1)}$ at low energies; a smooth transition between these two regions (e.g., by a spline fit) can then be integrated to give K_0 over the whole range with an accuracy of the order of about 10%.

Finally, it remains to consider diffusion. For the sorts of approximate calculations we are considering here, the generalized Einstein relations as modified for resonant charge exchange (Section 6-4C) should suffice. The calculation is thereby reduced to the estimation of $\tilde{A}$, which in turn entails the estimation of $\tilde{Q}^{(2)}$. Since charge exchange does not affect the even-index cross sections, we can use one of the methods described in Section 5-2F for estimating the momentum-transfer cross section, Q_D or $Q^{(1)}$. The method of an effective collision diameter is suitable. To review the method, a characteristic distance is chosen at which the potential energy is equal to some definite fraction of the initial relative kinetic energy, and this distance determines the cross section. For the cross section $\bar{Q}^{(2)}$, this estimation scheme takes the form

$$|V(d_2)| = \gamma_2 \varepsilon, \tag{6-5-51a}$$

$$\bar{Q}^{(2)} \approx \pi d_2^2. \tag{6-5-51b}$$

For inverse power potentials of the form $V(r) = \pm C_n/r^n$, this method yields the result

$$\bar{Q}^{(2)}(\text{approx.}) \approx \pi(C_n/\gamma_2^{\pm}\varepsilon)^{2/n}, \tag{6-5-52}$$

with different values of γ_2 for repulsive $(+)$ and attractive $(-)$ potentials. The empirical choices of $\gamma_2^+ = 0.3$ and $\gamma_2^- = 0.2$ give the results in Table 6-5-1, which is a comparison with the accurate values of Higgins and Smith (1968). The agreement is quite good.

Table 6-5-1   Accuracy of the Effective Collision Diameter Approximation for the Cross Section $\bar{Q}^{(2)a}$ Needed to Estimate Ion Temperatures, for the Potentials $V(r) = \pm C_n/r^n$

	$\bar{Q}^{(2)}$(approx.)$/\bar{Q}^{(2)}$	
n	Repulsion	Attraction
2	1.053	1.172
3	1.012	1.011
4	0.986	0.969
6	0.968	0.958
8	0.964	0.968
10	0.964	0.978
25	0.978	1.007
50	0.988	1.009
∞	1.000	1.000

[a]Corresponding results for $\bar{Q}^{(1)}$ appear in Table 5-2-11.

The agreement is also good for the case of an exponential repulsion, $V(r) = V_0 \exp(-r/\rho)$. The results are compared in Table 6-5-2 with the accurate calculations of Monchick (1959). For this potential the ratio $\bar{Q}^{(2)}$(approx.)$/\bar{Q}^{(2)}$ is a function of energy, rather than a constant as is the case for inverse-power potentials.

The reason for taking this much trouble over the estimation of $\tilde{A}$ is that $\bar{Q}^{(2)}$ and $\bar{Q}^{(1)}$ have quite different energy dependences when resonant charge exchange occurs. Thus $\tilde{A}$ depends rather strongly on T_{eff} and must be estimated with some care.

Table 6-5-2   Accuracy of the Effective Collision Diameter Approximation for the Cross Section $\bar{Q}^{(2)a}$ Needed to Estimate Ion Temperatures, for the Potential $V(r) = V_0 \exp(-r/\rho)$

ε/V_0	$\bar{Q}^{(2)}$(approx.)$/\bar{Q}^{(2)}$
0	1.000
10^{-4}	0.959
10^{-3}	0.956
10^{-2}	0.958
10^{-1}	0.986

[a]Corresponding results for $\bar{Q}^{(1)}$ appear in Table 5-2-12.

C. Ion Transfer

Having seen the important effects of resonant charge exchange on ion mobility and diffusion, we may expect that any process by which charge passes easily from an ion to a neutral will qualitatively and quantitatively affect mobility and diffusion. In particular, the transfer of an ionic fragment rather than an electron should be such a process. In this section we present two examples of ion transfer, one well-established and the other probable. Many others are suspected. Unfortunately, no real theory exists in usable form for the description of ion-transfer collisions. Such collisions are rearrangement collisions, indeed chemical reactions, and constitute a much more difficult problem than does electron transfer. With electron transfer the Born–Oppenheimer approximation allows the separation of the electronic and nuclear motions, and reduces the rearrangement to a simple splitting of energy states.

The first example has already been mentioned in Section 6-1D, and concerns the transfer of a He^+ ion during a collision between He_2^+ and He,

$$He_2^+ + He = He + He_2^+, \qquad (6\text{-}5\text{-}53)$$

the process being resonant if the incident He_2^+ is in its ground state ($^2\Sigma_u$). It was long thought that a helium ion of mobility $20 \ cm^2/V\text{-}s$ at room temperature was He_2^+ in its ground state, and theoretical calculations neglecting ion transfer were consistent with this interpretation (Geltman, 1953; Mason and Schamp, 1958). But later work showed that there were two ions of mass 8 having the mobilities of 16 and $20 \ cm^2/V\text{-}s$ (Madson et al., 1965). The now generally accepted explanation of this situation is that the faster ion is a metastable excited state $(He_2^+)^*$, probably the $^4\Sigma_u$ state arising from the interaction of He^+ (^2S) with an excited He (^3S) atom, for which ion transfer to ground-state He (^1S) would be nonresonant (Beaty et al., 1966). The results appear in Fig. 6-1-6, along with those for He^+ in He. They are all reasonably consistent with theoretical calculations except for ground-state He_2^+, for which no calculations exist for the reasons mentioned above. For this system the measurements are distinctly lower than those for $(He_2^+)^*$, and in fact lie below the polarization limit in the temperature range shown.

As an historical aside, the study of the mobility of helium ions in helium constitutes a complicated but interesting story, which has been reviewed by Massey (1971, Sec. 3.2.1). It was originally believed that the fast ion ($20 \ cm^2/V\text{-}s$) was in fact He^+, since this mobility is not far from the polarization limit of $21.7 \ cm^2/V\text{-}s$. But this value turned out to be completely inconsistent with the pioneering calculation of Massey and Mohr (1934) including resonant charge exchange, which gave only $12 \ cm^2/V\text{-}s$. For a time suspicion centered on the quantum theory of atomic collisions, which was still in its infancy. The full story provides some interesting lessons on the difficulties involved in obtaining unambiguous results for ion mobilities.

The second example concerns the mobility of H_3^+ ions in H_2, where a proton transfer is possible:

$$H_3{}^+ + H_2 = H_2 + H_3{}^+. \tag{6-5-54}$$

This is perhaps not exactly resonant because of some mismatch in equilibrium internuclear distances, but it is nevertheless highly probable. Mason and Vanderslice (1959) determined the potential energy between $H_3{}^+$ and H_2, partly from theory and partly from analysis of measurements of the elastic scattering of beams of $H_3{}^+$ ions by H_2 gas, and used this potential to calculate the mobility on the assumption that no ion transfer occurred. They obtained a mobility of about $22 \, cm^2/V\text{-}s$ at room temperature. Varney (1960) suggested that the mobility should be much lower because of proton transfer. Subsequent mass spectrographic identification of the ions involved by Barnes et al. (1961) showed that the hydrogen ion of room-temperature mobility 11 to $12 \, cm^2/V\text{-}s$, which was the one usually observed in mobility experiments, was indeed $H_3{}^+$.

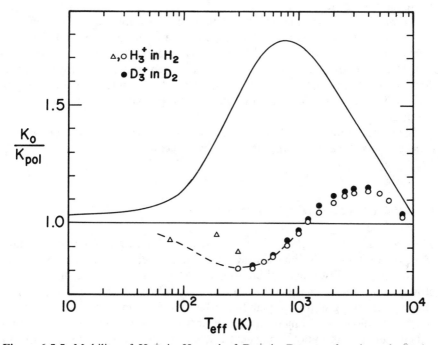

Figure 6-5-5. Mobility of $H_3{}^+$ in H_2 and of $D_3{}^+$ in D_2 as a function of effective temperature, showing the effect of proton or deuteron exchange. The solid curve is calculated from a potential with neglect of ion exchange (Mason and Vanderslice, 1959); the measured result is lower by nearly a factor of 2 at room temperature. The dashed curve has no theoretical status other than to link the measured points with the polarization limit. At very high temperatures the effect of ion exchange appears to diminish and the results approach each other. Notice that the scaling rules still apply. The points are taken from: $\triangle$ Chanin (1961); $\bigcirc$, $\bullet$ Ellis et al. (1976a).

Extensive measurements as a function of E/N were reported a few years later for mass-identified H_3^+ ions in H_2 (Albritton et al., 1968) and for mass-identified D_3^+ ions in D_2 (Miller et al., 1968). These can be used to illustrate how the whole temperature dependence of the mobility is altered by proton (or deuteron) transfer. The results are shown in Fig. 6-5-5. They are analogous to those of Fig. 6-1-6 for He_2^+ in He, but the T_{eff} scale is now logarithmic because a much greater range is covered. The solid curve is a slightly improved version of the Mason and Vanderslice calculation without proton transfer, using an $(8, 6, 4)$ potential with $\varepsilon_0 = 0.0366\,\text{eV}$, $r_m = 2.97\,\text{Å}$, and $\gamma = 0.2$ (Viehland et al., 1975). The experimental points are from the survey by Ellis et al. (1976a), plus a few earlier low-temperature results of Chanin (1961). The polarization limits are $K_{pol} = 14.01\,\text{cm}^2/\text{V-s}$ for H_3^+ in H_2 and $9.91\,\text{cm}^2/\text{V-s}$ for D_3^+ in D_2. Not only is the mobility lower by nearly a factor of 2 at room temperature by virtue of the proton transfer, but the whole temperature dependence of the mobility curve is altered. At very high temperatures it appears that repulsive collisions without ion transfer begin to dominate, and the calculated and experimental results approach each other.

Although proton transfer is a plausible mechanism for the results shown in Fig. 6-5-5, there remains the possibility that the initial drop in mobility with increasing temperature is caused by a strong valence attraction between H_3^+ and H_2. The ion H_5^+ is stable and has been experimentally identified (Saporoschenko, 1965; Massey, 1971, Sec. 3.5.1). Measurements on H_3^+ in D_2 or D_3^+ in H_2 to look for isotope scrambling would probably be required to settle the question.

The effects of ion transfer are thus seen to be analogous to those of electron transfer, even though a quantitative theory is lacking.

6-6. POLYATOMIC SYSTEMS

The two- and three-temperature theories discussed in Sections 6-2 and 6-3 have a major limitation: the ion-neutral interaction potential must be spherically symmetric and all ion-neutral collisions must be elastic. The results are thus strictly valid only for atomic ions (or electrons) in atomic gases, and only at electric field strengths low enough that excitation does not occur. In this section we present the general kinetic theory of drift-tube experiments involving both polyatomic ions and polyatomic neutrals. A restricted preview of such a theory was given for the case of vanishing field strength in Section 6-1-F on the one-temperature theory. The basic viewpoint for arbitrarily strong fields is not essentially different from this, but the methods of solution differ. That is, the Boltzmann equation is replaced by an analogous kinetic equation that takes into account the nonspherical interactions and inelastic collisions, and this new kinetic equation is solved by moment methods similar to those used in the two- and three-temperature theories.

A. Kinetic and Moment Equations

The simplest approach is to use the Wang Chang–Uhlenbeck–de Boer (Wang Chang et al., 1964) equation (WUB equation) in place of the Boltzmann equation. Almost all discussions to date have taken the WUB equation as their starting point (Lin et al., 1979b; Kumar et al., 1980; Kumar, 1980; Viehland et al., 1981; Weinert, 1983). This kinetic equation ignores all quantum-mechanical interference effects and replaces the elastic differential cross section $\sigma(\theta, \varepsilon)$ by an inelastic cross section $\sigma(\alpha\beta, \alpha'\beta'; \theta, \phi, \varepsilon)$ that describes transitions from internal states α and β before collision to internal states α' and β' after collision (Wang Chang et al., 1964). The interference effects arise between degenerate internal states, and the WUB equation discards these effects by the use of degeneracy-averaged cross sections. Each internal quantum state is treated as a separate species in a multicomponent mixture, and is described by its own Boltzmann-like equation. Thus the ions are described by a set of distribution functions $f^{(\alpha)}(\mathbf{r}, \mathbf{v}, t)$, one for each internal state α, and the neutrals by a set $F_j^{(\beta)}(\mathbf{r}, \mathbf{V}, t)$, one for each internal state β of neutral species j. Neglecting as usual all ion-ion interactions, we can write the WUB equation as

$$\frac{\partial f^{(\alpha)}}{\partial t} + \mathbf{v} \cdot \mathbf{V}_r f^{(\alpha)} + \mathbf{a} \cdot \mathbf{V}_v f^{(\alpha)} \equiv \frac{Df^{(\alpha)}}{Dt}$$

$$= \sum_j \sum_{\alpha'} \sum_{\beta\beta'} \iint [f^{(\alpha')}(\mathbf{v}')F_j^{(\beta')}(\mathbf{V}_j') - f^{(\alpha)}(\mathbf{v})F_j^{(\beta)}(\mathbf{V}_j)]$$

$$\times |\mathbf{v} - \mathbf{V}_j|\sigma_j(\alpha\beta, \alpha'\beta'; \theta, \phi, \varepsilon)\, d\mathbf{\Omega}_j\, d\mathbf{V}_j, \tag{6-6-1}$$

where

$$\mathbf{a} = e\mathbf{E}/m, \tag{6-6-2a}$$

$$\varepsilon = \tfrac{1}{2}\mu_j|\mathbf{v} - \mathbf{V}_j|^2, \tag{6-6-2b}$$

$$\mu_j = mM_j/(m + M_j), \tag{6-6-2c}$$

$$d\mathbf{\Omega} = \sin\theta\, d\theta\, d\phi, \tag{6-6-2d}$$

and the other symbols are as previously defined. This is just the Boltzmann equation of (5-3-5) for a species α, generalized to include transformations into and out of other species α' via binary collisions. Because the ions are present only in trace amounts the velocity distrubution functions of the neutrals can be taken as equilibrium Maxwellian distributions at the gas temperature T:

$$F_j^{(\beta)}(\mathbf{V}_j) = N_j(M_j/2\pi kT)^{3/2}Z_j^{-1}\exp[-(\tfrac{1}{2}M_jV_j^2 + \varepsilon_j^\beta)/kT], \tag{6-6-3}$$

where

$$Z_j = \sum_\beta \exp(-\varepsilon_j^\beta/kT) \tag{6-6-4}$$

is the internal partition function for neutral species j.

The WUB approach is semiclassical, in that the internal degrees of freedom are treated quantum mechanically but the translational degrees of freedom are treated classically. Somewhat surprisingly, this approach could lead to a practical difficulty via an inconsistency. Given a model of a nonspherical interaction and of internal degrees of freedom, it is usually easier to calculate the needed cross sections classically than semiclassically or quantum mechanically. If this is done, then for the sake of consistency a purely classical kinetic theory should be used. Strangely, the WUB semiclassical results do *not* necessarily reduce to a proper classical limit. The reason lies in the so-called "inverse collisions," the collisions that constitute the source term in the collision operator of the kinetic equation. The WUB theory assumes that these terms can be found by a time-reversal argument: a time-reversed collision restores the particles to their initial states. This assumption is not correct if the internal degrees of freedom are degenerate, as are rotational degrees of freedom, for instance. A more subtle argument is needed for the collisional source terms in a purely classical treatment, as was shown for dilute neutral gases by Taxman (1958).

In other words, the use of degeneracy-averaged cross sections in the semiclassical WUB treatment can foreclose the attainment of a proper classical limit by introducing an incorrect treatment of inverse collisions. The result is that the semiclassical and classical expressions for transport coefficients differ by terms that would be exactly zero if inverse collisions existed (Taxman, 1958). For neutral gases these terms turn out to be extremely small, but this is no guarantee that such terms will also be small for ion-neutral systems in a strong electric field.

In view of the foregoing considerations, Viehland (1986) recast the moment theory of polyatomic systems in a completely classical form for the case of nonvibrating (rigid rotor) diatomic ions and neutrals. The extra terms caused by the lack of inverse collisions are explicitly exhibited, but no numerical calculations of their magnitude have yet been reported.

In the remainder of this section we confine the discussion to the results based on the WUB kinetic equation. No qualitative conclusions are changed by a strictly classical treatment, but the numerical magnitudes of the extra terms must eventually be investigated.

Since we are interested in diffusion only within the framework of Fick's law, we carry out just a first-order expansion in the ion density gradient,

$$f^{(\alpha)} = f_0^{(\alpha)} - \left[f_x^{(\alpha)} \frac{\partial}{\partial x} + f_y^{(\alpha)} \frac{\partial}{\partial y} + f_z^{(\alpha)} \frac{\partial}{\partial z} \right] \ln n. \qquad (6\text{-}6\text{-}5)$$

This is the same as (6-1-46) of the one-temperature theory or (6-3-1) of the three-temperature theory, except that it applies separately to each internal state α of the ionic species. The normalization conditions therefore involve a summation over α,

$$\sum_{\alpha} \int f_0^{(\alpha)} \, d\mathbf{v} = n, \qquad \sum_{\alpha} \int f_{x,y,z}^{(\alpha)} \, d\mathbf{v} = 0, \qquad (6\text{-}6\text{-}6)$$

where $f_0^{(\alpha)}$ is the distribution function for state α in the spatially homogeneous case, and as usual we assume a uniform electric field in the z-direction. We need to consider moments for the spatially homogeneous reference case and for the spatially inhomogeneous case,

$$\langle \psi \rangle^{(0)} \equiv n^{-1} \sum_\alpha \int \psi^{(\alpha)} f_0^{(\alpha)} \, d\mathbf{v}, \tag{6-6-7a}$$

$$\langle \psi \rangle^{(x,y,z)} \equiv n^{-1} \sum_\alpha \int \psi^{(\alpha)} f_{x,y,z}^{(\alpha)} \, d\mathbf{v}. \tag{6-6-7b}$$

From these definitions we see that the mobility and diffusion coefficients are given by the expressions

$$\langle v_z \rangle^{(0)} = v_d = KE, \qquad D_L = \langle v_z \rangle^{(z)}; \tag{6-6-8a}$$

$$\langle v_x \rangle^{(0)} = \langle v_y \rangle^{(0)} = 0, \qquad D_T = \langle v_x \rangle^{(x)} = \langle v_y \rangle^{(y)}. \tag{6-6-8b}$$

These are the same as (6-1-48) of the one-temperature theory, except that the moment definitions now include a summation over α.

To solve the WUB equation we use a moment method, as in the case of elastic collisions. We multiply (6-6-1) by any (reasonable) function $\psi^{(\alpha)}(\mathbf{v})$, integrate over $\mathbf{v}$, sum over α, and apply the density-gradient expansion (6-6-5) to separate the result into spatially homogeneous and spatially inhomogeneous sets of equations. To make these equations more tractable, we make the quasi-steady-state assumptions that are reasonable for drift tubes, and which were discussed in Section 5-3B. In particular, we assume that ion properties other than number density are independent of position in the apparatus, and that all time derivatives are negligible except $\partial n/\partial t$. The physical justification for these assumptions is that ion momentum, energy, and other moments can be dissipated locally by collisions with neutral molecules, but ion mass cannot (except by chemically reactive collisions). These assumptions also imply that transitions into and out of the internal states occur fairly readily. We thus give up any description of transients, relaxation, and other time-dependent effects, which are dealt with in Chapter 8. We can eliminate $\partial n/\partial t$ by use of the equation of continuity, the moment equation with $\psi^{(\alpha)} = 1$,

$$(\partial n/\partial t) + \langle \mathbf{v} \rangle^{(0)} \cdot \nabla_r n = 0. \tag{6-6-9}$$

The other moment equations are simplified by integration by parts and use of the inverse collision property of the collision term (Monchick et al., 1963),

$$|\mathbf{v} - \mathbf{V}_j| \sigma_j(\alpha\beta, \, \alpha'\beta'; \, \theta, \, \phi, \, \varepsilon) \, d\mathbf{\Omega}_j \, d\mathbf{V}_j \, d\mathbf{v}$$

$$= |\mathbf{v}' - \mathbf{V}_j'| \sigma_j(\alpha'\beta', \, \alpha\beta; \, \theta', \, \phi', \, \varepsilon') \, d\mathbf{\Omega}_j' \, d\mathbf{V}_j' \, d\mathbf{v}'. \tag{6-6-10}$$

The resulting three sets of moment equations formally look just like the corresponding equations for elastic collisions:

$$(eE/m)\langle \partial\psi/\partial v_z\rangle^{(0)} = N \sum_j x_j \langle J_j\psi\rangle^{(0)}, \tag{6-6-11a}$$

$$(eE/m)\langle \partial\psi/\partial v_z\rangle^{(x)} + \langle v_x\psi\rangle^{(0)} = N \sum_j x_j \langle J_j\psi\rangle^{(x)}, \tag{6-6-11b}$$

$$(eE/m)\langle \partial\psi/\partial v_z\rangle^{(z)} + [\langle v_z\psi\rangle^{(0)} - \langle v_z\rangle^{(0)}\langle \psi\rangle^{(0)}] = N \sum_j x_j \langle J_j\psi\rangle^{(z)}, \tag{6-6-11c}$$

where the linear collision operator J_j is defined by

$$J_j\psi^{(\alpha)}(\mathbf{v}) \equiv N_j^{-1} \sum_{\alpha'} \sum_{\beta\beta'} \iint F_j^{(\beta)}(\mathbf{V}_j)[\psi^{(\alpha)}(\mathbf{v}) - \psi^{(\alpha')}(\mathbf{v}')]$$

$$\times |\mathbf{v} - \mathbf{V}_j|\sigma_j(\alpha\beta, \alpha'\beta'; \theta, \phi, \varepsilon)\, d\mathbf{\Omega}_j\, d\mathbf{V}_j. \tag{6-6-12}$$

These moment equations are really operator equations, and to solve them we must choose a complete set of basis functions that are orthogonal over some zero-order distribution function, expand the collision terms $J_j\psi^{(\alpha)}$ in terms of this set, and then devise some scheme for approximating the solutions of the resulting infinite set of algebraic equations for the moments. The choice of basis functions is of course crucial to the success of this procedure.

Previous experience with the case of elastic collisions provides the guidance for the following procedural details, in which we follow the development given by Viehland et al. (1981). The spatially uniform ion distribution function $f_0^{(\alpha)}$ is approximated by a zero-order function $g^{(\alpha)}$, which serves as the weight function for the orthogonality of the basis functions. The choice of $g^{(\alpha)}$ determines the basis functions via their orthogonality. Presumably, the successive approximations to the solutions of the moment equations will converge more rapidly the more that $g^{(\alpha)}$ resembles $f_0^{(\alpha)}$, in some sense. Two simplifying assumptions enter at this point. The first is that the distribution functions and basis functions do not depend on any of the angular momenta associated with the ions, but only on their energy. This simplification is distinct from the neglect of quantum-mechanical interference terms in the WUB equation, and is known to produce errors in the transport coefficients of neutral gases of a few percent at most, and usually less than one percent (Beenakker and McCourt, 1970). There is not much incentive to improve on this assumption until the experimental accuracy for ion transport coefficients is improved to the order of 0.1%, inasmuch as a large dose of messy algebra would be involved.

The second assumption concerns the internal energy distribution of the ions. Since the electric field can transfer energy into the internal states of an ion only by the indirect mechanism of ion-neutral collisions, it seems reasonable to assume that the internal energy distribution can be characterized by a single "internal-state temperature," T_i. That is, the ion basis temperature is given an independent internal-energy component, and we write

$$g^{(\alpha)}(\mathbf{v}, \varepsilon_i^\alpha) = h(\mathbf{v})Z_i^{-1} \exp(-\varepsilon_i^\alpha/kT_i), \tag{6-6-13}$$

where Z_i is the internal partition function,

$$Z_i = \sum_\alpha \exp(-\varepsilon_i^\alpha/kT_i). \tag{6-6-14}$$

The function $h(\mathbf{v})$ represents the zero-order approximation for the translational energy distribution, and can be chosen by analogy with the two- and three-temperature theories for the elastic-collision case.

The basis functions that correspond to the zero-order internal energy distribution assumed in (6-6-13) are the so-called Wang Chang–Uhlenbeck polynomials, $R_s(\varepsilon_i^\alpha/kT_i)$, a discrete set of nonclassical orthogonal polynomials that are generated as needed from the orthogonality condition (Wang Chang et al., 1964),

$$\sum_\alpha R_s(x^\alpha)R_{s'}(x^\alpha)\exp(-x^\alpha) = Z_i\delta_{ss'}, \tag{6-6-15a}$$

where

$$x^\alpha \equiv \varepsilon_i^\alpha/kT_i. \tag{6-6-15b}$$

The first two of these polynomials are

$$R_0(x^\alpha) = 1, \tag{6-6-16a}$$

$$R_1(x^\alpha) = N_1(x^\alpha - \bar{x}), \tag{6-6-16b}$$

where

$$\bar{x} = Z_i^{-1}\sum_\alpha x^\alpha \exp(-x^\alpha), \tag{6-6-16c}$$

and N_1 is a normalization constant. This is as far as we can go without choosing the translational basis functions. We make two choices, corresponding to the two- and three-temperature theories.

B. Spherical Basis Functions

The spherical basis functions of the two-temperature theory for elastic collisions provide a reasonable description of ion mobility, but fail for ion diffusion if $m > M$ (Viehland and Mason, 1978). We therefore consider only the spatially homogeneous case here. We also first consider only a single neutral gas, and save the easy extension to mixtures until later. With these restrictions we write

$$h(\mathbf{v}) = n(m/2\pi kT_b)^{3/2}\exp(-w_b^2), \tag{6-6-17a}$$

$$w_b^2 \equiv mv^2/2kT_b, \tag{6-6-17b}$$

where T_b is the basis temperature for the translational energy (not necessarily equal to the internal-state temperature T_i). The corresponding spherical basis functions are

$$\psi_{lmrs}^{(\alpha)} = w_b^l S_{l+1/2}^{(r)}(w_b^2) Y_l^m(\theta, \phi) R_s(\varepsilon_i^\alpha/kT_i), \qquad (6\text{-}6\text{-}18)$$

where $S_{l+1/2}^{(r)}(w_b^2)$ are the usual Sonine (generalized Laguerre) polynomials, and $Y_l^m(\theta, \phi)$ are the spherical harmonics.

We next use the completeness properties of the basis functions to expand the collision terms in the moment equations (6-6-11). Since the collision operators J_j are spherical operators, no coupling of l indices occurs in this expansion and the matrix elements are independent of the m indices. The results are

$$J_j \psi_{lmrs}^{(\alpha)} = \sum_{r's'} a_j(l; rs; r's') \psi_{lmr's'}^{(\alpha)}, \qquad (6\text{-}6\text{-}19)$$

where

$$a_j(l; rs, r's') \equiv (\psi_{lmr's'}, J_j \psi_{lmrs})/(\psi_{lmr's'}, \psi_{lmr's'}), \qquad (6\text{-}6\text{-}20)$$

and the inner products now include a summation over α in the definition,

$$(A, B) \equiv \sum_\alpha \int g^{(\alpha)} A^{(\alpha)\dagger} B^{(\alpha)} \, d\mathbf{v}. \qquad (6\text{-}6\text{-}21)$$

The operator moment equations are thereby reduced to algebraic moment equations. In particular, (6-6-11a) for the spatially homogeneous moments becomes

$$\mathscr{E}[(l + m)(l + \tfrac{1}{2} + r)\langle \psi_{(l-1)mrs} \rangle^{(0)} - (l + 1 - m)\langle \psi_{(l+1)m(r-1)s} \rangle^{(0)}]$$

$$= (l + \tfrac{1}{2}) \sum_{r'=0}^{\infty} \sum_{s'=0}^{\infty} \gamma(l; rs; r's')\langle \psi_{lmr's'} \rangle^{(0)}, \qquad (6\text{-}6\text{-}22)$$

which is the analog of (6-2-4) for elastic collisions. It is understood that any ψ_{lmrs} with negative indices is zero. The reduced field strength and matrix elements are defined as

$$\mathscr{E} \equiv \left[\left(\frac{m}{2kT_b} \right)^{1/2} \frac{e}{ma(1; 00; 00)} \right] \frac{E}{N}, \qquad (6\text{-}6\text{-}23)$$

$$\gamma(l; rs; r's') \equiv a(l; rs; r's')/a(1; 00; 00). \qquad (6\text{-}6\text{-}24)$$

We wish to solve (6-6-22) for the drift velocity,

$$v_d = (2kT_b/m)^{1/2}\langle \psi_{1000} \rangle^{(0)}, \qquad (6\text{-}6\text{-}25)$$

subject to the normalization condition

$$\langle \psi_{0000} \rangle^{(0)} = 1. \qquad (6\text{-}6\text{-}26)$$

We will also need solutions for higher moments in order to identify T_{eff}.

The calculations for the matrix elements $a(l; rs; r's')$ are tedious and complicated, somewhat more so than for elastic collisions, but they can finally be expressed in terms of linear combinations of irreducible collision integrals (Viehland et al., 1981). The matrix elements can be completely determined from knowledge of the three temperatures T, T_b, and T_i for a given ion-neutral system of known masses and cross sections. The original literature should be consulted for details. All we do here is introduce a compact notation for the irreducible collision integrals, in the hope of reducing awkwardness. For a given function $A(\mathbf{v})$ we define a collision integral $\langle\langle A \rangle\rangle_j$ for collisions of the ions with neutral species j as follows:

$$\langle\langle A \rangle\rangle_j \equiv \pi^{-3/2} Z_j^{-1} Z_i^{-1} \sum_{\alpha\alpha'} \sum_{\beta\beta'} \iint A(\mathbf{v}) \exp[-\gamma_j^2 - (\varepsilon_i^\alpha/kT_i) - (\varepsilon_j^\beta/kT)]$$

$$\times |\mathbf{v} - \mathbf{V}_j| \sigma_j(\alpha\beta; \alpha'\beta'; \theta, \phi, \varepsilon)\, d\mathbf{\Omega}_j\, d\gamma_j, \tag{6-6-27}$$

where

$$\gamma_j^2 \equiv \varepsilon/k(T_{\text{eff}})_j, \tag{6-6-28a}$$

$$(T_{\text{eff}})_j \equiv (mT + M_j T_b)/(m + M_j), \tag{6-6-28b}$$

in which ε is the relative collision energy defined by (6-6-2b), and Z_j and Z_i are the internal partition functions defined by (6-6-4) and (6-6-14), respectively.

1. Truncation-Iteration Scheme

A scheme based on the same arguments as used for the two-temperature theory of Section 6-2A seems to work satisfactorily, although only a very few computational tests have been carried out. The value of T is taken as given, and then initial values of T_b and T_i are chosen to match the average kinetic and internal energies,

$$\tfrac{1}{2}m\langle v^2 \rangle^{(0)} = \tfrac{3}{2}kT_b, \tag{6-6-29a}$$

$$\langle \varepsilon_i \rangle^{(0)} = Z_i^{-1} \sum_{\alpha} \varepsilon_i^\alpha \exp(-\varepsilon_i^\alpha/kT_i), \tag{6-6-29b}$$

which are equivalent to the conditions,

$$\langle \psi_{0010} \rangle^{(0)} = \langle \psi_{0001} \rangle^{(0)} = 0. \tag{6-6-29c}$$

That is, the moment equations are solved in lowest order. The values of T_b, T_i, and $\mathscr{E}$ found in lowest order are then kept fixed, and the moment equations truncated according to the scheme

$$\langle \psi_{10rs} \rangle^{(0)} = 0 \qquad \text{if } l + r + s \geqslant n + 1, \tag{6-6-30}$$

in the nth approximation. In each approximation v_d, $\langle v^2 \rangle$, $\langle \varepsilon_i \rangle$, and any other moments of interest are found by solving the truncated moment equations, and n is increased until satisfactory convergence is obtained (or patience and computer time run out). The irreducible collision integrals are functions of the internal temperature T_i and of an effective translational temperature, defined by (6-6-28b), as well as of the gas temperature T. The fact that three temperatures enter is clear from the exponential term in the integrand of (6-6-27) defining the $\langle\langle A \rangle\rangle_j$.

More specifically, the first approximation according to the above scheme ($n = 1$) reduces the moment equations to

$$\gamma(0; 01; 00) = 0, \tag{6-6-31a}$$

$$\mathscr{E} = \langle \psi_{1000} \rangle^{(0)}, \tag{6-6-31b}$$

$$\mathscr{E} \langle \psi_{1000} \rangle^{(0)} = -\tfrac{1}{2}\gamma(0; 10; 00). \tag{6-6-31c}$$

These equations determine v_d and $\langle v^2 \rangle$, as discussed next.

2. Mobility and Ion Energy

From (6-6-31b) and the definition of $\mathscr{E}$ given in (6-6-23) we obtain the expression for the drift velocity in pure gas j,

$$v_d = \frac{eE}{2\mu_j N \langle\langle \gamma_z(\gamma_z - \gamma_z') \rangle\rangle_j}. \tag{6-6-32}$$

This formula can be given a plausible physical interpretation by noticing that it involves a momentum-transfer collision integral. This integral is the average value of the product of the momentum along the field direction before a collision and the momentum change along this direction due to the collision. This is consistent with the physical arguments given in Section 5-2A, where the drift velocity of an ion swarm was calculated as the average velocity that results when the gain of momentum from the electric field is balanced at steady state by the loss of momentum to the neutral gas due to collisions.

More interesting is the close resemblance of (6-6-32) to the corresponding two-temperature formula for elastic collisions, and to the corresponding one-temperature formula for inelastic collisions. To show this we rewrite the collision integral in (6-6-32) in terms of the more familiar $\bar{\Omega}^{(1,1)}$, and obtain

$$[K]_1 = \frac{3e}{16N} \left[\frac{2\pi}{\mu_j k (T_{\text{eff}})_j} \right]^{1/2} \frac{1}{\bar{\Omega}_j^{(1,1)}}, \tag{6-6-33}$$

where $(T_{\text{eff}})_j$ is given in terms of T and T_b by (6-6-28b). This formula has exactly the same external appearance as (6-2-12) of the two-temperature theory, but the collision integral $\bar{\Omega}_j^{(1,1)}$ is now more complicated and depends on T_i and T as well as on $(T_{\text{eff}})_j$:

$$\bar{\Omega}_j^{(1,1)} \equiv (Z_j Z_i)^{-1/2} \sum_{\alpha\alpha'} \sum_{\beta\beta'} \exp[-(\varepsilon_i^\alpha/kT_i) - (\varepsilon_j^\beta/kT)]$$

$$\times \int_0^\infty \gamma_j^3 \exp(-\gamma_j^2)\, d\gamma_j \int_0^{2\pi} d\phi \int_0^\pi (\gamma_j^2 - \gamma_j\gamma_j' \cos\theta)$$

$$\times \sigma_j(\alpha\beta;\, \alpha'\beta';\, \theta,\, \phi,\, \varepsilon)\, \sin\theta\, d\theta. \tag{6-6-34}$$

This reduces to the inelastic one-temperature formula of (6-1-79) and (6-1-80a) if $T_i = (T_{\text{eff}})_j = T$.

In the one-temperature case the main difference in $\bar{\Omega}^{(1,1)}$ due to inelastic collisions, other than the expected summations over internal states, is the appearance of $\gamma_j\gamma_j' \cos\theta$ in place of $\gamma_j^2 \cos\theta$. As discussed in Section 6-1F, this difference appears to be minor, as evidenced by the fact that there is nothing remarkable about the diffusion coefficients of molecular gases as compared to atomic gases. But this is not true for ions at high field strengths, because the values of T_{eff} and T_i are no longer equal to T and may depend strongly on inelastic collisions. Thus the main effect of inelastic collisions on ion mobility is probably indirect, occurring through the ion temperatures. The suspicion is that the dependence of $\bar{\Omega}^{(1,1)}$ on T_{eff} is more important than its dependence on T_i, although not much information is available on this point. The main evidence comes from comparing the mobilities of atomic and molecular ions in the same molecular gas, as discussed in Section 6-6B4. We now turn to the question of the ion temperatures.

The total ion energy is characterized by two temperatures, a translational basis temperature T_b (or T_{eff}) and an internal energy T_i. We find the translational temperature from (6-6-31c); after some algebra we obtain

$$\tfrac{3}{2}kT_b[1 + (M_j/m)\xi_j] = \tfrac{3}{2}kT(1 - \xi_j) + \tfrac{1}{2}mv_d^2 + \tfrac{1}{2}M_j v_d^2, \tag{6-6-35}$$

where ξ_j is a dimensionless ratio characterizing fractional energy loss due to inelastic collisions. It can be written as the ratio of the collision integral for inelastic energy loss to that for momentum transfer,

$$\xi_j = \frac{\langle\langle \gamma^2 - (\gamma')^2 \rangle\rangle_j}{6\langle\langle \gamma_z(\gamma_z - \gamma_z') \rangle\rangle_j} = -\frac{\langle\langle \varepsilon_i^\alpha + \varepsilon_j^\beta - \varepsilon_i^{\alpha'} - \varepsilon_j^{\beta'} \rangle\rangle_j}{6\langle\langle \gamma_z(\gamma_z - \gamma_z') \rangle\rangle_j k(T_{\text{eff}})_j}. \tag{6-6-36}$$

This result is the generalization to inelastic collisions of the famous Wannier formula for the mean kinetic energy of the ions, which has been encountered several times already in Chapters 5 and 6; when $\xi_j = 0$ we recover the Wannier formula,

$$\tfrac{3}{2}kT_b = \tfrac{1}{2}m\langle v^2 \rangle = \tfrac{3}{2}kT + \tfrac{1}{2}mv_d^2 + \tfrac{1}{2}Mv_d^2.$$

The effect of inelastic collisions on this energy equation is thus entirely through

ξ_j, which from its definition (6-6-36) above is a function of $(T_{eff})_j$ and (more weakly, it is suspected) the temperatures T_i and T, just as is $\bar{\Omega}_j^{(1,1)}$.

The effective temperature $(T_{eff})_j$ is found by substituting (6-6-35) for T_b into the definition (6-6-28b); the result is

$$\tfrac{3}{2}k(T_{eff})_j[1 + (M_j/m)\xi_j] = \tfrac{3}{2}kT + \tfrac{1}{2}M_j v_d^2. \tag{6-6-37}$$

The mass ratio M_j/m appears in both (6-6-35) for T_b and (6-6-37) for T_{eff} because the energy loss due to inelastic collisions is unequally shared by ion and neutral in the laboratory system according to their mass ratio.

Several limiting cases for T_b and T_{eff} are worth noting. We have already remarked that Wannier's formula is recovered when $\xi_j = 0$. When the electric field is small, $\xi_j \to 0$ because of detailed balance (see below), $v_d \to 0$ because $E \to 0$, and therefore $T_{eff} \to T$. For heavy ions ($M_j/m \ll 1$), the effect of ξ_j disappears from the left-hand sides of (6-6-35) and (6-6-37) because of the mass ratio; the expression for T_{eff} is then the same as for elastic collisions only, although that for T_b becomes the same only at higher fields where kT is negligible compared to the field energy. However, for very light ions, especially electrons, the effect of ξ_j is amplified by the factor M_j/m and the effect of inelastic collisions can be quite pronounced. More detailed analysis for electrons confirms this expectation (Lin et al., 1979b). Roughly speaking, inelastic collisions are not expected to produce any striking new effects on the mobility of ions, but may do so for electrons. The case of negative differential conductivity for electrons is a case in point (Section 6-4D). The main known effect of inelastic collisions for ions is a breakdown in the scaling rule of T and E/N combining into the single variable T_{eff}, as illustrated in Section 6-6B4.

So much for T_b and T_{eff}. The internal temperature T_i characterizes the ionic internal energy that has been gained indirectly from the field by means of inelastic ion-neutral collisions. It is to be found from (6-6-31a), which yields the relation

$$\langle\!\langle \varepsilon_i^\alpha - \varepsilon_i^{\alpha'} \rangle\!\rangle_j = 0. \tag{6-6-38}$$

In accordance with our assumed steady-state conditions, this relation defines T_i as the temperature that results when the difference between the pre- and postcollision ion energies is zero on the average. In general, this average energy balance depends on the cross sections for inelastic collisions as compared to the elastic cross sections, and has to be determined by brute-force calculation in all but two special cases. These special cases occur when *either* the ion *or* the neutral is atomic and has no active internal degrees of freedom. The case of atomic ions is of course trivial, since they have no T_i at all. The case of molecular ions in an atomic gas is less obvious but of great practical importance. At steady state, the internal temperature must be equal to the effective translational temperature:

$$T_i = T_{eff}, \qquad \text{when } \varepsilon_j^\beta = 0. \tag{6-6-39}$$

The proof follows by setting $\varepsilon_j^\beta = 0$ in (6-6-38) and employing a detailed balance argument (Viehland et al., 1981). The relation has been experimentally verified for the rotational energy of N_2^+ ions in He (Duncan et al., 1983).

The equality of T_i and T_{eff} for molecular ions in an atomic gas can be given a simple physical interpretation. Energy is fed into the internal degrees of freedom of the ions by collisions with the structureless neutrals; the source of the internal energy is thus the translational motion. Energy leaks out of both the internal and translational degrees of freedom of the ions only through the translational (recoil) motion of the neutrals. Since the energy leak is the same for both forms of energy, and since the internal energy is fed only by the translational motion, it is not so surprising to find a relation like (6-6-39) at steady state. In contrast, with molecular neutrals the internal and translational energies of the ions have two leakage channels, and can leak into the internal degrees of freedom of the neutrals at different rates, depending on the details of the inelastic cross sections. The steady-state energy balance will thus depend on the cross sections, and no general relation like (6-6-39) can be obtained. Similarly, no simple result for T_i is obtained in a mixture of atomic neutral gases because of the diversity of energy leakage channels, as discussed in Section 6-6B5.

The potential importance of the relation (6-6-39) is connected with the use of drift tubes to study ion-molecule reaction rates. It offers a means to vary independently, within limits, the reactive internal and translational temperatures of the ions to see how the reaction rate depends separately on these temperatures. Ion-molecule reactions are discussed in Chapter 8.

3. Convergence of Approximations

Owing to a general lack of knowledge about cross sections for inelastic ion-neutral collisions, very few calculations have been made to test the convergence of the truncation-iteration scheme for the spherical basis functions discussed in this section. Comparison with experiment is essentially useless unless the cross sections are known. Viehland et al. (1981) have used a very simple model, whose properties could be calculated independently by Monte Carlo simulation, to test for convergence. The model has the following properties:

1. Structureless ions in a two-state neutral gas of the same mass ($M = m$).
2. Cold gas ($T = 0$). The neutrals are therefore in their ground state, and the thermal energy of the ions is negligible compared to their field energy ($kT \ll Mv_d^2$).
3. Isotropic differential cross sections (rigid-sphere-like).
4. Inelastic collisions occur only above a threshold energy corresponding to the relative speed v_{th}; an inelastic collision causes the translational energy of the ion-neutral pair to decrease by exactly the threshold energy in the center-of-mass system.

The differential cross sections are therefore given by

$$\text{elastic:} \quad \sigma_{el} = \sigma_0 \quad \text{(constant),} \quad (6\text{-}6\text{-}40a)$$

$$\text{inelastic:} \quad \sigma_{in} = \begin{cases} 0 & \text{for } v_r < v_{th}, \\ \sigma_0 & \text{for } v_r \geqslant v_{th}. \end{cases} \quad (6\text{-}6\text{-}40b)$$

The results of the calculations are shown in Figs. 6-6-1 and 6-6-2 for the drift velocity and the ion energy, in terms of dimensionless quantities involving the threshold velocity, v_{th}, the ion acceleration, $a = eE/m$, and the mean free path between ion-neutral collisions,

$$\lambda \equiv 1/NQ_0, \quad \text{where } Q_0 = 4\pi\sigma_0. \quad (6\text{-}6\text{-}41)$$

Figure 6-6-1 shows the reduced drift velocity as a function of reduced field strength, as calculated in first and fifth approximations. The fifth approximation

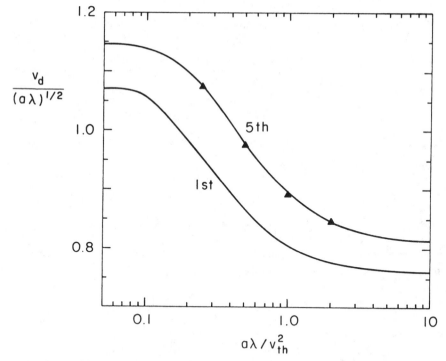

Figure 6-6-1. Reduced ion drift velocity as a function of reduced field strength for the isotropic two-state inelastic collision model discussed in the text, with $m = M$ and $T = 0$. The curves are the first and fifth approximations of the moment solutions with spherical basis functions, and the triangles are a Monte Carlo simulation (0.5% accuracy). There is a smooth transition from a constant value at low fields, where all collisions are elastic, to a lower constant value at high fields, where half the collisions are inelastic.

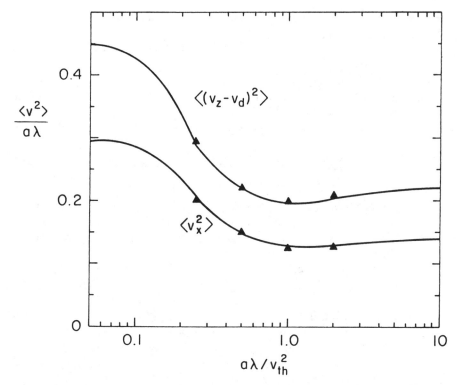

Figure 6-6-2. Reduced ion energies as a function of reduced field strength for the same system as in Fig. 6-6-1. The curves are the fifth moment approximation, and the triangles are a Monte Carlo simulation (2% accuracy). There is a smooth, but nonmonotonic, transition from the value for all elastic collisions to the value for one-half inelastic collisions, as in Fig. 6-6-1.

has converged within the 0.5% accuracy of the Monte Carlo simulation; the convergence error of the first approximation is thus seen to be about 10% in this case. In the absence of inelastic collisions, the reduced drift velocity in these units should be 1.1467 (Wannier, 1953; Skullerud, 1976). Inelastic collisions provide another channel for energy and momentum loss, and Fig. 6-6-1 shows that the reduced drift velocity gradually falls from this value to a lower value of about 0.81.

Figure 6-6-2 shows the reduced ion energies as a function of reduced field strength, as calculated in fifth approximation. This is essentially a plot of T_T and T_L vs. E/N. There is a smooth but nonmonotonic transition from the low-field constant values of 0.454 and 0.293 (Wannier, 1953; Skullerud, 1976) to high-field constant values of about 0.22 and 0.14, respectively. Agreement with the Monte Carlo simulation is excellent.

Viehland et al. (1981) also tested the convergence for the ion diffusion coefficients, since spherical basis functions are still usable when $m = M$ in the elastic collision case. There was agreement within the accuracy of the Monte Carlo simulation, which was only 10% for diffusion. The simple first-order GER using the accurate calculated values of T_T and T_L also gave agreement within the 10% uncertainty level.

These simple model calculations certainly do not comprehensively demonstrate the accuracy of moment calculations when inelastic collisions occur, but they at least indicate that the mere introduction of inelastic collisions is not likely to cause pathological convergence difficulties.

4. Temperature and Field Dependence of Mobility

Experimental data (Ellis et al., 1976a, 1978, 1984) indicate that the temperature and field dependence of the mobilities of molecular systems are qualitatively similar to those of atomic systems. The foregoing theoretical results are thus not called upon to account for any striking new phenomena. Lacking information on inelastic cross sections, the only comparison of theory with experiment that we can make is through the scaling rule that combines the two variables T and E/N into the single variable T_{eff}. That is, we can test whether zero-field measurements of K_0 as a function of T can be made to coincide with measurements of K_0 as a function of E/N by means of a plot of K_0 vs. T_{eff}. The success of this scaling rule for atomic systems was illustrated in Fig. 5-2-1.

To avoid any worry about the role of the internal temperature T_i, we first consider an atomic ion in a molecular gas. The only suitable system is Cl^- ions in N_2, for which $K(0)$ as a function of T has been measured by Eisele et al. (1981), and K has been measured as a function of E/N at 300 K by Viehland and Fahey (1983) and by Byers et al. (1983). The measurements have been smoothed and summarized in the compilation of Ellis et al. (1984). Since the energy-loss ratio ξ is unknown, we cannot use T_{eff} from (6-6-37); we therefore define an "elastic" effective temperature as

$$\tfrac{3}{2}kT_{\text{eff}}^{\text{el}} \equiv \tfrac{3}{2}kT + \tfrac{1}{2}Mv_d^2, \tag{6-6-42}$$

which really includes some inelastic collision effects through the measured values of v_d. For the zero-field measurements $T_{\text{eff}}^{\text{el}}$ is the same as T, but for the field-dependent measurements $T_{\text{eff}}^{\text{el}}$ is smaller than the true T_{eff} of (6-6-37) by the factor $[1 + (M/m)\xi]$. Thus the two types of measurements of K_0 vs. $T_{\text{eff}}^{\text{el}}$ should not coincide except at very low E/N, unless it happens that $(M/m)\xi \ll 1$. The results shown for Cl^- in N_2 in Fig. 6-6-3 fulfill this expectation: there is agreement at low E/N, but increasing divergence as E/N increases. Moreover, the two sets of measurements can be brought into coincidence by a scale factor applied to the temperature axis *only*. This scale factor must be equal to $T_{\text{eff}}/T_{\text{eff}}^{\text{el}}$, and determines ξ through comparison of (6-6-37) and (6-6-42):

$$\xi = \frac{m}{M}\left(\frac{T_{\text{eff}}}{T_{\text{eff}}^{\text{el}}} - 1\right). \tag{6-4-43}$$

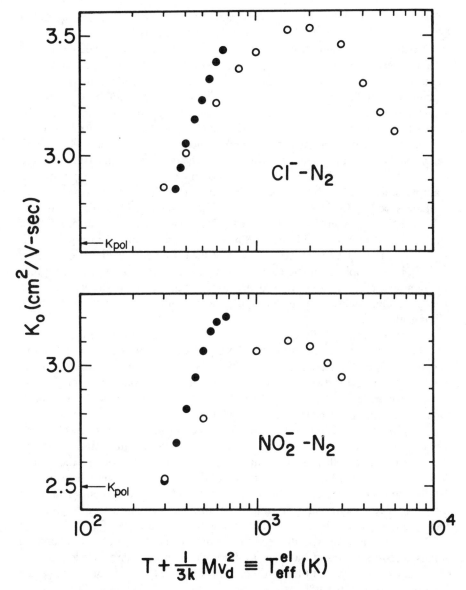

Figure 6-6-3. Scaling-rule plot of K_0 as a function of the "elastic" effective temperature for molecular systems, showing the effects of inelastic collisions. The filled circles are from zero-field measurements as a function of temperature, and the open circles are from measurements as a function of E/N at fixed temperature, all from the survey of Ellis et al. (1984). The failure of scaling is in contrast to the results of Fig. 5-2-1 for atomic systems. For a structureless ion like Cl^-, the results can be quantitatively interpreted in terms of an energy-loss ratio ζ according to (6-6-43). But for a molecular ion like NO_2^- an unknown internal ion temperature T_i also appears, as signaled by the fact that the two sets of measurements have maxima of different heights.

Values of ξ calculated in this way range from zero up to about 2.6 at $E/N = 130$ Td or $T_{\text{eff}}^{\text{el}} = 1900$ K (Viehland and Fahey, 1983).

The situation becomes more complicated when we consider molecular ions in molecular gases, because of the appearance of T_i. For the zero-field measurements T_{eff} and T_i are both the same as T, but the field-dependent measurements depend on both T_{eff} and T_i, and the relation between them is unknown. It is thus possible that the two kinds of measurements for such a system *cannot* be brought into coincidence in a plot of K_0 vs. $T_{\text{eff}}^{\text{el}}$ by a scale factor applied to the temperature axis only, because of the dependence of $\bar{\Omega}^{(1,1)}$ and hence K_0 on T_i as well as on T_{eff}. A clear sign of this would be for the two kinds of measurements to show maxima of different heights in a plot of K_0 vs. $T_{\text{eff}}^{\text{el}}$. An experimental example is provided by NO_2^- ions in N_2, for which $K(0)$ as a function of T has been measured by Eisele et al. (1980), and K has been measured as a function of E/N at 300 K by Viehland and Fahey (1983). The measurements have been smoothed and summarized by Ellis et al. (1984), and are shown in Fig. 6-6-3 as K_0 vs. $T_{\text{eff}}^{\text{el}}$. It is clear that the two sets of measurements cannot possibly be brought into complete coincidence by a temperature scale factor alone, because the high-temperature values of $K(0)$ are larger than the maximum values of the field-dependent measurements. Evidently the dependence of $\bar{\Omega}^{(1,1)}$ and K_0 on T_i is implicated.

A convincing test of the above interpretation would be provided by similar sets of measurements on the mobility of NO_2^- ions in a monatomic gas (say Ar). In this case $T_i = T_{\text{eff}}$ according to (6-6-39), so that $\bar{\Omega}^{(1,1)}$ and K_0 would depend only on T_{eff}, and the two sets of measurements again could be brought into coincidence by a scale factor applied to the temperature axis alone. Unfortunately, no such measurements are available.

The general conclusion to be drawn from the foregoing discussion is that ion mobility data for molecular systems contain substantial information on the effects of inelastic collisions and anisotropic potentials. Moreover, the theory allows at least a qualitative interpretation of the data without knowledge of any details about cross sections. Much further experimental and theoretical work will be needed to unravel quantitative details. Further discussion is given by Viehland and Fahey (1983), and in a review by Viehland (1984b).

The most important specific result from the theory that does not depend on details of cross sections is that $T_i = T_{\text{eff}}$ for molecular ions in an atomic gas. This somewhat surprising result finds direct application in the study of ion-molecule reactions in drift tubes (Chapter 8).

5. Mixtures

As usual in moment theories, the moment equations for a mixture have the same form as for a single gas, but the matrix elements $a_j(l; rs; r's')$ for single gas j are replaced by the linear combinations,

$$b(l; rs; r's') \equiv \sum_j x_j a_j(l; rs; r's').$$

$$(6-6-44)$$

In first approximation the ion drift velocity in a mixture is thus rather simple in appearance:

$$v_d = \frac{eE}{2mN}\left[\sum_j \frac{x_j M_j}{m + M_j}\langle\langle\gamma_z(\gamma_z - \gamma_z')\rangle\rangle_j\right]^{-1}, \qquad (6\text{-}6\text{-}45)$$

or in terms of the more familiar collision integral $\bar{\Omega}^{(1,1)}$ the mobility is

$$[K]_1 = \frac{3e}{16N}\left(\frac{2\pi}{mk}\right)^{1/2}\left[\sum_j x_j\left(\frac{M_j}{m + M_j}\right)^{1/2}(T_{\text{eff}})_j^{1/2}\bar{\Omega}_j^{(1,1)}\right]^{-1}. \qquad (6\text{-}6\text{-}46)$$

At first glance this result appears to imply that the mobility in a mixture obeys Blanc's law, but we have seen previously that this is not true even for atomic systems unless the electric field strength is vanishingly small. The reason is that the translational temperature of the ions is different in a mixture than in any of the single gases, even if E, N, and T are the same, and even if all the collisions are elastic. The mixture problem is inherently complicated by this fact, and the results are messy implicit equations. To complete this set of implicit equations we need expressions for the ion temperatures, and in polyatomic systems we also have the ion internal temperature T_i to deal with. The following expressions deal only with the first approximation.

The expression for the ion translational (basis) temperature in a mixture can be made to resemble (6-6-35) for a single gas by the use of suitable averages,

$$\tfrac{3}{2}kT_b(1 + \langle M\xi\rangle_{\text{mix}}/m) = \tfrac{3}{2}kT(1 - \langle\xi\rangle_{\text{mix}}) + \tfrac{1}{2}mv_d^2 + \tfrac{1}{2}\langle M\rangle_{\text{mix}}v_d^2, \qquad (6\text{-}6\text{-}47)$$

where the notation $\langle\ \rangle_{\text{mix}}$ signifies

$$\langle X\rangle_{\text{mix}} \equiv \sum_j \omega_j X_j \Big/ \sum_j \omega_j, \qquad (6\text{-}6\text{-}48a)$$

with weight factors ω_j defined as

$$\omega_j \equiv x_j M_j\langle\langle\gamma_z(\gamma_z - \gamma_z')\rangle\rangle_j/(m + M_j)^2. \qquad (6\text{-}6\text{-}48b)$$

The quantity X_j takes on the values $M_j\xi_j$, ξ_j, and M_j, respectively, in (6-6-47). Except for constant factors, which cancel out on normalization, the ω_j can also be written as

$$\omega_j = \frac{x_j}{m + M_j}\left(\frac{M_j}{m + M_j}\right)^{1/2}(T_{\text{eff}})_j^{1/2}\bar{\Omega}_j^{(1,1)}. \qquad (6\text{-}6\text{-}48c)$$

The expression for $(T_{\text{eff}})_j$ in a mixture is found by substitution of (6-6-47) into the definition (6-6-28b), and is

$$\tfrac{3}{2}k(T_{\text{eff}})_j \left(1 + \frac{\langle M\xi \rangle_{\text{mix}}}{m} \right) = \tfrac{3}{2}kT \left(1 + \frac{\langle M\xi \rangle_{\text{mix}} - M_j \langle \xi \rangle_{\text{mix}}}{m + M_j} \right)$$

$$+ \tfrac{1}{2}M_j \frac{m + \langle M \rangle_{\text{mix}}}{m + M_j} v_d^2, \tag{6-6-49}$$

which reduces to (6-6-37) in the case of a single gas, as it should.

In the atomic case, considerable simplification of the implicit equations could be achieved by Taylor-series expansions in terms of the difference between $(T_{\text{eff}})_j$ in the mixture and in the single gas j at the same values of E/N and T. This simplification was carried through in detail for the momentum-transfer theory in Section 5-2D, and exactly the same expansion procedure holds in the one-temperature theory (Section 6-1G) and in the two-temperature theory (Section 6-2F). In the molecular case, however, we have the additional quantities ξ_j and T_i to contend with in such an expansion. These quantities will also be different in the mixture than in any of the single gases, and no corresponding simple expression has been obtained.

The ion internal temperature T_i in a mixture is to be found from the relation,

$$\sum_j x_j \langle\langle \varepsilon^\alpha - \varepsilon^{\alpha'} \rangle\rangle_j = 0. \tag{6-6-50}$$

Unlike the case of a single gas, no simple result like (6-6-39) is obtained for T_i from this expression even if all the neutral gas species are atomic. The reason is that there is a different effective translational temperature $(T_{\text{eff}})_j$ for each neutral species in the mixture, and there is no a priori way to tell how the energy loss of the ion on collision will partition between its internal and translational modes unless relative cross sections for such energy loss are known for each neutral species. In other words, T_i in a mixture cannot be determined without knowledge of cross sections, even if the mixture is entirely made up of monatomic gases; there are too many channels for energy gain and loss.

C. Cartesian Basis Functions

Experience with atomic systems suggests that an anisotropic displayed distribution is needed to describe diffusion if $m > M$ (Lin et al., 1979c; Viehland and Lin, 1979). We therefore use the form of the three-temperature theory to represent the translational velocity distribution and write

$$h(\mathbf{v}) = n[m/2\pi k T_b^{(T)}][m/2\pi k T_B^{(L)}]^{1/2} \times \exp(-w_x^2 - w_y^2 - w_z^2), \tag{6-6-51a}$$

where

$$w_x^2 \equiv mv_x^2/2kT_b^{(T)}, \qquad w_y^2 \equiv mv_y^2/2kT_b^{(T)}, \tag{6-6-51b}$$

$$w_z^2 \equiv m(v_z - v_{\text{dis}})^2/2kT_b^{(L)}, \tag{6-6-51c}$$

in which $T_b^{(T)}$ and $T_b^{(L)}$ are the transverse and longitudinal basis temperatures for the translational energy, and v_{dis} is the displacement velocity. As usual, we consider only a single neutral gas at first, and postpone the extension to mixtures. The corresponding Cartesian basis functions are then Hermite polynomials,

$$\psi_{pqrs}^{(\alpha)} = H_p(w_x)H_q(w_y)H_r(w_z)R_s(\varepsilon_i^\alpha/kT_i), \qquad (6\text{-}6\text{-}52)$$

where $R_s(x^\alpha)$ are the same Wang Chang–Uhlenbeck polynomials that were used with the spherical basis functions.

We again use the completeness properties of the basis functions to expand the collision terms in the moment equations (6-6-11)

$$J_j\psi_{pqrs}^{(\alpha)} = \sum_{p'q'r's'} a_j(pqrs; p'q'r's')\psi_{p'q'r's'}^{(\alpha)}, \qquad (6\text{-}6\text{-}53)$$

where the matrix elements are defined as

$$a_j(pqrs; p'q'r's') \equiv (\psi_{p'q'r's'}, J_j\psi_{pqrs})/(\psi_{p'q'r's'}, \psi_{p'q'r's'}). \qquad (6\text{-}6\text{-}54)$$

The operator moment equations are thereby reduced to algebraic moment equations.

In particular, from (6-6-11a) we obtain the equation for the spatially homogeneous moments,

$$2r\mathscr{E}\langle\psi_{pq(r-1)s}\rangle^{(0)} = \sum_{p'q'r's'} \gamma(pqrs; p'q'r's')\langle\psi_{p'q'r's'}\rangle^{(0)}, \qquad (6\text{-}6\text{-}55)$$

which is the analog of (6-3-11) for elastic collisions. It is understood that any ψ_{pqrs} with negative indices is zero. The reduced field strength and matrix elements are defined as

$$\mathscr{E} \equiv \left[\frac{m}{2kT_b^{(L)}}\right]^{1/2}\left[\frac{e}{ma(0010; 0000)}\right]\frac{E}{N}, \qquad (6\text{-}6\text{-}56)$$

$$\gamma(pqrs; p'q'r's') \equiv a(pqrs; p'q'r's')/a(0010; 0000). \qquad (6\text{-}6\text{-}57)$$

We wish to solve (6-6-55) for the drift velocity,

$$v_d = v_{dis} + [kT_b^{(L)}/2m]^{1/2}\langle\psi_{0010}\rangle^{(0)}, \qquad (6\text{-}6\text{-}58)$$

subject to the normalization condition

$$\langle\psi_{0000}\rangle^{(0)} = 1. \qquad (6\text{-}6\text{-}59)$$

With spherical basis functions the experimental cylindrical symmetry in a drift tube is described in a natural way, but with Cartesian basis functions we must describe this symmetry by imposing the following special conditions:

$$\gamma(pqrs; p'q'r's') = \gamma(qprs; q'p'r's'), \qquad (6\text{-}6\text{-}60a)$$

$$\gamma(pqrs; p'q'r's') = 0 \text{ if } (p + p') \text{ or } (q + q') \text{ is odd}; \qquad (6\text{-}6\text{-}60b)$$

$$\langle \psi_{pqrs} \rangle^{(0)} = \langle \psi_{qprs} \rangle^{(0)}, \qquad (6\text{-}6\text{-}60c)$$

$$\langle \psi_{pqrs} \rangle^{(0)} = 0 \text{ if } p \text{ or } q \text{ is odd}. \qquad (6\text{-}6\text{-}60d)$$

The equations for the spatially inhomogeneous moments are obtained from (6-6-11b) and (6-6-11c),

$$2r\mathscr{E}\langle \psi_{pq(r-1)s} \rangle^{(x,z)} + h_{pqrs}^{(x,z)} = \sum_{p'q'r's'} \gamma(pqrs; p'q'r's')\langle \psi_{p'q'r's'} \rangle^{(x,z)}, \qquad (6\text{-}6\text{-}61)$$

where

$$h_{pqrs}^{(x)} \equiv \left[\frac{kT_b^{(T)}}{2m} \right]^{1/2} \frac{\langle \psi_{(p+1)qrs} \rangle^{(0)} + 2p\langle \psi_{(p-1)qrs} \rangle^{(0)}}{Na(0010; 0000)}, \qquad (6\text{-}6\text{-}62a)$$

$$h_{pqrs}^{(z)} \equiv \left[\frac{kT_b^{(L)}}{2m} \right]^{1/2} \frac{\langle \psi_{pq(r+1)s} \rangle^{(0)} + 2r\langle \psi_{pq(r-1)s} \rangle^{(0)} - \langle \psi_{0010} \rangle^{(0)}\langle \psi_{pqrs} \rangle^{(0)}}{Na(0010; 0000)}. \qquad (6\text{-}6\text{-}62b)$$

These are the analogs of (6-3-12) and of (6-3-15a) and (6-3-15b) for elastic collisions. We wish to solve these equations for the diffusion coefficients,

$$D_T = [kT_b^{(T)}/2m]^{1/2}\langle \psi_{1000} \rangle^{(x)}, \qquad (6\text{-}6\text{-}63a)$$

$$D_L = [kT_b^{(L)}/2m]^{1/2}\langle \psi_{0010} \rangle^{(z)}, \qquad (6\text{-}6\text{-}63b)$$

subject to the normalization conditions

$$\langle \psi_{0000} \rangle^{(x)} = \langle \psi_{0000} \rangle^{(z)} = 0, \qquad (6\text{-}6\text{-}64)$$

and to symmetry conditions like (6-6-60).

As might be expected, the calculations for the matrix elements $a(pqrs; p'q'r's')$ are tedious and complicated, but they can finally be expressed in terms of linear combination of irreducible collision integrals (Viehland et al., 1981). These collision integrals $\langle\langle A \rangle\rangle_j$ are formally defined in the same way as those that arise with the spherical basis functions, namely by (6-6-27), but the integration variable γ is defined differently. Instead of the definition of γ given by (6-6-28) for the spherical basis functions, the definition of γ for the Cartesian basis functions reflects the anisotropy and displacement of the zero-order distribution function, as follows:

$$\varepsilon = (\gamma_x^2 + \gamma_y^2)kT_{\text{eff}}^{(T)} + (\gamma_z + \tilde{v}_{\text{dis}})^2 kT_{\text{eff}}^{(L)}, \qquad (6\text{-}6\text{-}65)$$

where

$$T_{\text{eff}}^{(T,L)} \equiv [mT + MT_b^{(T,L)}]/(m + M), \tag{6-6-66a}$$

$$\tilde{v}_{\text{dis}} \equiv [\mu v_{\text{dis}}^2/2kT_{\text{eff}}^{(L)}]^{1/2}. \tag{6-6-66b}$$

These definitions are the same as those used with the three-temperature theory for elastic collisions in Section 6-3A.

1. Truncation-Iteration Scheme

A suggested scheme is based on the three-temperature theory of Section 6-3A. After some numerical experimentation, Viehland et al. (1981) suggested that an effective truncation scheme for (6-6-55) and (6-6-61) would be to set the following moments to zero in an nth approximation:

$$\langle \psi_{pqrs} \rangle^{(0)} = 0 \qquad \text{for } p + q + r + 2s \geq n + 2, \tag{6-6-67a}$$

$$\langle \psi_{pqrs} \rangle^{(x,z)} = 0 \qquad \text{for } p + q + r + 2s \geq n + 1. \tag{6-6-67b}$$

The value of T is taken as given, and initial values of $T_b^{(T)}$, $T_b^{(L)}$, T_i, and v_{dis} are chosen to satisfy the equations

$$\tfrac{1}{2}m\langle v_x^2 \rangle^{(0)} = \tfrac{1}{2}m\langle v_y^2 \rangle^{(0)} = \tfrac{1}{2}kT_b^{(T)}, \tag{6-6-68a}$$

$$\tfrac{1}{2}m\langle (v_z - v_{\text{dis}})^2 \rangle^{(0)} = \tfrac{1}{2}kT_b^{(L)}, \tag{6-6-68b}$$

$$\langle \varepsilon_i \rangle^{(0)} = Z_i^{-1} \sum_\alpha \varepsilon_i^\alpha \exp(-\varepsilon_i^\alpha/kT_i), \tag{6-6-68c}$$

$$\langle v_z \rangle^{(0)} = v_d = v_{\text{dis}}, \tag{6-6-68d}$$

which are equivalent to the conditions

$$\langle \psi_{0001} \rangle^{(0)} = \langle \psi_{0010} \rangle^{(0)} = \langle \psi_{2000} \rangle^{(0)} = \langle \psi_{0020} \rangle^{(0)} = 0. \tag{6-6-68e}$$

These conditions are equivalent to a solution of the moment equations in lowest order. The values of $T_b^{(T)}$, $T_b^{(L)}$, T_i, v_{dis}, and $\mathscr{E}$ found in this lowest order are then kept fixed, and higher-order equations ($n > 1$) are solved for v_d, $\langle v_x^2 \rangle$, $\langle v_z^2 \rangle$, $\langle \varepsilon_i \rangle$, D_T, D_L, and any other moments of interest until satisfactory convergence is obtained with increasing n.

Specifically, in the first approximation ($n = 1$) the moment equations are reduced to

$$\mathscr{E} = \tfrac{1}{2}, \tag{6-6-69a}$$

$$\gamma(2000; 0000) = \gamma(0020; 0000) = \gamma(0001; 0000) = 0, \tag{6-6-69b}$$

and

$$\gamma(1000;\ 1000)\langle\psi_{1000}\rangle^{(x)} = h^{(x)}_{1000}, \tag{6-6-70a}$$

$$\gamma(0010;\ 0010)\langle\psi_{0010}\rangle^{(z)} = h^{(z)}_{0010}. \tag{6-6-70b}$$

The relations (6-6-69) determine E/N, $T_b^{(T)}$, $T_b^{(L)}$, and T_i, respectively, and the relations (6-6-70) determine D_T and D_L, respectively, in terms of the independent variables T and $v_{dis} = v_d$, as discussed next.

2. First Approximations

By analogy with the elastic-collision case, we expect that the first approximations with the Cartesian basis functions will usually be more accurate than the corresponding results with the spherical basis functions. However, there have not been any numerical calculations made to test this expectation, or to check the convergence of the proposed truncation-iteration scheme. But some interesting results already emerge in just the first approximation, to which we confine the following discussion.

a. Mobility and Diffusion. The expression for v_d looks much like the corresponding formula (6-6-32) obtained with spherical basis functions. Little can be added to the discussion already given. Of more interest are the ion diffusion coefficients and their relation to the mobility. In first approximation the diffusion coefficients are given by the expressions

$$[D_T]_1 = \frac{kT_b^{(T)}}{mNa(1000;\ 1000)}, \tag{6-6-71a}$$

$$[D_L]_1 = \frac{kT_b^{(L)}}{mNa(0010;\ 0010)}. \tag{6-6-71b}$$

In this form it is not obvious that the diffusion coefficients obey the simple generalized Einstein relations (Viehland et al., 1981). To obtain the GER it is much better to introduce differential mobilities and proceed as in Section 6-4. In place of (6-4-13) we obtain the relations

$$\frac{D_T}{K_T^x} = \frac{\langle\psi_{1000}\rangle^{(x)}_D}{\langle\psi_{1000}\rangle^{(x)}_E}, \tag{6-6-72a}$$

$$\frac{D_L}{K_L^z} = \frac{\langle\psi_{0010}\rangle^{(z)}_D}{\langle\psi_{0010}\rangle^{(z)}_E}, \tag{6-6-72b}$$

and in first approximation these yield the usual GER (Waldman and Mason, 1982, unpublished calculations), as given by (6-4-2), say. This provides a solid pedigree for the GER in the WUB equation, and rationalizes the fact that the

GER seem to work fairly well for atomic ions in simple molecular gases (Section 6-4D).

The inclusion of inelastic collisions thus does not change the formal appearance of the simple first-order GER; the effects of the inelastic collisions appear only implicitly in the coefficients themselves, namely D_T, D_L, K, K', T_T, and T_L. The extension to a second order of approximation to find the analogs of the corrections Δ_T and Δ_L, which were obtained in Section 6-4 for the case of elastic collisions, has not been carried out. This extension is probably essential for electrons in molecular gases, as suggested by the phenomenon of negative differential conductivity (Section 6-4D).

Of course, even the first-order GER are experimentally useless unless some information is available on the ion temperatures, to which we turn next.

b. *Ion Temperatures.* The ion temperatures T_T and T_L are defined in terms of components of the mean ion energy, and in the first approximation they are equal to the basis temperatures $T_b^{(T)}$ and $T_b^{(L)}$, as given by (6-6-68a) and (6-6-68b). Although expressions for these quantities can be found in terms of collision integrals based on the Cartesian basis functions, it is more satisfactory to relate them to the total translational temperature T_b and the fractional energy-loss parameter ξ of (6-6-36). That is, it is better to put the expressions for the ion temperatures in terms of moments of the *spherical* basis functions in order to recapture T_b and ξ. This procedure was followed by Viehland et al. (1981), whose results can be written in the following form:

$$kT_{T,L}[1 + (M/m)\xi] = kT(1 - \xi) + \zeta_{T,L}Mv_d^2, \qquad (6\text{-}6\text{-}73)$$

where

$$\zeta_T = \frac{(m + M)(A^* - \tfrac{5}{3}\xi)}{5m + 3MA^*}, \qquad (6\text{-}6\text{-}74a)$$

$$\zeta_L = \frac{5m - (2m - M)A^* - \tfrac{1}{3}\xi(5m - 10M + 9MA^*)}{5m + 3MA^*}. \qquad (6\text{-}6\text{-}74b)$$

For $\xi = 0$ these formulas reduce to those of the one- and two-temperature theories, as given by (6-1-89a) and (6-1-89b), for instance. But when inelastic collisions occur we must include ξ. Similarly, the quantity A^* is the generalization to inelastic collisions of the usual ratio of collision integrals, $\bar{\Omega}^{(2,2)}/\bar{\Omega}^{(1,1)}$. It is defined in terms of the generalized spherical-basis collision integrals as

$$A_j^* \equiv \frac{5\langle\!\langle(3\gamma_z^2 - \gamma^2)[\gamma_z^2 - (\gamma_z')^2]\rangle\!\rangle_j}{12\langle\!\langle\gamma_z(\gamma_z - \gamma_z')\rangle\!\rangle_j}, \qquad (6\text{-}6\text{-}75)$$

for neutral species j. It is readily verified that these formulas reduce to (6-6-35) for T_b on substitution into the expression

$$\tfrac{3}{2}kT_b = \tfrac{1}{2}k(2T_T + T_L) + \tfrac{1}{2}mv_d^2, \tag{6-6-76}$$

which follows exactly from the definitions of the ion temperatures in terms of ion energies.

However, it is *not* true that $(2T_T + T_L) = T_{eff}$, as is the case for elastic collisions; the relation turns out to be

$$2T_T + T_L = T_{eff}(1 - \xi), \tag{6-6-77}$$

which follows by comparison of the formulas above with formula (6-6-37) for T_{eff}.

This is as far as the results with Cartesian basis functions have been carried. It remains only to indicate the extension for mixtures of neutral gases.

3. Mixtures

As usual, the moment equations for a mixture can be obtained from those for a single gas by replacement of the matrix elements $a_j(pqrs; p'q'r's')$ by the linear combinations,

$$b(pqrs; p'q'r's') = \sum_j x_j a_j(pqrs; p'q'r's'). \tag{6-6-78}$$

Although this gives a method of calculation, not much insight is obtained. The only simple results that have been extracted from the formalism are expressions for the ion temperatures. As with single gases, these expressions have been written in terms of averages in the spherical-basis system, and have been made to resemble the corresponding formulas for single gases. Thus in place of (6-6-73) we have

$$\langle kT_{T,L}\rangle_{\mathrm{mix}}(1 + \langle M\xi\rangle_{\mathrm{mix}}/m) = kT(1 - \langle \xi\rangle_{\mathrm{mix}}) + \langle \zeta_{T,L}\rangle_{\mathrm{mix}}\langle M\rangle_{\mathrm{mix}}v_d^2, \tag{6-6-79}$$

where the notation $\langle\ \rangle_{\mathrm{mix}}$ has been defined by (6-6-48) as far as $M\xi$, ξ, and M are concerned. For $\langle \zeta_{T,L}\rangle_{\mathrm{mix}}$ the expressions are (Viehland et al., 1981)

$$\langle \zeta_T\rangle_{\mathrm{mix}} = \frac{(m + \langle M\rangle_{\mathrm{mix}})(\langle MA^*\rangle_{\mathrm{mix}} - \tfrac{5}{3}\langle M\xi\rangle_{\mathrm{mix}})}{\langle M\rangle_{\mathrm{mix}}(5m + 3\langle MA^*\rangle_{\mathrm{mix}})}, \tag{6-6-80a}$$

$$\langle \zeta_L\rangle_{\mathrm{mix}} =$$

$$\frac{5m\langle M\rangle_{\mathrm{mix}} - (2m - \langle M\rangle_{\mathrm{mix}})\langle MA^*\rangle_{\mathrm{mix}} - \tfrac{1}{3}\langle M\xi\rangle_{\mathrm{mix}}(5m - 10\langle M\rangle_{\mathrm{mix}} + 9\langle MA^*\rangle_{\mathrm{mix}})}{\langle M\rangle_{\mathrm{mix}}(5m + 3\langle MA^*\rangle_{\mathrm{mix}})}.$$

$$\tag{6-6-80b}$$

The quantity $\langle MA^* \rangle_{\text{mix}}$ is defined according to (6-6-48). If $\xi = 0$, these formulas reduce to those of the one- and two-temperature theories, as given by (6-1-91) and (6-1-92).

D. Final Remarks

The results of this section show how inelastic collisions can be introduced into the kinetic theory of drift-tube experiments in a systematic way. There have been very few applications of these results to date, largely because of the lack of information on inelastic cross sections. The collision dynamics of inelastic processes can be quite complicated to calculate, and this appears to be the biggest bottleneck for applications at present. In a sense this is a different problem than the purely kinetic-theory problem considered here, which can be considered as essentially solved. Simple calculations (e.g., Figs. 6-6-1 and 6-6-2) and simple scaling arguments suggested by the theory (e.g., Fig. 6-6-3) indicate that ion transport data for molecular systems contain substantial information about inelastic processes and nonspherical ion-neutral interactions, but the way to extract such information is not yet clear. Extensive numerical exploration with simple models will probably be the first step, and this awaits advances in the computation of inelastic collision dynamics.

REFERENCES

Albritton, D. L., T. M. Miller, D. W. Martin, and E. W. McDaniel (1968). Mobilities of mass-identified H_3^+ and H^+ ions in hydrogen, *Phys. Rev.* **171**, 94–102.

Alievskiĭ, M. Ya., and V. M. Zhdanov (1969). Transport and relaxation phenomena in polyatomic gas mixtures, *Soviet Phys. JETP* (English Transl.) **28**, 116–121 [*Zh. Eksp. Teor. Fiz.* **55**, 221–232 (1968)].

Arthurs, A. M., and A. Dalgarno (1960). The mobilities of ions in molecular gases, *Proc. R. Soc. London* **A256**, 552–558.

Baker, G. A., Jr. (1975). *Essentials of Padé Approximants*, Academic Press, New York.

Baraff, G. A. (1964). Maximum anisotropy approximation for calculating electron distributions: Application to high field transport in semiconductors, *Phys. Rev.* **133**, A26–A33.

Barnes, W. S., D. W. Martin, and E. W. McDaniel (1961). Mass spectrographic identification of the ion observed in hydrogen mobility experiments, *Phys. Rev. Lett.* **6**, 110–111.

Beaty, E. C., J. C. Browne, and A. Dalgarno (1966). Ion mobilities in helium, *Phys. Rev. Lett.* **16**, 723–724.

Beenakker, J. J. M., and F. R. McCourt (1970). Magnetic and electric effects on transport properties, *Ann. Rev. Phys. Chem.* **21**, 47–72.

Biondi, M. A., and L. M. Chanin (1961). Blanc's law—Ion mobilities in helium-neon mixtures, *Phys. Rev.* **122**, 843–847.

Braglia, G. L. (1981). On the accuracy of the iterative method for swarm transport coefficients, *J. Chem. Phys.* **74**, 2990–2992.

Byers, M. S., M. G. Thackston, R. D. Chelf, F. B. Holleman, J. R. Twist, G. W. Neeley, and E. W. McDaniel (1983). Mobilities of Tl^+ ions in Kr and Xe, Li^+ in Kr and Xe, and Cl^- in N_2, *J. Chem. Phys.* **78**, 2796–2797.

Chanin, L. M. (1961). Temperature variation of ionic mobilities in hydrogen, *Phys. Rev.* **123**, 526–529.

Chapman, S., and T. G. Cowling (1970). *The Mathematical Theory of Non-Uniform Gases*, 3rd ed., Cambridge University Press, London.

Cohen, J. S., and B. Schneider (1975). Collisions of Ne* (3s) and Ne^+ with Ne: Excitation and charge transfer, elastic scattering, and diffusion, *Phys. Rev.* **A11**, 884–892.

Dalgarno, A. (1958). The mobilities of ions in their parent gases, *Philos. Trans. R. Soc. London* **A250**, 426–439.

Dickinson, A. S. (1968a). The mobility of He^+ ions in He, *J. Phys.* **B1**, 387–394.

Dickinson, A. S. (1968b). The mobility of protons and alpha particles in helium, *J. Phys.* **B1**, 395–401.

Dickinson, A. S., and M. S. Lee (1978). The mobility of protons in helium, *J. Phys.* **B11**, L377–L379.

Duncan, M. A., V. M. Bierbaum, G. B. Ellison, and S. R. Leone (1983). Laser-induced fluorescence studies of ion collisional excitation in a drift field: Rotational excitation of N_2^+ in helium, *J. Chem. Phys.* **79**, 5448–5456.

Eisele, F. L., M. G. Thackston, W. M. Pope, I. R. Gatland, H. W. Ellis, and E. W. McDaniel (1977). Experimental test of the generalized Einstein relation for Cs^+ ions in molecular gases: H_2, N_2, O_2, CO, and CO_2, *J. Chem. Phys.* **67**, 1278–1279.

Eisele, F. L., M. D. Perkins, and E. W. McDaniel (1980). Mobilities of NO_2^-, NO_3^-, and CO_3^- in N_2 over the temperature range 217-675 K, *J. Chem. Phys.* **73**, 2517–2518.

Eisele, F. L., M. D. Perkins, and E. W. McDaniel (1981). Measurement of the mobilities of Cl^-, $NO_2^- \cdot H_2O$, $NO_3^- \cdot H_2O$, $CO_3^- \cdot H_2O$, and $CO_4^- \cdot H_2O$ in N_2 as a function of temperature, *J. Chem. Phys.* **75**, 2473–2475.

Ellis, H. W., R. Y. Pai, E. W. McDaniel, E. A. Mason, and L. A. Viehland (1976a). Transport properties of gaseous ions over a wide energy range, *At. Data Nucl. Data Tables* **17**, 177–210.

Ellis, H. W., M. G. Thackston, R. Y. Pai, and E. W. McDaniel (1976b). Longitudinal diffusion coefficients of Rb^+ ions in He, Ne, Ar, H_2, N_2, O_2, and CO_2, *J. Chem. Phys.* **65**, 3390–3391.

Ellis, H. W., E. W. McDaniel, D. L. Albritton, L. A. Viehland, S. L. Lin, and E. A. Mason (1978). Transport properties of gaseous ions over a wide energy range: Part 2, *At. Data Nucl. Data Tables* **22**, 179–217.

Ellis, H. W., M. G. Thackston, E. W. McDaniel, and E. A. Mason (1984). Transport properties of gaseous ions over a wide energy range: Part 3, *At. Data Nucl. Data Tables* **31**, 113–151.

Evans, D. J. (1977). Transport properties of homonuclear diatomics: 1. Dilute gases, *Mol. Phys.* **34**, 103–112.

Evans, D. J., and R. O. Watts (1976). A theoretical study of transport coefficients in benzene vapor, *Mol. Phys.* **32**, 995–1015.

Fahr, H., and K. G. Müller (1967). Ionenbewegung unter dem Einfluss von Umladungsstössen, *Z. Phys.* **200**, 343–365.

Ferziger, J. H., and H. G. Kaper (1972). *Mathematical Theory of Transport Processes in Gases*, North-Holland, Amsterdam.

Firsov, O. B. (1951). Resonant charge transfer of ions during slow collisions (in Russian), *Zh. Eksp. Teor. Fiz.* **21**, 1001–1008.

Gatland, I. R., W. F. Morrison, H. W. Ellis, M. G. Thackston, E. W. McDaniel, M. H. Alexander, L. A. Viehland, and E. A. Mason (1977). The Li^+-He interaction potential, *J. Chem. Phys.* **66**, 5121–5125.

Geltman, S. (1953). The mobility of helium molecular ions in helium, *Phys. Rev.* **90**, 808–816.

Grew, K. E., and T. L. Ibbs (1952). *Thermal Diffusion in Gases*, Cambridge University Press, London.

Grew, K. E., and T. L. Ibbs (1962). *Thermodiffusion in Gasen*, VEB Deutschen Verlag der Wissenschaften, Berlin.

Hahn, H., and E. A. Mason (1972). Field dependence of gaseous-ion mobility: Theoretical tests of approximate formulas, *Phys. Rev.* **A6**, 1573–1577.

Hariharan, P. C., and V. Staemmler (1976). Potential energy curve of $^1\Sigma^+$ Li^+/He, *Chem. Phys.* **15**, 409–414.

Hassé, H. R. (1926). Langevin's theory of ionic mobility, *Philos. Mag.* **1**, 139–160.

Heiche, G. (1967). Charge exchange and mobility of atomic ions in their parent gases, *U.S. Naval Ordnance Lab. Rep. NOL TR 67-150*.

Heiche, G., and E. A. Mason (1970). Ion mobilities with charge exchange, *J. Chem. Phys.* **53**, 4687–4696.

Helm, H. (1975). The mobilities of Kr^+ $(^2P_{3/2})$ and Kr^+ $(^2P_{1/2})$ in krypton at 295 K, *Chem. Phys. Lett.* **36**, 97–99.

Helm, H. (1976a). The mobilities and equilibrium reactions of helium ions in helium at 77 K, *J. Phys.* **B9**, 1171–1189.

Helm, H. (1976b). The mobilities of atomic krypton and xenon ions in the $^2P_{1/2}$ and $^2P_{3/2}$ state in their parent gas, *J. Phys.* **B9**, 2931–2943.

Helm, H. (1977). The cross section for symmetric charge exchange of He^+ in He at energies between 0.3 and 8 eV, *J. Phys.* **B10**, 3683–3697.

Helm, H., and M. T. Elford (1977a). The influence of fine-structure splitting on the mobility of atomic neon ions in neon, *J. Phys.* **B10**, 983–991.

Helm, H., and M. T. Elford (1977b). The mobility of Ar^+ ions in argon and the effect of spin-orbit coupling, *J. Phys.* **B10**, 3849–3851.

Herzberg, G. (1950). *Spectra of Diatomic Molecules*, 2nd ed., Van Nostrand Reinhold, New York.

Higgins, L. D., and F. J. Smith (1968). Collision integrals for high temperature gases, *Mol. Phys.* **14**, 399–400.

Hirschfelder, J. O., C. F. Curtiss, and R. B. Bird (1964). *Molecular Theory of Gases and Liquids*, Wiley, New York.

Holstein, T. (1952). Mobilities of positive ions in their parent gases, *J. Phys. Chem.* **56**, 832–836.

Holstein, T. (1955). Mobility of positive ions in gas mixtures, *Phys. Rev.* **100**, 1230 (A).

Hoselitz, K. (1941). The mobility of alkali ions in gases: 5. Temperature measurements in the inert gases, *Proc. R. Soc. London* **A177**, 200–204.

Howorka, F., F. C. Fehsenfeld, and D. L. Albritton (1979). H^+ and D^+ ions in He: Observations of a runaway mobility, *J. Phys.* **B12**, 4189–4197.

Iinuma, K., M. Takebe, Y. Satoh, and K. Seto (1982). Numerical calculations of generalized mobilities derived from n-6-4 potentials including core sizes, *Technol. Rep. Tohoku Univ.* **47**, 139–158.

James, D. R., E. Graham, G. R. Akridge, I. R. Gatland, and E. W. McDaniel (1975). Longitudinal diffusion of K^+ ions in He, Ne, Ar, H_2, NO, O_2, CO_2, N_2 and CO, *J. Chem. Phys.* **62**, 1702–1705.

Kihara, T. (1953). The mathematical theory of electrical discharges in gases: B. Velocity-distribution of positive ions in a static field, *Rev. Mod. Phys.* **25**, 844–852.

Kleban, P., and H. T. Davis (1978). Electron drift and diffusion in polyatomic gases: Calculations for CH_4, CD_4, and related models, *J. Chem. Phys.* **68**, 2999–3006.

Koizumi, T., T. Tsurugai, and I. Ogawa (1987). State dependence of mobilities for Kr^{++} in Kr at 88 K, *J. Phys. Soc. Jpn.* **56**, 17–20.

Kolos, W. (1976). Long- and intermediate-range interaction in the three lowest sigma states of the HeH^+ ion, *Int. J. Quantum Chem.* **10**, 217–224.

Kolos, W., and J. M. Peek (1976). New ab initio potential curve and quasibound states of HeH^+, *Chem. Phys.* **12**, 381–386.

Kumar, K. (1977). Relation between mobility and diffusion in swarms, *Proc. 13th Int. Conf. Phenomena Ionized Gases*, Phys. Soc. G.D.R., Berlin, Part 2, pp. 747–748.

Kumar, K. (1980). Matrix elements of the Boltzmann collision operator in a basis determined by an anisotropic Maxwellian weight function including drift, *Aust. J. Phys.* **33**, 449–468.

Kumar, K., and R. E. Robson (1973). Mobility and diffusion: 1. Boltzmann equation treatment for charged particles in a neutral gas, *Aust. J. Phys.* **26**, 157–186.

Kumar, K., H. R. Skullerud, and R. E. Robson (1980). Kinetic theory of charged particle swarms in neutral gases, *Aust. J. Phys.* **33**, 343–448.

Landau, L. D., and E. M. Lifshitz (1977). *Quantum Mechanics: Non-Relativistic Theory*, 3rd ed., Pergamon, London.

Langevin, P. (1905). Une formule fondamentale de théorie cinétique, *Ann. Chim. Phys.* **5**, 245–288. A translation is given by E. W. McDaniel, *Collision Phenomena in Ionized Gases*, Wiley, New York, 1964, App. 2, pp. 701–726.

Lin, S. L., and E. A. Mason (1979). Influence of resonant charge transfer on ion mobility, *J. Phys.* **B12**, 783–789.

Lin, S. L., I. R. Gatland, and E. A. Mason (1979a). Mobility and diffusion of protons and deuterons in helium—A runaway effect, *J. Phys.* **B12**, 4179–4188.

Lin, S. L., R. E. Robson, and E. A. Mason (1979b). Moment theory of electron drift and diffusion in neutral gases in an electrostatic field, *J. Chem. Phys.* **71**, 3483–3498.

Lin, S. L., L. A. Viehland, and E. A. Mason (1979c). Three-temperature theory of gaseous ion transport, *Chem. Phys.* **37**, 411–424.

Lindenfeld, M. J., and B. Shizgal (1979). Matrix elements of the Boltzmann collision operator for gas mixtures, *Chem. Phys.* **41**, 81–95.

Loeb, L. B. (1960). *Basic Processes of Gaseous Electronics*, 2nd ed., University of California Press, Berkeley, Calif.

McDaniel, E. W., and E. A. Mason (1973). *The Mobility and Diffusion of Ions in Gases*, Wiley, New York.

McDaniel, E. W., and J. T. Moseley (1971). Tests of the Wannier expressions for diffusion coefficients of gaseous ions in electric fields, *Phys. Rev.* **A3**, 1040–1044.

McDaniel, E. W., and L. A. Viehland (1984). The transport of slow ions in gases: Experiment, theory, and applications, *Phys. Rep.* **110**, 333–367.

Madson, J. M., H. J. Oskam, and L. M. Chanin (1965). Ion mobilities in helium, *Phys. Rev. Lett.* **15**, 1018–1020.

Maitland, G. C., V. Vesovic, and W. A. Wakeham (1981). The inelastic contributions to transport collision integrals in the infinite-order sudden approximation, *Mol. Phys.* **42**, 803–815.

Märk, E., and T. D. Märk (1984). Transverse ion diffusion in gases, in *Swarms of Ions and Electrons in Gases*, W. Lindinger, T. D. Märk, and F. Howorka, Eds., Springer-Verlag, New York, pp. 60–86.

Marrero, T. R., and E. A. Mason (1972). Gaseous diffusion coefficients, *J. Phys. Chem. Ref. Data* **1**, 3–118.

Mason, E. A. (1957a). Higher approximations for the transport properties of binary gas mixtures: 1. General formulas, *J. Chem. Phys.* **27**, 75–84.

Mason, E. A. (1957b). Higher approximations for the transport properties of binary gas mixtures: 2. Applications, *J. Chem. Phys.* **27**, 782–790.

Mason, E. A., and L. Monchick (1962). Heat conductivity of polyatomic and polar gases, *J. Chem. Phys.* **36**, 1622–1639.

Mason, E. A., and H. W. Schamp, Jr. (1958). Mobility of gaseous ions in weak electric fields, *Ann. Phys. N.Y.* **4**, 233–270.

Mason, E. A., and J. T. Vanderslice (1959). Mobility of hydrogen ions (H^+, $H_2{}^+$, $H_3{}^+$) in hydrogen, *Phys. Rev.* **114**, 497–502.

Mason, E. A., J. T. Vanderslice, and J. M. Yos (1959). Transport properties of high-temperature multicomponent gas mixtures, *Phys. Fluids* **2**, 688–694.

Mason, E. A., R. J. Munn, and F. J. Smith (1966). Thermal diffusion in gases, *Adv. At. Mol. Phys.* **2**, 33–91.

Mason, E. A., H. O'Hara, and F. J. Smith (1972). Mobilities of polyatomic ions in gases: Core model, *J. Phys.* **B5**, 169–176.

Massey, H. S. W. (1971). *Slow Collisions of Heavy Particles*, 2nd ed., Vol. 3 of *Electronic and Ionic Impact Phenomena*, H. S. W. Massey, E. H. S. Burhop, and H. B. Gilbody, Eds., Oxford University Press, London.

Massey, H. S. W., and C. B. O. Mohr (1934). Free paths and transport phenomena in gases and the quantum theory of collisions: 2. The determination of the laws of force between atoms and molecules, *Proc. R. Soc. London* **A144**, 188–205.

Massey, H. S. W., and R. A. Smith (1933). The passage of positive ions through gases, *Proc. R. Soc. London* **A142**, 142–172.

Miller, T. M., J. T. Moseley, D. W. Martin, and E. W. McDaniel (1968). Reactions of H^+ in H_2 and D^+ in D_2; Mobilities of hydrogen and alkali ions in H_2 and D_2 gases, *Phys. Rev.* **173**, 115–123.

Monchick, L. (1959). Collision integrals for the exponential repulsive potential, *Phys. Fluids* **2**, 695–700.

Monchick, L., K. S. Yun, and E. A. Mason (1963). Formal kinetic theory of transport phenomena in polyatomic gas mixtures, *J. Chem. Phys.* **39**, 654–669.

Monchick, L., A. N. G. Pereira, and E. A. Mason (1965). Heat conductivity of polyatomic and polar gases and gas mixtures, *J. Chem. Phys.* **42**, 3241–3256.

Monchick, L., R. J. Munn, and E. A. Mason (1966). Thermal diffusion in polyatomic gases: A generalized Stefan–Maxwell diffusion equation, *J. Chem. Phys.* **45**, 3051–3058.

Monchick, L., S. I. Sandler, and E. A. Mason (1968). Thermal diffusion in polyatomic gases: Nonspherical interactions, *J. Chem. Phys.* **49**, 1178–1184.

Mott, N. F., and H. S. W. Massey (1965). *The Theory of Atomic Collisions*, 3rd ed., Oxford University Press, London.

Munn, R. J., E. A. Mason, and F. J. Smith (1964). Some aspects of the quantal and semiclassical calculation of phase shifts and cross sections for molecular scattering and transport, *J. Chem. Phys.* **41**, 3978–3988.

Nyeland, C., L. L. Poulsen, and G. D. Billing (1984). Rotational relaxation and transport coefficients for diatomic gases: Computations on nitrogen, *J. Phys. Chem.* **88**, 1216–1221.

Orient, O. J. (1967). The measurement of the temperature dependence of atomic and molecular ion mobilities in helium, *Can. J. Phys.* **45**, 3915–3922.

Orient, O. J. (1971). The mobility of mass-identified protons in helium gas, *J. Phys.* **B4**, 1257–1266.

Orient, O. J. (1972). The mobility of mass-identified D^+ ions in helium gas, *J. Phys.* **B5**, 1056–1058.

Pai, R. Y., H. W. Ellis, and E. W. McDaniel (1976). The generalized Einstein relation—Application to Li^+ and Na^+ ions in hydrogen gas, *J. Chem. Phys.* **64**, 4238–4239.

Parker, G. A., and R. T. Pack (1978). Rotationally and vibrationally inelastic scattering in the rotational IOS approximation. Ultrasimple calculation of total (differential, integral, and transport) cross sections for nonspherical molecules, *J. Chem. Phys.* **68**, 1585–1601.

Patterson, P. L. (1970a). Mobilities of negative ions in SF_6, *J. Chem. Phys.* **53**, 696–704.

Patterson, P. L. (1970b). Temperature dependence of helium-ion mobilities, *Phys. Rev.* **A2**, 1154–1164.

Petrović, Z. L. (1986). The application of Blanc's law to the determination of the diffusion coefficients for thermal electrons in gases, *Aust. J. Phys.* **39**, 237–247.

Petrović, Z. L., R. W. Crompton, and G. N. Haddad (1984). Model calculations of negative differential conductivity in gases, *Aust. J. Phys.* **37**, 23–35.

Robson, R. E. (1972). A thermodynamic treatment of anisotropic diffusion in an electric field, *Aust. J. Phys.* **25**, 685–693.

Robson, R. E. (1976). On the generalized Einstein relations for gaseous ions in an electrostatic field, *J. Phys.* **B9**, L337–L339.

Robson, R. E. (1984). Generalized Einstein relation and negative differential conductivity in gases, *Aust. J. Phys.* **37**, 35–44.

Robson, R. E., and K. Kumar (1973). Mobility and diffusion: 2. Dependence on experimental variables and interaction potentials for alkali ions in rare gases, *Aust. J. Phys.* **26**, 187–201.

Sandler, S. I., and J. S. Dahler (1967). Kinetic theory of loaded spheres: 4. Thermal diffusion in a dilute-gas mixture of D_2 and HT, *J. Chem. Phys.* **47**, 2621–2630.

Sandler, S. I., and E. A. Mason (1967). Thermal diffusion in a loaded sphere-smooth sphere mixture: A model for ^{4}He-HT and ^{3}He-HD, *J. Chem. Phys.* **47**, 4653–4658.

Sandler, S. I., and E. A. Mason (1968). Kinetic-theory deviations from Blanc's law of ion mobilities, *J. Chem. Phys.* **48**, 2873–2875.

Saporoschenko, M. (1965). Mobility of mass-analyzed H^+, H_3^+, and H_5^+ ions in hydrogen gas, *Phys. Rev.* **139**, A349–A351.

Schruben, D. L., and D. W. Condiff (1973). External fields in the kinetic theory of gases, *J. Chem. Phys.* **59**, 306–323.

Sejkora, G., P. Girstmair, H. C. Bryant, and T. D. Märk (1984). Transverse diffusion of Ar^+ and Ar^{2+} in Ar, *Phys. Rev.* **A29**, 3379–3387.

Shanks, D. (1955). Non-linear transformations of divergent and slowly convergent sequences, *J. Math & Phys.* **34**, 1–42.

Sinha, S., and J. N. Bardsley (1976). Symmetric charge transfer in low-energy ion-atom collisions, *Phys. Rev.* **A14**, 104–113.

Sinha, S., S. L. Lin, and J. N. Bardsley (1979). The mobility of He^+ ions in He, *J. Phys.* **B12**, 1613–1622.

Skullerud, H. R. (1969). Diffusion of gaseous ions in strong electric fields under the influence of charge-transfer collisions, *J. Phys.* **B2**, 86–90.

Skullerud, H. R. (1972). Mobility, diffusion and interaction potential for potassium ions in argon, *Tech. Rep. EIP 72-3*, Physics Dept., Norwegian Inst. Tech., Trondheim.

Skullerud, H. R. (1973). Mobility, diffusion and interaction potential for potassium ions in argon, *J. Phys.* **B6**, 918–928.

Skullerud, H. R. (1974). On the theory of electron diffusion in electrostatic fields in gases, *Aust. J. Phys.* **27**, 195–209.

Skullerud, H. R. (1976). On the relation between the diffusion and mobility of gaseous ions moving in strong electric fields, *J. Phys.* **B9**, 535–546.

Skullerud, H. R. (1984). On the calculation of ion swarm properties by velocity moment methods, *J. Phys.* **B17**, 913–929.

Skullerud, H. R., T. Eide, and T. Stefánsson (1986). Transverse diffusion of lithium ions in helium, *J. Phys.* **D19**, 197–208.

Smirnov, B. M. (1967). Ion mobility in a neutral host gas of the same species, *Soviet Phys. Tech. Phys.* (English Transl.) **11**, 1388–1393 [*Zh. Tekhn. Fiz.* **36**, 1864–1871 (1966)].

Smith, F. J. (1967). Low energy elastic and resonant exchange cross sections between complex atoms, *Mol. Phys.* **13**, 121–130.

Smith, F. T. (1965). Classical and quantal scattering: 1. The classical action, *J. Chem. Phys.* **42**, 2419–2426.

Taxman, N. (1958). Classical theory of transport phenomena in dilute polyatomic gases, *Phys. Rev.* **110**, 1235–1239.

Thackston, M. G., M. S. Byers, F. B. Holleman, R. D. Chelf, J. R. Twist, and E. W. McDaniel (1983). Longitudinal diffusion coefficients and test of the generalized Einstein relation for Tl^+ ions in Kr and Xe, Li^+ in Kr and Xe, and Cl^- in N_2, *J. Chem. Phys.* **78**, 4781–4782.

Thomson, G. M., J. H. Schummers, D. R. James, E. Graham, I. R. Gatland, M. R. Flannery, and E. W. McDaniel (1973). Mobility, diffusion, and clustering of K^+ ions in gases, *J. Chem. Phys.* **58**, 2402–2411.

Varney, R. N. (1960). Mobility of hydrogen ions, *Phys. Rev. Lett.* **5**, 559–560.

Viehland, L. A. (1982). Gaseous ion transport coefficients, *Chem. Phys.* **70**, 149–156.

Viehland, L. A. (1983). Interaction potentials for Li^+-rare gas systems, *Chem. Phys.* **78**, 279–294.

Viehland, L. A. (1984a). Interaction potentials for the alkali ion-rare gas systems, *Chem. Phys.* **85**, 291–305.

Viehland, L. A. (1984b). Internal-energy distribution of molecular ions in drift tubes, in *Swarms of Ions and Electrons in Gases*, W. Lindinger, T. D. Märk, and F. Howorka, Eds., Springer-Verlag, New York, pp. 27–43.

Viehland, L. A. (1986). Classical kinetic theory of drift tube experiments involving molecular ion-neutral systems, *Chem. Phys.* **101**, 1–16.

Viehland, L. A., and D. W. Fahey (1983). The mobilities of NO_3^-, NO_2^-, NO^+, and Cl^- in N_2: A measure of inelastic energy loss, *J. Chem. Phys.* **78**, 435–441.

Viehland, L. A., and M. Hesche (1986). Transport properties for systems with resonant charge transfer, *Chem. Phys.* **110**, 41–54.

Viehland, L. A., and S. L. Lin (1979). Application of the three-temperature theory of gaseous ion transport, *Chem. Phys.* **43**, 135–144.

Viehland, L. A., and E. A. Mason (1975). Gaseous ion mobility in electric fields of arbitrary strength, *Ann. Phys. N.Y.* **91**, 499–533.

Viehland, L. A., and E. A. Mason (1978). Gaseous ion mobility and diffusion in electric fields of arbitrary strength, *Ann. Phys. N.Y.* **110**, 287–328.

Viehland, L. A., E. A. Mason, and J. H. Whealton (1974). Mean energy distribution of gaseous ions in electrostatic fields, *J. Phys.* **B7**, 2433–2439.

Viehland, L. A., E. A. Mason, W. F. Morrison, and M. R. Flannery (1975). Tables of transport collision integrals for $(n, 6, 4)$ ion-neutral potentials, *At. Data Nucl. Data Tables* **16**, 495–514.

Viehland, L. A., S. L. Lin, and E. A. Mason (1981). Kinetic theory of drift-tube experiments with polyatomic species, *Chem. Phys.* **54**, 341–364.

Volz, D. J., H. J. Schummers, R. D. Laser, D. W. Martin, and E. W. McDaniel (1971). Mobilities and longitudinal diffusion coefficients of mass-identified potassium ions and positive nitric oxide ions in nitric oxide, *Phys. Rev.* **A4**, 1106–1109.

Waldman, M., and E. A. Mason (1981). Generalized Einstein relations from a three-temperature theory of gaseous ion transport, *Chem. Phys.* **58**, 121–144.

Waldman, M., E. A. Mason, and L. A. Viehland (1982). Influence of resonant charge transfer on ion diffusion and generalized Einstein relations, *Chem. Phys.* **66**, 339–349.

Waldmann, L. (1965). Quantum-theoretical transport equations for polyatomic gases, in *Statistical Mechanics of Equilibrium and Non-equilibrium*, J. Meixner, Ed., North-Holland, Amsterdam, pp. 177–191.

Waldmann, L. (1968). Kinetic theory of dilute gases with internal degrees of freedom, in *Fundamental Problems in Statistical Mechanics*, Vol. 2, E. G. D. Cohen, Ed., North-Holland, Amsterdam, pp. 276–305.

Waldmann, L. (1973). On kinetic equations for particles with internal degrees of freedom, *Acta Phys. Austriaca Suppl.* **10**, 223–246.

Wang Chang, C. S., G. E. Uhlenbeck, and J. de Boer (1964). The heat conductivity and viscosity of polyatomic gases, *Stud. Stat. Mech.* **2**, 241–268.

Wannier, G. H. (1953). Motion of gaseous ions in strong electric fields, *Bell Syst. Tech. J.* **32**, 170–254.

Weinert, U. (1978). Matrix elements of the linearized collision operator for multi-temperature gas-mixtures, *Z. Naturforsch.* **33a**, 480–492.

Weinert, U. (1980). Thermal diffusion of ions drifting in an electric field through a neutral gas, *Phys. Fluids* **23**, 1518–1525.

Weinert, U. (1982). Multi-temperature generalized moment method in Boltzmann transport theory, *Phys. Rep.* **91**, 297–399.

Weinert, U. (1983). Gaussian basis functions in kinetic theory: 1. Incorporation of inelastic processes, *Physica* **121A**, 150–174.

Weinert, U., and E. A. Mason (1980). Generalized Nernst-Einstein relations for nonlinear transport coefficients, *Phys. Rev.* **A21**, 681–690.

Whealton, J. H., and E. A. Mason (1974). Transport coefficients of gaseous ions in an electric field, *Ann. Phys. N.Y.* **84**, 8–38.

Whealton, J. H., E. A. Mason, and R. E. Robson (1974). Composition dependence of ion-transport coefficients in gas mixtures, *Phys. Rev.* **A9**, 1017–1020.

Wood, H. T. (1971). General expression for the quantum transport cross sections, *J. Chem. Phys.* **54**, 977–979.

7

INTERACTION POTENTIALS AND TRANSPORT COEFFICIENTS

The relations connecting mobilities and diffusion coefficients with the ion-neutral interaction potential $V(r)$ have been developed in considerable detail in Chapters 5 and 6. In this chapter we consider how to use these relations in practical ways. There are three main uses for these relations:

1.. Scaling rules
2. Calculation of transport coefficients from $V(r)$
3. Determination of $V(r)$ from transport measurements—the inversion problem.

Scaling rules are by far the easiest to implement because they involve very little computation, and have already been used throughout Chapters 5 and 6. The scaling of T and E/N into the single variable T_{eff} is the most important rule, illustrated for a variety of systems in Figs. 5-2-1 (K^+ in He, Ne, Ar), 5-2-2 (e^- in He, Ne, Ar), 6-1-2 (H^+, D^+, Li^+ in He), 6-1-5 (SF_5^-, SF_6^-, $(SF_6)SF_6^-$, $(SF_6)_2SF_6^-$ in SF_6), 6-1-6 (He^+, He_2^+ in He), 6-5-5 (H_3^+ in H_2, D_3^+ in D_2), and 6-3-3 (Cl^-, NO_2^- in N_2). The less important mass scaling rule for isotopes is illustrated in Figs. 6-1-2 and 6-5-5, with hydrogen and deuterium. We have nothing further to add in this chapter on the subject of scaling rules.

The calculation of transport coefficients from a given $V(r)$ is straightforward, but usually requires the aid of a computer (unless the computer results are already available in the form of standard tables). There are two aspects to this application. The first is to test some independently known $V(r)$, as calculated from a theoretical model, for instance, in order to assess its accuracy. Or if such a $V(r)$ has been determined from other kinds of measurements, such as ion-beam scattering or spectroscopic results, then the calculation tests the consistency of these measurements with the transport measurements. The second aspect is to estimate needed transport coefficients in the face of meager information, by making educated semiempirical guesses about the nature and magnitude of $V(r)$.

The third use of transport measurements, to determine $V(r)$, is potentially the most important application of the theory. It is also the most difficult, both conceptually and computationally. It is an example of a large number of inversion problems that occur throughout physics and chemistry. Substantial progress has been made on this inversion problem in recent years, although many intriguing questions remain.

All the real applications that we can properly discuss should refer to spherical ion-atom potentials. The reason is that present computing power is not yet able to cope routinely with nonspherical potentials and inelastic collisions. But the day may not be far distant when such computations become routinely feasible, and the results for spherical interactions should then supply a useful guide. However, in some cases it seems reasonable at present to treat the interactions between polyatomic species via an effective spherical potential, or by some related simple approximation, and we will therefore make a few comments on nonspherical interactions in this connection.

We begin with a summary of computational methods for evaluating transport cross sections and collision integrals when $V(r)$ is given. Next we give a very brief sketch of the nature of ion-neutral interactions, to serve as background for the subsequent discussion. This subsequent discussion consists of three parts. The first part illustrates the use of ion mobility and diffusion data to test potentials obtained from a variety of sources. The second part illustrates how reasonable values of transport coefficients can be estimated when only fragmentary information is available. The third and last part is essentially a status report on the fundamental inversion problem.

7-1. CALCULATION OF TRANSPORT CROSS SECTIONS FROM INTERACTION POTENTIALS

Whatever moment method is adopted for solving the Boltzmann equation, the numerical computation of transport coefficients eventually comes down to first evaluating a set of transport cross sections, usually

$$Q^{(l)}(\varepsilon) = 2\pi \int_0^\pi (1 - \cos^l\theta)\sigma(\theta, \varepsilon) \sin\theta \, d\theta, \qquad (7\text{-}1\text{-}1)$$

or sometimes

$$\sigma^{(l)}(\varepsilon) = 2\pi \int_0^\pi [1 - P_l(\cos\theta)]\sigma(\theta, \varepsilon) \sin\theta \, d\theta \qquad (7\text{-}1\text{-}2)$$

(see Section 5-3E). The one- and two-temperature theories average the $Q^{(l)}$ over energies to form the collision integrals $\bar{\Omega}^{(l,s)}$ defined by (6-1-10), which are then combined to form the needed matrix elements. The three-temperature theory averages the $Q^{(l)}$ to form the collision integrals $[p, q]^{(l)}$ defined by (6-3-18), which are used to form matrix elements. Other versions of moment theory, not described in this book, use socalled interaction integrals that are essentially matrix elements of the $\sigma^{(l)}$ with respect to Sonine polynomials (Kumar, 1967; Ness and Robson, 1985).

In other words, the cross sections are the crucial connections between the interaction and the transport coefficients. Their evaluation is often the hardest part of the whole computation, and is certainly much harder than the simple integration needed to obtain collision or interaction integrals. The rest of the computation is mostly just matrix algebra. Except for very simple forms of interaction, the calculation of transport cross sections must be done numerically, and this can be a fairly formidable task even for elastic collisions and spherical interactions. In this section we review such calculations, but leave the problems of inelastic collisions and nonspherical interactions to future generations of computing machines. We first consider the case where classical mechanics affords an adequate description of ion-atom collisions, and then the case where quantum effects cannot be ignored. We finally note the modifications that must be introduced when resonant charge exchange occurs.

A. Classical Calculations

Given the potential, two successive sets of integrations are needed to obtain cross sections. The first integrations determine the deflection angle in an ion-atom collision as a function of impact parameter b and relative energy ε:

$$\theta(b, \varepsilon) = \pi - 2b \int_{r_0}^{\infty} \left[1 - \frac{b^2}{r^2} - \frac{V(r)}{\varepsilon} \right]^{-1/2} \frac{dr}{r^2}, \tag{7-1-3}$$

where the distance of closest approach r_0 is the outermost root of

$$1 - \frac{b^2}{r_0^2} - \frac{V(r_0)}{\varepsilon} = 0. \tag{7-1-4}$$

It is more efficient and accurate to work in an impact parameter formulation with $\theta(b, \varepsilon)$ than to calculate the differential cross section $\sigma(\theta, \varepsilon)$,

$$\sigma(\theta, \varepsilon) \sin \theta = b|d\theta/db|^{-1}. \tag{7-1-5}$$

The reason is that $\sigma(\theta, \varepsilon)$ has singularities (rainbows) in certain regions, which are troublesome to handle by numerical methods.

The second set of integrations yields $Q^{(l)}(\varepsilon)$; in the impact parameter formulation this is a weighted average over impact parameters,

$$Q^{(l)}(\varepsilon) = 2\pi \int_0^{\infty} (1 - \cos^l \theta) b \, db. \tag{7-1-6}$$

Numerical integration for $Q^{(l)}$ is troublesome when θ becomes large during orbiting collisions at lower energies. Numerical integration to find $\theta(b, \varepsilon)$ is also difficult in orbiting regions.

These two sets of integrations can be numerically formidable without the aid of high-speed computing machinery, and have a long history that has been reviewed in some detail in the treatises of Hirschfelder et al. (1964, Sec. 8.4) and McDaniel and Mason (1973, Sec. 6-1). Most of the difficulties can now be handled routinely by a widely-used computer program developed by O'Hara and Smith (1970, 1971) and modified by Neufeld and Aziz (1972) and Viehland (1982). Another program based on the method of Barker et al. (1964) is readily available in the monograph by Maitland et al. (1981, App. 12). However, these programs cannot always cope with peculiar features of some potentials, such as a long-range maximum or an extra minimum, and a more elaborate program has been developed to deal with such problems by Rainwater et al. (1982).

Calculations of the $\bar{\Omega}^{(l,s)}$ for several standard forms of ion-atom potentials have been enshrined as numerical tables, reducing the computational problem to simple interpolation. The most extensive are those for the $(n, 6, 4)$ potential given in (6-1-62) (Viehland et al., 1975) and the $(12, 4)$ core model of (6-1-63) (Mason et al., 1972). Unfortunately, these tabulations do not include the transport cross sections, and so can be used only in conjunction with the one- and two-temperature moment theories. We have prepared tables of reduced transport cross sections for $(n, 6, 4)$ and $(n, 4)$ core models, which are given in Appendix II. These tables can be used with the three-temperature moment theory, or other versions of moment theory that use collision integrals other than the $\bar{\Omega}^{(l,s)}$. The tables are also useful in connection with calculations involving the Lorentz model mentioned in Section 5-3A, and the socalled two-term approximation mentioned in Section 5-3D4, in which energy averages of reciprocal transport cross sections appear. The reduced cross sections in Appendix II are defined as

$$Q^{(l)*}(\varepsilon^*) \equiv \bar{Q}^{(l)}(\varepsilon)/\pi r_m^2, \qquad (7\text{-}1\text{-}7a)$$

$$\varepsilon^* \equiv \varepsilon/\varepsilon_0, \qquad (7\text{-}1\text{-}7b)$$

where r_m and ε_0 are, as usual, the position and depth of the potential minimum.

If the potential is orientation dependent, a complete scattering calculation is usually prohibitively difficult, and some sort of approximation scheme must be used. The simplest method ignores inelastic collisions and merely averages the potential over all orientations to convert it to a spherically symmetric form. Although easy, this procedure may discard too much; for example, a dipole potential averages exactly to zero, and so all dipole effects are discarded by averaging over orientations. Various weight factors for the angular averaging have been suggested, such as a Boltzmann energy or free-energy average, but these have no real theoretical status. The simplest method with a reasonable physical basis fixes the relative orientation during the whole collision, so that the potential depends only on separation, evaluates the cross sections for a number of orientations, and then averages the cross sections or collision integrals over orientations (Monchick and Mason, 1961). The physical reasoning behind this averaging procedure is that the deflection angle in a collision is determined mostly by the interaction around the distance of closest approach, in which the orientation probably varies little during the collision. In mathematical language, most of the contribution to the integral in (7-1-3) comes from the vicinity of the lower limit, r_0. This procedure has been applied mostly to orientation-dependent interactions between neutral molecules, and not to ion-neutral interactions. It seems to give reasonable results with only a moderate computational effort. Somewhat more elaborate methods that have been proposed (Parker and Pack, 1978; Curtiss and Tonsager, 1985) also have not been applied to ion-neutral collisions.

B. Semiclassical and Quantal Calculations

A preliminary discussion of quantum effects was given in Section 6-1E, where the role of the scattering phase shift was outlined. In the calculation of transport cross sections, the phase shift in a quantal calculation is analogous to the deflection angle in a classical calculation, and the summation over the phase shifts is analogous to the integration over classical impact parameters. However, the labor involved is much greater in the quantum-mechanical calculations. For one thing, an additional parameter enters, which corresponds to the de Broglie wavelength λ, a quantity implicitly taken as zero in a classical calculation. For another, the first integration for $\delta_l(\kappa)$ in the quantum-mechanical case is much more laborious than the first integration for $\theta(b, \varepsilon)$ in the classical case.

The first integration determines the phase shifts as a function of the angular-momentum quantum number l and the wave number of relative motion, defined as

$$\kappa = \mu v/\hbar = 2\pi/\lambda = (2\mu\varepsilon)^{1/2}/\hbar, \qquad (7\text{-}1\text{-}8)$$

by numerical integration of the radial wave equation,

$$\frac{d^2 G_l(r)}{dr^2} + \kappa^2 \left[1 - \frac{(l+1)}{\kappa r^2} - \frac{V(r)}{\varepsilon} \right] G_l(r) = 0, \qquad (7\text{-}1\text{-}9)$$

where $l = 0, 1, 2, \ldots$ is the angular-momentum quantum number. In principle, integration is to be carried out from the origin to a value of r large enough for $G_l(r)$ to acquire its asymptotic form,

$$G_l(r) \sim \sin\left(\kappa r - \frac{l\pi}{2} + \delta_l \right), \qquad (7\text{-}1\text{-}10)$$

from which δ_l is identified. This numerical integration for $\delta_l(\kappa)$ is the most arduous part of the calculation. In addition to the numerical labor, difficulties develop from the same features that cause classical orbiting. A number of laborsaving tricks have been devised, some of which are discussed by Munn et al. (1964) and by Buckingham et al. (1965), but no generally available computer programs are available at present.

The transport cross sections are obtained by summing trigonometric functions of the phase shifts over the angular-momentum quantum numbers l, according to the formulas already given in Section 6-1E; this corresponds to the integration over impact parameters in the classical case. For reference, the momentum-transfer cross section is given by the expression

$$\bar{Q}^{(1)}(\kappa) = \frac{4\pi}{\kappa^2} \sum_{l=0}^{\infty} (l+1) \sin^2(\delta_l - \delta_{l+1}), \qquad (7\text{-}1\text{-}11)$$

and the other transport cross sections are given by equations (6-1-68*b*), (6-1-68*c*), and (6-1-68*d*).

The final integrations for collision integrals or interaction integrals are the same as in the classical case, although some extra difficulty occurs because of the quantum-mechanical orbiting resonances in the transport cross sections.

The foregoing description corresponds to a full quantum-mechanical calculation. If the de Broglie wavelength λ is small compared to the scale or range of the potential, then much numerical effort can be saved by judicious use of the semiclassical JWKB approximation for the phase shifts, given by (6-1-70). However, this approximation cannot be used for all l and κ, because there are always regions where it fails, no matter how small λ is. In fact, a thorough use of semiclassical approximations for the transport cross sections has the unexpected result of completely throwing out *all* quantum effects, as explained in Section 6-1E. A discussion of the regions where approximations are safe and unsafe can be found in McDaniel and Mason (1973, Sec. 6-1B).

Unlike the classical case, no computer programs for calculating quantal transport cross sections have been published.

C.   Charge-Exchange Calculations

The effect of resonant charge exchange on ion mobility and diffusion was discussed in some detail in Section 6-5. The type of numerical work needed in a calculation of the transport cross sections from the interaction potentials is the same as that described in the preceding section, except that phase shifts must be calculated for two or more potentials instead of for just one. There is, of course, no complete counterpart to the classical calculations described in Section 7-1A, since resonant charge exchange is not describable classically. If, however, the exchange probability P_{ex} is taken as known from some other source, then the semiclassical description given in Section 6-5B1 is the classical analogue.

Various approximations that are useful for making reasonable estimates with little labor were discussed in Section 6-5B5 and need not be elaborated here.

7-2. NATURE OF ION-NEUTRAL INTERACTIONS

We almost never know the true interaction potential completely. Instead we must be satisfied with mathematical models that we hope mimic the true potential in a reasonable way, or with approximate and limited results obtained by inversion of experimental data. In either case it is very helpful to have some information, however limited or fragmentary, on the nature and magnitude of the potential. The purpose of this section is to supply a brief overview of the information available, which is mostly in the form of knowledge about asymptotic behavior. We do not go into much detail because a number of excellent reviews and treatises are available (Hirschfelder et al., 1964; Hirschfelder, 1967; Margenau and Kestner, 1969; Certain and Bruch, 1972; Maitland et al., 1981; Gray and Gubbins, 1984). It is convenient, although arbitrary, to divide the interactions into long-range, short-range, and intermediate-range potentials.

A. Long-Range Potentials

The long-range contributions to the interaction are usually taken to be the asymptotic forms given by theory as series in r^{-n}, and can be conveniently divided into three parts,

$$V(\text{long range}) = V(\text{polarization}) + V(\text{dispersion}) + V(\text{electrostatic}) \quad (7\text{-}2\text{-}1)$$

The polarization or induction potential arises primarily from the interaction of the ionic charge with the multipole moments it induces in the neutral atom or molecule. The dominant long-range ion-neutral interaction is usually the r^{-4} ion-induced dipole potential, as was pointed out in Section 6-1D, but the ion also induces quadrupole and higher moments. If the neutral is not spherically symmetric, the interaction may be angle dependent because of the anisotropy of the polarizability. Further small contributions to the polarization potential occur if either the ion or the neutral has permanent dipole or higher multipole moments; these contributions all involve the interaction of a permanent multipole on one partner with an induced multipole on the other. All of the polarization interactions are entirely classical and do not involve quantum effects. The only way in which quantum mechanics would enter the picture would be through *ab initio* calculations of properties such as polarizabilities and multipole moments, but the best values of such properties usually come from experiment.

The dispersion potential, first clearly recognized by London (1930a, b) and by Eisenschitz and London (1930), is fundamentally quantum-mechanical in nature, but has a simple semiclassical interpretation. The electron distribution of an ion or molecule undergoes quantum-mechanical fluctuations that give rise to transient multipole moments, which induce in-phase transient moments in another molecule or ion. The electrostatic interactions between these two sets of moments give rise to the attractive dispersion energy, which may be angle dependent because of anisotropy of the polarizability.

The electrostatic potential arises from the interactions between the charge on the ion and the permanent multipole moments of the neutral, plus interactions arising from any permanent multipole moments of the ion. These interactions are entirely classical.

For reference, we give expressions for the first few terms of the asymptotic expressions for the long-range potentials. Summaries of numerical values of the parameters occurring in these expressions are given in Appendix III for some common ions, atoms, and molecules. We restrict ourselves here to formulas for cylindrically symmetric ions and neutrals, which are fairly simple; in general, multipoles and multipole polarizabilities are tensors, and the formulas for their interactions are complicated (Buckingham, 1967). We use the following notation:

$$\alpha_d = \text{dipole polarizability}$$

$$\kappa = \text{anisotropy of } \alpha_d$$

$$\alpha_q = \text{quadrupole polarizability}$$

$$\mu = \text{dipole moment}$$

$$\Theta = \text{quadrupole moment}$$

The leading polarization term is the charge-induced dipole interaction,

$$V_{\text{pol}}(e, \text{ind } \mu) = -\frac{e^2\bar{\alpha}_d}{2r^4}[1 + \kappa(3\cos^2\theta - 1)], \qquad (7\text{-}2\text{-}2)$$

where θ is the angle that the molecular axis of the neutral makes with the line drawn from the center of the ion to the center of the neutral. The average dipole polarizability $\bar{\alpha}_d$ and the anisotropy of the polarizability κ (not to be confused with wavenumber, for which we have used the same symbol) are defined in terms of the dipole polarizabilities parallel and perpendicular to the molecular axis as

$$\bar{\alpha}_d = \tfrac{1}{3}(\alpha_d^{\|} + 2\alpha_d^{\perp}), \qquad (7\text{-}2\text{-}3)$$

$$\kappa = (\alpha_d^{\|} - \alpha_d^{\perp})/3\bar{\alpha}_d. \qquad (7\text{-}2\text{-}4)$$

The angle-dependent term in (7-2-2) is usually ignored in transport calculations because of the difficulty of evaluating the appropriate cross sections (Section 7-1A). There are two contributions to α_d for molecules, an electronic part due to the adjustment of the electron positions in the electric field of the ion, and an atomic part due to the adjustment of the relative positions of the nuclei. At high frequencies (e.g., optical frequencies) only the electronic contribution is active, but in low-frequency processes such as collisions both are active. Gislason et al. (1977) have shown experimentally that the atomic contributions to α_d are important in collisions of K^+ beams with CF_4 and SF_6 molecules, and should be included in the polarization interaction. Since swarm experiments involve slow ion-neutral collisions, the atomic contribution should be included in α_d; it can be determined from dielectric-constant measurements.

The next polarization term is the charge-induced quadrupole interaction; averaged over all orientations, this is

$$\bar{V}_{pol}(e, \text{ind } \Theta) = -\frac{e^2 \bar{\alpha}_q}{2r^6}, \qquad (7\text{-}2\text{-}5)$$

where $\bar{\alpha}_q$ is the average quadrupole polarizability. This term can make a significant contribution in many ion-neutral interactions (Margenau, 1941; Mason and Schamp, 1958); numerical examples are given later in this section.

The next such polarization term would be the charge-induced octopole interaction, which varies as r^{-8}. This is almost always neglected.

Additional polarization terms arise from permanent multipole moments of the ion or neutral. The leading such term is the interaction between a permanent dipole on one partner and the dipole it induces in the other; averaged over all orientations, this is

$$\bar{V}_{pol}(\mu, \text{ind } \mu) = -\frac{1}{r^6}(\mu_n^2 \bar{\alpha}_{di} + \mu_i^2 \bar{\alpha}_{dn}), \qquad (7\text{-}2\text{-}6)$$

where the subscript i denotes ion and the subscript n denotes neutral molecule. Note that this interaction has the same r^{-6} behavior as the charge-induced quadrupole interaction. Further polarization terms vary as still higher powers of r, and we neglect them here.

Available values of α_d, κ, α_q, and μ for some common ions and neutral molecules are given in Appendix III. The values of α_d for molecules include both electronic and atomic contributions. When values are not available, they can sometimes be estimated by means of simple model calculations, which have been summarized by McDaniel and Mason (1973, Sec. 6-2-A).

We next turn to the dispersion interaction, which operates even between spherically symmetric ions and neutrals. The long-range potential between the fluctuating multipoles and the moments they induce can be written as a series,

$$V(\text{dispersion}) = -\frac{C^{(6)}}{r^6} - \frac{C^{(8)}}{r^8} - \cdots, \qquad (7\text{-}2\text{-}7)$$

in which the first term represents the fluctuating dipole-induced dipole interactions, the second term represents the fluctuating dipole-induced quadrupole interactions, and so on. In keeping with our neglect of r^{-8} and higher terms in the polarization potential, we will confine our discussion to the $C^{(6)}$ coefficient, and indeed just to its average over all orientations, $\bar{C}^{(6)}$. Quite accurate values of $\bar{C}^{(6)}$ for many interacting pairs have been obtained by a combination of quantum theory with dielectric, optical, and other data (Starkschall and Gordon, 1971; Tang et al., 1976; Zeiss and Meath, 1977; Standard and Certain, 1985). An excellent way of consolidating and summarizing the available information on dispersion interactions is to use the approximate Slater–Kirkwood (1931) formula for $\bar{C}^{(6)}$ as a correlation formula. The formula for the interaction of particles 1 and 2 is

$$\bar{C}_{12}^{(6)} = \frac{3}{2} \frac{\bar{\alpha}_{d1}\bar{\alpha}_{d2}}{(\bar{\alpha}_{d1}/N_1)^{1/2} + (\bar{\alpha}_{d2}/N_2)^{1/2}} e^2 a_0^5, \qquad (7\text{-}2\text{-}8)$$

where $\bar{\alpha}_{d1}$ and $\bar{\alpha}_{d2}$ are in units of a_0^3 and N_1 and N_2 are the number of equivalent electron oscillators in the particles. By choosing N_1 and N_2 to reproduce the accurately known values of $\bar{C}_{11}^{(6)}$ and $\bar{C}_{22}^{(6)}$, respectively, we can calculate the value of $\bar{C}_{12}^{(6)}$ with a remarkable accuracy of a few percent or better (Wilson, 1965; Kramer and Herschbach, 1970). That is, treatment of N_1 and N_2 as empirical parameters is equivalent to use of the combination rule,

$$\bar{C}_{12}^{(6)} = \frac{2\bar{C}_{11}^{(6)}\bar{C}_{22}^{(6)}}{(\bar{\alpha}_{d2}/\bar{\alpha}_{d1})\bar{C}_{11}^{(6)} + (\bar{\alpha}_{d1}/\bar{\alpha}_{d2})\bar{C}_{22}^{(6)}}. \qquad (7\text{-}2\text{-}9)$$

Even more remarkably, screening-constant calculations suggest that the value of N is the same for an isoelectronic series—that is, that Ar, K^+, Cl^-, and Ca^{2+} all have the same N value. This assumption gives values of $\bar{C}^{(6)}$ accurate within a few percent (Koutselos and Mason, 1986).

Values of N for a number of atoms, molecules, and ions are given in Appendix III. Except for atoms in excited electronic states, these empirical values of N are rather close to the number of electrons in the outermost shell of the atom or ion. For molecules the value of N is fairly close to the number of electrons involved in the chemical bonds. These two rough rules enable crude estimates of $\bar{C}^{(6)}$ to be made by (7-2-8) from knowledge of dipole polarizabilities alone.

We turn finally to the electrostatic interactions. The leading electrostatic terms result from the interaction of the charge of the ion with the permanent multipole moments of the neutral. It is not very well known that such terms can arise even with atomic neutrals. An atom with degenerate s and p states (e.g., an excited H atom) interacts with an ion as if it had a permanent dipole moment, leading to an r^{-2} term in the potential (Hirschfelder and Meath, 1967; LeRoy, 1973, p. 117). Atoms in P states (e.g., B, C, O, F, Cl, Br, I) have permanent quadrupole moments and interact with ions with a long-range r^{-3} potential (LeRoy, 1973, p. 118). More familiar are the interactions of the ionic charge with the permanent multipoles of a neutral molecule, which are entirely orientation-dependent and vanish when averaged over all orientations (Hirschfelder et al.,

1964, Sec. 1.3). They have so far played no important role in mobility and diffusion calculations, presumably because of the difficulty of calculating the transport cross sections. For cylindrically symmetric neutrals the first two terms are

$$V_{el}(e, \mu) = -\frac{e\mu}{r^2}\cos\theta, \tag{7-2-10}$$

$$V_{el}(e, \Theta) = +\frac{e\Theta}{2r^3}(3\cos^2\theta - 1). \tag{7-2-11}$$

The latter term has been invoked to account for apparently low mobilities of some ions in N_2 gas (Revercomb and Mason, 1975). Other terms arise from the interactions of the permanent multipole moments of the ion with those of the neutral. The leading such term is the dipole-dipole interaction,

$$V_{el}(\mu, \mu) = -\frac{\mu_i \mu_n}{r^3}[2\cos\theta_i \cos\theta_n - \sin\theta_i \sin\theta_n \cos(\phi_i - \phi_n)], \tag{7-2-12}$$

where θ_i and θ_n are the angles made by the ionic and molecular dipole axes, respectively, with the line drawn between centers, and ϕ_i and ϕ_n are the azimuthal rotation angles about the line of centers. The next terms are the dipole-quadrupole interactions, which vary as r^{-4}; the ion dipole-molecular quadrupole term is

$$V_{el}(\mu, \Theta) = +\frac{3\mu_i \Theta_n}{2r^4}[(3\cos^2\theta_n - 1)\cos\theta_i$$
$$- 2\sin\theta_i \sin\theta_n \cos\theta_n \cos(\phi_i - \phi_n)]. \tag{7-2-13}$$

A similar term for the ion quadrupole-molecular dipole interaction is obtained by interchange of the subscripts i and n.

Some values of μ and Θ for neutral molecules are given in Appendix III. It should be noted that the definition of Θ we are using here is equal to one-half that used by Hirschfelder et al. (1964). No reliable values of either μ or Θ for molecular ions appear to be available.

The expressions given in this section, plus the numerical values of parameters tabulated in Appendix III, allow estimates to be made of long-range interactions that should be adequate for use in connection with mobility and diffusion calculations. To give an idea of relative magnitudes we show in Table 7-2-1 the ratio of the two r^{-6} coefficients (ion-induced quadrupole polarization and dipole-dipole dispersion) for some simple ion-atom pairs. It can be seen that neither term always dominates the other. For small ions (e.g., Be^{2+} and Li^+) the polarization energy dominates, whereas for polarizable ions (e.g., H^-, Cl^-, Cs^+) the dispersion energy dominates. In Table 7-2-2 we show the ratio of the

Table 7-2-1 Comparison of the Charge-Induced Quadrupole Polarization Energy with the Dipole-Dipole Dispersion Energy for Some Ion-Atom Pairs[a]

	$\frac{1}{2}e^2\alpha_q/C^{(6)}$		
	He	Ar	Xe
Be^{2+}	51	175	397
Li^+	4.1	14	30
H^-	0.037	0.098	0.17
Ca^{2+}	1.2	3.9	8.4
K^+	0.21	0.64	1.3
Cl^-	0.066	0.19	0.37
Ba^{2+}	0.50	1.5	3.1
Cs^+	0.090	0.27	0.54
I^-	0.039	0.11	0.21

[a]Data from Appendix III.

Table 7-2-2 Comparison of the Total r^{-6} Energy with the r^{-4} Polarization Energy for Some Ion-Atom Pairs, Taken at the Position of the Potential Energy Minimum[a]

	$[\frac{1}{2}e^2\alpha_q + C^{(6)}]/\frac{1}{2}e^2\alpha_d r_m^2$		
	He	Ar	Xe
Li^+	0.17	0.24	0.36
K^+	0.35	0.38	0.33
Cl^-		0.80	0.73
Cs^+	0.67	0.50	0.37
I^-		0.97	

[a]Data from Appendix III.

total r^{-6} energy to the r^{-4} polarization energy, at the position of the potential minimum, for a few ion-atom pairs. We can conclude from these results that it would seldom be safe to ignore the r^{-6} energy completely.

B. Short-Range Potentials

Accurate information on short-range interactions is scarcer than for the long-range interactions. Their quantum-mechanical origin is clear enough but accurate *ab initio* calculations with the Schrödinger equation are difficult. When an ion is brought close enough to an atom or molecule that their electronic charge clouds can overlap, large distortions are produced because of the requirements of the Pauli exclusion principle. If the pair originally had closed electronic shells, their electrons tend to avoid each other according to the Pauli principle, and there is a decrease of charge density in the region between them. This decrease reduces the screening of the nuclear charges from each other, and the net effect is one of repulsion between the pair. If the original pair did not have closed shells, there may be an increase in charge density between them because of electron spin pairing, leading to the formation of a chemical bond. Short-range repulsive forces thus have the same origin as chemical bonds and are sometimes called overlap or valence forces. They are also sometimes called exchange forces because any wavefunctions used to calculate them must be suitably antisymmetrized with respect to electron exchange.

One general feature seems to emerge from a large number of calculations of short-range repulsive potentials, whether based on elaborate *ab initio* quantum-mechanical calculations, on crude approximations thereof, or on outright models such as an electron-gas model or a delta-function model for the Coulomb interactions. This feature is that the repulsive potential can usually be represented over a large range of separations by a simple exponential function,

$$V_{\text{rep}}(r) = Ae^{-ar}, \tag{7-2-14}$$

where A and a are constants. This form is indeed suggested by approximate quantum-mechanical calculations, but the large range over which it is empirically found to be accurate is a bit surprising. Somewhat more elaborate forms are sometimes used as curve-fitting devices, e.g. using two exponentials or taking A to be a simple polynomial. For mathematical convenience $V_{\text{rep}}(r)$ is sometimes represented by an inverse power function,

$$V_{\text{rep}}(r) \approx \frac{B}{r^n}, \tag{7-2-15}$$

where B and n are constants. This is often found empirically to be satisfactory over a limited range of r, but it has no particular theoretical basis.

There are no simple methods known by which the short-range parameters A and a (or B and n) can be estimated from properties of the isolated ions and atoms. This situation stands in sad contrast to the case of the long-range parameters discussed in Section 7-2A, which can be determined from dielectric, optical, and other data. The best that can be done at present is to take all the information available on V_{rep} for various systems from various sources and try to correlate all these V_{rep} in terms of known ionic and atomic properties, using a minimum number of empirical parameters. We describe one such correlation scheme here, as devised by Sondergaard and Mason (1975). There is every reason to expect that even more accurate and comprehensive correlation schemes can be devised in the future.

The correlation scheme is based on models and empirical observations. It begins by adopting three features of a simple one-dimensional model that replaces the nuclear potential wells by delta-function wells (a so-called delta-function model), as follows:

1. V_{rep} is of exponential form, (7-2-14).
2. The preexponential constant A for the interaction of two like atoms or ions is proportional to N, the number of electrons in one atom or ion.
3. The constant A is proportional to the square of the constant a in the exponent.

According to these three features, (7-2-14) can be rewritten as

$$V_{rep}(r) \propto Na^2 e^{-ar}. \tag{7-2-16}$$

A fourth feature is taken from a suggestion of Butterfield and Carlson (1972),

4. The parameter a is inversely proportional to a softness parameter ρ that describes the charge density in the outer region of one of the isolated atoms or ions. That is, the outer atomic charge density is proportional to $\exp(-r/\rho)$, where r is the distance from the nucleus.

A fifth feature is empirical and comes from the examination of a number of measurements and calculations of V_{rep} by Sondergaard and Mason:

5. The constant of proportionality in (7-2-16) is not universal, but can be divided into separate families corresponding to closed s-shells, closed p-shells, neutrals, and ions.

Thus the result for a homonuclear interaction can finally be written in the form

$$V_{rep}(r) = \alpha N(\beta/\rho)^2 e^{-\beta r/\rho}, \tag{7-2-17}$$

where α and β are empirical parameters found by fitting known results. The parameter β seems to be universal, and the parameter α falls in family groups. The softness parameters ρ are taken from Hartree–Fock–Slater calculations on single ions and atoms. Numerical values of the parameters obtained by Sondergaard and Mason are collected in Appendix III.

For heteronuclear interactions the simplest procedure is to apply a combination rule to the results for the homonuclear systems summarized in Appendix III. It happens that a geometric-mean combination rule works surprisingly well,

$$(V_{12})_{rep} = [(V_{11})_{rep}(V_{22})_{rep}]^{1/2}, \tag{7-2-18}$$

or in terms of the parameters A and a,

$$A_{12} = (A_{11}A_{22})^{1/2}, \tag{7-2-19a}$$

$$a_{12} = \tfrac{1}{2}(a_{11} + a_{22}). \tag{7-2-19b}$$

This combination rule can be given some theoretical basis, but the reasons are not very strong (Mason and Monchick, 1967). However, it has been tested by direct experiment a number of times, for noble gas interactions determined from scattering experiments with fast atomic beams (reviewed by Jordan et al., 1972).

Another combination rule, probably somewhat more accurate than the geometric-mean rule, is based on physical arguments about the distortion of the individual charge clouds of the interacting pair. The original suggestion is due to Gilbert (1968) and was elaborated by F. T. Smith (1972), leading to the expressions

$$V_{12} = A_{12}e^{-r/\rho_{12}}, \tag{7-2-20}$$

$$\rho_{12} = \tfrac{1}{2}(\rho_{11} + \rho_{22}), \tag{7-2-21a}$$

$$\left(\frac{A_{12}}{\rho_{12}}\right)^{2\rho_{12}} = \left(\frac{A_{11}}{\rho_{11}}\right)^{\rho_{11}} \left(\frac{A_{22}}{\rho_{22}}\right)^{\rho_{22}}. \tag{7-2-21b}$$

In many cases the two sets of combination rules lead to quite similar results, despite their different appearances.

It should be emphasized that the correlation above for the repulsive short-range potentials is considerably less accurate than are the analogous relations for the attractive long-range potentials. In particular, direct measurements or accurate calculations are usually to be preferred if available.

Electron spin pairing can lead to strongly attractive valence forces. This type of interaction can often be empirically represented by sums and differences of exponential functions; an example is the Morse potential, the difference of two exponentials,

$$V_{val} = D_e[e^{-2\beta(r-r_e)} - 2e^{-\beta(r-r_e)}], \tag{7-2-22}$$

where D_e is the dissociation energy, r_e is the equilibrium bond length, and β is a parameter related to the fundamental bond vibrational frequency. Many other similar empirical potential functions have been used, primarily for spectroscopic purposes (Steele et al., 1962). Usually many molecular states result from the interaction of two ground-state atoms or ions if electron pairing is possible (Mason and Monchick, 1967).

C. Intermediate-Range Potentials

Short-range interactions depend essentially on the overlap of electronic charge clouds, whereas long-range interactions are independent of such overlap, depending rather on the correlation of electron motions caused by Coulombic effects. In the intermediate range we have to worry about both effects together, and it would be theoretically inconsistent simply to add the long-range and the short-range potentials together and hope for the best in between. Not only would it be inconsistent, but a spurious singularity would occur at $r = 0$ if we added any long-range terms in r^{-n} to an exponential repulsion such as (7-2-14). Despite this admonition, surprisingly reasonable results are sometimes obtained by such a simpleminded addition, especially if a parameter is left adjustable (McDaniel and Mason, 1973, Sec. 6-4; Alvarez-Rizzatti and Mason, 1975). But better results can be obtained with the aid of a little theoretical guidance, as described below.

If the long-range part of the potential is *not* expanded into the asymptotic form of a series of terms in r^{-n}, and overlap effects are included, the result can be represented as the original asymptotic form multiplied by damping functions (Kreek and Meath, 1969; Koide et al., 1981; and other papers referred to therein). That is, a term in r^{-n} of V(long range) is modified as follows:

$$\frac{C_n}{r^n} \rightarrow \frac{C_n}{r^n} f_n(r), \qquad (7\text{-}2\text{-}23)$$

where $f_n(r)$ is a damping function. Of course, such a result is only formal and is useless in a practical sense if $f_n(r)$ has to be determined by elaborate quantum-mechanical calculations, but the structure of $f_n(r)$ is usually so simple that it can be easily approximated. It is obvious on physical grounds that $f_n(0) = 0$ and $f_n(\infty) = 1$, and exploratory calculations on simple systems suggest that $f_n(r)$ shifts smoothly from 0 to 1 over a fairly limited range of r. Thus $f_n(r)$ can be represented by a simple empirical form, and by multiplying the terms of V(long range) by suitable $f_n(r)$ and adding the products to $V_{\text{rep}}(r)$, it is possible to obtain good results for the complete $V(r)$ over the whole range of interest (Ahlrichs et al., 1977; Tang and Toennies, 1984; and many other papers referred to therein). It even happens that the same damping function can often be used for all the important terms of V(long range). As an example, one useful simple form is

$$f_n(r) = \begin{cases} \exp\left[-(c_n\dfrac{r_m}{r} - 1)^2 \right] & \text{for } r < c_n r_m, \\ 1 & \text{for } r > c_n r_m, \end{cases} \qquad (7\text{-}2\text{-}24)$$

where r_m is the position of the potential minimum. Ahlrichs et al. (1977) have found that the single value of $c_n = 1.28$ works well for all the dispersion terms of some simple systems.

In summary, reasonable estimates of ion-neutral interactions can often be obtained by combining the repulsive branch of the potential with the asymptotic attractive branch, modified by damping functions. There are many papers on this subject, especially for neutral-neutral interactions, and this section represents only a brief overview of the subject and an entrance to the literature. A useful review with a similar purpose has been given by Scoles (1980).

7-3. TESTS OF POTENTIALS

In this section we give some illustrations of the use of ion transport measurements to test potentials that have been obtained in some independent way. We select three sources of potentials to be tested: quantum-mechanical calculations, electron-gas model calculations, and ion-beam scattering measurements.

A. Quantum-Mechanical Calculations

In principle, ion-atom potentials can be calculated *ab initio* from quantum mechanics plus values of a few fundamental physical constants, and computers are now powerful enough to make many such calculations practical, especially for simple systems with only a small number of electrons. For systems with many electrons, however, such calculations are impractical without recourse to simplifications and approximations, which need to be tested.

A few examples have already been exhibited in Chapter 6. Quantum-mechanical potentials for H^+ (or D^+) + He and for Li^+ + He were used to calculate the mobility curves shown in Fig. 6-1-2, which are in good agreement with the experimental mobilities. Similar results are shown in Fig. 6-1-6 for the interaction He^+ + He, which involves two potential curves and the phenomenon of resonant charge exchange. All of these systems involve no more than four electrons.

Another simple four-electron system for which *ab initio* quantum-mechanical calculations are available is H^- + He. Olson and Liu (1978) have made self-consistent-field (SCF) calculations of the interaction potential, and later a presumably more accurate configuration-interaction (CI) calculation (Olson and Liu, 1980). These *ab initio* potentials have been tested by using them to calculate the mobility of H^- ions in He gas as a function of E/N at 300 K (Viehland et al., 1981). The attractive well for H^- + He is so shallow that the transport coefficients at 300 K are completely dominated by the repulsive wall of the potential; we therefore expect the mobility to decrease with increasing E/N, with the rate of decrease being a rough measure of the steepness of the repulsion (see Sections 6-1D and 6-2C). Figure 7-3-1 compares the calculated mobilities with the experimental ones (McFarland et al., 1973; Ellis et al., 1976). It is apparent that the SCF potential is too large, because it yields a low mobility, whereas the CI potential gives much better agreement. The measurements probably do not extend to higher E/N because of the onset of electron detachment from H^- in energetic collisions with He.

The somewhat similar closed-shell system Cl^- + Ar has a much deeper attractive well than does H^- + He, but the much larger number of electrons makes only SCF calculations practicable (Olson and Liu, 1978). Transport calculations are no harder, however, and the mobility results are shown in Fig. 7-3-2 (Viehland et al., 1981), compared with the experimental data (Dotan and Albritton, 1977; Thackston et al., 1979; Ellis et al., 1978). The distinct maximum in the experimental mobilities shows that the attractive well is important in this

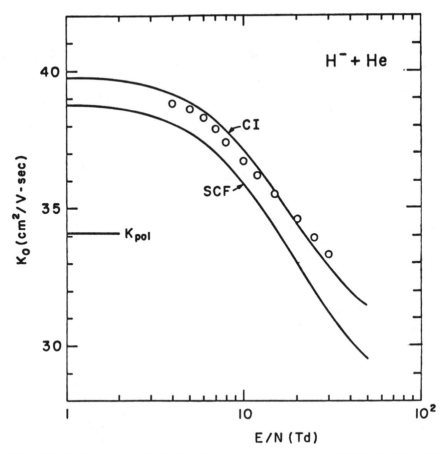

Figure 7-3-1. Comparison of calculated and measured standard mobilities for $H^- + He$ as a function of E/N at 300 K, as a test of *ab initio* quantum-mechanical potentials (Viehland et al., 1981). The curves are based on the self-consistent-field and configuration-interaction potentials of Olson and Liu (1978, 1980), and the measurements are those of McFarland et al. (1973) as summarized by Ellis et al. (1976).

energy range. It is also apparent from the figure that the mobility probes the repulsive wall of the potential only for E/N greater than about 200 Td. In this region the results are similar to those for $H^- + He$, namely that the SCF potential is somewhat too large but has about the right steepness. The well region, however, is given rather poorly by the SCF calculation, as shown by the poor agreement between the calculations and measurements below about 100 Td. The SCF potential gives a mobility maximum that is much too high and that occurs at a value of E/N that is too low by at least a factor of 2. From the discussion in Section 6-1D and 6-2C we can surmise that the latter feature means that the calculated potential well is too shallow by about a factor of 2. The first feature probably means roughly that the "width" of the potential well is

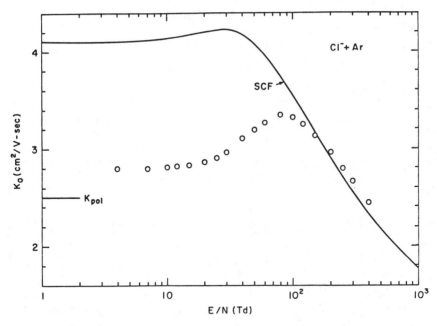

Figure 7-3-2. Comparison of calculated and measured standard mobilities for Cl^- + Ar as a function of E/N at 300 K, as a test of an *ab initio* quantum-mechanical potential (Viehland et al., 1981). The curve is based on the self-consistent-field potential of Olson and Liu (1978), and the measurements are those of Dotan and Albritton (1977) and of Thackston et al. (1979) as summarized by Ellis et al. (1978).

too great, or that the curvature at the minimum is too small. The SCF calculations were, of course, not designed to give the potential well accurately; they describe only the r^{-4} polarization component of the attractive part of the potential, and there are other important contributions to the interaction in the well region, as was discussed in Section 7-2.

As a final example we consider the interaction of a ^4S ground-state O^+ ion with a ground-state Ar atom, which leads to the $X^4\Sigma^-$ molecular potential-energy curve. This system has special interest because the mobility data exhibit a rare minimum as a function of E/N (Dotan et al., 1976; Ellis et al., 1976). An *ab initio* potential curve has been obtained from a polarization plus configuration-interaction (POL-CI) calculation by Guest et al. (1979), which is in good agreement with earlier scattering measurements of Ding et al. (1977). The mobility measurements probe a somewhat different portion of the potential than do the scattering measurements, however, and have been used by Viehland and Mason (1982) to test the calculated potential. The results are shown in Fig. 7-3-3, which also includes a plot of the calculated *ab initio* potential. The measurements above about 130 Td test the potential below about 3 Å, which can be seen to cover the region of the potential well. In this region the agreement is good. At the lowest values of E/N, however, the agreement is not as good,

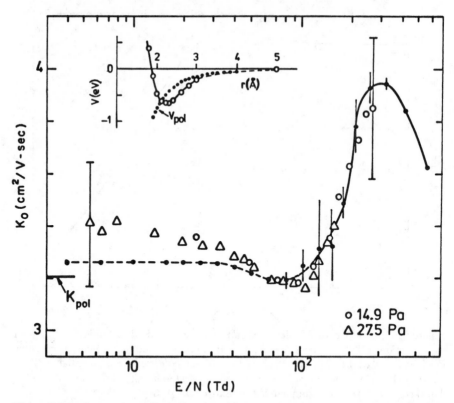

Figure 7-3-3. Comparison of calculated and measured standard mobilities for $O^+ + Ar$ as a function of E/N at 300 K, as a test of an *ab initio* quantum-mechanical potential (Viehland and Mason, 1982). The potential on which the calculations are based is shown in the insert (Guest et al., 1979). The black mobility dots with bars represent calculated points with estimated convergence errors. Experimental points are represented by open circles and triangles with error bars, as measured at two pressures (Dotan et al., 1976). That part of the mobility curve shown dashed ($E/N < 130$ Td) corresponds to the dashed tail of the potential in the insert ($r > 3$ Å). The fact that the total potential shows more attraction than $V_{pol} = -e^2\alpha_d/2r^4$ at large r is believed responsible for the minimum in the mobility curve.

although both calculations and measurements agree in showing a minimum. This region corresponds to the tail of the potential, where the *ab initio* calculations are sparse and probably less accurate. The discrepancy between the calculated and experimental mobilities at the lowest E/N, although systematic, is just at the limit of estimated maximum experimental error. Only minor adjustment of the tail region of the potential would probably be sufficient to produce a satisfactory result.

The particular feature of the potential that is responsible for the minimum in the mobility curve is interesting, and can be surmised from the (12, 6, 4) potential model calculations shown in Fig. 6-1-4. For this model a minimum in the mobility curve develops as r^{-6} attraction energy is added to the long-range r^{-4} polarization energy. Thus the essential feature appears to be that the potential should show *more* attraction than $V_{pol}(e, ind \mu)$ as the particles approach each other from infinity. The insert in Fig. 7-3-3 shows that this is indeed the case for O^+-Ar. This view is supported by the further model calculations shown in Fig. 6-1-5, in which the extra attraction at large r is simulated by shifting a (12, 4) potential to larger distances by the addition of a rigid core.

B. Electron-Gas Model Calculations

Because of the complexity of full *ab initio* quantum-mechanical calculations of interaction potentials, especially for systems with many electrons, there have been many attempts to devise simpler calculation procedures (or models) that would yield accurate results with less expenditure of effort. One very successful procedure treats the electrons in the interacting systems in a statistical way as an electron gas, taking the electron distributions of the isolated atoms or ions as given by independent Hartree–Fock self-consistent-field calculations. Long-range polarization and dispersion energies can also be incorporated in the model, as discussed by Waldman and Gordon (1979), who give references to earlier work.

Ion transport data are very useful for testing such procedures or models, and in this section we give, as an example, the results of such a test by Gatland et al. (1977) on the 12 combinations of Li^+, Na^+, K^+, and Rb^+ with He, Ne, and Ar as calculated by Waldman and Gordon. The results are shown in Fig. 7-3-4, with

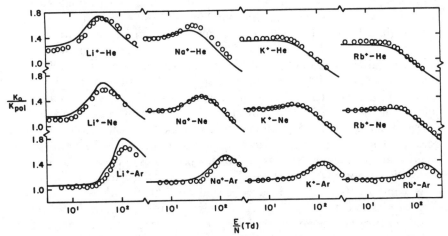

Figure 7-3-4. Comparison of calculated and measured standard ion mobilities for alkali ions in noble gases as a function of E/N at 300 K, as a test of electron-gas model calculations of the potentials (Gatland et al., 1977). The diameters of the circles representing the measurements indicate the experimental uncertainty.

the experimental mobilities taken from the survey of Ellis et al. (1976). Three features stand out in the figure. First, the agreement is remarkably good, showing that the electron-gas model produces potentials of good accuracy. Second, the agreement is nevertheless not wholly within the experimental uncertainty, showing that there is further information on the potential still available in the mobility data. Third, definite patterns can be seen in the deviations, either with a fixed ion and varying inert gas (vertically in the figure), or with a fixed inert gas and varying ion (horizontally in the figure). These patterns suggest that a closer analysis of the relation between mobility and potential would be useful in assessing a potential model, and in extracting the maximum amount of information about the potential from mobility data. We postpone such an analysis to Section 7-5, in which we deal with the direct determination of the potential from mobility data, except for one remark below concerning a possible principle of corresponding states for the mobilities of alkali ions in noble gases.

A simple principle of corresponding states predicts that the curves of K_0 vs. T_{eff} would have the same *shape* for all the alkali ion–noble gas pairs, differing only in scale factors on the K_0 and T_{eff} axes. Such a correspondence would presumably be a reflection of the fact that the potentials all had the same shape, differing only in scale factors for energy and distance; that is, the potentials would be the same for different systems except for the values of ε_0 and r_m. Such a principle holds for the transport properties of the noble gases and all their mixtures to a remarkable accuracy (Kestin et al., 1972; Najafi et al., 1983). Since alkali ions have the same electronic structure as noble gases, the possibility of a similar principle of corresponding states for alkali ions in noble gases seems reasonable. A suitable test would be to see whether a proper choice of well-depth parameters ε_0 would scale the results for all the systems onto a single curve of K_0/K_{pol} vs. $kT_{\text{eff}}/\varepsilon_0$. The results shown in Fig. 7-3-4 show immediately that this is impossible, because the heights of the mobility maxima relative to K_{pol} are different; that is, $(K_0/K_{\text{pol}})_{\max}$ does not have a universal value for all the systems. Detailed analysis in terms of specific potential models only serves to confirm this conclusion (Takebe, 1983). A little reflection on the nature of the potentials reveals the reason. The interaction between two noble gas atoms depends equally on the properties of both atoms, and so might conceivably have a universal shape, but the interaction between an alkali ion and a noble gas atom does not have this pairwise character. As was discussed in Section 7-2, the polarization interaction depends only on the properties of the atom and not on the particular ion involved (since all the ions have the same charge), but the other components of the potential depend on both the ion and atom. For example, the potentials for $Li^+ + Ar$ and $Cs^+ + Ar$ have identical polarization tails, but the rest of the potentials differ markedly because Cs^+ is a much larger and more polarizable ion than Li^+. Thus the potentials inherently have different shapes and cannot be scaled together by varying ε_0 and r_m.

C. Ion-Beam Scattering

Information on ion-neutral potentials can be obtained through measurements of the scattering of ion beams by gas targets. A substantial body of such data has been acquired over the past ten years or so, and it is important to check whether these results are consistent with swarm measurements of ion transport coefficients. The easiest procedure is to take an ion-atom potential deduced from ion-beam scattering (usually by adjusting the parameters of an assumed potential model), and use it to calculate ion transport coefficients that can be compared with swarm measurements. We give here a sampling of such comparisons.

The easiest beam experiments to analyze are those in which the ion beam has a large enough kinetic energy that the scattering probes only the repulsive branch of the ion-atom potential. A simple exponential expression for $V(r)$, such as given by (7-2-14), then usually suffices to fit the scattering data, and it is straightforward to use this potential to calculate mobilities and diffusion coefficients for comparison with swarm data. The energy ranges covered by the two types of measurements usually do not coincide, but the high-energy range of the transport results overlaps the low-energy range of the beam results. An example of the resulting comparison has already been shown in Fig. 5-2-1, for the mobility of K^+ ions in He, Ne, and Ar. Viehland and Mason (1984a, b) have made an extensive comparison for most of the available high-energy beam data involving closed-shell alkali and halide ions on noble gas atoms. These beam experiments had kinetic energies in the keV range and measured integral cross sections by beam attenuation. A representative example is shown in Fig. 7-3-5, for the mobilities of Cl^- and Br^- ions in He and Ne. The overall agreement was fairly good for most of the systems studied, and the probable causes of the discrepancies could usually be assigned with some confidence because of the high redundancy in the data.

Similar integral cross-section measurements at lower energies have been performed by Gislason and coworkers (Polak-Dingels et al., 1982; Rajan and Gislason, 1983; Budenholzer et al., 1977, 1983). These experiments probe down into the well region of the potential, and so their analysis is more complicated. Potential models were chosen and their parameters adjusted to fit the scattering measurements; initially $(n, 6, 4)$ models were used, but were later replaced by more flexible Morse–spline–van der Waals functions. A very thorough comparison of these beam potentials with ion transport data has been carried out by Viehland (1983, 1984). He not only used the beam potentials to calculate transport coefficients for comparison with experiment, but also compared the beam potentials themselves with potentials that he obtained by direct inversion of mobility data (as described in Section 7-5). The positions and depths of the potential wells found by both types of measurements are given in Table III-7. Overall, the agreement of the beam potentials with transport data is quite good, but there are some discrepancies for systems with shallow potential wells

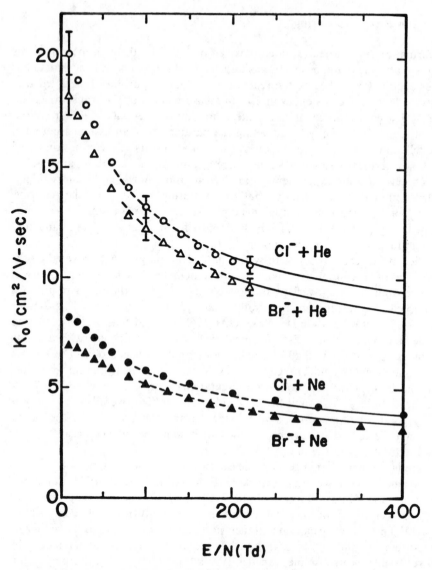

Figure 7-3-5. Comparison of calculated and measured standard ion mobilities for the halide ions Cl^- and Br^- in He and Ne as a function of E/N at 300 K, as a test of ion-beam scattering measurements (Viehland and Mason, 1984a). The beam results are based on integral cross-section measurements in the keV range, and probe only the repulsive branch of the ion-atom potential (Kita et al., 1976). The measured mobilities are from the surveys of Ellis et al. (1978, 1984).

(specifically, for Na$^+$, K$^+$, Cs$^+$ in He, and K$^+$, Cs$^+$ in Ne). Viehland (1984) believes that these discrepancies are due to "some systematic feature of the scattering measurements made at very low energies that has not been properly taken into account in the data analysis."

As a final example, we consider an ion-atom potential obtained from differential, rather than integral, cross-section measurements. The differential scattering of F$^-$ ions by Xe atoms was measured by deVreugd et al. (1979) over the angular range of 0 to 6 mrad with ion beams of energies from 250 to 500 eV. They fitted their results by means of a potential of the form

$$V(r) = Ae^{-ar} - \frac{C}{r^4}, \qquad (7\text{-}3\text{-}1)$$

with parameters $C = 371.4\,\text{eV-}a_0^4$ taken from the known polarizability of Xe,

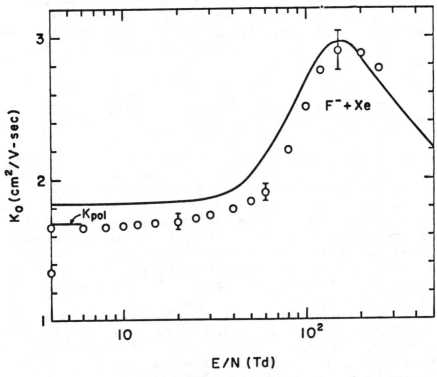

Figure 7-3-6. Comparison of calculated and measured standard ion mobilities for F$^-$ + Xe as a function of E/N at 300 K, as a test of ion-beam scattering measurements (Viehland and Mason, 1981). The beam results are based on differential cross-section measurements (deVreugd et al., 1979), and the measured mobilities are from the survey of Ellis et al. (1978).

$A = 5538\,\text{eV}$ taken from an electron-gas model calculation on the isoelectronic system Ne + Xe by Cohen and Pack (1974), and $a = 1.94a_0^{-1}$ adjusted to fit the scattering cross sections. This potential has been tested by using it to calculate the transport coefficients of F^- in Xe for comparison with swarm experiments (Viehland and Mason, 1981). The mobility results are shown in Fig. 7-3-6, with experimental mobilities taken from the survey of Ellis et al. (1978). The agreement is remarkably good, especially since the (exp, 4) model of (7-3-1) would ordinarily be considered fairly crude for representing mobility data. Both the position and the height of the mobility maximum are given correctly. Roughly speaking, the correct position means that the potential well depth is correct, and the correct height means that the "width" of the well is also approximately correct. The discrepancies of about 10% at low values of E/N indicate that the shape of the potential well, especially its outer branch, is not represented so accurately by the (exp, 4) model.

In summary, calculations of ion transport coefficients, especially mobilities, give a straightforward method for testing proposed potentials, and for locating any deficiencies found. The calculations are even fairly quick and easy if a suitable computer program is available (see Section 7-1). The phenomenon of resonant charge exchange is no obstacle to this sort of test, as demonstrated by the semiclassical calculations of Viehland and Hesche (1986) on the noble gas ions in their parent gases.

7-4. ESTIMATION OF TRANSPORT COEFFICIENTS FROM MEAGER DATA

In the preceding sections we discussed the calculation of ion transport coefficients when the ion-neutral potential was rather accurately known. It frequently happens that the interaction is known only poorly or must even be guessed. In this section we show by a few examples how transport coefficients can be estimated in the face of meager information. The methods are also useful when the ion-neutral potential is accurately known, but only approximate values are needed for the transport coefficients, for which the full computational methods discussed in Section 7-1 are too much trouble. The examples in this section indicate the sort of accuracy that can be obtained in estimates of transport properties.

A number of simplifications can be adopted when we are aiming at only a modest level of accuracy. First, we might as well use the simplest possible kinetic-theory approximation. For the mobility this is the first approximation of the two-temperature theory. We expect this to be accurate within about a 10% level (see the discussion in Section 6-2B2). Second, in calculating the needed collision integrals from some approximate $V(r)$, we should take advantage of available tabulations and not repeat all the numerical integrations discussed in Section 7-1. That is, we can use some reasonable model for $V(r)$, adjust its parameters to suit the system at hand, and then just look up the numerical results in available tables. Third, we should calculate ion diffusion coefficients directly from mobilities via GER. The two-temperature theory is inaccurate at high fields for the diffusion of heavy ions ($m > M$), and the three-temperature theory, insofar as it goes beyond the GER, is too much trouble. For most systems we can expect the simple GER to be accurate on a 10 to 20% level.

A. Summary of Formulas and Tables

Let us summarize the formulas and tables to be used. The standard mobility is given by

$$K_0 \approx \frac{3e}{16N_0}\left(\frac{2\pi}{\mu kT_{\text{eff}}}\right)^{1/2}\frac{1}{\bar{\Omega}^{(1,1)}(T_{\text{eff}})},\qquad (7\text{-}4\text{-}1)$$

which is just (6-2-25) in first approximation. In practical units this becomes

$$K_0 \approx \frac{1.85 \times 10^4}{T_{\text{eff}}^{1/2}\bar{\Omega}^{(1,1)}}\left(\frac{m+M}{mM}\right)^{1/2}\qquad \text{cm}^2/\text{V-s},\qquad (7\text{-}4\text{-}2)$$

where m and M are in atomic mass units (g/mol or dalton), T_{eff} is in K, and $\bar{\Omega}^{(1,1)}$ is in Å^2. The effective temperature is given by

$$\tfrac{3}{2}kT_{\text{eff}} \approx \tfrac{3}{2}kT + \tfrac{1}{2}Mv_d^2,\qquad (7\text{-}4\text{-}3)$$

which is (6-2-26) in first approximation. In practical units this becomes

$$T_{\text{eff}} \approx T + 2.89 \times 10^{-4}MK_0^2(E/N)^2,\qquad (7\text{-}4\text{-}4)$$

where T_{eff} and T are in K, M is in atomic mass units, K_0 is in cm^2/V-s, and E/N is in Td. Ordinarily, one would calculate K_0 as a function of T_{eff} from (7-4-2) and then convert to K_0 as a function of E/N by (7-4-4).

The ion diffusion coefficients, if needed, are best calculated from the parameterized GER expressions of Section 6-4B; in practical units the formulas (6-4-22) for $D_{T,L}$ become

$$ND_T \approx 2.32 \times 10^{15}K_0T_T\left(1 + \Delta_T\frac{K'}{1+K'}\right),\qquad (7\text{-}4\text{-}5a)$$

$$ND_L \approx 2.32 \times 10^{15}K_0T_L[1 + (1 + \Delta_L)K'],\qquad (7\text{-}4\text{-}5b)$$

where $ND_{T,L}$ are in cm^{-1} s^{-1}, K_0 is in cm^2/V-s, $T_{T,L}$ are in K, and $\Delta_{T,L}$ and K' are dimensionless. Numerical values of the quantities Δ_T and Δ_L, which depend primarily on the ion/neutral mass ratio, are given in Table 6-4-1; in many calculations they can safely be neglected. The ion temperatures are given by (6-4-

20), which in practical units is

$$T_{T,L} \approx T + 8.68 \times 10^{-4} \zeta_{T,L} M K_0^2 (E/N)^2 (1 + \beta_{T,L} K'), \qquad (7\text{-}4\text{-}6)$$

the units being the same as in (7-4-4) for T_{eff}. Numerical values of the dimensionless quantities $\beta_{T,L}$ are given in Table 6-4-1. The dimensionless quantities $\zeta_{T,L}$ are given by (6-4-17)

$$\zeta_T = \frac{(m + M)\tilde{A}}{4m + 3M\tilde{A}}, \qquad (7\text{-}4\text{-}7a)$$

$$\zeta_L = \frac{4m - (2m - M)\tilde{A}}{4m + 3M\tilde{A}}. \qquad (7\text{-}4\text{-}7b)$$

The dimensionless quantity $\tilde{A}$ is available in Table 6-4-2 for systems whose interactions resemble those of alkali ions with noble-gas atoms. For other systems $\tilde{A}$ can be found from A^* and C^*, dimensionless quantities available from the same calculations or tabulations used to find $\bar{\Omega}^{(1,1)}$. The relation is given by (6-4-23), which is equivalent to

$$\tilde{A} \approx \tfrac{2}{3} A^*/C^*. \qquad (7\text{-}4\text{-}8)$$

In a pinch, a rough estimate of $\tilde{A} \approx 0.9$ can be used (corresponding to $A^* \approx 1.1$ and $C^* \approx 0.8$), *except* if resonant charge exchange is important (a case we discuss separately below).

It remains only to find $K' \equiv d \ln K_0/d \ln (E/N)$. It should *not* be found by differentiation of the calculated K_0, but rather from the value of C^*, by means of the formulas (6-2-17) and (6-2-18), which yield

$$\frac{K'}{1 + K'} \approx -\frac{T_{\text{eff}} - T}{T_{\text{eff}}} (6C^* - 5). \qquad (7\text{-}4\text{-}9)$$

It is usually not safe to estimate C^* here as a rough numerical constant, as it is in (7-4-8) above, because it appears in a sensitive way in the group $(6C^* - 5)$, which is a rather variable quantity (see, e.g., Figs. 6-1-8 and 6-1-9).

The foregoing formulas supply the complete recipe for approximate calculations of ion transport coefficients from $V(r)$.† Tabulations of $\bar{\Omega}^{(1,1)}$, A^*, and C^* for various model forms of $V(r)$ are available, as already mentioned in Section 7-1A. Although all of the formulas and tabulations apply strictly only to spherically symmetric interactions, they nevertheless usually give reasonable results for many molecular species. The special precautions required in the case of resonant charge exchange are considered separately below.

---

†The astute reader may have noticed that (7-4-4) for T_{eff} is not quite consistent with (7-4-6) for $T_{T,L}$, and should include a factor $(1 + \beta K')$, where β is given by (6-4-26). This correction is usually not worth making since it has only a minor influence on the results.

The most important step in a calculation is to find the nature of the short-range ion-neutral interaction. This part of the interaction completely controls the transport coefficients at high T_{eff}, and determines the general nature of their behavior at intermediate T_{eff} (see Sections 6-1D and 6-2C). The three major short-range interactions are repulsion, valence attraction, and resonant charge exchange. Once the nature of the interaction is known, the general behavior of the transport coefficients is thereby determined, and only a few pieces of information are needed to pin down the scale of this behavior. We always assume that the dipole polarizability of the neutral is known, so that the value of K_{pol} can be calculated from (6-1-61). This forms a sort of fiducial mark from which the rest of the behavior as a function of T_{eff} is measured. Some examples follow. These examples are limited to calculations of mobilities, since the GER for D_T and D_L have already been examined in Section 6-4B. Comparisons with experimental data then indicate the sorts of errors introduced by the physical assumptions made, in addition to the convergence errors of about 10%.

B. Short-Range Repulsion

This type of interaction can often be inferred from structural considerations: the short-range interaction between species with closed electronic shells is repulsive. Another clue is supplied if the room-temperature mobility is higher than K_{pol}. We have already given several examples in connection with other matters—some molecular ions in SF_6 (Fig. 6-1-5), $He_2{}^+$ in He (Fig. 6-1-6), and F^- in Xe (Fig. 7-3-6). We briefly review these systems before considering others.

The crudest sorts of estimates are represented by the ions $SF_5{}^-$, $SF_6{}^-$, $(SF_6)SF_6{}^-$, and $(SF_6)_2SF_6{}^-$ in SF_6. These are bulky ions in a bulky neutral gas, so a core model, such as given by (6-1-63), would seem to be a good guess. This model has three parameters, ε_0, r_m, and a, but we can consider one as effectively known because we know the coefficient of the r^{-4} polarization term. It is easiest to eliminate r_m by considering the ratio K_0/K_{pol}, which is related to tabulated quantities for the (12, 4) core model by

$$\frac{K_0}{K_{pol}} = \frac{1.80(1 - a^*)^2}{(T^*)^{1/2}\Omega^{(1,1)*}}, \tag{7-4-10}$$

where $a^* = a/r_m$, $T^* = kT/\varepsilon_0$, and $\Omega^{(1,1)*} = \bar{\Omega}^{(1,1)}/\pi r_m^2$. (The numerical factor depends on the particular core model used—see Table II-9 for the relevant formulas.) A very rough estimate of the well depth ε_0 can be obtained by taking it equal to the r^{-4} polarization energy at a separation corresponding to the ion-neutral diameter d_{12},

$$\varepsilon_0 \approx e^2\bar{\alpha}_d/2d_{12}^4, \tag{7-4-11}$$

where d_{12} is estimated from individual ion and neutral diameters,

$$d_{12} \approx \tfrac{1}{2}(d_1 + d_2). \tag{7-4-12}$$

Patterson (1970) estimated these diameters as follows. The diameter of neutral SF_6 was taken to be 5.51 Å on the basis of second virial coefficient measurements; this is the value of the parameter σ of a (12, 6) potential used to fit the measurements. The diameters of the SF_5^- and SF_6^- ions were assumed to be about the same as that of neutral SF_6. A value twice as big, 11.0 Å, was assumed for the $(SF_6)SF_6^-$ ion, and for the $(SF_6)_2SF_6^-$ ion a value of 11.9 Å was estimated for an equilateral triangular array of SF_6 spheres. Thus both scales are fixed for a plot of K_0/K_{pol} vs. kT_{eff}/ε_0. Only one parameter, a^*, remains to be found in order to estimate mobilities. Here some of the mobility measurements themselves were used to find a^*, after which the other mobilities could be estimated. The measured mobilities at room temperature all lie slightly below the polarization limit, which suggests a value of $a^* \approx 0.2$ on the basis of the (12, 4) core model (Mason et al., 1972). This value of a^* gives a reasonable representation of the temperature (or field) dependences of the mobilities of all these molecular ions, as Fig. 6-1-5 shows. The representation is, in fact, surprisingly good in view of the crudeness of the parameter estimations. The main reason is that the temperature variation of K_0 is rather flat for these ions, so that the results are insensitive to the crude values estimated for the parameter ε_0 via (7-4-11).

A similar analysis for a number of large polyatomic ions in various neutral gases was carried out by Patterson (1972). Since no calculations for core models were available at that time, Patterson used the (12, 6, 4) model, which happens to give similar mobility curves (compare Figs. 6-1-4 and 6-1-5). Unfortunately, unreasonably large values of the r^{-6} coefficients were required to fit the data, owing to the unrealistic nature of this potential for large polyatomic ions. More extensive recent measurements on various SF_6 ions in SF_6 gas have been interpreted with (9, 6, 4) potentials rather than core potentials, leading to quite large values of the r^{-6} coefficients (Brand and Jungblut, 1983).

A more sophisticated mobility estimate is represented by He_2^+ in He, a system with a rather interesting history that was discussed in Section 6-5C. In order to throw further light on the puzzle presented by the mobilities of various helium ions in He gas, Geltman (1953) made approximate *ab initio* quantum-mechanical calculations of the potential energy between He_2^+ and He, so as to obtain a mobility based on first principles. His mobility calculations were rather rough, however, and were later refined by Mason and Schamp (1958), who fitted a (12, 6, 4) model to Geltman's average potential, taking $\gamma = 0.25$, $\varepsilon_0 = 0.099$ eV, and $r_m = 1.93$ Å. The resulting mobility is shown as the uppermost curve in Fig. 6-1-6. It is in astonishingly good agreement (no doubt somewhat fortuitous) with the later measurements of Orient (1967) on the (presumed) metastable $^4\Sigma_u(He_2^+)^*$ ion. The amount of computational work involved in finding $V(r)$ in a case like this is very large, and would hardly ever be undertaken unless the system had some special interest.

It is worth remarking that even a very elaborate calculation will give poor results if an essential piece of the physics is left out. In particular, the above calculations did not really apply to ground-state He_2^+ as they were intended to do, because the phenomenon of ion transfer in collisions was not considered. As was shown in Fig. 6-1-6, ion transfer significantly affects the mobility. Another example of the effect of (presumed) ion transfer occurs with $H_3{}^+$ in H_2; here the effect is even larger, as discussed in Section 6-5C and shown in Fig. 6-5-5.

The system F^- in Xe represents a mobility estimate more typical than the very crude estimates for polyatomic ions in SF_6 or the elaborate calculations for $He_2{}^+$ in He. The short-range repulsion was estimated by a combination of molecular-beam experiments and approximate quantum-mechanical calculations, and combined with the r^{-4} polarization attraction in the form of an (exp, 4) potential model. From this potential the mobility was calculated, in good agreement with experiment, as was shown in Fig. 7-3-6. This calculation involved a full numerical integration on a computer because no tables of collision integrals exist for the (exp, 4) model; the effort was justified because the aim was to test the potential by means of mobility data. However, if the aim had been merely to use the potential to estimate mobilities, the best procedure would have been to use an $(n, 4)$ model to fit the potential and then simply look up the collision integrals in the published tables of Viehland et al. (1975). The fitting is easily done, beginning by using the proper potential-well parameters, $\varepsilon_0 = 0.28$ eV and $r_m = 5.35a_0$. The value of n is then found from the condition that the two potentials must match at the minimum and have the same r^{-4} coefficient; this translates into $n = ar_m$, where a is the exponential parameter of the repulsion, or $n = (1.94)(5.35) = 10.4$. To avoid interpolation in the tables, it would have been satisfactory to use a (10, 4) model with the correct ε_0 and r_m, or to adjust either ε_0 or r_m slightly to keep the r^{-4} coefficient and hence K_{pol} correct.

As final examples we pick two systems for which good measurements of K_0 are available, but we pretend to have only limited information for purposes of illustration.

We first give an updated treatment of O^+ in He, a system studied in earlier estimates of ion mobilities for air constituents (Mason, 1970). We assume that one zero-field mobility measurement is available at 300 K, and combine this with calculated values of the attractive r^{-4} and r^{-6} coefficients to determine the parameters of some $(n, 6, 4)$ potential models,

$$V(r) = \frac{n\varepsilon_0}{n(3 + \gamma) - 12(1 + \gamma)} \left[\frac{12}{n}(1 + \gamma)\left(\frac{r_m}{r}\right)^n - 4\gamma\left(\frac{r_m}{r}\right)^6 - 3(1 - \gamma)\left(\frac{r_m}{r}\right)^4 \right],$$

$$(7\text{-}4\text{-}13)$$

where n, γ, ε_0, and r_m are the parameters. Since there is insufficient information to determine all the parameters (three pieces of information and four parameters), we use two different values of n to illustrate the effect on the results. From (7-2-2) and the quantities listed in Appendix III we calculate the coefficient of the charge-induced dipole energy to be $C_{ind}^{(4)} = 1.476$ eV-Å^4, which yields $K_{pol} = 17.1$ cm^2/V-s. There are two contributions to the $C^{(6)}$ coefficient, the charge-induced quadrupole energy and the dispersion energy. From (7-2-5) we calculate $C_{ind}^{(6)}/C_{ind}^{(4)} = 0.14 \text{ Å}^2$, and from (7-2-8) we calculate $C_{dis}^{(6)}/C_{ind}^{(4)} = 1.46 \text{ Å}^2$, using the tables of Appendix III. From the compilation of Ellis et al. (1976) we take $K_0 = 22.5 \text{ cm}^2/\text{V-s}$ at 300 K. Picking the parameter n arbitrarily, we find that the ratio $K_0/K_{pol} = 1.32$ at 300 K determines T^* and hence ε_0 for a given value of γ; the value of $C_{ind}^{(4)}$ then determines r_m. We try a series of values of γ until we find the one that gives about the correct value of $C^{(6)}/C_{ind}^{(4)}$. For $n = 8$ we find $\gamma = 0.2$, $\varepsilon_0 = 0.063 \text{ eV}$, $r_m = 1.92 \text{ Å}$, and for $n = 12$ we find $\gamma = 0.2$, $\varepsilon_0 = 0.037 \text{ eV}$, $r_m = 2.40 \text{ Å}$. Thus all the potential parameters are obtained and K_0 is straightforwardly calculated from the tables of $\Omega^{(1,1)*}(T^*)$ given by Viehland et al. (1975). The results are compared in Fig. 7-4-1 with the measurements summarized by Ellis et al. (1976). The agreement is within about 10% up to 600 K, but at higher temperatures the uncertainty about the short-range repulsion (i.e., the value of the parameter n) leads to poorer agreement, the error increasing to about 25% at 3000 K. One more piece of suitable information

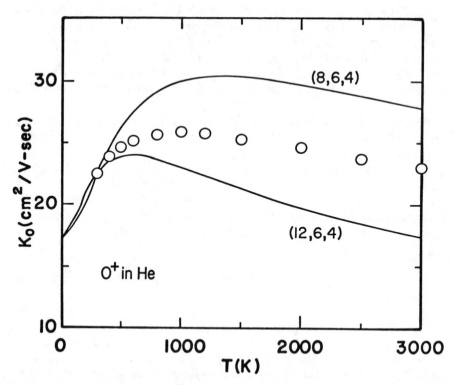

Figure 7-4-1. Example of estimation methods for ion mobilities involving short-range repulsion. Input information consists of calculated values of the $C^{(4)}$ and $C^{(6)}$ coefficients, and one measured mobility at 300 K.

would be sufficient to make a much improved estimate of K_0 that is accurate at high temperatures.*

The system $Cl^- + N_2$ constitutes our final example involving short-range repulsion. Many of the complications of this real system must perforce be ignored in a simple estimation procedure. In particular, we ignore the effects of inelastic collisions, which were illustrated for this system in Fig. 6-6-3, and we ignore the interaction between the ionic charge and the permanent quadrupole moment of N_2 as described by (7-2-11). We estimate the short-range repulsion from ion-beam scattering measurements, and calculate values of $C^{(4)}$ and $C^{(6)}$. Using Appendix III we calculate $C_{ind}^{(4)} = 12.5$ eV-Å^4, which yields $K_{pol} = 2.64$ cm^2/V-s and $C_{dis}^{(6)} = 82$ eV-Å^6. To calculate $C_{ind}^{(6)}$ we need the quadrupole polarizability of N_2, which is not known. It can, however, be estimated by a relation obtained from a harmonic-oscillator model (Margenau, 1941; McDaniel and Mason, 1973, pp. 250–251),

$$\alpha_q \approx \tfrac{3}{2}\alpha_d^2 \frac{h\nu}{e^2 f},$$ (7-4-14a)

where $h\nu$ is the oscillator frequency or some mean excitation energy, and f is the oscillator strength. The usual simple guess takes $h\nu$ equal to the ionization potential and $f = 1$, which yields

$$\alpha_q(\text{Å}^5) \approx 0.104[\alpha_d(\text{Å}^3)]^2 I \text{ (eV)}.$$ (7-4-14b)

From this we calculate $C_{ind}^{(6)} = 20$ eV-Å^6.

Beams of fast Cl^- ions have not been scattered in N_2, but have been scattered in Ar (Kita et al., 1976). Results with beams of fast K^+ ions (Amdur et al., 1972, 1973) suggest that N_2 and Ar behave somewhat similarly as scatterers, so we estimate V_{rep} for $Cl^- + N_2$ as

$$V_{rep}(Cl^- + N_2) \approx V_{rep}(Cl^- + Ar)\frac{V_{rep}(K^+ + N_2)}{V_{rep}(K^+ + Ar)}$$

$$= 3.14 \times 10^{-3}e^{-3.16r} \text{ eV}, \quad 1.95 \leqslant r \leqslant 2.81 \text{ Å}.$$ (7-4-15)

Fitting this result with an r^{-8} potential at 2.8 Å in order to use tabulated values of $\Omega^{(1,1)*}$, we finally obtain

$$V(Cl^- + N_2) \approx \frac{1.8 \times 10^3}{r^8} - \frac{102}{r^6} - \frac{12.5}{r^4} \text{ eV}, \quad r \text{ in Å}.$$ (7-4-16)

This is a standard (8, 6, 4) potential, which we represent with rounded parameter

---

*Note added in proof: The mobility data for O^+ in He have recently been used to test *ab initio* calculations of the potential (Simpson et al., 1987).

values of $\gamma = 0.4$, $\varepsilon_0 = 0.059$ eV, $r_m = 3.51$ Å, leading to the results shown in Fig. 7-4-2. The calculated mobilities are too high by about 10 to 20%, which is not too bad considering the approximations made.

Because the inclusion of the charge-permanent quadrupole interaction would be expected to lower the calculated mobilities, especially at the lower temperatures, it is interesting to make a rough numerical estimate of this interaction. We use an averaging procedure to convert the orientation-dependent potential to an effective spherical potential, including a Boltzmann factor to allow for the fact that some orientations are energetically favored over others (Hirschfelder et al., 1964, pp. 27–28),

$$\langle V(r, T) \rangle = \frac{\int_0^\pi V(r, \theta) \exp(-V/kT) \sin \theta \, d\theta}{\int_0^\pi \exp(-V/kT) \sin \theta \, d\theta}. \qquad (7\text{-}4\text{-}17)$$

Inserting $V_{el}(e, \Theta)$ from (7-2-11), expanding the exponential in a power series, and integrating term by term, we obtain a high-temperature approximation (Revercomb and Mason, 1975),

$$\langle V_{el}(e, \Theta) \rangle = -\frac{1}{5kT} \frac{e^2 \Theta^2}{r^6} + \cdots. \qquad (7\text{-}4\text{-}18)$$

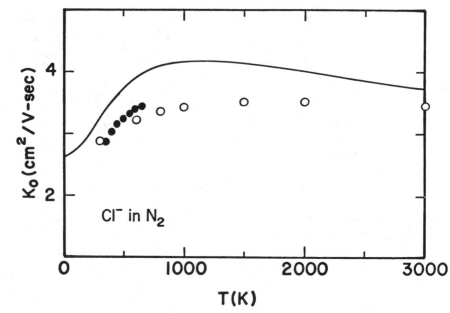

Figure 7-4-2. Example of estimation methods for ion mobilities involving short-range repulsion. Input information consists of calculated values of the $C^{(4)}$ and $C^{(6)}$ coefficients, and some ion-beam scattering data. Inelastic collisions and the ion-permanent quadrupole interaction are ignored. (The data are the same as shown in Fig. 6-6-3.)

The result is thus a net attractive r^{-6} term with a temperature-dependent coefficient. For $Cl^- + N_2$ we calculate the coefficient to be $4.09 \times 10^4/T$ eV-Å⁶, which gives 136 eV-Å⁶ at 300 K and 41 eV-Å⁶ at 1000 K. These would be very substantial additions to the value of 102 eV-Å⁶ used in (7-4-16). We therefore suspect that much of the discrepancy shown in Fig. 7-4-2 can be attributed to the $V_{el}(e, \Theta)$ interaction. Unfortunately, the orientation dependence of this interaction makes it very difficult to include in any simple estimation of mobility. We just have to include it in the estimate of uncertainty.

C. Resonant Charge Exchange

This phenomenon can always be suspected to be important for an ion in its parent gas. As was discussed in Section 6-5, the effect of charge exchange is to convert glancing collisions into apparently nearly head-on collisions, thereby greatly increasing the momentum-transfer cross section. The temperature dependence of the mobility is thus profoundly affected, as was illustrated in Fig. 6-1-6 for He^+ in He. A full theoretical calculation is quite complicated because of the multiple potential energy curves involved (Section 6-5B2), and the calculation of the theoretical curve in Fig. 6-1-6 required a substantial effort (Sinha et al., 1979). However, if direct beam measurements of the charge-exchange cross section are available, the mobility can be straightforwardly calculated from the relation $Q^{(1)} \approx 2Q_{ex}$ (Section 6-5B5). If Q_{ex} is fitted by an expression of the form

$$Q_{ex}^{1/2} = a_1 - a_2 \ln \varepsilon, \tag{7-4-19}$$

then the integration for $\bar{\Omega}^{(1,1)}$ yields (Gradshteyn and Ryzhik, 1980, p. 578)

$$\bar{\Omega}^{(1,1)} \approx 2[a_1 + a_2\psi(3) - a_2 \ln kT]^2 + 2a_2^2\psi'(3), \tag{7-4-20}$$

where $\psi(3) = 0.9228$ is a digamma function and $\psi'(3) = 0.3949$ is a trigamma function (Abramowitz and Stegun, 1964, pp. 258, 260). The units of kT in (7-4-20) must be the same as those of ε in (7-4-19) if $\bar{\Omega}^{(1,1)}$ is to be in the same units as Q_{ex}.

We wish to illustrate this estimation procedure for He^+ in He, for which full theoretical calculations are available for comparison, and for N_2^+ in N_2, for which full calculations are hardly practicable.

For He^+ in He we use the results obtained by Rundel et al. (1979) with a merging-beam technique for collision energies between 0.1 and 187 eV,

$$Q_{ex}^{1/2}(\text{Å}) = 5.09 - 0.299 \ln W, \tag{7-4-21}$$

where $W = 2\varepsilon$ is the laboratory collision energy in eV. Substituting this expression into (7-4-20), we obtain the mobility results shown in Fig. 7-4-3. At low temperatures the mobility is no longer dominated by charge exchange, and

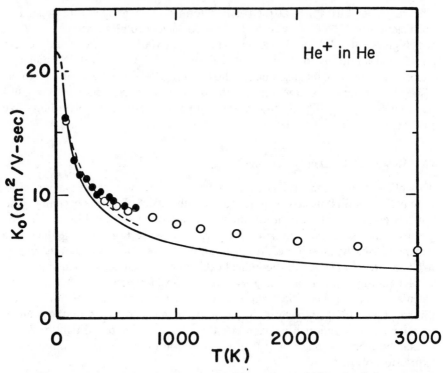

Figure 7-4-3. Example of estimation methods for ion mobilities involving resonant charge exchange. Input information consists only of beam measurements of Q_{ex}, leading to the calculated solid curve, which is connected graphically to the polarization limit at low temperatures by the dashed section. The dashed curve running up to about 700 K represents the full theoretical calculation of Sinha et al. (1979). (The data are the same as shown in Fig. 6-1-6.)

a simple graphical connection is made to the calculated limit of $K_{pol} = 21.6 \, \text{cm}^2/\text{V-s}$ at $T = 0$. The experimental data and the full theoretical calculations shown in the figure for comparison are the same as those in Fig. 6-1-6, but Fig. 7-4-3 extends to much higher effective temperatures. The agreement is quite reasonable, but probably ought to be better at the highest temperatures. Whether this discrepancy is to be attributed to actual inconsistency between the beam and swarm measurements, or to the use of just the first approximation in the calculations, cannot be decided without more elaborate calculations. It is known that the peculiar scattering due to resonant charge exchange strongly distorts the ion velocity distribution function at high fields, and increases the convergence error in the first approximation formulas (Lin and Mason, 1979).

The example of N_2^+ in N_2 constitutes a fairly severe test of the estimation method, inasmuch as the charge exchange may not be exactly resonant because the bond lengths of N_2^+ and N_2 are not identical. Values of Q_{ex} have been measured by Stebbings et al. (1963) from 30 to 10,000 eV, and by Nichols and

Witteborn (1966) from 0.5 to 17 eV. Above about 3 eV both sets of measurements are accurately represented by the expression,

$$Q_{\text{ex}}^{1/2}(\text{Å}) = 6.48 - 0.24 \ln W, \qquad (7\text{-}4\text{-}22)$$

where $W = 2\varepsilon$ is the laboratory collision energy in eV. Substituting this expression into (7-4-20), we obtain the mobility results shown in Fig. 7-4-4. At low temperatures a graphical connection (shown dashed) is made to the limiting value of $K_{\text{pol}} = 2.81$ cm^2/V-s. The agreement with the experimental data is quite reasonable. The larger discrepancies at the lower temperatures can be attributed to deviations of Q_{ex} from the logarithmic expression at energies below $W = 3$ eV, which were observed by Nichols and Witteborn. However, the increase they observed at their very lowest energies was far too large to be consistent with the mobility data, and may not be real.

Thus the estimation of mobilities with resonant charge exchange can be quite successful, despite the multiplicity of potentials. The reason is the accuracy of the relation $Q^{(1)} \approx 2Q_{\text{ex}}$, and the accuracy of the logarithmic representation of $Q_{\text{ex}}^{1/2}$ according to (7-4-19), which permits reliable extrapolation to low energies. This is the good news about charge exchange. The bad news is that the estimation of the diffusion coefficients requires knowledge of the dimensionless quantity $\tilde{A}$,

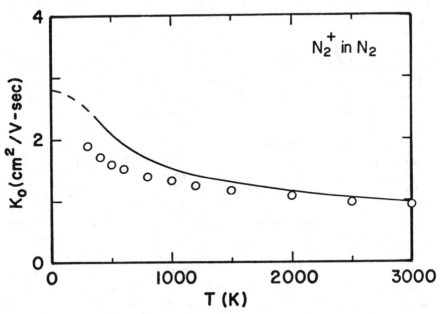

Figure 7-4-4. Example of estimation methods for ion mobilities involving resonant charge exchange. Input information consists only of beam measurements of Q_{ex}, leading to the calculated solid curve, which is connected graphically to the polarization limit at low temperatures by the dashed section. The data are from the survey of Ellis et al. (1976).

which in turn requires knowledge of the individual potentials, as discussed in Section 6-4C. The reason the news is bad is that $\tilde{A}$ is approximately proportional to $Q^{(2)}/Q^{(1)}$, and charge exchange as such does not affect $Q^{(2)}$, although it greatly increases $Q^{(1)}$. Thus $\tilde{A}$ appears to be abnormally low compared to systems without charge exchange. In the absence of knowledge about the individual potential-energy curves involved, we can still make an estimate by means of the very simple charge-exchange model given by (6-4-32). This model involves a parameter δ, lying between 0 and 1, for the fractional contribution of charge exchange. From the relation $Q^{(1)} \approx 2Q_{ex}$ we can infer that the probability of charge exchange in a collision is about $\frac{1}{2}$, and accordingly take $\delta \approx \frac{1}{2}$. From (6-4-33) we then estimate $\tilde{A}$ to be

$$\tilde{A} = \tfrac{2}{3}(1 - \delta) \approx \tfrac{1}{3}. \qquad (7\text{-}4\text{-}23)$$

In other words, in a pinch we take $\tilde{A} \approx \frac{1}{3}$. That this is a reasonable guess can be checked from the detailed calculations for He^+ in He summarized in Table 6-4-10, where $\tilde{A}$ is found to be about 0.37 over a range of field strengths.

D. Valence Attraction

Valence attraction usually causes the tail of the potential to show a stronger attraction than the r^{-4} polarization energy, and thereby produce a minimum in K_0 vs. T_{eff}. An example of a weak minimum was already exhibited for O^+ in Ar in Section 7-3A (see Fig. 7-3-3). However, valence attraction may make the potential well depth so large that the minimum in K_0 is displaced to very high temperatures, and may not even be seen. The result is that K_0 may appear to fall steadily downward from K_{pol}. An example of this behavior was shown in Fig. 6-1-2 for H^+ in He, which has a well depth of about 2.0 eV.

If the decrease of K_0 from K_{pol} is gradual and extends to high temperatures, accurate values of the potential well depth and position are not needed in order to make reasonable estimates of the mobility. Fairly crude values can suffice. We illustrate this here for two systems, H^+ in H_2 and N^+ in N_2. Both H_3^+ and N_3^+ are known to be stable species with fairly large binding energies.

Although the interaction of H^+ with H_2 is actually known rather accurately (Giese and Gentry, 1974), for purposes of illustration we will use only rough estimates. From some approximate quantum-mechanical calculations and some measurements of the elastic scattering of H^+ ions in H_2 gas, Mason and Vanderslice (1959) estimated $\varepsilon_0 \approx 2.7$ eV and $r_m \approx 1.5$ Å for H^+ in H_2. The coefficient of the r^{-4} polarization term is readily calculated to be 5.81 eV-Å^4 from the data in Appendix III. In order to use tabulated values of collision integrals, and thus avoid a large machine computation, we adopt the $(n, 6, 4)$ model as a purely empirical representation of the real potential, requiring only that the r^{-4} coefficient be correct in order to keep the correct limit of $K_{pol} = 18.1$ cm^2/V-s. Since the repulsion is important only at extremely high effective temperatures, we arbitrarily pick $n = 12$, which thereby fixes the parameter γ to be 0.72. It is then straightforward to calculate K_0 from the tables of Viehland et al. (1975), with the results shown in Fig. 7-4-5. The calculated mobilities are systematically too high by 10 to 20% over the range from 300 to 3000 K, which seems reasonable in view of the crudeness of the estimation procedure.

Less is known about the interaction of N^+ with N_2 so we proceed partly by analogy with H^+ in H_2, using a (12, 6, 4) potential model with $\gamma = 0.72$. From the data in Appendix III we calculate $K_{pol} = 3.44$ cm^2/V-s. Only one more piece of information is then required to determine the mobility. We take the zero-field mobility at 300 K as known to be $K_0 = 3.01$ cm^2/V-s. The resulting mobility curve is shown in Fig. 7-4-6. The corresponding potential parameters are $\varepsilon_0 \approx 0.7$ eV and $r_m \approx 2.5$ Å. The agreement with the experimental data is quite good, largely because we have forced agreement to occur at 300 K.

In short, mobilities with valence attraction can be estimated on the basis of only crude information, provided that the valence attraction is strong enough. The reason is that the valence attraction causes only a gradual decrease of K_0 from its K_{pol} limit up to quite high effective temperatures.

The foregoing examples show how mobilities can be estimated with fair

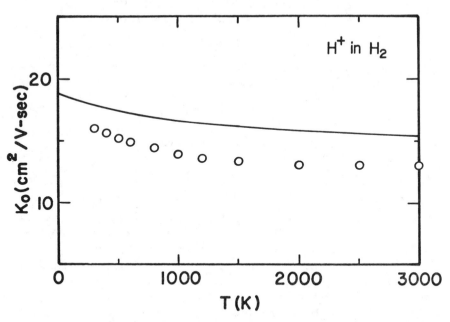

Figure 7-4-5. Example of estimation methods for ion mobilities involving valence attraction. Input information consists of the calculated $C^{(4)}$ coefficient, and rough estimates of the depth and position of the potential well from beam measurements and quantum-mechanical calculations. The data are from the survey of Ellis et al. (1976).

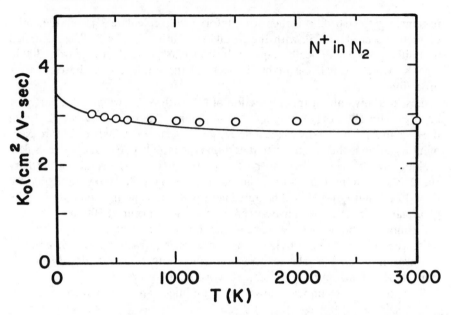

Figure 7-4-6. Example of estimation methods for ion mobilities involving valence attraction. Input information consists of the calculated $C^{(4)}$ coefficient, one measured mobility at 300 K, and the assumption that the potentials for $H^+ + H_2$ and $N^+ + N_2$ have the same shape. The data are from the survey of Ellis et al. (1976).

accuracy on the basis of meager data and educated guessing. We now turn to the more fundamental problem of determining $V(r)$ by direct inversion of transport measurements.

7-5. DIRECT DETERMINATION OF POTENTIALS: INVERSION PROBLEM

Science is full of inversion problems, essentially because what we can measure seldom directly tells us what we really want to know. What we want to know is usually buried under various averagings, convolutions, and so on. The problem then is to deconvolute the measurements to find the quantity of real interest. Examples abound—particle interactions from particle scattering measurements in physics, reaction mechanisms from reaction rate measurements in chemistry, tomography in biology and medicine, and so on.

In many cases (maybe most) there is no unambiguous or unique answer to the inversion problem—too much information has been lost through the averaging performed by the experimental measurements. The difficulty is not that the inversion problem has *no* answer, but rather that it has *too many* answers. Some way must be found to narrow down the field of possible answers allowed by the observations. This is usually done by introducing enough additional information about the nature of the answer to make it unique. Such additional information may have a sound theoretical basis, or it may be nothing but a plausible guess (a "model"). In either case, the traditional way of introducing the extra information in the present case is to assume a definite mathematical form for $V(r)$, containing a number of parameters that can be adjusted to produce a fit of the experimental data. This was the method used to determine ion-neutral potentials from the limited data on the zero-field mobility as a function of gas temperature (Mason and Schamp, 1958). It has also proved quite successful in determining atom-atom potentials from a variety of experimental data; a good example is the case of the Ar-Ar potential studied by Aziz and Chen (1977). A weak point of this method is that the results with one assumed form give little guidance for the selection of an improved assumption. Another weak point in the method is that it is often difficult to find a sufficiently flexible mathematical form for the potential without introducing too many adjustable parameters. Too many parameters may force one to spend an inordinate amount of time (and money) wandering around in multiparameter space seeking a minimum in the deviations, only to find that the minimum has a large flat bottom so that neither the parameters nor the potential is well-determined (Buck, 1975, pp. 334–336). Such a fate is avoided in the case of neutral gases by the use of independently determined values of the first three long-range dispersion coefficients, $C^{(6)}$, $C^{(8)}$, and $C^{(10)}$. This trick is not available for ion transport data, and the great range of T_{eff} covered by the experiments points to a serious problem with too many parameters. This situation has led to a search for direct inversion schemes, in which $V(r)$ is determined directly from the experimental data without any explicit assumption being made about the functional form of the potential.

In the present case we wish to invert measurements of K_0 as a function of T and E/N to find the ion-atom potential $V(r)$. This is very much like the older analogous problem for neutral gases of trying to invert measurements of the viscosity or diffusion coefficient as a function of T to find $V(r)$. It is easy to show that these inversions cannot be unique in the general case, simply by exhibiting a counter example: none of these transport measurements can distinguish between an attractive and a repulsive inverse-power potential (Kestin and Mason, 1973). Even if this sort of limitation could be overcome by the introduction of extra information about $V(r)$, the prospects for direct inversion still do not appear promising at first glance, because three layers of averaging intervene between $V(r)$ and $K_0(T, E/N)$. Nevertheless, despite the apparently dismal prospects for success, direct inversions of transport coefficients *can* be carried out, provided only that some fairly weak constraints are imposed on the shape of $V(r)$. This section is devoted to a description of the inversion procedure for the mobility, together with a physical rationalization for its success, and a statement of the accompanying constraints on $V(r)$.

A. Fundamental Equations

To make matters explicit we write down once again the set of equations that must be dealt with. We use the formulas from the two-temperature theory, which makes the description a little simpler, but the treatment in terms of the three-temperature theory is not essentially different. The kinetic-theory results are summarized by (6-2-25) and (6-2-26), which we write as

$$K_0 = \frac{3e}{16N_0} \left(\frac{2\pi}{\mu k T_{\text{eff}}}\right)^{1/2} \frac{1+\alpha}{\bar{\Omega}^{(1,1)}(T_{\text{eff}})}, \qquad (7\text{-}5\text{-}1)$$

$$\tfrac{3}{2}k T_{\text{eff}} = \tfrac{3}{2}kT + \tfrac{1}{2}Mv_d^2(1+\beta'), \qquad (7\text{-}5\text{-}2)$$

where α and β' represent all the higher-order correction terms. The desired information on $V(r)$ is contained in $\bar{\Omega}^{(1,1)}(T_{\text{eff}})$, which is the quantity to be directly inverted. Since α and β' are usually small, we can extract $\bar{\Omega}^{(1,1)}(T_{\text{eff}})$ from $K_0(T, E/N)$ by iteration during the inversion procedure, which itself is iterative in nature. We will therefore discuss just the inversion of $\bar{\Omega}^{(1,1)}(T_{\text{eff}})$ and not bother here with any further details about α and β'.

The three layers of integration that intervene between $\bar{\Omega}^{(1,1)}(T_{\text{eff}})$ and $V(r)$ are as follows:

$$\bar{\Omega}^{(1,1)}(T_{\text{eff}}) = \frac{1}{2(kT_{\text{eff}})^3} \int_0^\infty \bar{Q}^{(1)}(\varepsilon) \exp(-\varepsilon/kT_{\text{eff}})\varepsilon^2 \, d\varepsilon, \qquad (7\text{-}5\text{-}3)$$

$$\bar{Q}^{(1)}(\varepsilon) = 2\pi \int_0^\infty (1 - \cos\theta)b \, db, \qquad (7\text{-}5\text{-}4)$$

$$\theta(b, \varepsilon) = \pi - 2b \int_{r_0}^\infty \left[1 - \frac{b^2}{r^2} - \frac{V(r)}{\varepsilon}\right]^{-1/2} \frac{dr}{r^2}. \qquad (7\text{-}5\text{-}5)$$

These are the classical-mechanical relations, to which we confine the discussion. The quantum-mechanical relations tend to wash out the details of $V(r)$ even more, owing to the nonzero value of the deBroglie wavelength. Given $V(r)$, all possible angles of deflection in a collision are first calculated as a function of collision energy and impact parameter, by integration over the trajectories according to (7-5-5). No essential information is lost at this point. It is understood how to reconstruct $V(r)$ if $\theta(b, \varepsilon)$ is known—or its quantum-mechanical analog, $\delta_l(\kappa)$ (Child, 1974, pp. 82–84; Buck, 1975, pp. 338–340; Pauly, 1979, Sec. 7.2.2; Wheeler, 1976). The deflection angles are then averaged over impact parameters according to (7-5-4) to produce the momentum-transfer cross section $\bar{Q}^{(1)}(\varepsilon)$. It is at this point that massive loss of information would be expected. Some further loss of information would be expected from the final averaging of $\bar{Q}^{(1)}(\varepsilon)$ over a distribution of energies according to (7-5-3) to produce $\bar{\Omega}^{(1,1)}(T_{\text{eff}})$.

We first describe the inversion procedure, and then consider how it manages to succeed through such apparently severe averaging.

B. Inversion Method

The idea behind the method is based on the approximate procedure for making quick estimates of cross sections that was described in Section 5-2F. The key point is that the energy-averaged cross section represented by $\bar{\Omega}^{(1,1)}$ at a particular value of T_{eff} is determined by the interaction of an ion and atom at a particular separation (or at most a small range of separations). This separation, which we call $\bar{r}$, is the one at which $V(\bar{r})$ is approximately equal to kT_{eff}, so that

$$\bar{\Omega}^{(1,1)}(T_{\text{eff}}) = \pi\bar{r}^2, \tag{7-5-6a}$$

$$V(\bar{r}) = G \cdot kT_{\text{eff}}, \tag{7-5-6b}$$

where G is a quantity of order unity. Clearly this procedure can operate in either direction—given $V(r)$ we can calculate $\bar{\Omega}^{(1,1)}(T_{\text{eff}})$, or inversely. In an inversion, a measured value of $\bar{\Omega}^{(1,1)}$ gives $\bar{r}$, and the temperature T_{eff} at which it was measured gives V at that $\bar{r}$, provided that G is known. Thus each pair of experimental numbers, $\bar{\Omega}^{(1,1)}$ and T_{eff}, translate directly into a pair of calculated numbers, $\bar{r}$ and V, which represent one point on the potential energy curve.

The above procedure, in whichever direction it is operated, obviously depends on knowing G. For quick estimates of average cross sections, it is adequate to take G as a constant, albeit a different constant for repulsive and attractive potentials [Section 5-2F; see also (6-5-52) in Section 6-5B5]. But a constant G is much too crude to be acceptable for an accurate inversion method. However, the relations (7-5-6) can obviously be made exact for any particular $V(r)$ by a suitable selection of G, which will then depend on some complicated way on $V(r)$ and T_{eff}. A little dimensional analysis shows that G must be independent of the distance scale parameter r_m of the potential, and depend only on a dimensionless temperature and a number of dimensionless parameters that characterize the shape of $V(r)$,

$$G = G(T^*, \gamma_1, \gamma_2, \ldots), \tag{7-5-7a}$$

$$T^* = kT_{\text{eff}}/\varepsilon_0, \tag{7-5-7b}$$

where ε_0 is the depth of the potential well (or some other suitable energy scale parameter), and $\gamma_1, \gamma_2 \ldots$ are the dimensionless shape parameters. Examples of the latter are the parameters n and γ of the $(n, 6, 4)$ potential model given by (7-4-13). The remarkable finding that makes a direct inversion scheme possible is that G depends mainly on T^*, and only weakly on the shape parameters—at least for potentials with a short-range repulsion, a long-range attraction, and a single smooth minimum between the two. This feature of G was first noticed for the viscosity of neutral gases by Gough et al. (1972), and later adapted to ion mobilities by Viehland et al. (1976).

Examples of the inversion function $G(T^*)$ are shown in Fig. 7-5-1 for several $(n, 6, 4)$ potentials. Although the potentials are rather different in detail (for example, their curvatures at their minima range over nearly a factor of 2), their

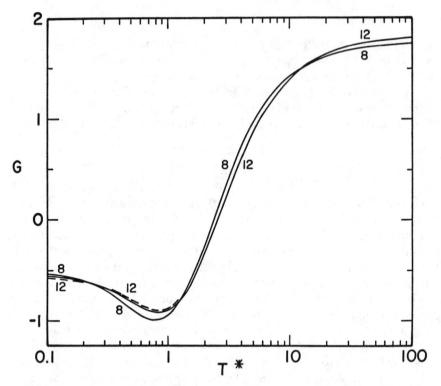

Figure 7-5-1. The inversion function G for mobility as a function of $T^* = kT_{\text{eff}}/\varepsilon_0$ for several $(n, 6, 4)$ potentials. The solid curves are for $n = 8$ and $n = 12$ with $\gamma = 0$ (no r^{-6} term), and the dashed curve is for $n = 12$, $\gamma = 0.4$.

inversion functions are quite close. This means that an inversion procedure can be devised that cycles through the insensitive function $G(T^*)$, as follows.

1. The inversion is started by fitting some convenient potential model to the data. This gives an initial value of ε_0 and a first approximation for $G(T^*)$. A simple way to carry out this initial fit is to plot the experimental K_0/K_{pol} vs. $\ln T_{\text{eff}}$ and compare it with plots of theoretical K_0/K_{pol} vs. $\ln T^*$. The experimental plot should be superposable on one of the theoretical plots by translation along the temperature axis, and the amount of translation directly gives $\ln (\varepsilon_0/k)$.

2. The small correction terms α and β' are calculated and used to convert the experimental $K_0(T, E/N)$ to $\bar{\Omega}^{(1,1)}(T_{\text{eff}})$.

3. Using the starting values of ε_0 and $G(T^*)$, the $\bar{\Omega}^{(1,1)}(T_{\text{eff}})$ data are inverted point-by-point according to (7-5-6). This yields a better approximation to $V(r)$ than the model potential of step 1, but will have virtually the same value of ε_0.

4. The new $V(r)$ is used to calculate a new inversion function $G(T^*)$, and to obtain improved values of the corrections α and β'.

5. The process is repeated starting from step 2 until the results converge (i.e., become stable). This will be the best potential obtainable with the particular value of ε_0 chosen.

6. The best value of ε_0 is then found by search; that is, a new value of ε_0 is chosen and the iteration process is repeated. This is not such a blind search as it might seem. Each calculation of a new inversion function $G(T^*)$ is equivalent to a calculation of the mobility itself, so that a comparison with the experimental data can be made at each step of the iteration. The fit near the maximum of K_0 is moderately sensitive to the value of ε_0 and can guide the search.

7. The best value of ε_0 is chosen according to two criteria. One is, of course, the fit of the mobility data. The other is the rate of convergence of the iteration. Convergence is rather sensitive to ε_0; a poor value of ε_0 leads to slow convergence in step 5. An uncertainty of 1% in the experimental data eventually leads to an uncertainty of about 5% in ε_0.

Four comments on the foregoing inversion scheme are in order. First, the data must include the maximum in K_0 if ε_0 is to be determined. If the data start at a value of T_{eff} beyond the maximum, then only the repulsive wall of $V(r)$ is probed by the measurements. If the data end at a value of T_{eff} before the maximum is reached, then only the attractive tail of $V(r)$ is probed by the measurements. In either of these cases, the precise value of ε_0 used in the iteration is unimportant and does not affect the final $V(r)$, which appears only as a numerical table covering the region probed by the measurements.

The foregoing point is illustrated on the generalized mobility curve shown in Fig. 7-5-2, in which the effective temperatures corresponding to various parts of $V(r)$ are indicated.

Second, a straightforward way to test the inversion scheme is to invert "simulated data" calculated from some assumed potential, and compare the inverted potential with the exact original. This was done by Viehland et al. (1976) with a potential corresponding approximately to the interaction of Li^+ with He; three iterations with the two-temperature theory were sufficient to produce convergence, including the value of ε_0. Later experience with other systems has shown that this convergence happened to be unusually fast, but convergence has always been achieved.

Third, the inversion scheme obviously requires extensive machine computation, far beyond the scope of hand calculations. Numerical techniques for the efficient direct computation of transport properties from potentials to a predetermined level of accuracy have been described by Viehland (1982), for both the two- and three-temperature theories. In addition, Viehland (1983) has described a computer program for the direct inversion procedure. Potentials for a large number of real systems have now been determined by the direct inversion of mobility data, and compared with other potentials obtained by independent means (Viehland, 1983, 1984; Gatland, 1984; Kirkpatrick and Viehland, 1985). Agreement is generally satisfactory. In many cases the potentials from mobilities

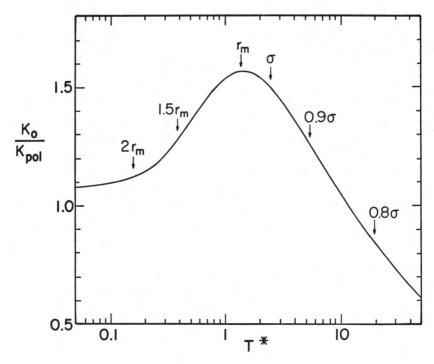

Figure 7-5-2. Generalized curve of ion mobility as a function of effective temperature, $T^* = kT_{\text{eff}}/\varepsilon_0$. The arrows indicate the regions dominated by the various parts of $V(r)$—the potential minimum at r_m, the zero of the potential at σ, and so on. [The curve shown actually corresponds to a (12, 4) potential model.]

are the most extensive and reliable ones available at the present time. The main features of most of these potentials are given in Appendix III.

The fourth and final comment is that nothing in the inversion scheme limits the procedure to mobilities or to ion-atom interactions. The original ideas were in fact developed for atom-atom interactions, and only later modified on the basis of the experience gained from the inversion of ion mobilities. In particular, atom-atom potentials, including well depths, can be determined by direct inversion of gas viscosity data alone, which is usually the most accurately known transport property. The quantity $\eta/T^{5/6}$ as a function of T, where η is the gas viscosity, has a maximum and plays the same role as K_0 as a function of T_{eff}. The factor of $T^{5/6}$ suppresses a low-temperature singularity in the collision integral caused by the long-range r^{-6} interaction, just as the definition of mobility suppresses a low-temperature singularity caused by the r^{-4} interaction. This technique has been developed by Boushehri et al. (1978) and by Maitland and Wakeham (1978a), and tested with "simulated data." A great many noble gas interactions have been determined by this method (Maitland and Wakeham, 1978b).

We now proceed to a justification of the inversion method, with some attention to hidden assumptions and inherent limitations, especially the question of what might happen if a different assumption about the general shape of the potential is made.

C. Justification and Limitations

We first give some arguments to explain how information on $V(r)$ can survive the three layers of integration that lead to $\bar{\Omega}^{(1,1)}$. Second, we make the arguments somewhat more quantitative by drawing a connection between the inversion method for $\bar{\Omega}^{(1,1)}$ and established inversion methods for deflection angles. We next point out how the method is related to established techniques for solving integral equations. Finally, we discuss the limitations of the method, insofar as they are known. Many questions still remain unanswered about direct inversion. Our discussion largely follows the expositions of Maitland et al. (1978, 1985a).

We consider first the final integration over energies to obtain $\bar{\Omega}^{(1,1)}$ according to (7-5-3). The weight factor over which the cross section $\bar{Q}^{(1)}$ is averaged is $\varepsilon^2 \exp(-\varepsilon/kT_{\text{eff}})$, which has a maximum at $\varepsilon = 2kT_{\text{eff}}$. This weight factor will wash out fine details of $\bar{Q}^{(1)}$, but will leave its main features unchanged. This is in fact the case for *all* the $\bar{\Omega}^{(l,s)}$ collision integrals, which have general weight factors of $\varepsilon^{s+1} \exp(-\varepsilon/kT_{\text{eff}})$, and we can define a mean collision energy $\bar{\varepsilon}_{ls}$ such that

$$\bar{\Omega}^{(l,s)}(T_{\text{eff}}) = \bar{Q}^{(l)}(\bar{\varepsilon}_{ls}), \qquad (7\text{-}5\text{-}8a)$$

where the value of $\bar{\varepsilon}_{ls}$ will occur approximately at the maximum of the weight function,

$$\bar{\varepsilon}_{ls} \approx (s + 1)kT_{\text{eff}}. \qquad (7\text{-}5\text{-}8b)$$

Thus the energy averaging does little to degrade the information on $V(r)$ that is contained in a particular cross section $\bar{Q}^{(l)}$. At a given T_{eff} the collisions that determine $\bar{\Omega}^{(l,s)}$ are essentially those that determine a cross section $\bar{Q}^{(l)}$ at a particular energy $\bar{\varepsilon}_{ls}$. These collisions are described by deflection angles that are given by $\theta(b, \bar{\varepsilon}_{ls})$ according to (7-5-5).

Turning next to the deflection angle $\theta(b, \bar{\varepsilon}_{ls})$, we see from (7-5-5) that the integration for this angle apparently averages $V(r)$ over all distances from the distance of closest approach, r_0, out to infinity. However, the integrand of (7-5-5) goes to infinity at r_0, so that most of the contribution to θ comes from $V(r)$ in a region near r_0. Thus we see that the integration for θ does little to degrade the information on $V(r)$—at a given b and $\bar{\varepsilon}$, θ is essentially determined by $V(r_0)$. The effects of $V(r)$ on θ at larger r can be taken into account quantitatively, but we need not discuss that refinement here.

We are finally left with the averaging over all trajectories (impact parameters) to determine $\bar{Q}^{(1)}$ according to (7-5-4). At first glance we would expect massive loss of information from this integration, and the reason we do not is due to a curious trade-off between the importance of large deflection angles and the three-dimensional nature of space. The factor $(1 - \cos \theta)$ in the integrand of (7-5-4) shows that large deflection angles are the important ones in determining $\bar{Q}^{(1)}$, because the factor becomes very small when θ is small. (In fact, it is this factor that keeps the classical momentum-transfer cross section from diverging for potentials with long-range tails that nominally extend to infinity; the classical total scattering cross section *does* diverge for such potentials.) The three-dimensional nature of space is responsible for the factor of $b \, db$ in (7-5-4)— this factor becomes $b^{d-2} \, db$ for a space of d dimensions. The extra factor of b means that collisions with large impact parameters count more than those with small impact parameters in three dimensions, other things being equal. The reason, of course, is simply that many more collisions can occur statistically with large b than with small b. Thus the integrand of (7-5-4) is small when b is small because of the factor $b \, db$. But the integrand is also small when b is large because then θ is small and the factor $(1 - \cos \theta)$ is even smaller. Qualitatively, then, the main contribution to $\bar{Q}^{(1)}$ must come from a restricted range of impact parameters, neither too small nor too large. Only in such a restricted range can the two requirements of large deflection angle *and* large impact parameter be satisfied together. It is this feature that prevents the massive loss of information expected from averaging over all possible trajectories. This behavior holds for all the cross sections $\bar{Q}^{(l)}$, incidentally.

We can make the foregoing argument somewhat more quantitative by considering the integrand of $\bar{Q}^{(1)}$ in more detail, and this enables us to draw a connection with known inversion methods for deflection angles. For this purpose we use essentially the "random-angle" approximation of Hahn and Mason (1971), which was discussed in Section 5-2F. There it was pointed out that a reasonable approximation to $\bar{Q}^{(1)}$ could be obtained by replacing the factor $(1 - \cos \theta) = 2 \sin^2(\theta/2)$ by its mean value of 1 from $b = 0$ out to some cutoff value b_c, and by zero for $b > b_c$. In this approximation we obtain

$$\bar{Q}^{(1)} \approx \pi b_c^2. \qquad (7\text{-}5\text{-}9)$$

Of course, this approximation can always be made exact by a suitable choice of b_c; the remarkable fact is that the value of θ corresponding to b_c is nearly constant, independent of energy and of the details of $V(r)$, even though the value of b_c does depend on these quantities. In other words, the dominant collisions that determine $\bar{Q}^{(1)}$ at *any* energy are those that have a cutoff deflection angle θ_c. The value of b_c corresponding to θ_c depends on ε and $V(r)$, but collisions with $b > b_c$ do not matter because $V(r)$ is too weak and θ is too small, whereas collisions with $b < b_c$ give virtually the same results for all $V(r)$. Thus the essential feature of $V(r)$ that determines $\bar{Q}^{(1)}$ at a specific ε is the value of the distance of closest approach r_c that leads to a deflection angle of θ_c.

Just why θ_c should be nearly constant is a bit of a mystery. Even more of a mystery is why the Firsov (1951) approximation for θ_c works so well. Firsov suggested, in a somewhat different connection involving charge exchange, that a factor of $\sin^2 g(b)$ in an integrand could be replaced by its mean value of 1/2 out to b_c and by zero beyond b_c, with the value of b_c chosen so that $|g(b_c)| = 1/\pi$ (see Section 6-5B5). This leads to $|\theta_c/2| = 1/\pi$ for $\bar{Q}^{(1)}$ and $|\theta_c| = 1/\pi$ for $\bar{Q}^{(2)}$. But even without understanding the reason for the success of this particular approximation, we can see how it relates the inversion procedures for transport coefficients to known inversion methods for $\theta(b, \varepsilon)$. As mentioned in Section 7-5A, $V(r)$ can be reconstructed if $\theta(b, \varepsilon)$ is known, either as $\theta(b)$ at fixed ε or as $\theta(\varepsilon)$ at fixed b (or at fixed angular momentum). These possibilities correspond to two independent slices of the deflection-angle surface in $(\theta, b\ \varepsilon)$ space. Such a surface is shown schematically in Fig. 7-5-3, but as $(|\theta|, b, \varepsilon)$ since experiments do not distinguish between positive and negative θ. The fact that transport coefficients correspond to a nearly constant value of θ_c means that they represent a third independent slice of the deflection-angle surface, namely a slice in $(|\theta|, b, \varepsilon)$ space that is parallel to the b, ε plane and that gives a curve relating b_c and ε at the constant value of θ_c. It is reasonable to expect that almost any nontrivial slice in $(|\theta|, b, \varepsilon)$ space is sufficient for a reconstruction of $V(r)$, and hence that transport coefficients contain roughly the same information as $\theta(b)$ at fixed ε or $\theta(\varepsilon)$ at fixed b.

This expectation can be put into more rigorous form by first formally inverting (7-5-5) for $\theta(b, \varepsilon)$ to obtain (Maitland et al., 1978, 1985a),

$$\frac{V}{\varepsilon} = 1 - \exp[-2I(z, \varepsilon)], \qquad (7\text{-}5\text{-}10a)$$

$$\bar{r} = z \exp[I(z, \varepsilon)], \qquad (7\text{-}5\text{-}10b)$$

where

$$I(z, \varepsilon) = \frac{1}{\pi} \int_z^\infty \frac{\theta(b, \varepsilon)}{(b^2 - z^2)^{1/2}} \, db. \qquad (7\text{-}5\text{-}11)$$

These expressions show how to obtain $V(\bar{r})$ if $\theta(b, \varepsilon)$ is known. For an initial choice of ε, the integral $I(z, \varepsilon)$ is found as a function of z by integration of $\theta(b, \varepsilon)$ over impact parameters, thereby giving both V/ε and $\bar{r}$ as functions of z. By elimination of z, a pair of numbers $\bar{r}, V$ is obtained corresponding to the initial choice of ε. The process is repeated for other choices of ε, thereby generating $V(\bar{r})$. This inversion scheme for $\theta(b, \varepsilon)$ can be converted to a more practical one for $\bar{\Omega}^{(1,1)}$ by using (7-5-6) and (7-5-8) to write

$$\bar{\Omega}^{(1,1)}(T_{\text{eff}}) = \pi \bar{r}^2 = z^2 \exp[2I(z, \bar{\varepsilon})], \qquad (7\text{-}5\text{-}12a)$$

$$\frac{V(\bar{r})}{\bar{\varepsilon}} = 1 - \exp[-2I(z, \bar{\varepsilon})] \equiv \tfrac{1}{2}G, \qquad (7\text{-}5\text{-}12b)$$

where

$$\bar{\varepsilon} = 2kT_{\text{eff}}.$$

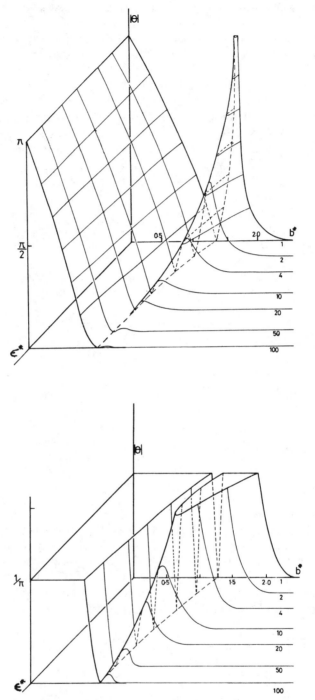

Figure 7-5-3. The three-dimensional ($|\theta|$, b, ε) surface, showing slices at constant ε (above) and at constant θ_c (below). The surface corresponds to a (12, 6) potential model, with $\varepsilon^* = \varepsilon/\varepsilon_0$ and $b^* = b/\sigma$.

This is a set of integral equations that can be solved by standard iterative techniques, one of which is the one described in Section 7-5B. More details on these techniques are given by Maitland et al. (1985a). Maitland et al. (1985b, 1986) have also discussed how the method might be extended to nonspherical potentials.

The foregoing arguments provide a theoretical justification for the direct inversion of data on mobilities or other transport coefficients. It remains to consider the limitations of the method. Clearly there are some limitations, because we know that such an inversion cannot in general be unique. Here we have no firm mathematical results, only some indications based on numerical experimentation. It seems obvious that we cannot recover a potential unless it is rather smoothly varying, since the averaging inherent in $\bar{\Omega}^{(1,1)}$ must limit the amount of detailed structure in $V(r)$ that can be recovered. For example, inversion of data corresponding to a square-well potential yields an inverted potential that scarcely resembles the original square well (Viehland and Mason, unpublished calculations). Outside of that, the most important factor that determines whether the true $V(r)$ is obtained in the inversion procedure appears to be the choice of the initial approximation used to start the iteration. A sufficient condition for success is that the initial choice must have the same general shape as the true $V(r)$. The minimal definition of "same general shape" seems to be the same number of zero crossing points of the potential and its first derivative, the force (Maitland et al., 1978). If this is not the case, then the form of the inversion function $G(T^*)$ is essentially altered and the iteration may not converge to the correct result. Incidentally, it is still not properly understood why $G(T^*)$ is so insensitive to details as long as $V(r)$ has the proper "general shape."

As an illustration, we show in Fig. 7-5-4 the effect of choosing an initial approximation for $V(r)$ of the wrong shape (Maitland et al., 1978). This particular example was calculated for the case of the viscosity of a single neutral gas, but the same considerations apply to ion mobility. The solid curve shows the true potential, a Lennard-Jones (12, 6) model, which was used to generate simulated experimental data. To invert these simulated data, a (12, 6, 3) potential having an outer maximum was chosen as the initial approximation; this potential is shown as the dashed curve. Although the maximum was reduced in amplitude and moved to slightly larger separations during the first three iterations, further iteration failed to remove it and only the repulsive wall of the true (12, 6) potential was recovered satisfactorily. Of course, the inverted potential is rather peculiar looking, which in practice might serve as an indication of an improper initial choice.

From a practical point of view it thus seems important to try a number of different functions as initial approximations for $V(r)$, and to monitor the convergence carefully. At present only a few sufficient conditions for the meaning of "same general shape" are known; essentially nothing is yet known concerning necessary conditions. The whole subject of inversion methods is full of unanswered questions and interesting problems.

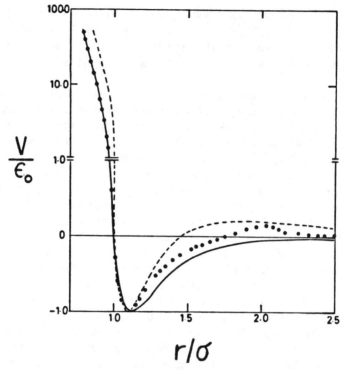

Figure 7-5-4. Effect of an improper initial choice of $V(r)$ on the inversion procedure. The solid curve is a (12, 6) potential model representing the true $V(r)$, and the dashed curve is a (12, 6, 3) model used as an initial choice. The filled circles are the third-iteration potential obtained from inverting viscosity data for the (12, 6) potential.

REFERENCES

Abramowitz, M., and I. A. Stegun (1964). *Handbook of Mathematical Functions*, U.S. National Bureau of Standards, Washington, D.C.

Ahlrichs, R., R. Penco, and G. Scoles (1977). Intermolecular forces for simple systems, *Chem. Phys.* **19**, 119–130.

Alvarez-Rizzatti, M., and E. A. Mason (1975). Charge-transfer energy in closed-shell ion-atom interactions, *J. Chem. Phys.* **63**, 5290–5295.

Amdur, I., J. E. Jordan, K.-R. Chien, L. W.-M. Fung, R. L. Hance, E. Hulpke, and S. E. Johnson (1972). Scattering of fast potassium ions by helium, neon, and argon, *J. Chem. Phys.* **57**, 2117–2121.

Amdur, I., J. E. Jordan, L. W.-M. Fung, L. J. F. Hermans, S. E. Johnson, and R. L. Hance (1973). Scattering of fast potassium ions by small molecules, *J. Chem. Phys.* **59**, 5329–5332.

Aziz, R. A., and H. H. Chen (1977). An accurate intermolecular potential for argon, *J. Chem. Phys.* **67**, 5719–5726.

Barker, J. A., W. Fock, and F. Smith (1964). Calculation of gas transport properties and the interaction of argon atoms, *Phys. Fluids* **7**, 897–903.

Boushehri, A., L. A. Viehland, and E. A. Mason (1978). Direct determination of interaction potentials from gas viscosity measurements alone, *Chem. Phys.* **28**, 313–318.

Brand, K. P., and H. Jungblut (1983). The interaction potentials of SF_6 ions in SF_6 parent gas determined from mobility data, *J. Chem. Phys.* **78**, 1999–2007.

Buck, U. (1975). Elastic scattering, *Adv. Chem. Phys.* **30**, 313–388.

Buckingham, A. D. (1967). Permanent and induced molecular moments and long-range intermolecular forces, *Adv. Chem. Phys.* **12**, 107–142.

Buckingham, R. A., J. W. Fox, and E. Gal (1965). The coefficients of viscosity and thermal conductivity of atomic hydrogen from 1 to 400°K, *Proc. R. Soc. London* **A284**, 237–251.

Budenholzer, F. E., E. A. Gislason, and A. D. Jorgensen (1977). Determination of potassium ion-rare gas potentials from total cross section measurements, *J. Chem. Phys.* **66**, 4832–4846.

Budenholzer, F. E., E. A. Gislason, and A. D. Jorgensen (1983). Comment on "Recent determinations of potassium ion-rare gas potentials," *J. Chem. Phys.* **78**, 5279–5280.

Butterfield, C., and E. H. Carlson (1972). Ionic soft sphere parameters from Hartree–Fock–Slater calculations, *J. Chem. Phys.* **56**, 4907–4911.

Certain, P. R., and L. W. Bruch (1972). Intermolecular forces, in *MTP International Review of Science, Physical Chemistry Series One*, Vol. 1, *Theoretical Chemistry*, W. Byers-Brown, Ed., Butterworth, London, Chap. 4, pp. 113–165.

Child, M. S. (1974). *Molecular Collision Theory*, Academic, New York.

Cohen, J. S., and R. T. Pack (1974). Modified statistical method for intermolecular potentials. Combining rules for higher van der Waals coefficients, *J. Chem. Phys.* **61**, 2372–2382.

Curtiss, C. F., and M. W. Tonsager (1985). Atom-diatomic molecule kinetic theory cross sections, *J. Chem. Phys.* **82**, 3795–3801.

de Vreugd, C., R. W. Wijnaendts van Resandt, and J. Los (1979). The well depths of XeF^- and $XeCl^-$ from differential scattering measurements, *Chem. Phys. Lett.* **65**, 93–94.

Ding, A., J. Karlau, and J. Weise (1977). The potential of $Ar\text{-}O^+(^4\Sigma^-)$, *Chem. Phys. Lett.* **45**, 92–95.

Dotan, I., and D. L. Albritton (1977). Mobilities of F^-, Cl^-, Br^-, and I^- ions in argon, *J. Chem. Phys.* **66**, 5238–5240.

Dotan, I., W. Lindinger, and D. L. Albritton (1976). Mobilities of various mass-identified positive and negative ions in helium and argon, *J. Chem. Phys.* **64**, 4544–4547.

Eisenschitz, R., and F. London (1930). Über das Verhältnis der van der Waalsschen Kräfte zu den homöopolaren Bindungskräften, *Z. Phys.* **60**, 491–527.

Ellis, H. W., R. Y. Pai, E. W. McDaniel, E. A. Mason, and L. A. Viehland (1976). Transport properties of gaseous ions over a wide energy range, *At. Data Nucl. Data Tables* **17**, 177–210.

Ellis, H. W., E. W. McDaniel, D. L. Albritton, L. A. Viehland, S. L. Lin, and E. A. Mason (1978). Transport properties of gaseous ions over a wide energy range: Part 2, *At. Data Nucl. Data Tables* **22**, 179–217.

Ellis, H. W., M. G. Thackston, E. W. McDaniel, and E. A. Mason (1984). Transport properties of gaseous ions over a wide energy range: Part 3, *At. Data Nucl. Data Tables* **31**, 113–151.

Firsov, O. B. (1951). Resonant charge transfer of ions during slow collisions (in Russian), *Zh. Eksp. Teor. Fiz.* **21**, 1001–1008.

Gatland, I. R. (1984). Determination of ion-atom potentials from mobility experiments, in *Swarms of Ions and Electrons in Gases*, W. Lindinger, T. D. Märk, and F. Howorka, Eds., Springer-Verlag, New York, pp. 44–59.

Gatland, I. R., L. A. Viehland, and E. A. Mason (1977). Tests of alkali ion-inert gas interaction potentials by gaseous ion mobility experiments, *J. Chem. Phys.* **66**, 537–541.

Geltman, S. (1953). The mobility of helium molecular ions in helium, *Phys. Rev.* **90**, 808–816.

Giese, C. F., and W. R. Gentry (1974). Classical trajectory treatment of inelastic scattering in collisions of H^+ with H_2, HD, and D_2, *Phys. Rev.* **A10**, 2156–2173.

Gilbert, T. L. (1968). Soft-sphere model for closed-shell atoms and ions, *J. Chem. Phys.* **49**, 2640–2642.

Gislason, E. A., F. E. Budenholzer, and A. D. Jorgensen (1977). The ion-induced dipole potential: A caveat concerning molecular polarizabilities, *Chem. Phys. Lett.* **47**, 434–435.

Gough, D. W., G. C. Maitland, and E. B. Smith (1972). The direct determination of intermolecular potential energy functions from gas viscosity measurements, *Mol. Phys.* **24**, 151–161.

Gradshteyn, I. S., and I. M. Ryzhik (1980). *Tables of Integrals, Series, and Products*, 4th ed., Academic, New York.

Gray, C. G., and K. E. Gubbins (1984). *Theory of Molecular Fluids*, Vol. 1, *Fundamentals*, Oxford University Press, Oxford.

Guest, M. F., A. Ding, J. Karlau, J. Weise, and L. H. Hillier (1979). Potential interactions between O^+ and rare gases, *Mol. Phys.* **38**, 1427–1444.

Hahn, H., and E. A. Mason (1971). Random-phase approximation for transport cross sections, *Chem. Phys. Lett.* **9**, 633–635.

Hirschfelder, J. O., Ed. (1967). *Advances in Chemical Physics*, Vol. 12, *Intermolecular Forces*, Wiley, New York.

Hirschfelder, J. O., and W. J. Meath (1967). The nature of intermolecular forces, *Adv. Chem. Phys.* **12**, 3–106.

Hirschfelder, J. O., C. F. Curtiss, and R. B. Bird (1964). *Molecular Theory of Gases and Liquids*, Wiley, New York.

Jordan, J. E., E. A. Mason, and I. Amdur (1972). Molecular beams in chemistry, in *Physical Methods of Chemistry*, Vol. 1, Part 3D, A. Weissberger and B. W. Rossiter, Eds., Wiley, New York, Chap. 6, pp. 365–446.

Kestin, J., and E. A. Mason (1973). Transport properties in gases (comparison between theory and experiment), in *Transport Phenomena—1973, AIP Conf. Proc. No. 11*, J. Kestin, Ed., American Institute of Physics, New York, pp. 137–192.

Kestin, J., S. T. Ro, and W. A. Wakeham (1972). An extended law of corresponding states for the equilibrium and transport properties of the noble gases, *Physica* **58**, 165–211.

Kirkpatrick, C. C., and L. A. Viehland (1985). Interaction potentials for the halide ion-rare gas systems, *Chem. Phys.* **98**, 221–231.

Kita, S., K. Noda, and H. Inouye (1976). Repulsive potentials for $Cl^- - R$ and $Br^- - R$ (R = He, Ne, and Ar) derived from beam experiments, *J. Chem. Phys.* **64**, 3446–3449.

Koide, A., W. J. Meath, and A. R. Allnatt (1981). Second order charge overlap effects and damping functions for isotropic atomic and molecular interactions, *Chem. Phys.* **58**, 105–119.

Koutselos, A., and E. A. Mason (1986). Correlation and prediction of dispersion coefficients for isoelectronic systems, *J. Chem. Phys.* **85**, 2154–2160.

Kramer, H. L., and D. R. Herschbach (1970). Combination rules for van der Waals force constants, *J. Chem. Phys.* **53**, 2792–2800.

Kreek, H., and W. J. Meath (1969). Charge-overlap effects. Dispersion and induction forces, *J. Chem. Phys.* **50**, 2289–2302.

Kumar, K. (1967). The Chapman–Enskog solution of the Boltzmann equation: A reformulation in terms of irreducible tensors and matrices, *Aust. J. Phys.* **20**, 205–252.

LeRoy, R. J. (1973). *Molecular Spectroscopy*, Vol. 1, *A Specialist Periodical Report*, The Chemical Society, London, pp. 113–176.

Lin, S. L., and E. A. Mason (1979). Influence of resonant charge transfer on ion mobility, *J. Phys.* **B12**, 783–789.

London, F. (1930a). Über einige Eigenschaften und Anwendungen der Molekularkräfte, *Z. Phys. Chem.* **B11**, 222–251.

London, F. (1930b). Zur Theorie und Systematik der Molekularkräfte, *Z. Phys.* **63**, 245–279.

McDaniel, E. W., and E. A. Mason (1973). *The Mobility and Diffusion of Ions in Gases*, Wiley, New York.

McFarland, M., D. L. Albritton, F. C. Fehsenfeld, E. E. Ferguson, and A. L. Schmeltekopf (1973). Flow-drift technique for ion mobility and ion-molecule reaction rate constant measurements: 1. Apparatus and mobility measurements, *J. Chem. Phys.* **59**, 6610–6619.

Maitland, G. C., and W. A. Wakeham (1978a). Direct determination of intermolecular potentials from gaseous transport coefficients alone: Part 1. The method, *Mol. Phys.* **35**, 1429–1442.

Maitland, G. C., and W. A. Wakeham (1978b). Direct determination of intermolecular potentials from gaseous transport coefficients alone: Part 2. Application to unlike monatomic interactions, *Mol. Phys.* **35**, 1443–1469.

Maitland, G. C., E. A. Mason, L. A. Viehland, and W. A. Wakeham (1978). A justification of methods for the inversion of gas transport coefficients, *Mol. Phys.* **36**, 797–816.

Maitland, G. C., M. Rigby, E. B. Smith, and W. A. Wakeham (1981). *Intermolecular Forces. Their Origin and Determination*, Oxford University Press, Oxford.

Maitland, G. C., V. Vesovic, and W. A. Wakeham (1985a). The inversion of thermophysical properties: 1. Spherical systems revisited, *Mol. Phys.* **54**, 287–300.

Maitland, G. C., V. Vesovic, and W. A. Wakeham (1985b). The inversion of thermophysical properties: 2. Non-spherical systems explored, *Mol. Phys.* **54**, 301–319.

Maitland, G. C., M. Mustafa, V. Vesovic, and W. A. Wakeham (1986). The inversion of thermophysical properties: 3. Highly anisotropic interactions, *Mol. Phys.* **57**, 1015–1033.

Margenau, H. (1941). On the forces between positive ions and neutral molecules, *Philos. Sci.* **8**, 603–613.

Margenau, H., and N. R. Kestner (1969). *Theory of Intermolecular Forces*, Pergamon, New York.

Mason, E. A. (1970). Estimated ion mobilities for some air constituents, *Planet. Space Sci.* **18**, 137–144.

Mason, E. A., and L. Monchick (1967). Methods for the determination of intermolecular forces, *Adv. Chem. Phys.* **12**, 329–387.

Mason, E. A., and H. W. Schamp (1958). Mobility of gaseous ions in weak electric fields, *Ann. Phys. N.Y.* **4**, 233–270.

Mason, E. A., and J. T. Vanderslice (1959). Mobility of hydrogen ions (H^+, H_2^+, H_3^+) in hydrogen, *Phys. Rev.* **114**, 497–502.

Mason, E. A., H. O'Hara, and F. J. Smith (1972). Mobilities of polyatomic ions in gases: Core model, *J. Phys.* **B5**, 169–176.

Monchick, L., and E. A. Mason (1961). Transport properties of polar gases, *J. Chem. Phys.* **35**, 1676–1697.

Munn, R. J., E. A. Mason, and F. J. Smith (1964). Some aspects of the quantal and semiclassical calculation of phase shifts and cross sections for molecular scattering and transport, *J. Chem. Phys.* **41**, 3978–3988.

Najafi, B., E. A. Mason, and J. Kestin (1983). Improved corresponding states principle for the noble gases, *Physica* **119A**, 387–440.

Ness, K. F., and R. E. Robson (1985). Interaction integrals in the kinetic theory of gases, *Transp. Theory Stat. Phys.* **14**, 257–290.

Neufeld, P. D., and R. A. Aziz (1972). Program ACQN to calculate transport collision integrals adapted to run on IBM computers, *Comput. Phys. Commun.* **3**, 269–271.

Nichols, B. J., and F. C. Witteborn (1966). Measurements of resonant charge exchange cross sections in nitrogen and argon between 0.5 and 17 eV, *NASA Tech. Note* **D-3265**.

O'Hara, H., and F. J. Smith (1970). The efficient calculation of the transport properties of a dilute gas to a prescribed accuracy, *J. Comput. Phys.* **5**, 328–344.

O'Hara, H., and F. J. Smith (1971). Transport collision integrals for a dilute gas, *Comput. Phys. Commun.* **2**, 47–54.

Olson, R. E., and B. Liu (1978). Self-consistent-field potential energies for the ground negative-ion and neutral states of HeH, ArH, and ArCl, *Phys. Rev.* **A17**, 1568–1574.

Olson, R. E., and B. Liu (1980). Interactions of H and H^- with He and Ne, *Phys. Rev.* **A22**, 1389–1394.

Orient, O. J. (1967). The measurement of the temperature dependence of atomic and molecular ion mobilities in helium, *Can. J. Phys.* **45**, 3915–3922.

Parker, G. A., and R. T. Pack (1978). Rotationally and vibrationally inelastic scattering in the rotational IOS approximation. Ultrasimple calculation of total (differential, integral, and transport) cross sections for nonspherical molecules, *J. Chem. Phys.* **68**, 1585–1601.

Patterson, P. L. (1970). Mobilities of negative ions in SF_6, *J. Chem. Phys.* **53**, 696–704.

Patterson, P. L. (1972). Mobilities of polyatomic ions, *J. Chem. Phys.* **56**, 3943–3947.

Pauly, H. (1979). Elastic scattering cross sections: 1. Spherical potentials, in *Atom-Molecule Collision Theory*, R. B. Bernstein, Ed., Plenum, New York, pp. 111–199.

Polak-Dingels, P., M. S. Rajan, and E. A. Gislason (1982). Determination of lithium ion-rare gas potentials from total cross section measurements, *J. Chem. Phys.* **77**, 3983–3993.

Rainwater, J. C., P. M. Holland, and L. Biolsi (1982). Binary collision dynamics and numerical evaluation of dilute gas transport properties for potentials with multiple extrema, *J. Chem. Phys.* **77**, 434–447.

Rajan, M. S., and E. A. Gislason (1983). Determination of cesium ion-rare gas potentials from total cross section measurements, *J. Chem. Phys.* **78**, 2428–2437.

Revercomb, H. E., and E. A. Mason (1975). Theory of plasma chromatography/gaseous electrophoresis—A review, *Anal. Chem.* **47**, 970–983.

Rundel, R. D., D. E. Nitz, K. A. Smith, M. W. Geis, and R. F. Stebbings (1979). Resonant charge transfer in $He^+ - He$ collisions studied with the merging-beams technique, *Phys. Rev.* **A19**, 33–42.

Scoles, G. (1980). Two-body, spherical, atom-atom, and atom-molecule interaction energies, *Ann. Rev. Phys. Chem.* **31**, 81–96.

Simpson, R. W., R. G. A. R. Maclagan, and P. W. Harland (1987). Interaction potentials and mobility calculations for the HeO^+ system, *J. Chem. Phys.* **87**, 5419–5424.

Sinha, S., S. L. Lin, and J. N. Bardsley (1979). The mobility of He^+ ions in He, *J. Phys.* **B12**, 1613–1622.

Slater, J. C., and J. G. Kirkwood (1931). The van der Waals forces in gases, *Phys. Rev.* **37**, 682–697.

Smith, F. T. (1972). Atomic distortion and the combining rule for repulsive potentials, *Phys. Rev.* **A5**, 1708–1713.

Sondergaard, N. A., and E. A. Mason (1975). Delta-function model for short-range interatomic forces: Correlation scheme for closed-shell atoms and ions, *J. Chem. Phys.* **62**, 1299–1302.

Standard, J. M., and P. R. Certain (1985). Bounds to two- and three-body long-range interaction coefficients for S-state atoms, *J. Chem. Phys.* **83**, 3002–3008.

Starkschall, G., and R. G. Gordon (1971). Improved error bounds for the long-range forces between atoms, *J. Chem. Phys.* **54**, 663–673.

Stebbings, R. F., B. R. Turner, and A. C. H. Smith (1963). Charge transfer in oxygen, nitrogen, and nitric oxide, *J. Chem. Phys.* **38**, 2277–2279.

Steele, D., E. R. Lippincott, and J. T. Vanderslice (1962). Comparative study of empirical internuclear potential functions, *Rev. Mod. Phys.* **34**, 239–251.

Takebe, M. (1983). The generalized mobility curve for alkali ions in rare gases: Clustering reactions and mobility curves, *J. Chem. Phys.* **78**, 7223–7226.

Tang, K. T., and J. P. Toennies (1984). An improved simple model for the van der Waals potential based on universal damping functions for the dispersion coefficients, *J. Chem. Phys.* **80**, 3726–3741.

Tang, K. T., J. M. Norbeck, and P. R. Certain (1976). Upper and lower bounds of two- and three-body dipole, quadrupole, and octupole van der Waals coefficients for hydrogen, noble gas, and alkali atom interactions, *J. Chem. Phys.* **64**, 3063–3074.

Thackston, M. G., F. L. Eisele, W. M. Pope, H. W. Ellis, and E. W. McDaniel (1979). Mobility of Cl^- ions in Ne, Ar, and Kr, *J. Chem. Phys.* **70**, 3996–3997.

Viehland, L. A. (1982). Gaseous ion transport coefficients, *Chem. Phys.* **70**, 149–156.

Viehland, L. A. (1983). Interaction potentials for Li^+ — rare gas systems, *Chem. Phys.* **78**, 279–294.

Viehland, L. A. (1984). Interaction potentials for the alkali ion-rare gas systems, *Chem. Phys.* **85**, 291–305.

Viehland, L. A., and M. Hesche (1986). Transport properties for systems with resonant charge exchange, *Chem. Phys.* **110**, 41–54.

Viehland, L. A., and E. A. Mason (1981). Well depths of XeF^- and $XeCl^-$ from ion transport data, *Chem. Phys. Lett.* **83**, 298–300.

Viehland, L. A., and E. A. Mason (1982). Test of the $ArO^+(X^4\Sigma^-)$ potential by ion mobility data, *Mol. Phys.* **47**, 709–712.

Viehland, L. A., and E. A. Mason (1984a). Repulsive interactions of closed-shell ions with He and Ne atoms: Comparison of beam and transport measurements, *J. Chem. Phys.* **80**, 416–422.

Viehland, L. A., and E. A. Mason (1984b). Repulsive interactions of closed-shell ions with Ar, Kr, and Xe atoms: Comparison of beam and transport measurements, *J. Chem. Phys.* **81**, 903–908.

Viehland, L. A., E. A. Mason, W. F. Morrison, and M. R. Flannery (1975). Tables of transport collision integrals for (n, 6, 4) ion-neutral potentials, *At. Data Nucl. Data Tables* **16**, 495–514.

Viehland, L. A., M. M. Harrington, and E. A. Mason (1976). Direct determination of ion-neutral molecule interaction potentials from gaseous ion mobility measurements, *Chem. Phys.* **17**, 433–441.

Viehland, L. A., E. A. Mason, and S. L. Lin (1981). Test of the interaction potentials of H^- and Br^- ions with He atoms and of Cl^- ions with Ar atoms, *Phys. Rev.* **A24**, 3004–3009.

Waldman, M., and R. G. Gordon (1979). Scaled electron gas approximation for intermolecular forces, *J. Chem. Phys.* **71**, 1325–1339.

Wheeler, J. A. (1976). Semiclassical analysis illuminates the connection between potential and bound states and scattering, in *Studies in Mathematical Physics*, E. H. Lieb, B. Simon, and A. S. Wightman, Eds., Princeton University Press, Princeton, N.J., pp. 351–422.

Wilson, J. N. (1965). On the London potential between pairs of rare-gas atoms, *J. Chem. Phys.* **43**, 2564–2565.

Zeiss, G. D., and W. J. Meath (1977). Dispersion energy constants $C_6(A, B)$, dipole oscillator strength sums and refractivities for Li, N, O, H_2, N_2, O_2, NH_3, H_2O, NO and N_2O, *Mol. Phys.* **33**, 1155–1176.

8

SPECIAL TOPICS AND
APPLICATIONS OF THE THEORY

This chapter is a compendium of topics related to the experimental techniques and results described in Chapters 1 through 4 and to the theory discussed in Chapters 5 and 6. It is by no means complete, and is meant merely to serve as an introduction to the scope of the subject of gaseous ion transport and some of the topics touched by it.

8-1. ION MOBILITY SPECTROMETRY (PLASMA CHROMATOGRAPHY)

It is fairly obvious that ions of different mobilities can be separated by a time-of-drift technique, provided that they do not interact too strongly with each other. Electrophoresis is a long-established procedure for large ions in aqueous solution, and the same principles should apply to gaseous ions as well. In fact, the method was used by Powell and Brata (1932; see also Tyndall, 1938, pp. 56–57) to resolve the lithium isotopes ^{6}Li and ^{7}Li using Ar drift gas, and to search for "ekacaesium" (element 87) using Xe drift gas. The latter search was of course unsuccessful, since element 87 (francium) does not occur in nature. The method was not developed into a technique for quantitative chemical analysis until about 1970, however, when instrumentation was developed that operated at atmospheric pressure and was capable of detecting and measuring trace concentrations of a wide variety of compounds. It would lead us too far afield to discuss in any detail this application, which now has a fairly extensive literature in the field of analytical chemistry, but we give below a very brief account of its main features. The reader wishing more details can find them in a recent collection of review articles published in book form (Carr, 1984).

The basic experimental procedure is as follows. The sample gas to be analyzed is mixed with a carrier gas (such as purified N_2) and exposed to an ionization source, usually the beta-emitter ^{63}Ni. Primary ions and electrons are produced in the carrier gas, and start a sequence of reactions that quickly lead to a few simple molecular ions. These molecular ions in turn undergo ion-molecule reactions with the trace sample compound to form the stable ion or ions to be measured in the drift tube. The sample ions formed in this reaction region are extracted and introduced into the drift region, a drift tube designed to operate at about atmospheric pressure. The drift gas is the same as the carrier gas to avoid further ion-molecule reactions. The transit times of the ions through the drift region, which are of the order of milliseconds, are recorded to give an output of ion current vs. time. This output resembles a gas chromatogram with a millisecond time scale (hence the name "plasma chromatography"), and can be used in much the same way to detect and "fingerprint" molecules.

Because of the high pressure of operation, E/N is small and almost all instruments operate in the low-field region.

The procedure can be remarkably sensitive. Some organic molecules can be detected in air at concentrations of only 1 part in 10^{12}. The capability of the instrument can be greatly extended by preceding it with a conventional gas or liquid chromatograph, and by following it with a mass spectrometer for further ion identification. Additional variations are also possible, such as different ion sources and different drift gases. Some of these are only beginning to be developed (Carr, 1984).

The feature that makes plasma chromatography somewhat unusual as an analytical technique, as compared to aqueous electrophoresis, for instance, is the existence of the highly developed kinetic theory that connects gaseous ion mobility to details of ion-molecule interactions in the drift tube. The technique can therefore be carried well beyond the empirical "fingerprint" stage. Kinetic theory immediately shows how the transit time (proportional to K^{-1}) varies with the mass, size, and charge state of the ion,

$$K^{-1} \propto \mu^{1/2} \bar{\Omega}^{(1,1)}/z, \qquad (8\text{-}1\text{-}1)$$

where z is the number of elementary charges on the ion. The application of this formula in plasma chromatography is, however, rather different than one might expect from the discussions in Chapters 5, 6, and 7, or even from the lithium isotope resolution mentioned at the beginning of this section. For a series of small (atomic) ions of the same z in the same drift gas, the main variation of K comes from μ rather than from $\bar{\Omega}^{(1,1)}$, which in fact is exactly constant in the polarization limit. But for a series of *large* ions such as are encountered in plasma chromatography, μ is nearly constant and the mobility is controlled by $\bar{\Omega}^{(1,1)}$, since $\mu \approx M$ when $m \gg M$. It is an empirical fact that the mobilities of large ions in air or nitrogen show a definite trend with mass; this can only be due to trends in $\bar{\Omega}^{(1,1)}$. Since, as a rough rule of thumb, large molecules tend to be heavier than small molecules, we can expect that the observed transit time should give an approximate measure of the ion mass.

This is indeed the case, and a number of useful semiempirical correlations of ion mass with transit time have been given. An example is given in Fig. 8-1-1

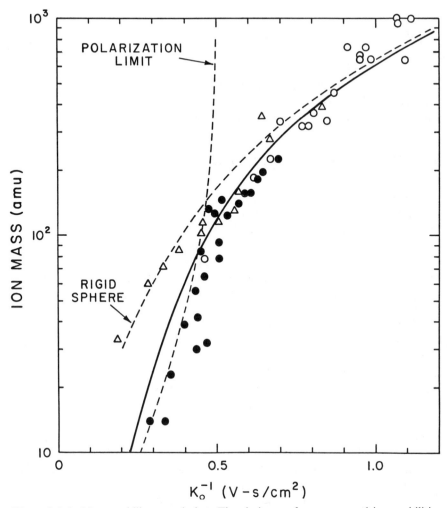

Figure 8-1-1. Mass-mobility correlation. The circles are from measured ion mobilities, the filled circles representing ions mass-identified by mass spectrometer and the open circles representing educated guesses. The triangles are from vapor diffusion coefficients. The solid curve is calculated for rigid spheres with an added polarization potential—the $(\infty, 4)$ model of (6-1-57)—the sphere volumes being taken proportional to ionic and molecular masses. The dashed curve marked "rigid sphere" is similarly calculated, but without the polarization potential.

(Revercomb and Mason, 1975). Such correlations are by no means perfect, of course, and the observed scatter can be attributed to the fact that $\bar{\Omega}^{(1,1)}$ depends somewhat on the details of molecular structure and not simply on mass. Much smaller scatter can be expected within a chemically homologous series, and this expectation has been confirmed experimentally by Griffin et al. (1973), the scatter being reduced by about an order of magnitude (e.g., from 20% to 2%).

What plasma chromatography does determine unambiguously is the quantity $\mu^{1/2}\bar{\Omega}^{(1,1)}/z$, which is essentially ion "size" through $\bar{\Omega}^{(1,1)}$. Ion size is obviously an identifying characteristic in its own right, and does not necessarily have to be linked to ion mass to be useful.

In summary, it is interesting that the transport properties of ions in gases are the basis of a sensitive technique for the quantitative analysis of trace materials. The review edited by Carr (1984) can be consulted for details.

8-2. MACRO-IONS (AEROSOL PARTICLES)

If we proceed in size from the large molecular ions of plasma chromatography to still larger ions, we enter the realm of smokes, aerosols, and other charged particulate matter. Their transport properties in gases are often of great practical importance (Thomson and Thomson, 1928, pp. 187–189; Hidy and Brock, 1970, Chap. 3). Such macro-ions may have diameters that are comparable to, or even larger than, the molecular mean free path of the gas, and therefore behave much differently than the molecular ions with which the rest of this book is concerned. Multiple collisions become important, and all the theoretical discussions based on binary collisions break down. The primary problem is therefore to work out how the mobility behaves as the particle size becomes comparable to the mean free path (or how the diffusion coefficient behaves, since most applications refer to the low-field region where the Nernst–Townsend–Einstein relation holds). That is, the problem is to find how the mobility depends on the Knudsen number, the ratio of the mean free path to the particle dimension.

It is easy to see why molecular ions and macro-ions behave so differently by thinking about the drag force exerted on the ion by the gas, which must on the average be equal and opposite to the force exerted by the electric field. The mobility will be inversely proportional to this drag. For ions of molecular dimensions the drag is due to the collisions of individual gas molecules with the ion. The drag is therefore proportional to the molecular density N, and to the ion-molecule cross section, as was discussed in detail in Section 5-1 and 5-2. But for very large macro-ions, the gas behaves like a continuum fluid and creeps around the surface of the particle in viscous flow. The drag is then proportional to the gas viscosity and to the particle diameter (or some other linear dimension), rather than to the square of the diameter as in a cross section. For large spherical particles of radius R the drag force, without slip, is given by Stokes's law,

$$F_{\text{drag}} = 6\pi R \eta v_d, \tag{8-2-1}$$

where η is the gas viscosity and v_d is the drift velocity of the particle. At steady state the drag force balances the force zeE from the electric field, and since

$v_d = KE$ by definition, we see that the mobility of a spherical macro-ion is given by

$$K = ze/6\pi R\eta. \qquad (8\text{-}2\text{-}2)$$

Because η is independent of gas pressure (density), so is K, whereas the mobility of an ion of molecular dimensions is inversely proportional to pressure (density).

Thus the mobility of a macro-ion is very different from that of a molecular ion—it is independent of N rather than inversely proportional to N; it varies inversely as the first power of the particle dimension rather than as the second power; and it varies directly with the cross section of the neutral molecules (through η), rather than in some inverse fashion as might be expected on simple physical grounds.

Because particulate matter encountered in practice is likely to be neither very large nor very small compared to the molecular mean free path, the challenge is to work out how one extreme behavior changes smoothly into the other, since the mechanisms are so different in the two extremes. We begin by noting that the breakdown of the binary collision assumption implies the breakdown of the E/N scaling law— the mobilities of macro-ions may depend on E and N separately and not just on their ratio. We therefore consider only the vanishing-field limit in which we do not have to worry about the dependence of the mobility on E.

Even in this low-field limit, no general theoretical formula connecting the two extreme results was known until comparatively recently, although several limiting expressions and empirical interpolation formulas had been available for many years. A brief summary of these, as well as of more recent theoretical calculations, has been given by Annis et al. (1972). The initial deviations from Stokes's law are given by the Stokes–Cunningham formula,

$$K = \frac{ze}{6\pi R\eta}\left(1 + A\frac{\lambda}{R}\right), \qquad (8\text{-}2\text{-}3)$$

where A is a constant, approximately unity, that depends on the law of reflection of the molecules from the surface of the sphere, and λ is the mean free path of the gas molecules, to be calculated from the gas viscosity,

$$\eta = \tfrac{1}{2}NM\bar{V}\lambda, \qquad (8\text{-}2\text{-}4)$$

in which $\bar{V} = (8kT/\pi M)^{1/2}$ is the mean molecular speed. The ratio λ/R is the Knudsen number. Somewhat surprisingly, this formula gives the correct inverse density dependence of K in the molecular limit where $\lambda \gg R$, but its other predictions for this limit are wrong. It is also inaccurate in the transition region where $\lambda \sim R$.

An empirical interpolation formula devised by Millikan in connection with his famous oil-drop experiments is still one of the best results available

(Millikan, 1923; see also Knudsen, 1950, pp. 32–33),

$$K = \frac{ze}{6\pi R\eta}\left[1 + \frac{\lambda}{R}(A + Be^{-cR/\lambda})\right],\tag{8-2-5}$$

where A, B, and c are constants. This expression reduces to the Stokes–Cunningham formula (8-2-3) for small λ/R, the coefficient B takes care of the proper limit for large λ/R, and the exponential with the parameter c takes care of the transition region. No theoretical value for c is known, and usually all three parameters A, B, c are empirically adjusted to fit experimental data.

More recently, a theoretical interpolation formula has been derived on the basis of the so-called "dusty-gas model," in which the particle is treated as one component of a binary gas mixture, and the molecular and viscous flows are assumed to be additive (Annis et al., 1972). Variation of the mole fraction of "dust" corresponds to variation of the Knudsen number. The result is

$$K = \frac{ze}{6\pi R\eta}\left(1 + A\frac{\lambda}{R}\frac{1 + \beta\lambda/R}{1 + \beta'\lambda/R}\right),\tag{8-2-6}$$

where A, β, and β' are constants. The constant A is the same as in the Millikan formula, and β and β' can be chosen so that the fit to experimental data is virtually the same.

The results of (8-2-5) and (8-2-6) are compared with each other and with some oil-drop data in Fig. 8-2-1 in terms of a customary "slip factor," equal to $6\pi R\eta K$ in the present notation. (For this purpose the charge ze has been absorbed into the definition of K.) The agreement is quite good. Notice that the measurements cover nearly four orders of magnitude in λ/R.

For ions it is convenient to have (8-2-5) and (8-2-6) explicit in the molecular-ion and macro-ion limits, and explicit in the gas pressure rather than the mean free path, even though the relation between the two is easily found from (8-2-4). We suppose that the mobility K is always reported as a standard mobility, $K_0 = NK/N_0$, calculated as if K were indeed inversely proportional to N. In the molecular-ion (low-pressure) limit K_0 is a true constant, which we denote as $K_0^0 \equiv \lim_{p\to 0} K_0$, but in the transition region K_0 depends on p. In the macro-ion (high-pressure) limit K itself is a constant independent of pressure, which we denote as K^∞. Then (8-2-5) and (8-2-6) take the forms

$$K_0 = K^\infty\frac{p}{N_0 kT} + K_0^0[1 - \delta(1 - e^{-p/\pi})],\tag{8-2-7a}$$

$$K_0 = K^\infty\frac{p}{N_0 kT} + K_0^0\left[1 - \frac{\delta(p/\pi)}{1 + (p/\pi)}\right],\tag{8-2-7b}$$

where $N_0 = 2.687 \times 10^{19}$ molecules/cm³ is the standard gas density, and δ and π are constants. The constant δ is dimensionless, and π has the dimensions of

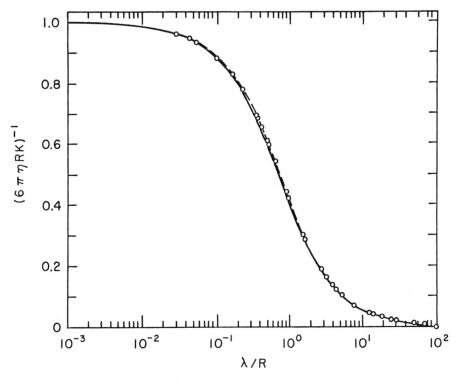

Figure 8-2-1. Reciprocal of the "slip factor," $6\pi R\eta K$, for oil droplets in air as a function of Knudsen number, λ/R. The circles are experimental results for watchmaker's oil. The dashed curve is Millikan's formula, (8-2-5), and the solid curve is (8-2-6).

pressure. They are related to the constants of (8-2-5) and (8-2-6) as follows:

$$\delta = \frac{B}{A + B} = \frac{\beta - \beta'}{\beta}, \qquad (8\text{-}2\text{-}8a)$$

$$\pi/p = \lambda/cR = \beta'\lambda/R. \qquad (8\text{-}2\text{-}8b)$$

The first two terms on the right of (8-2-7), the terms without the δ, correspond to the Stokes–Cunningham formula, but are arranged to give both K_0^0 and K^∞ correctly as limiting values. The third term, involving δ and π, is important only in the transition region; the corrections introduced by this term are of the order of 10%. To the extent that this term is negligible, we see that the mobility is simply the sum of its high- and low-pressure limits.

The expectation is that the limitation to spherical particles has been absorbed into the quantities K_0^0, K^∞, δ, and π, so that (8-2-7) apply to macro-ions of arbitrary shape. That is, the four parameters may depend on the shape and size of the macro-ion, but this is of no consequence as long as the parameters must be determined experimentally—the explicit pressure dependence of K_0 is still given correctly.

It is of interest to extend (8-2-7) to gas mixtures, and results for the mobility of macro-ions in gas mixtures have been obtained by the dusty-gas model (Annis et al., 1973). In the low-pressure limit we expect to recover Blanc's law (Section 5-2D1), and in the high-pressure limit we expect to recover Stokes's law with η as the viscosity of the gas mixture. This indeed is what is found; the low-pressure limit is

$$\frac{1}{K^0_{\text{mix}}} = \sum_j \frac{x_j}{K^0_j}, \tag{8-2-9}$$

where both K^0_{mix} and K^0_j are normalized to the usual standard gas density, and the high-pressure limit is

$$K^\infty_{\text{mix}} = ze/6\pi R\eta_{\text{mix}}, \tag{8-2-10}$$

where K^∞_{mix} is independent of pressure because η_{mix} is. The full expression valid at all pressures is

$$K_{\text{mix}} = K^\infty_{\text{mix}} \frac{p}{N_0 kT} + \left\{ \sum_j \frac{x_j}{K^0_j} \left[1 - \frac{\delta_j(p/\pi_j)}{1 + (p/\pi_j)} \right]^{-1} \right\}^{-1}, \tag{8-2-11}$$

where K_{mix} and K^0_j are normalized to standard gas density, and δ_j and π_j are the constants of (8-2-7) for the mobility in single gas j. To the extent that the terms involving the δ_j are negligible, we see that the mobility of a macro-ion in a gas mixture is still just the sum of its limiting values at high and low pressures. That is, K_{mix} at fixed gas composition is nearly linear in the pressure.

The composition dependence of K_{mix} is less simple than it might appear from (8-2-11), because the gas viscosity η_{mix}, which appears in K^∞_{mix}, has a fairly complicated composition dependence (Chapman and Cowling, 1970, Sec. 12.4). But if this part of the composition dependence is subtracted out, the remaining composition dependence is simple. To demonstrate this, we rearrange (8-2-11) as follows:

$$\frac{1}{K_{\text{mix}} - K^\infty_{\text{mix}}(p/N_0 kT)} = \sum_j \frac{x_j}{K^0_j} \left[1 - \frac{\delta_j(p/\pi_j)}{1 + (p/\pi_j)} \right]^{-1}. \tag{8-2-12}$$

That is, $[K_{\text{mix}} - K^\infty_{\text{mix}}(p/N_0 kT)]^{-1}$ is linear in the mole fractions of the neutral gases. This expression has the appearance of a sort of extended form of Blanc's law as given by (8-2-9); the K_{mix} on the left-hand side is modified by a term involving K^∞_{mix}, and the summation involving the K^0_j on the right-hand side is modified by multiplying each term by a correction factor near unity. (It is important to remember that K_{mix} and K^0_j are normalized to standard gas density, and that K^∞_{mix} is independent of gas density.)

The mixture formulas for macro-ions have never been tested because of lack of data.

To summarize, the formulas given here encompass ions ranging in size from ordinary atomic and molecular ions to large charged aerosol particles. The main restriction is that the electric field is weak. The formulas show how the mobility (and diffusion coefficient) of an ion varies with both neutral gas pressure and gas composition as the Knudsen number varies between 0 and ∞.

8-3. RUNAWAY IONS

We now come to a topic in which supposedly accurate moment solutions of the Boltzmann equation fail to converge, and thereby signal the onset of the new physical phenomenon of ion runaway. Despite this mathematical failure, the basic physics of the phenomenon can be discussed with the aid of the simple momentum-transfer theory presented in Section 5-2A.

The runaway phenomenon occurs when the ion-neutral momentum-transfer cross section decreases with increasing collision energy so rapidly that the ions are unable to lose enough momentum by collisions to achieve a steady-state drift velocity. As a result, the ions are continuously accelerated by the electric field, and no mobility properly exists; the drift tube becomes a gassy accelerator.

Although a runaway phenomenon was known for electrons both in plasmas (Shkarofsky et al., 1966, Secs. 8–10) and in neutral gases (Cavalleri and Paveri-Fontana, 1972), the analogous phenomenon for ions came as somewhat of a surprise. It was first encountered for H^+ (and D^+) ions in He, which so far remains the only system in which it has been experimentally observed, although calculations suggest that many systems should exhibit runaway at sufficiently large values of E/N. The existence of an extremely accurate *ab initio* potential for H^+ + He encouraged Lin et al. (1979a) to calculate its transport coefficients, inasmuch as the calculations could in principle cover a greater range with greater accuracy than experimental measurements. Unexpectedly, the calculations using the two-temperature theory were plagued by apparently poor convergence for values of E/N greater than about 25 Td. The "corrections" to the first-order approximation became large and varied erratically in successive orders of approximation, so that solutions of the moment equations became unstable. This problem of unstable solutions was finally traced to the softness of the repulsive wall for the H^+ + He interaction, which led to a rapid decrease of the momentum-transfer cross section with increasing collision energy. At high energies collisions are inadequate to prevent continued acceleration and hence runaway. Since the runaway begins to occur for ions in the high-energy tail of the velocity distribution, the phenomenon appears mathematically as instability in the higher-order moment solutions, because the higher moments sample the tail more effectively.

Because of this theoretical prediction, Howorka et al. (1979) made a successful search for ion runaway with both H^+ and D^+ ions in He. They found several experimental indications of runaway. There was a precipitous increase in the apparent mobility above a certain "onset" value of E/N, and this onset value scaled with ion mass as predicted by theory. The arrival-time spectrum of the ions was also abnormally broadened, in accord with a predicted runaway of the ion diffusion coefficients, and showed an anomalous "early toe" corresponding to the acceleration of the higher-energy ions.

To illustrate the numerical accuracy achieved in the moment-theory calculations, we show in Fig. 8-3-1 the zero-field mobility as a function of temperature as obtained by two completely independent calculations. (This is essentially an expanded version of part of Fig. 6-1-2.) The classical calculations of Lin et al. (1979a) are shown as a solid curve, and the quantum-mechanical calculations of Dickinson and Lee (1978) are shown as individual points. Except for the lowest temperatures, where genuine quantum effects occur (see Section 6-1E), the agreement is within the estimated uncertainty of 0.2% for each calculation. The convergence is very rapid—the first approximation is accurate within 0.2% below 800 K, and within 3% at 5000 K. The quantum deviations amount to only about 3% at 10 K and are negligible above about 50 K. The steady decrease of K_0 from its polarization limit at $T = 0\,\mathrm{K}$ is caused by attractive interactions of shorter range than the r^{-4} polarization potential, as

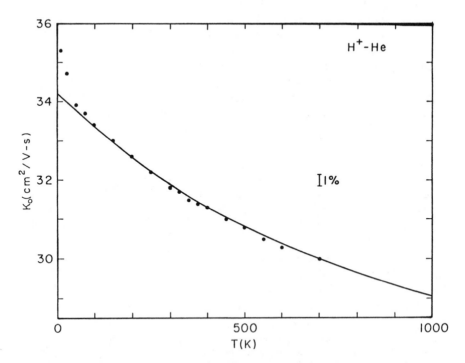

Figure 8-3-1. Zero-field mobility of H^+ in He as a function of temperature as calculated classically (Lin et al., 1979a; solid curve) and quantum mechanically (Dickinson and Lee, 1978; filled circles).

discussed in Section 6-1D. The r^{-6} charge-induced quadrupole energy appears first, followed by the strong valence attraction that leads to a stable HeH^+ molecular ion.

The rather mundane behavior shown in Fig. 8-3-1 changes dramatically when K_0 is calculated as a function of E/N at fixed T, as shown in Fig. 8-3-2.

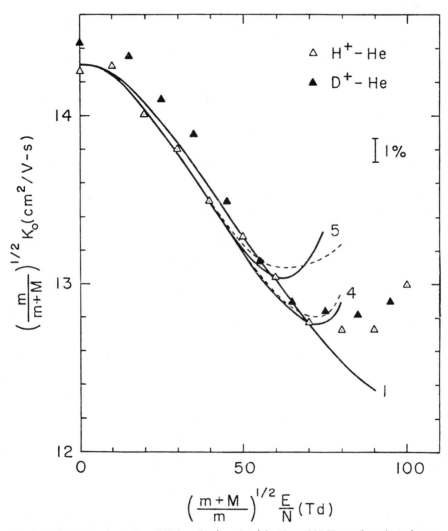

Figure 8-3-2. Mass-scaled mobilities of H^+ and D^+ in He at 300 K as a function of mass-scaled field strength, showing the onset of runaway. The curves are calculations from the two-temperature theory, full curves for H^+ and dashed curves for D^+ (Lin et al., 1979a). The numbers by the curves refer to the order of approximation. The triangles are smoothed values taken from the measurements of Howorka et al. (1979), whose experimental uncertainty is about 4%.

Everything appears normal at low E/N, but the convergence becomes poorer as E/N is increased, the apparent mobility suddenly increases rapidly, and finally the solutions of the moment equations become unstable. The axes of Fig. 8-3-2 have been mass-scaled, as discussed below, to bring the results for H^+ and D^+ together. The experimental data of Howorka et al. (1979) follow the scaling well within the estimated experimental uncertainty, and agree in absolute value with the calculations for E/N below the onset of runaway.

The onset of runaway as shown in Fig. 8-3-2 occurs at somewhat different values of E/N for the calculations and the measurements. This difference is not surprising, since the calculations and experiments are sensitive in different degrees to the high-velocity tail of the distribution function, where the runaway ions originate.

Approximate scaling rules that account for the change of the K_0 vs. E/N curve when the ion mass is changed can readily be obtained from the first-approximation formulas of the two-temperature theory (as was discussed in Section 6-2C), or even from the momentum-transfer theory given in Section 5-2A. The following mass-scaling argument is a paraphrase of that given in Section 5-2A. From the expression (6-2-25) for K we see that $\mu^{1/2}K_0$ vs. T_{eff} should give a single curve for both H^+ and D^+ in He, where μ is the ion-neutral reduced mass. The formula (6-2-26) for T_{eff} shows that T_{eff} in such a plot can be replaced by $(T_{\text{eff}} - T)$ if T is fixed and E/N is varied. But $(T_{\text{eff}} - T)$ is proportional to $Mv_d^2 = MK^2E^2$; since μK^2 depends only on T_{eff}, it then follows that E^2/μ depends only on T_{eff} if T is fixed. We conclude that $\mu^{1/2}K_0$ vs. $\mu^{-1/2}E/N$ is a single curve for both H^+ and D^+ in He at fixed T. Since M is fixed, it is more convenient to use the dimensionless quantity $m/(m + M)$ in place of μ, as is done in Fig. 8-3-2. The scaling is exact in the first approximation, but deviations can be seen in the higher approximations in Fig. 8-3-2, as the differences between the solid and dashed curves.

The runaway phenomenon can be understood rather simply by the momentum-transfer arguments of Chapter 5. We can expect runaway whenever the average momentum gained from the field exceeds that lost by collisions,

$$eE > \mu v_d \bar{v}, \tag{8-3-1}$$

where the mean collision frequency $\bar{v}$ is

$$\bar{v} = N\bar{v}_r Q^{(1)} \approx N(2\bar{\varepsilon}/\mu)^{1/2}Q^{(1)}, \tag{8-3-2}$$

in which $Q^{(1)}$ is the momentum-transfer cross section and we have taken $\bar{\varepsilon} \approx \frac{1}{2}\mu\bar{v}_r^2$. To eliminate v_d from (8-3-1) we use the energy-balance equation,

$$\bar{\varepsilon} = \frac{3}{2}kT + \frac{1}{2}Mv_d^2, \tag{8-3-3}$$

and neglect the thermal energy; the average momentum loss by collisions thereby becomes

$$\mu v_d \bar{v} \approx 2N \left(\frac{m}{m+M} \right)^{1/2} \bar{\varepsilon} Q^{(1)}. \tag{8-3-4}$$

Substituting this expression into (8-3-1) we can write an approximate criterion for runaway as

$$\frac{e}{2} \left(\frac{m+M}{m} \right)^{1/2} \frac{E}{N} > \bar{\varepsilon} Q^{(1)}. \tag{8-3-5}$$

Runaway will actually occur even sooner because of the high-energy ions in the tail of the distribution.

The criterion for runaway can be shown graphically by a plot of $\varepsilon Q^{(1)}$ vs. ε. If the curve has an absolute maximum, it will always be possible to choose a value of E/N large enough to meet the criterion (8-3-5), and runaway will occur. But if the curve continues to rise, no runaway is possible. Such a plot is shown in Fig. 8-3-3 for H^+–He; the curve never rises higher than about 110×10^{-17} eV-cm^2. This behavior can be contrasted with that of an (8, 4) potential having the same depth and position of the potential well, shown as a dashed curve in Fig. 8-3-3. (This is a larger version of the curve shown in Fig. 6-3-1.) After a first local minimum near $\varepsilon = 1$ eV, the (8, 4) curve continues to rise, whereas the H^+–He curve shows a second maximum at about $\varepsilon = 10^2$ eV and then falls off. This fall-off is caused by the softness of the repulsive wall of the potential.

The shape of the $\varepsilon Q^{(1)}$ curve can be correlated with the potential in the following way. At low energies the collisions are dominated by the r^{-4} polarization potential, for which $Q^{(1)}$ varies as $\varepsilon^{-1/2}$. Thus $\varepsilon Q^{(1)}$ rises until a shorter-ranged repulsion manifests itself. There is initially a partial cancellation, leading to the first local maximum, and then the repulsion begins to dominate the scattering, leading to a minimum and subsequent rise in $\varepsilon Q^{(1)}$. Usually, this rise continues to very high energies, but for H^+–He the short-range repulsion becomes so soft that $Q^{(1)}$ falls more rapidly than ε^{-1} at large ε. The result is a second local maximum in $\varepsilon Q^{(1)}$ and a subsequent fall.

A criterion for the *absence* of any runaway can therefore be stated as follows: at high energies $Q^{(1)}$ must decrease less rapidly than ε^{-1}.

Notice that the mass scaling for runaway given by (8-3-5) is the same as the general mass scaling for E/N shown in Fig. 8-3-2. We thus expect the mass-scaling rules to hold even after the onset of runaway.

Since runaway ions continue to accelerate, the mobility concept becomes meaningless, although an "apparent" drift velocity for a particular apparatus can always be measured in terms of the length of the drift tube L divided by the transit time t. The resulting apparent mobility will then depend separately on E and N rather than just their ratio E/N, as well as on L. Although Howorka et al. (1979) pointed out this effect, they were not able to vary N over enough of a range to check it experimentally. In terms of the usual experimental procedures,

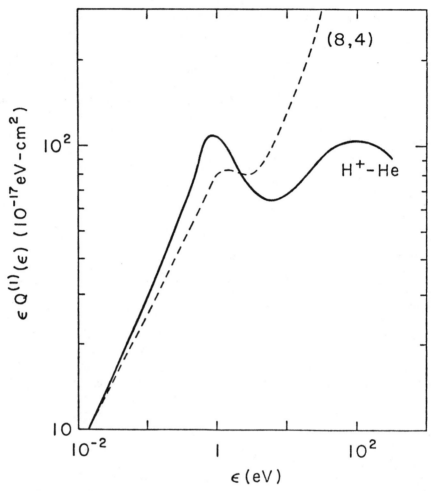

Figure 8-3-3. Collisional momentum loss as a function of collision energy, showing the conditions for runaway with H^+ in He. The full curve is calculated for the actual H^+–He potential, and the broken curve shown for comparison is calculated for an (8, 4) potential.

the effect appears as a dependence of the apparent reduced mobility on N at fixed E/N. That is, if K_0 is plotted against E/N for a series of runs at different constant values of N, the points for the different runs fall on a single curve at low E/N, before the onset of runaway, but separate into a different curve for each N after runaway begins. Such an effect was found by Moruzzi and Kondo (1980). What was surprising about their results, however, was the strong dependence of the apparent K_0 on N at fixed E/N. Even if the ions accelerated all the way through the drift tube, the density dependence would only be as $N^{1/2}$, whereas for a true steady drift there would be no dependence on N at all. Experimentally, a dependence on N was observed that did not seem to fall between the extreme limiting cases.

This seemingly anomalous dependence on N can be qualitatively explained by noting that an induction time is required before the ions acquire a speed sufficient to reach runaway, assuming that the ions are injected into the drift tube with zero velocity (Waldman and Mason, 1981). The apparent mobility as a function of N thus starts out along a curve corresponding to an eventual steady-state drift that would be independent of N, but somewhere runaway sets in and the apparent mobility shifts over to an accelerator-like curve approaching an $N^{1/2}$ behavior. The transition region from a drift curve to an accelerator curve may show a strong local dependence on N. To demonstrate this behavior, Waldman and Mason (1981) used a simple model in which the collision frequency v_m as a function of v_d is constant until a critical speed v_c is reached, after which v_m becomes abruptly zero. Although this model represents reality only crudely, the expectation is that it embodies the essential qualitative features. They then used this model for v_m in an average equation of motion for the time dependence of the ion swarm velocity,

$$\frac{dv_d}{dt} = \frac{eE}{m} - v_m v_d, \qquad (8\text{-}3\text{-}6a)$$

$$v_m = (\mu/m)\bar{v}, \qquad (8\text{-}3\text{-}6b)$$

where $\bar{v}$ is the collision frequency as we have usually defined it. The mass factor occurs because of the conversion from center-of-mass coordinates to laboratory coordinates. This equation of motion could be integrated to find the time to reach v_c and the transit time through the drift tube, from which the apparent mobility could be found. Notice that for a steady state (8-3-6a) reduces to $eE = \mu\bar{v}v_d$, the usual equation of momentum balance.

The results are shown in Fig. 8-3-4 in terms of a dimensionless reduced mobility K_0^* and number density N^*, defined as

$$K_0^* \equiv \frac{L/t}{v_\infty}, \qquad (8\text{-}3\text{-}7)$$

$$N^* \equiv \frac{vL}{v_\infty}, \qquad (8\text{-}3\text{-}8)$$

where $v_\infty \equiv eE/\mu v$ is the asymptotic drift velocity that would result if v were constant without a cutoff at v_c. There is a family of curves of K_0^* vs. N^* indexed by a dimensionless ratio R,

$$R \equiv v_c/v_\infty. \qquad (8\text{-}3\text{-}9)$$

This family of curves represents the density dependence of K_0 at fixed E/N, with each value of R corresponding to a different value of E/N. For each curve E/N is fixed and so v_∞ is constant, but v_∞ increases as the value of E/N is increased. Thus R decreases as E/N is increased, and $R = 0$ corresponds to runaway

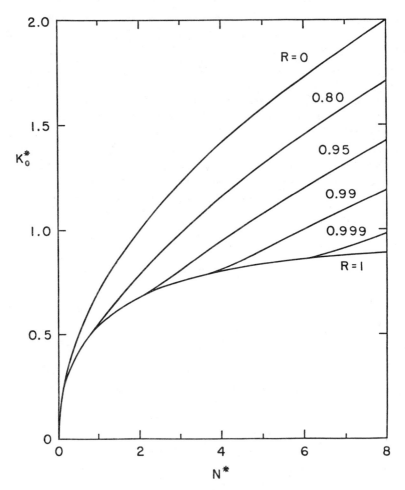

Figure 8-3-4. Reduced apparent mobility K_0^* as a function of reduced gas density N^* at fixed E/N, showing the density dependence of ion runaway. The runaway ratio R indexes the value of E/N; $R = 0$ refers to the pure accelerator case, and $R = 1$ to the true steady-state drift case.

throughout the entire length of the drift tube. Similarly, $R = 1$ corresponds to no runaway, and the attainment of a true steady state.

The important feature to note is that some of the curves exhibit a rise faster than N over a range of N^*—that is, they show regions of positive curvature. These correspond to a strong dependence of K_0 on N at fixed E/N, as observed by Moruzzi and Kondo (1980). These regions disappear as E/N is increased (i.e., as R is decreased). This last feature has not been tested experimentally.

Another feature predicted by the model is the dependence of runaway on L. Notice that because v_∞ is a constant quantity when E/N is fixed, the density N^*

as defined by (8-3-8) is proportional to NL. Thus the model predicts that N and L enter as the single variable NL. A similar result from gas-discharge physics is known as Paschen's law (Thomson and Thomson, 1933, pp. 486–490). This feature of runaway has not been tested experimentally, either.

In summary, the phenomenon of ion runaway is explained rather well by simple momentum-transfer arguments. The full moment theory in its usual form cannot deal with such a transient phenomenon, which is excluded by its basic postulates; it can only signal the existence of a new physical feature by its failure to converge. Runaway is not likely to have any practical applications—it is more likely to be something to avoid because of its possible catastrophic effects.

8-4. RELAXATION IN DRIFT TUBES

Our discussions of ion transport have so far dealt almost exclusively with steady-state conditions, with the notable exception of the runaway phenomenon of the preceding section, for which the moment methods failed. Moment methods can nevertheless deal with a limited class of time-dependent phenomena, and we shall discuss several in the remainder of this chapter. These are phenomena in which processes are sufficiently slow that the ion velocity distribution manages to maintain a form that is not too drastically different from the gaussian forms assumed for $f^{(0)}$ in the one-, two-, and three-temperature theories. These might be characterized as quasi-steady-state phenomena. The parameters of $f^{(0)}$, such as the basis temperature T_b, may vary with time, but the approximately gaussian mathematical form remains intact. This condition clearly fails with runaway, where a large high-velocity tail develops in the distribution.

In this section we consider some transient effects in drift tubes, such as the attainment of a steady drift velocity and mean ion energy when ions are first injected. The usual hope is, of course, that such entrance relaxation effects will not spoil the measurements, but it is helpful to have a theory available that can describe such transients, so that errors can be estimated and perhaps corrected. Entrance transients may well be important for very heavy or very light (e.g., electrons) ions in short drift tubes. A similar transient effect occurs in drift-tube studies of ion-molecule reactions in which the ion residence time is increased by periodic reversal of the electric field (Heimerl et al., 1969; Johnsen and Biondi, 1972). A transient occurs at each reversal. We consider both kinds of transients in this section.

We follow the presentation of Lin et al. (1977) and first present a simple phenomenological theory, followed by a full kinetic theory. The lowest-order moment equations of the kinetic theory are the same as the phenomenological equations, but give precise microscopic definitions of the collision frequencies introduced in the phenomenological theory. Detailed experimental data on transients being very scarce, we use "simulated data" from Monte Carlo computer calculations for a few special cases in order to make comparisons with the theoretical predictions. The treatment is limited to elastic collisions, but it is easy to see how an extension to inelastic collisions might proceed.

The ideas behind the phenomenological theory are simple extensions of the momentum-transfer theory introduced in Section 5-2A. If the momentum acquired by the ions from the field is not balanced by momentum loss due to collisions, then the difference should lead to an increase in the average ion momentum,

$$m\frac{dv_d(t)}{dt} = eE - v_m m v_d(t),$$
(8-4-1)

where v_m is a mean collision frequency for momentum transfer. If we look back at the steady-state momentum-balance equation (5-2-8), we see that v_m is related to the usual mean collision frequency $\bar{v}$ by

$$v_m = (\mu/m)\bar{v}.$$
(8-4-2)

This is the same equation of motion as used in the preceding section to discuss the pressure dependence of ion runaway, namely (8-3-6).

Similarly, if the energy acquired by the ions from the field is not balanced by energy loss due to collisions, then the difference should lead to an increase in the average ion energy. However, it is only the ion energy *in excess of thermal energy* that is lost by collisions, so we write

$$\tfrac{1}{2}m\frac{\overline{dv^2(t)}}{dt} = eEv_d(t) - v_e[\tfrac{1}{2}m\overline{v^2(t)} - \tfrac{3}{2}kT],$$
(8-4-3)

where v_e is a mean collision frequency for energy transfer. A similar thermal term does not occur in the momentum equation (8-4-1) because the mean momentum of the neutrals is zero. If we look back at the steady-state energy-balance equation (5-2-19), we find that v_e is related to $\bar{v}$ by

$$v_e = \frac{2mM}{(m+M)^2}\bar{v} = \frac{2m}{m+M}v_m.$$
(8-4-4)

In practice, we must imagine that both v_m and v_e can depend on time, because the mean collision energy changes with time. It is only for the Maxwell model that the collision frequencies are constants. However, for illustration purposes we assume that v_m and v_e are constants in order to integrate the differential equations; the results are

$$v_d(t) - v_d(\infty) = [v_d(0) - v_d(\infty)]e^{-v_m t},$$
(8-4-5)

$$\overline{v^2(t)} - \overline{v^2(\infty)} = [\overline{v^2(0)} - \overline{v^2(\infty)}]e^{-v_e t}$$
$$+ \left(\frac{v_e}{v_e - v_m}\right)\left[1 - \frac{v_d(0)}{v_d(\infty)}\right]\left[\overline{v^2(\infty)} - \frac{3kT}{m}\right](e^{-v_e t} - e^{-v_m t}).$$
(8-4-6)

Roughly speaking, the decay of $v_d(t)$ is like that of a single radioactive species, whereas the behavior of $\overline{v^2(t)}$ is like that of a radioactive daughter of this species.

To construct a full kinetic theory of momentum and energy relaxation in drift tubes, we go back to the general development of moment equations outlined in Section 5-3B and reconsider some of the assumptions. The general equation describing the moments of any function $\psi(\mathbf{v})$ of the ion velocity was given by (5-3-12), which we repeat here,

$$\frac{\partial}{\partial t}(n\langle\psi\rangle) + \mathbf{V}_r \cdot (n\langle\mathbf{v}\psi\rangle) - \frac{ne}{m}\mathbf{E}\cdot\langle\mathbf{V}_v\psi\rangle = -n\sum_j N_j\langle J_j\psi\rangle, \qquad (8\text{-}4\text{-}7)$$

where J_j is the linear collision operator defined by (5-3-13). The usual assumption made in applying this equation to drift tubes is that the time derivatives of all velocity moments are small and can be neglected, except for $\partial n/\partial t$. We now wish to modify this assumption and keep at least the time derivatives of the moments corresponding to momentum and energy, which of course makes the equations more complicated. In order to reduce some of this complexity for the particular case we are treating here, we make an important simplification by assuming spatial homogeneity of the ions. That is, we neglect spatial derivatives in the moment equations and attribute all relaxation to the collisional terms alone. Although this assumption does not quite correspond to the way that real drift tubes operate, it is usually regarded as a reasonable first approximation, and was in fact made implicitly in the preceding phenomenological theory—that is, no convection or diffusion terms for momentum and energy appeared in (8-4-1) and (8-4-3). Physically, this assumption means that processes of a diffusive nature are much slower and hence less important than collisional processes in determining time development (except for $\partial n/\partial t$, since n cannot decay by collisions). This assumption then implies that $\partial n/\partial t = 0$, because the equation of continuity connects $\partial n/\partial t$ with spatial derivatives, which can be regarded as a result either of true spatial homogeneity or, better, of slow diffusion.

The general moment equation then simplifies to

$$\frac{\partial}{\partial t}\langle\psi\rangle = \frac{e}{m}\mathbf{E}\cdot\langle\mathbf{V}_v\psi\rangle - \sum_j N_j\langle J_j\psi\rangle. \qquad (8\text{-}4\text{-}8)$$

Notice that the dropping of the spatial gradient terms means that we are ignoring the effects of boundaries. The next step is to choose a set of basis functions and expand the collision term; for this purpose we choose the two-temperature Burnett functions summarized in Section 5-3D, namely

$$\psi_{lm}^{(r)} = w_b^l S_{l+1/2}^{(r)}(w_b^2) Y_l^m(\theta, \phi), \qquad (8\text{-}4\text{-}9a)$$

$$w_b^2 = mv^2/2kT_b, \qquad (8\text{-}4\text{-}9b)$$

where $S_{l+1/2}^{(r)}(w_b^2)$ are Sonine polynomials, $Y_l^m(\theta, \phi)$ are spherical harmonics, and

T_b is the ion basis temperature. The index m is always zero in the present application, so we henceforth omit it. For simplicity, we now consider only a single neutral gas in the drift tube; the case of a gas mixture is easily worked out, and the results are given by Lin et al. (1977). The expansion of the collision term then gives

$$J\psi_l^{(r)} = \sum_{s=0}^{\infty} a_{rs}(l)\psi_l^{(s)}, \tag{8-4-10}$$

where the matrix elements $a_{rs}(l)$ are eventually expressed as linear combinations of irreducible collision integrals (see, e.g., Table 6-2-1). The rest of the calculation then proceeds as for the steady-state case, except that T_b can vary with time because it represents the mean ion energy. The term $\langle \mathbf{V}_v\psi_l^{(r)} \rangle$ is evaluated through the recursion relation (6-1-3), and (8-4-8) becomes

$$\frac{d}{dt}\langle \psi_l^{(r)} \rangle + [\tfrac{1}{2}(l+r)\langle \psi_l^{(r)} \rangle - (l+\tfrac{1}{2}+r)\langle \psi_l^{(r-1)} \rangle]\frac{d\ln T_b}{dt}$$

$$= \frac{v_m\mathscr{E}}{(l+1/2)}[l(l+\tfrac{1}{2}+r)\langle \psi_{l-1}^{(r)} \rangle - (l+1)\langle \psi_{l+1}^{(r-1)} \rangle]$$

$$- v_m\sum_{s=0}^{\infty}\gamma_{rs}(l)\langle \psi_l^{(s)} \rangle, \tag{8-4-11}$$

where

$$\mathscr{E} \equiv \left(\frac{m}{2kT_b}\right)^{1/2}\frac{eE}{mv_m}, \tag{8-4-12}$$

$$v_m \equiv Na_{00}(1) = \tfrac{4}{3}N\frac{\mu}{m}\left(\frac{8kT_{\text{eff}}}{\pi\mu}\right)^{1/2}\bar{\Omega}^{(1,1)}(T_{\text{eff}}), \tag{8-4-13}$$

$$T_{\text{eff}} \equiv (mT + MT_b)/(m+M), \tag{8-4-14}$$

$$\gamma_{rs}(l) \equiv a_{rs}(l)/a_{00}(1). \tag{8-4-15}$$

Except for the time-dependent terms, which we have just introduced, this result is the same as the moment equations (6-2-4) for the steady state. We have chosen the definition of v_m to give the same correspondence between $\bar{v}$ and $\bar{\Omega}^{(1,1)}$ that was exhibited in (6-1-15a). This is an infinite set of coupled equations that must be solved by truncation and iteration.

The moment equations corresponding to momentum and energy relaxation are the two equations of the infinite set (8-4-11) with $l=1$, $r=0$, and $l=0$, $r=1$. For a first approximation we identify T_b as the mean ion energy, which is equivalent to taking $\langle \psi_0^{(1)} \rangle = 0$, and neglect all terms in the summation with

$s > r$. Recalling that normalization gives $\langle \psi_0^{(0)} \rangle = 1$, we obtain

$$\frac{d}{dt}\langle \psi_1^{(0)} \rangle + \tfrac{1}{2}\langle \psi_1^{(0)} \rangle \frac{d \ln T_b}{dt} = v_m \mathscr{E} - v_m \langle \psi_1^{(0)} \rangle, \qquad (8\text{-}4\text{-}16)$$

$$-\frac{3}{2}\frac{d \ln T_b}{dt} = -2v_m \mathscr{E} \langle \psi_1^{(0)} \rangle - v_m \gamma_{10}(0), \qquad (8\text{-}4\text{-}17)$$

with

$$\tfrac{3}{2}kT_b = \tfrac{1}{2}m\langle v^2 \rangle, \qquad (8\text{-}4\text{-}18)$$

$$\langle \psi_1^{(0)} \rangle = (m/2kT_b)^{1/2}v_d, \qquad (8\text{-}4\text{-}19)$$

$$\gamma_{10}(0) = -\frac{3m}{m + M}\frac{T_b - T}{T_b}. \qquad (8\text{-}4\text{-}20)$$

The result for $\gamma_{10}(0)$ follows from the formulas for the matrix elements in Table 6-2-1. This pair of moment equations is closed, in that only two unknowns are involved, $\langle \psi_1^{(0)} \rangle$ and T_b. In higher approximations, other moments occur in (8-4-16) and (8-4-17), and further equations must be included to obtain a closed set. We shall not bother with higher approximations here. When the expressions for T_b, $\langle \psi_1^{(0)} \rangle$, and $\gamma_{10}(0)$ are substituted back into (8-4-16) and (8-4-17), we obtain equations for $v_d(t)$ and $\langle v^2(t) \rangle$ that are identical in form to the phenomenological equations (8-4-1) and (8-4-3), provided that the collision frequency for energy transfer is defined as

$$v_e = -\frac{2}{3}\left(\frac{T_b}{T_b - T}\right)\gamma_{10}(0)v_m = \frac{2m}{m + M}v_m, \qquad (8\text{-}4\text{-}21)$$

the last equality following from the expression for $\gamma_{10}(0)$. In other words, the full kinetic theory reduces to the phenomenological momentum-transfer theory in first approximation. The only real difference is that the kinetic theory gives a more precise definition of v_m in terms of a collision integral for ion-neutral collisions, and also provides for the systematic calculation of corrections, if needed.

Because of the lack of detailed experimental information with which to test the foregoing theory, Lin et al. (1977) used computer-generated simulated data, obtained by following ion trajectories numerically by means of a Monte Carlo technique (Lin and Bardsley, 1975, 1977, 1978). Two different initial conditions were chosen:

1. The ion swarm was initially stationary, so that $v_d(0) = 0$ and $\tfrac{1}{2}m\langle v^2(0) \rangle = \tfrac{3}{2}kT$.

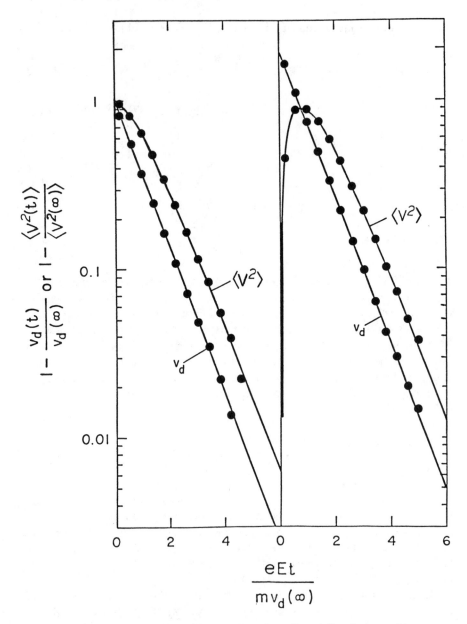

Figure 8-4-1. Semilogarithmic plots of the relaxation of ion drift velocity and ion energy as a function of dimensionless time, $eEt/mv_d(\infty)$, for the Maxwell model with $T = 0$ and $m = 4M$. The two sets of initial conditions are an initially stationary ion swarm and a field reversal, as described in the text. The full curves are calculated from moment theory and the filled circles are Monte Carlo computer simulations.

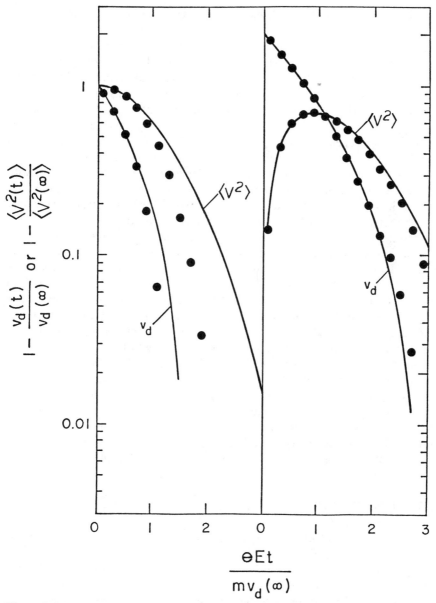

Figure 8-4-2. Same as Fig. 8-4-1, for the rigid-sphere model with $T = 0$ and $m = M$.

2. The ion swarm had previously reached a steady state with the electric field oriented in the opposite direction, so that $v_d(0) = -v_d(\infty)$ and $\langle v^2(0) \rangle = \langle v^2(\infty) \rangle$.

Two ion-neutral interactions were investigated, the Maxwell model and the rigid-sphere model.

The results for the Maxwell model are shown in Fig. 8-4-1, for a cold gas ($T = 0$) and $m = 4M$. For this model the collision frequencies v_m and v_e are constants and the first kinetic-theory approximation is exact, so that the agreement between theory and simulation should be nearly perfect. The agreement is indeed within the estimated 1% numerical accuracy of the Monte Carlo simulations. This is really a check on the simulations rather than a true test of the kinetic theory, since we believe that the latter is exact in this case.

For rigid spheres it is $\bar{\Omega}^{(1,1)}$ that is constant, and v_m and v_e depend on time through their dependence on T_{eff}. Then the differential equations for $v_d(t)$ and $\langle v^2(t) \rangle$ must be integrated numerically, and moreover are only first approximations. Results are shown in Fig. 8-4-2 for ions in a cold gas ($T = 0$) with $m = M$. Some discrepancies are now apparent between the theoretical curves and the Monte Carlo simulations, as is to be expected because the theory is only a first approximation, whereas the simulations are accurate to about 1%. It is interesting that the theory gives better agreement with the Monte Carlo simulations for the field-reversal situation than for the initially stationary ion swarm. This can be rationalized in terms of the anisotropy of the ion energy in the two situations. In the stationary-swarm case, the ion energy first flows into the direction parallel to the field, and then by collisions into the perpendicular direction. In the field-reversal case, the ion energy is already fully distributed into the perpendicular direction. It is thus plausible that the stationary-swarm case is more anisotropic, and hence less well described by the two-temperature theory, which does not have anisotropy built in from the beginning as the three-temperature theory does.

In summary, the foregoing calculations show that moment theory can handle at least some time-dependent phenomena. It is unlikely that measurements of transient effects in drift tubes will ever be of interest for their own sake, but these results show that the theory can be used to describe such transients, and at the least is useful for the estimation of errors resulting from neglect of transients.

8-5.   ION CYCLOTRON RESONANCE

Another time-dependent phenomenon that can be treated by moment methods is ion cyclotron resonance (ICR). In their simplest form, ICR experiments involve an ion swarm moving in a neutral gas (or gas mixture) under the influence of a constant magnetic field $\mathbf{B}$ and an oscillating electric field $\mathbf{E}(t)$ perpendicular to $\mathbf{B}$. The magnetic field causes the ions to rotate with a cyclotron frequency $\omega_c = eB/m$. If there were no collisions, the ions would absorb energy from the oscillating electric field as a sharp resonance around ω_c. Collisions with neutrals broaden the resonance to an approximately Lorentzian shape as a function of frequency ω, centered at ω_c. Study of the broadened line shape then offers possibilities for obtaining information about ion-neutral collisions. In particular, study of the ICR line shape as a function of gas temperature and amplitude of the electric field can in principle yield the same sort of information as the study of the drift velocity as a function of T and E in a drift tube. In this section we give a brief review of the kinetic theory of ICR line shape in the collision-dominated regime.

We keep the review brief because ICR has in fact not developed into a serious competitor for drift-tube experiments. The reason is probably historical—ICR was developed as an experimental technique much later than drift tubes were. Instead, ICR has been mostly exploited for the study of ion-molecule reactions by working in a nearly collision-free regime, where drift tubes are useless. Here ICR is essentially used as an analytical tool to monitor the decay of reactant ions and the growth of product ions by observation of their sharp resonance lines. Kinetic theory plays no role in the interpretation of such experiments, because collisions are rare events; these experiments are more analogous to beam experiments than to drift-tube experiments.

Reviews of ICR experimental techniques and applications have been given by Baldeschwieler (1968) and by Beauchamp (1971). The usual starting point for the theoretical interpretation of the results is a phenomenological equation of motion for the mean ion velocity, very much like the one used in the previous section to discuss relaxation in drift tubes; the main difference is the addition of an external force term due to the magnetic field,

$$m\frac{d}{dt}\bar{\mathbf{v}} = e(\mathbf{E} + \bar{\mathbf{v}} \times \mathbf{B}) - v_m m\bar{\mathbf{v}}, \qquad (8\text{-}5\text{-}1)$$

where v_m is a mean collision frequency for momentum transfer (Beauchamp, 1967). Other differences are that the mean velocity is now a three-dimensional vector $\bar{\mathbf{v}}$ rather than a one-dimensional drift velocity v_d, and $\mathbf{E}$ varies with time,

$$\mathbf{E} = \mathbf{E}_0 \sin \omega t. \qquad (8\text{-}5\text{-}2)$$

As might by now be anticipated, this equation of motion also appears as a first approximation in the two-temperature moment theory (Viehland et al., 1975b).

After (8-5-1) is solved for $\overline{\mathbf{v}(t)}$, the instantaneous power absorption per ion can be calculated from the relation

$$A(t, \omega) = e\mathbf{E} \cdot \bar{\mathbf{v}} = e\mathbf{E}_0 \cdot \overline{\mathbf{v}(t)} \sin \omega t. \qquad (8\text{-}5\text{-}3)$$

In most ICR experiments both ω and ω_c are much greater than both $(\omega - \omega_c)$ and v_m, so that an ion makes many revolutions between collisions, and the ions are initially nearly at rest. Under these conditions the general result can be shown to simplify to (Comisarow, 1971)

$$A(t, \omega) = A(\infty, \omega)\left\{1 - e^{-v_m t}\left[\cos(\omega - \omega_c)t - \frac{\omega - \omega_c}{v_m}\sin(\omega - \omega_c)t\right]\right\},$$

$$(8\text{-}5\text{-}4)$$

where the steady-state $(t \to \infty)$ power absorption is

$$A(\infty, \omega) = \frac{e^2 E_0^2 v_m}{4m[(\omega - \omega_c)^2 + v_m^2]}. \tag{8-5-5}$$

Although the collision-dominated quantity $A(\infty, \omega)$ is the item of main interest here, it is worth noting that (8-5-4) holds even in the collisionless region ($v_m = 0$ because $N = 0$), where it reduces to

$$A(t, \omega) \to \frac{e^2 E_0^2}{4m(\omega - \omega_c)} \sin(\omega - \omega_c)t. \tag{8-5-6}$$

If v_m is independent of E_0 and $(\omega - \omega_c)$, then $A(\infty, \omega)$ has a Lorentzian shape and v_m can be determined experimentally as equal to the half-width at half-height. That is, measurements of ICR absorption can be used to determine v_m and hence the average ion-neutral momentum-transfer cross section, which is essentially the same information that is obtained from drift-tube measurements. However, v_m in general does depend on E_0 and $(\omega - \omega_c)$ because it depends on the ion-neutral collision energy. The absorption line shape is then not Lorentzian, and a quantitative interpretation based on the phenomenological theory becomes suspect. It is only in the low-field limit, where the ion energy is thermal, that v_m can be expected to depend only on T and be independent of E_0 and $(\omega - \omega_c)$.

An accurate kinetic theory is thus necessary if ICR experiments are to yield detailed quantitative information on ion-neutral interactions. We present here a summary of the two-temperature moment theory developed by Viehland et al. (1975b). We proceed as usual to form moment equations, and then argue that spatial variations of the moments are small and that diffusion is slow in collision-dominated ICR experiments; this procedure yields the general moment equation,

$$\frac{\partial}{\partial t} \langle \psi \rangle = \frac{e}{m} \langle \mathbf{E} + \mathbf{v} \times \mathbf{B} \rangle \cdot \nabla_v \psi \rangle - \sum_j N_j \langle J_j \psi \rangle. \tag{8-5-7}$$

Except for the $\mathbf{v} \times \mathbf{B}$ term, this is the same as the general moment equation (8-4-8) for relaxation in drift tubes, and we can follow the same method of solution. Choosing the two-temperature Burnett functions as a basis set, expanding the collision term, and restricting ourselves to a single neutral gas, we reduce the result to

$$\frac{\partial}{\partial t} \langle \psi_l^{(r)} \rangle = v_m(\mathscr{E}_0 \sin \omega t + \mathscr{E}_z) \left\langle \frac{\partial \psi_l^{(r)}}{\partial w_{bz}} \right\rangle$$

$$+ v_m \mathscr{E}_y \left\langle \frac{\partial \psi_l^{(r)}}{\partial w_{by}} \right\rangle - \omega_c \left\langle w_{bz} \frac{\partial \psi_l^{(r)}}{\partial w_{bx}} - w_{bx} \frac{\partial \psi_l^{(r)}}{\partial w_{bz}} \right\rangle$$

$$- v_m \sum_{s=0}^{\infty} \gamma_{rs}(l) \langle \psi_l^{(s)} \rangle, \tag{8-5-8}$$

where $v_m \equiv N a_{00}(1)$, $\gamma_{rs}(l) \equiv a_{rs}(l)/a_{00}(1)$, and $\mathscr{E} \equiv (m/2kT_b)^{1/2}(eE/mv_m)$, exactly as in the preceding section. The derivative terms can be evaluated through recursion relations for Burnett functions, leaving us with an infinite set of coupled ordinary differential equations, which we can solve by truncation and iteration to find the mean ion velocity, represented by $\langle \psi_1^{(0)} \rangle$, and hence the power absorption.

In a first approximation we obtain an equation that is essentially the same as the phenomenological equation of motion (8-5-1); written out, this is a pair of coupled equations for the components of $\langle \psi_1^{(0)} \rangle$,

$$\frac{d}{dt} \langle \psi_{1z}^{(0)} \rangle = v_m(\mathscr{E}_0 \sin \omega t + \mathscr{E}_z) + \omega_c \langle \psi_{1x}^{(0)} \rangle - v_m \langle \psi_{1z}^{(0)} \rangle, \qquad (8\text{-}5\text{-}9a)$$

$$\frac{d}{dt} \langle \psi_{1x}^{(0)} \rangle = -\omega_c \langle \psi_{1z}^{(0)} \rangle - v_m \langle \psi_{1x}^{(0)} \rangle. \qquad (8\text{-}5\text{-}9b)$$

These equations can be solved formally without assuming that v_m is independent of the field strength or the frequency difference $(\omega - \omega_c)$. The result looks exactly like the phenomenological result given by (8-5-4) and (8-5-5). The essential difference, however, is that v_m now has a precise microscopic definition in terms of ion-neutral collisions, and can depend on E_0 and $(\omega - \omega_c)$.

To complete the description we need a specification of the parameter T_b of the two-temperature theory. As usual we identify it with the mean ion energy, which can be found in first approximation from another moment equation. The result is

$$\tfrac{3}{2} kT_b = \tfrac{1}{2} m \langle v^2 \rangle = \tfrac{3}{2} kT + \frac{m + M}{2mv_m} A(\infty, \omega). \qquad (8\text{-}5\text{-}10)$$

As before, v_m is a function of a single variable T_{eff} as given by (8-4-13) and (8-4-14). Substituting for $A(\infty, \omega)$ from (8-5-5), we find this effective temperature to be

$$kT_{\text{eff}} = kT + \frac{M}{3m} \frac{e^2 E_0^2}{4m[(\omega - \omega_c)^2 + v_m^2]}, \qquad (8\text{-}5\text{-}11)$$

with v_m given as a function of T_{eff} by (8-4-13).

Thus the quantity T_{eff} plays the same sort of role as in the case of steady-state drift-tube experiments, and represents the total random energy of the ions. It consolidates the dependence of ICR absorption line shape on the three variables T, E_0, and $(\omega - \omega_c)$ into a dependence on the single variable T_{eff}. Measurements of ICR line shape as a function of these three variables can thus yield the quantity $v_m(T_{\text{eff}})$, or equivalently the collision integral $\bar{\Omega}^{(1,1)}(T_{\text{eff}})$ for ion-neutral collisions. In principle, then, collision-dominated ICR measurements are capable of yielding the same information as drift-tube measurements. However, as remarked at the beginning of this section, ICR has not developed in this direction, and so we have little more to say on the subject of ICR.

Two final remarks are in order. Viehland et al. (1975b) have investigated the second approximation for the solution of (8-5-8), and concluded that the convergence errors are about the same for ICR line shapes as for drift velocities. They have also written down the explicit results for ICR in a mixture of neutral gases.

8-6. ION-MOLECULE REACTIONS

Reactions between ions and neutrals play important roles in many areas, such as upper-atmosphere chemistry, gas discharges, radiation chemistry, flames, and others. Such reactions can be studied in drift tubes by adding a small amount of reactive gas to the large excess of inert drift gas (buffer gas). The rate coefficient for the ion-molecule reaction can then be determined by measuring the attenuation of the ion current as a function of the added reactant gas concentration. Drift-tube measurements have been useful in covering a significant energy gap that exists between the thermal regime ending at about 600°C and the regime of ion beams beginning at several eV (Albritton, 1979).

A simplification in the interpretation occurs because reactive collisions are such rare events in a drift tube that the reverse reactions can be completely neglected. Therefore we need consider only the integral cross section $Q_R^*(\alpha\beta; \varepsilon_R)$ for reaction of an ion in internal state α with a molecule of reactant species R in internal state β, with collision energy ε_R. In contrast, for the buffer gas we must consider the full differential cross section $\sigma_B(\alpha\beta, \alpha'\beta'; \theta, \phi, \varepsilon)$, as discussed in Section 6-6. Note that the same neutral gas can serve as both a buffer and a reactant only if the reaction probability is very small.

In compensation, a complication exists because the reaction rate coefficient that is measured depends on the velocity distribution function of the ions, which is far from Maxwellian at high electric fields, and which depends on details of the ion-buffer gas collisions and is thus an unknown function of the temperature and electric field strength. That is, the measured rate coefficient k_R depends not only on the temperature T of the buffer and reactant gases, but also in an unknown way on E/N through the ion distribution function, as follows:

$$k_R(T, E/N) = (nN_R)^{-1} \sum_\alpha \sum_\beta \iint f^{(\alpha)}(\mathbf{v}, T, E/N) F_R^{(\beta)}(\mathbf{V}_R, T)$$

$$\times |\mathbf{v} - \mathbf{V}_R| Q_R^*(\alpha\beta; \varepsilon_R) \, d\mathbf{V}_R \, d\mathbf{v}, \qquad (8\text{-}6\text{-}1)$$

where the notation is the same as for polyatomic systems in Section 6-6. Although the distribution function $F_R^{(\beta)}(\mathbf{V}, T)$ of the reactant gas can be taken as Maxwellian, the distribution function $f^{(\alpha)}(\mathbf{v}, T, E/N)$ of the ions is completely unknown. The only simplification is that $f^{(\alpha)}$ can be considered to be determined entirely by collisions with the buffer gas molecules, provided that the reactant gas is present in the drift tube in only trace amounts. Mercifully, the buffer gas can always be experimentally chosen to be monatomic, thereby avoiding difficulties with the effect of the buffer-gas internal degrees of freedom on $f^{(\alpha)}$.

The most desirable outcome of drift-tube measurements of $k_R(T, E/N)$ would be the determination of $Q_R^*(\alpha\beta; \varepsilon_R)$. According to (8-6-1) this seems to require the solution of two very difficult problems. First, the ion distribution function $f^{(\alpha)}(\mathbf{v}, T, E/N)$ must be found, presumably by solution of the Boltzmann equation plus analysis of mobility measurements of the ion in the pure buffer gas. Second, $Q_R^*(\alpha\beta; \varepsilon_R)$ must be extracted from $k_R(T, E/N)$ by deconvolution of (8-6-1). It is entirely possible that neither of these problems has a unique solution in any particular case (Russ et al., 1975), or that small errors in k_R may be enormously amplified in Q_R^*. For example, Melton and Gordon (1969) have shown that the extraction of Q_R^* from even *thermal* rate data may be subject to large inaccuracies unless its dependence on ε is quite innocuous.

Despite these formidable difficulties, some real progress has been made on the full inversion problem. Lin and Bardsley (1977) showed how Monte Carlo methods could be used to calculate $f^{(\alpha)}$ from mobility data, and obtained velocity distributions for O^+ ions in He and in Ar buffer gases. The distributions are rather different, and were used by Albritton et al. (1977) to analyze their measurements on the reaction of O^+ ions with O_2, N_2, and NO. They succeeded in inverting their measurements in He buffer gas to determine Q_R^*, from which they calculated the rate coefficient to be expected in Ar buffer gas. The rate coefficients for the same reaction in the two buffer gases are rather different, even when E/N is scaled into a suitable center-of-mass kinetic energy. The agreement with the experimental measurements was very good, as is illustrated in Fig. 8-6-1. A very readable account of this work is given in a review article by Albritton (1979), and shows that the fundamental theoretical machinery can be made to work, at least in a few simple cases. Unfortunately, this direct approach requires rather heroic efforts for each individual case, and there is no assurance of its practicability in general. Indeed, some recent work on the reaction $O^+ + N_2$ using guided ion beam mass spectrometry casts doubt on the quantitative results of the earlier drift-tube studies, probably owing to difficulties with the ion distribution function (Burley et al., 1987).

In view of the foregoing difficulties, it is worth asking whether a result less ambitious than the complete determination of $Q_R^*(\alpha\beta; \varepsilon_R)$ might be useful, and more easily obtained. In particular, one is often happy just to know a *thermal* rate coefficient, corresponding to the rate coefficient that would be measured under conditions of near thermal equilibrium at some effective temperature. Such a rate coefficient may be all that is really needed to solve some problem involving an ion-molecule reaction. It represents just some sort of average over $f^{(\alpha)}$ and Q_R^*, and does not require complete knowledge of either one. Moment methods are particularly suited for finding such averages, and have been applied to the analysis of ion-molecule reactions by Viehland and Mason (1977, 1979) and by Viehland et al. (1981). Robson (1986) has given an extended form of momentum-transfer theory that includes reactive phenomena; as might be expected from earlier remarks, this corresponds to the first approximation of a more general moment theory.

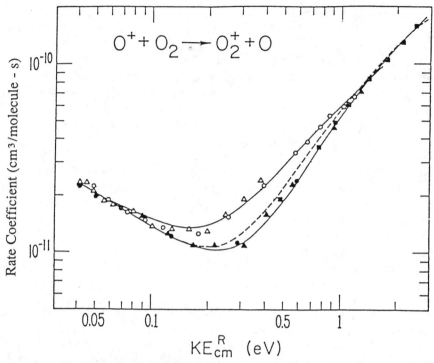

Figure 8-6-1. Comparison of rate coefficients measured by Albritton et al. (1977) in He buffer (filled symbols) and in Ar buffer (open symbols). The lower solid curve was fitted to the measurements using the O^+ in He distribution function calculated by Lin and Bardsley (1977), to find the reaction cross section. The upper solid curve was then predicted using the O^+ in Ar distribution function. The dashed curve is the thermal rate coefficient calculated from (8-6-2) for a temperature T_{eff}^R. The energy scale for KE_{cm}^R was determined according to (8-6-6).

It is even easy to guess, from the discussions in Chapters 5 and 6, what form the results should take in the first approximation of a moment theory. For example, if we ignore any internal degrees of freedom, then the thermal rate coefficient $k_R^{(0)}$ corresponding to a measured rate coefficient $k_R(T, E/N)$ is found from (8-6-1) to be

$$k_R(T, E/N) \approx k_R^{(0)}(T_{eff}^R) = \frac{2}{k T_{eff}^R} \left(\frac{2}{\pi \mu_R k T_{eff}^R} \right)^{1/2}$$

$$\times \int_0^\infty Q_R^*(\varepsilon_R) \exp(-\varepsilon_R / k T_{eff}^R) \varepsilon_R \, d\varepsilon_R, \qquad (8\text{-}6\text{-}2)$$

where

$$T_{eff}^R = (mT + M_R T_B)/(m + M_R), \qquad (8\text{-}6\text{-}3)$$

$$\mu_R = mM_R/(m + M_R).$$ (8-6-4)

This result follows by substitution of a Maxwellian distribution for $f^{(\alpha)}$ corresponding to an ion temperature T_B, and integration over the center-of-mass coordinates. Moreover, we can find T_B from the measured drift velocity of the ions in the pure buffer gas,

$$\tfrac{3}{2}kT_B = \tfrac{3}{2}kT + \tfrac{1}{2}mv_d^2 + \tfrac{1}{2}M_Bv_d^2,$$ (8-6-5)

from which the effective reaction temperature T_{eff}^R is found from (8-6-3) to be

$$\tfrac{3}{2}kT_{\text{eff}}^R = \tfrac{3}{2}kT + \tfrac{1}{2}M_Rv_d^2\frac{m + M_B}{m + M_R} = KE_{\text{cm}}^R,$$ (8-6-6)

where KE_{cm}^R is the mean kinetic energy between the ions and the reactant molecules in the center-of-mass frame.

In other words, to a first approximation we expect the measured rate coefficient $k_R(T, E/N)$ to correspond to a thermal rate coefficient $k_R^{(0)}(T_{\text{eff}}^R)$ at a temperature T_{eff}^R that can be calculated from the measured drift velocity.

If internal degrees of freedom are included, then the gas temperature T enters explicitly, through the distribution function for the internal states of the reactant molecule. In addition, a corresponding internal-state temperature T_i for the ions (Section 6-6) should enter explicitly. The formal expression for the thermal rate coefficient in first approximation is then

$$k_R^{(0)}(T_{\text{eff}}^R, T_i, T) = \frac{1}{Z_iZ_R}\frac{2}{kT_{\text{eff}}^R}\left(\frac{2}{\pi\mu_R kT_{\text{eff}}^R}\right)^{1/2}$$

$$\times \sum_\alpha \exp(-\varepsilon_i^\alpha/kT_i)\sum_\beta \exp(-\varepsilon_R^\beta/kT)$$

$$\times \int_0^\infty Q_R^*(\alpha\beta; \varepsilon_R)\exp(-\varepsilon_R/kT_{\text{eff}}^R)\varepsilon_R\, d\varepsilon_R,$$ (8-6-7)

where Z_i and Z_R are the internal partition functions for the ion and reactant, respectively. The expressions for T_{eff}^R and T_i are also more complicated, and are discussed further below.

To proceed beyond these fairly obvious first approximations, we need the full moment theory. That we must take this next step is indicated by the measurements of Albritton et al. (1977) on some reactions of O^+ ions in He and Ar buffer gases, already referred to, which do not completely scale together when reduced in terms of T_{eff}^R or KE_{cm}^R according to (8-6-6).

We begin as usual by writing down the appropriate kinetic equation, which is just the WUB equation (6-6-1) with an added term for the change in $f^{(\alpha)}$ due to the reactive collisions. With the neglect of reverse reactions leading to the reformation of primary ions, this equation is

$$\frac{Df^{(\alpha)}(\mathbf{v})}{Dt} + \sum_\beta \int f^{(\alpha)}(\mathbf{v})F_R^{(\beta)}(\mathbf{V}_R)|\mathbf{v} - \mathbf{V}_R|Q_R^*(\alpha\beta; \varepsilon_R)\, d\mathbf{V}_R$$

$$= \sum_{\alpha'} \sum_{\beta\beta'} \iint [f^{(\alpha')}(\mathbf{v}')F_B^{(\beta')}(\mathbf{V}_B') - f^{(\alpha)}(\mathbf{v})F_B^{(\beta)}(\mathbf{V}_B)]$$

$$\times |\mathbf{v} - \mathbf{V}_B|\sigma_B(\alpha\beta, \alpha'\beta'; \theta, \phi, \varepsilon)\, d\Omega\, d\mathbf{V}_B, \tag{8-6-8}$$

where the subscript R signifies reactant and the subscript B signifies buffer. The other symbols are as defined previously. This equation is written for only one reactant gas species and one buffer gas species. The generalization to mixtures of several species is straightforward, and we do not give it here. It has been given in detail by Viehland et al. (1981).

We now follow essentially the same procedure as outlined in Section 6-6 for polyatomic systems, with only minor modifications because of the presence of the reaction term. We first form general moment equations by multiplying (8-6-8) by any function $\psi^{(\alpha)}(\mathbf{v})$, and then integrating over $\mathbf{v}$ and summing over α. We assume spatially homogeneous and quasi-steady-state conditions. The first assumption means that we do not worry about any effects of gradients on the reaction, or at any rate suppose that these amount at most to minor corrections that are the responsibility of the experimenter. The second assumption means that we assume that all time derivatives are negligible except $\partial n/\partial t$. In the special case where $\psi^{(\alpha)}(\mathbf{v})$ is a constant, we obtain the equation of continuity for the ions, as modified by the reaction rate, which supplies a term for ion loss. For spatially homogeneous conditions (no gradients), this is simply the rate equation,

$$\frac{dn}{dt} = -k_R n N_R, \tag{8-6-9}$$

with k_R given by (8-6-1). Notice that $\partial n/\partial t$ has become the ordinary derivative dn/dt, because n is assumed not to depend on position. For later purposes it is convenient to rewrite (8-6-1) as a moment over a reaction operator,

$$k_R = \langle J_R^* 1 \rangle, \tag{8-6-10}$$

where the operator J_R^* is defined as

$$J_R^* \psi^{(\alpha)}(\mathbf{v}) = \sum_\beta \int F_R^{(\beta)}(\mathbf{V}_R)\psi^{(\alpha)}(\mathbf{v})|\mathbf{v} - \mathbf{V}_R|Q_R^*(\alpha\beta; \varepsilon_R)\, d\mathbf{V}_R. \tag{8-6-11}$$

In all of the rest of the moment equations, obtained when $\psi^{(\alpha)}$ is not just a constant, we can neglect the reactive collision term because $N_R \ll N_B$. This merely reproduces the moment equations (6-6-11a) for polyatomic systems without reaction. The only reason we cannot ignore the reactive collision term in the lowest moment equation, the one in which $\psi^{(\alpha)}$ is constant, is that the nonreactive collision term is identically zero in this case.

The foregoing results are only a somewhat mathematical way of saying that $f^{(\alpha)}$ is determined entirely by collisions with the buffer gas. The reaction term appears in the equation of continuity, but nowhere else, because it is the only thing that makes dn/dt nonzero in the absence of gradients.

The next step is to solve the general moment equations, as before, by choosing a set of basis functions in which to expand the collision term $J\psi^{(\alpha)}$. It is simplest to choose the spherical basis functions $\psi^{(\alpha)}_{lmrs}$ described in Section 6-6B. The expansion for the reactive collision term can then be written as

$$J_R^* \psi^{(\alpha)}_{lmrs} = \sum_{r's'} a_R^*(l; rs; r's') \psi^{(\alpha)}_{lmr's'}, \qquad (8\text{-}6\text{-}12)$$

where the matrix elements a_R^* are

$$a_R^*(l; rs; r's') \equiv (\psi_{lmr's'}, J_R^* \psi_{lmrs})/(\psi_{lmr's'}, \psi_{lmr's'}), \qquad (8\text{-}6\text{-}13)$$

with the inner products $(\cdot,\cdot)$ being defined by (6-6-21). Substituting the expression (8-6-12) into (8-6-10) for k_R, we obtain

$$k_R = \sum_{r's'} a_R^*(0; 00; r's') \langle \psi_{00r's'} \rangle. \qquad (8\text{-}6\text{-}14)$$

The first term gives the thermal rate coefficient $k_R^{(0)}$,

$$k_R^{(0)} = a_R^*(0; 00; 00), \qquad (8\text{-}6\text{-}15)$$

which follows because $\langle \psi_{0000} \rangle = 1$ by normalization. Higher approximations for k_R itself can then be written as

$$k_R = k_R^{(0)} \left[1 + \sum_{\substack{r's' \\ >0}} \gamma_R^*(0; 00; r's') \langle \psi_{00r's'} \rangle \right], \qquad (8\text{-}6\text{-}16)$$

where

$$\gamma_R^*(0; 00; r's') \equiv a_R^*(0; 00; r's')/a_R^*(0; 00; 00). \qquad (8\text{-}6\text{-}17)$$

Equation (8-6-16) gives the full moment solution for k_R. The coefficients γ_R^* can be calculated once Q_R^* is known, and the moments $\langle \psi_{00r's'} \rangle$ are found by solving the ion-buffer moment equations (in which the role of the reactive collisions is negligible). The solution of the ion-buffer part of the problem is also needed in order to find T_B, from which T_{eff}^R is calculated, as well as to find the internal temperature T_i of the ions.

At first glance the moment solution appears to save little, if any, effort, because we must still find Q_R^*. But this is not true. In first approximation we have

$k_R = k_R^{(0)}$, and it is not necessary to know Q_R^* explicitly. It is only necessary to find T_{eff}^R and T_i, and this is easy, at least in the first approximation. But higher approximations do seem to require knowledge of Q_R^*, which embroils us in the full inversion problem again. Nevertheless, this inversion can be avoided by approximating γ_R^* in terms of derivatives of k_R, as explained below. Similarly, the ion-buffer moments $\langle \psi_{00r's'} \rangle$ can be obtained from an approximate analysis of the mobility measurements. These approximations should be adequate if the moment series (8-6-16) converges rapidly enough.

Let us first consider the temperatures T_B, T_{eff}^R, and T_i. The results are much simpler if the buffer gas has no internal degrees of freedom; since there is no experimental difficulty in choosing the buffer to be He or Ar, for example, we assume that this is always done. Then the relation between T_B and the drift velocity of the ions is independent of any internal structure of the ions. At steady state the internal degrees of freedom of the ion are, so to speak, "filled up," and on the average have no direct influence on T_B. In more mathematical terms, the quantity ξ that characterizes energy loss due to inelastic collisions, given in equations (6-6-35) and (6-6-36), is zero for structureless neutrals. We can therefore take over the formulas already derived for T_B in Section 6-2; in a second approximation these are

$$\tfrac{3}{2}kT_B = \tfrac{1}{2}m\langle v^2 \rangle = \tfrac{3}{2}kT + \tfrac{1}{2}(m + M_B)v_d^2(1 + \beta K' + \cdots), \qquad (8\text{-}6\text{-}18)$$

where

$$\beta = \frac{mM_B(5 - 2A_B^*)}{5(m^2 + M_B^2) + 4mM_B A_B^*}. \qquad (8\text{-}6\text{-}19)$$

The dimensionless ratio $A_B^* \equiv \bar{\Omega}_B^{(2,2)}/\bar{\Omega}_B^{(1,1)}$, which is usually near unity, is a function only of the effective ion temperature relative to the buffer, T_{eff}^B, which is

$$\tfrac{3}{2}kT_{\text{eff}}^B = \tfrac{3}{2}kT + \tfrac{1}{2}M_B v_d^2(1 + \beta K' + \cdots). \qquad (8\text{-}6\text{-}20)$$

Any influence of the internal degrees of freedom of the ions is contained in v_d and K', which are experimentally measured quantities. The effective reaction temperature T_{eff}^R is then found from (8-6-3) to be

$$\tfrac{3}{2}kT_{\text{eff}}^R = KE_{\text{cm}}^R = \tfrac{3}{2}kT + \tfrac{1}{2}M_R v_d^2 \frac{m + M_B}{m + M_R}(1 + \beta K' + \cdots). \qquad (8\text{-}6\text{-}21)$$

Up to the present, the accuracy of measurement of k_R has not justified the use of the correction term $\beta K'$.

A more interesting role is played by T_i. When the buffer molecules are identical and structureless, the internal temperature of the ions becomes equal to their translational temperature, as was discussed in Section 6-6B2,

$$T_i = T_{\text{eff}}^B, \tag{8-6-22}$$

where T_{eff}^B is given by (8-6-20) above. But since T_{eff}^B is not the same as the effective reaction temperature T_{eff}^R, the relative values of T_i and T_{eff}^R can be manipulated by varying the mass of the buffer gas. In particular, by comparing (8-6-20) and (8-6-21), we find that (Viehland et al., 1981)

$$\frac{T_i - T}{T_{\text{eff}}^R - T} = \frac{T_{\text{eff}}^B - T}{T_{\text{eff}}^R - T} = \frac{M_B}{M_R} \frac{m + M_R}{m + M_B}. \tag{8-6-23}$$

This result suggests that unless $m \ll M_B$ and M_R, the effects of ion internal and translational energies on reaction rates can be at least partially separated by making measurements in two or more buffer gases.

The foregoing result is strikingly confirmed by some experimental work summarized in the review by Albritton (1979). The rate coefficient as a function of KE_{cm}^R for the reaction $O_2^+ + CH_4$ was found by Dotan et al. (1978) to be almost an order of magnitude higher in Ar buffer than in He buffer, except at nearly thermal energies. Since the difference seemed much larger than might be expected for differences in the translational energy distributions of the ions (i.e., the effect illustrated in Fig. 8-6-1), it was suggested that vibrational excitation of the O_2^+ ion by collisions with Ar was responsible. This is in the direction indicated by (8-6-23). The suggestion was tested by adding a small amount of O_2 to quench the O_2^+ vibrational excitation via resonant charge exchange, but not enough was added to perturb the translational energy distribution. The rate coefficient in He buffer was not affected by the added quenching gas, but that in Ar buffer was markedly decreased, thereby confirming that the higher rate in Ar buffer was probably due to vibrational excitation of O_2^+.

The manipulation of T_i relative to T_{eff}^R by use of different buffer gases has since been used to study the effect of internal excitation on a number of ion-molecule rate coefficients. In some cases the internal energy has essentially no effect (Albritton et al., 1983), but in other cases it has a pronounced effect (Durup-Ferguson et al., 1983, 1984). This gives important information on the mechanism of the reaction and on the shape of the potential energy surface for the reaction.

We now return to the problem of determining the thermal rate coefficient $k_R^{(0)}$ from the measured k_R, without becoming embroiled in the full inversion problem for Q_R^*; that is, in correcting for the effect of the ion translational energy distribution beyond what is included in KE_{cm}^R or T_{eff}^R. The motivation for this is twofold. First, knowledge of $k_R^{(0)}$ may be all that is really desired, and the inversion for Q_R^* may be difficult (or even impossible). Second, if information on Q_R^* is in fact needed, we now have it as an independent second problem of inverting the integral (8-6-2) for $k_R^{(0)}(T_{\text{eff}}^R)$ rather than the more difficult integral (8-6-1) for $k_R(T, E/N)$.

The procedure is to keep only the first few correction terms in (8-6-16), use recursion relations for the irreducible reaction collision integrals in order to find the matrix elements $\gamma_R^*(0; 00; r's')$, and to evaluate the $\langle\psi_{00r's'}\rangle$ by an analysis of mobility measurements in pure buffer gas. The details have been worked out only for atomic ions, for which all $s' = 0$ (Viehland and Mason, 1977), and at present we can only hope that the results apply at least qualitatively to molecular ions. The first correction term corresponds to $r' = 1$, $s' = 0$, but vanishes identically because $\langle\psi_{0010}\rangle = 0$ owing to the relation $\frac{3}{2}kT_B = \frac{1}{2}m\langle v^2\rangle$. Keeping only the first few nonzero correction terms, we can therefore write

$$k_R = k_R^{(0)}[1 + \gamma_R^*(0; 00; 20)\langle\psi_{0020}\rangle + \cdots]. \qquad (8\text{-}6\text{-}24)$$

The first matrix element is found from recursion relations to be

$$\gamma_R^*(0; 00; 20) = -\frac{4}{15}\left[1 - \left(\frac{m}{m + M_R}\right)\frac{T}{T_{\text{eff}}^R}\right]^2$$
$$\times\left[\frac{d \ln k_R^{(0)}}{d \ln T_{\text{eff}}^R} - \left(\frac{d \ln k_R^{(0)}}{d \ln T_{\text{eff}}^R}\right)^2 - \frac{d^2 \ln k_R^{(0)}}{d(\ln T_{\text{eff}}^R)^2}\right]. \qquad (8\text{-}6\text{-}25)$$

Further matrix elements can also be expressed in terms of various logarithmic derivatives of $k_R^{(0)}$, but we do not reproduce the formulas here.

The expressions for the moments $\langle\psi_{00r0}\rangle$ are rather complicated and we do not present them here; they are given by Viehland and Mason (1977). These moments depend only on the masses m and M_B, on the temperatures T and T_{eff}^B, and on various ratios of collision integrals. The first few such ratios are

$$A_B^* \equiv \bar{\Omega}_B^{(2,2)}/\bar{\Omega}_B^{(1,1)}, \qquad (8\text{-}6\text{-}26a)$$

$$C_B^* \equiv \bar{\Omega}_B^{(1,2)}/\bar{\Omega}_B^{(1,1)}, \qquad (8\text{-}6\text{-}26b)$$

$$E_B^* \equiv \bar{\Omega}_B^{(2,3)}/\Omega_B^{(2,2)}, \qquad (8\text{-}6\text{-}26c)$$

which are all approximately unity and depend only on T_{eff}^B and the ion-buffer interaction. The main features of the first few moments are illustrated in Fig. 8-6-2 for the somewhat extreme case of rigid spheres at high electric fields. It can be seen that the moments are generally less than unity in magnitude, and can even be zero for certain ion-buffer mass ratios. The fact that they are of comparable magnitude implies that the expansion for k_R converges, since the coefficients $\gamma_R^*(0; 00; r0)$ probably decrease suitably rapidly as the index r increases (Viehland and Mason, 1977).

The procedure for calculating $k_R^{(0)}$ from measured k_R is then as follows. The mobility data are first analyzed in terms of some reasonable potential model for the ion-buffer interaction, such as the $(n, 6, 4)$ model given by (6-1-62). The collision integral ratios of (8-6-26) are computed for the potential, and then the $\langle\psi_{00r0}\rangle$ are calculated from the formulas given by Viehland and Mason (1977).

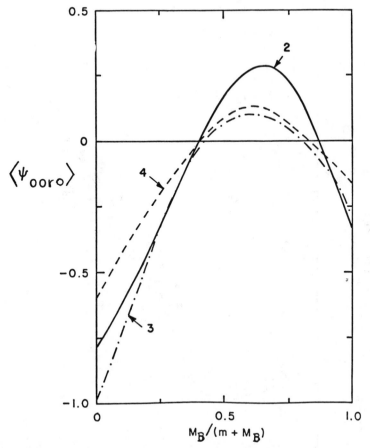

Figure 8-6-2. Ion-buffer moments $\langle\psi_{00r0}\rangle$, needed to correct measured k_R to equivalent thermal rate coefficients $k_R^{(0)}$, as a function of mass ratio for rigid spheres at high electric fields (cold gas limit). The first two moments are always $\langle\psi_{0000}\rangle = 1$ and $\langle\psi_{0010}\rangle = 0$.

Extensive numerical integration is required to find the collision integral ratios for a general potential (see Section 7-1), but the results of such integrations for the $(n, 6, 4)$ potential are available as convenient numerical tables (Viehland et al., 1975a). The coefficients $\gamma_R^*(0; 00; r0)$ are next calculated from (8-6-25) and similar formulas; iteration may be required if k_R and $k_R^{(0)}$ are much different. From these results $k_R^{(0)}$ follows directly from (8-6-24). As a guide to the estimation of convergence, the difference between k_R and $k_R^{(0)}$ is probably less than about 10% when the factor in the second brackets in (8-6-25)—that is, the factor containing the derivatives of $k_R^{(0)}$—is less than about 1/3.

The procedure above has been tested by Viehland and Mason (1977) for the reaction

$$O^+ + N_2 \to NO^+ + O,$$

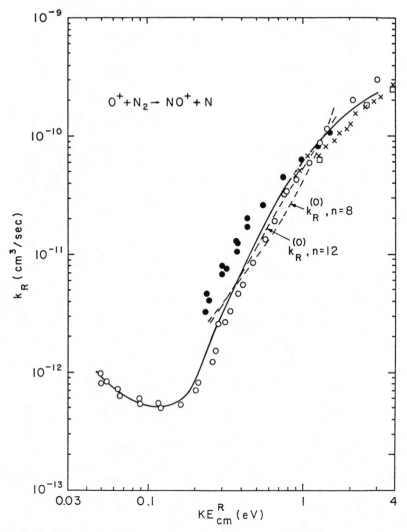

Figure 8-6-3. Rate coefficients for $O^+ + N_2$ as determined by three different measurements (Viehland and Mason, 1977). Open circles: drift-tube measurements in He buffer (McFarland et al., 1973a, b; Albritton et al., 1977); filled circles: drift-tube measurements in Ar buffer (Albritton et al., 1977); squares and crosses: ion-beam measurements (Rutherford and Vroom, 1971). The curves are thermal rate coefficients $k_R^{(0)}$ calculated from the measured k_R by moment theory: solid curves from the He buffer data; dashed curves from the Ar buffer data, using two different $(n, 6, 4)$ potentials for the O^+–Ar interaction.

as measured in He and in Ar buffer gases (McFarland et al., 1973a, b; Albritton et al., 1977). As a check, values of $k_R^{(0)}$ at high energies can be calculated from the ion-beam experiments of Rutherford and Vroom (1971). The results are shown in Fig. 8-6-3 as k_R vs. KE_{cm}^R; clearly the data for the two buffer gases do *not* scale to a single curve, just as was the case for the $O^+ + O_2$ reaction shown in Fig. 8-6-1. The solid curve represents the $k_R^{(0)}$ calculated from the He buffer data, for which the corrections are rather small and it is satisfactory to use only one correction term in (8-6-24). The corrections for the Ar buffer data are much larger, and involve at least two extra difficulties. First, it would be desirable to use additional correction terms; unfortunately, this is not feasible because higher logarithmic derivatives of $k_R^{(0)}$ are required, for which the experimental accuracy is not adequate. Second, the mobility data on O^+ in Ar did not extend to high enough field strengths to probe the upper end of the reaction energy range: from (8-6-20) and (8-6-21) we find that at the same value of E/N, $T_{eff}^B \approx 1.12 T_{eff}^R$ for Ar buffer, but $T_{eff}^B \approx 0.31 T_{eff}^R$ for He buffer. Despite these difficulties, the $k_R^{(0)}$ calculated from the Ar buffer measurements, shown as dashed curves, agree fairly well with those calculated from the He buffer measurements. (The experimental uncertainties in the data are about 30%.) It is also encouraging that the calculated $k_R^{(0)}$ are not overly sensitive to the potential chosen to represent the O^+–Ar interaction, as shown in Fig. 8-6-3. The drift-tube measurements are also in agreement with the beam data where they overlap each other.

We conclude that it is entirely feasible to extract high-temperature thermal rate coefficients $k_R^{(0)}$ from swarm measurements of k_R made at room temperature as a function of E/N. Moment methods provide a direct route for this calculation, and explicitly bypass the difficult problem of finding the ion energy distribution and of extracting the reaction cross section Q_R^* from the rate data.

We conclude this section with two observations. The first is that Figs. 8-6-1, 8-6-2, and 8-6-3 suggest that a buffer gas can be chosen so that the conversion of k_R to $k_R^{(0)}$ involves only a small correction. In fact, more detailed calculations show that a *mixture* of buffer gases can be chosen to make the correction essentially zero (Viehland and Mason, 1979). This scheme can work only with atomic ions, however, unless additional information on inelastic cross sections is available.

The second observation is that all the considerations in this section on the analysis of ion-molecule reactions in drift tubes apply, with only minor modifications, to reactions studied by ion cyclotron resonance techniques (see the discussion in Section 8-5). Details have been worked out for the collision-dominated ICR regime by Viehland and Mason (1976).

8-7. ELECTRON TRANSPORT

The topic of electron transport is a large subject in its own right. Less than fifteen years ago a survey of "modern developments" required a large book (Huxley and Crompton, 1974), and a similar survey today would require at least another such book. This section therefore attempts only to place the topic in the general context of charged-particle swarm phenomena, and to supply some important references.

Although, from the general standpoint of moment solutions of the Boltzmann equation, electrons are just very light atomic ions, the theory of electron transport has historically developed almost independently of the theory for ions. The reason is probably the one alluded to in Section 5-3D4—because $m/M \ll 1$ for electrons, the electron distribution function is nearly isotropic in velocity space when only elastic collisions occur, so that only two terms are needed in an expansion in spherical harmonics. This allows many simplifications that are not possible with heavier charged particles, such as replacement of the Boltzmann integral collision operator by a differential representation, and the fundamental connections between electron and ion transport become lost in the subsequent details. Important general reviews that emphasize the connections have been given by Braglia (1980), by Kumar et al. (1980), and by Kumar (1984). However, these simplifications are not generally possible when electrons undergo inelastic collisions, and the anisotropy of the distribution function becomes significant. It is largely the attempts to treat this anisotropy that have dominated recent theoretical work, and have brought the theories of electron and ion transport closer together.

The important theme that underlies both electron and ion transport, as well as much of kinetic theory in general, is the expansion of the distribution function in spherical harmonics. This expansion was briefly discussed in Section 5-3D4, and we repeat the relevant equation (5-3-35) here,

$$f(\mathbf{v}, \mathbf{r}, t) = n(\mathbf{r}, t) \sum_{l=0}^{\infty} \sum_{m=-l}^{l} f_{lm}(v) Y_l^m(\theta, \phi), \qquad (8\text{-}7\text{-}1)$$

where the $f_{lm}(v)$ are unknown functions of v and the $Y_l^m(\theta, \phi)$ are the spherical harmonics describing the angular dependence of $f(\mathbf{v})$ in velocity space. In practice only a finite number of terms can be dealt with, and the summation is usually truncated at some l_{max} (or else some equivalent assumption about the higher terms is made.) The two-term approximation takes $l_{max} = 1$, which is excellent for elastic collisions of electrons, but dubious for inelastic collisions. Most of the recent theoretical efforts have been concerned with the inclusion of more terms, and there is now quite an extensive literature on the subject. It has been surveyed and discussed by Robson and Ness (1986), in a paper concerned primarily with the general features of the hierachy of equations that result from using the expansion (8-7-1) in the Boltzmann equation.

The first accurate systematic use of the spherical-harmonic expansion (8-7-1) that could be used for electric fields of arbitrary strength and for electron-neutral cross sections of nearly arbitrary energy dependence was given by Lin et al. (1979b). The crucial new feature they introduced was the treatment of the functions $f_{lm}(v)$. They used an expansion in Sonine polynomials (not in itself a new idea), but with an arbitrary electron basis temperature T_b rather than the gas temperature T. In other words, they combined the approach of the two-temperature moment theory for ions discussed in Sections 5-3D2 and 6-2 with the spherical-harmonics expansion (8-7-1). They were thereby able to calculate accurate results, even for electrons in CH_4, a difficult system that shows the phenomenon of negative differential conductivity (described in Section 6-4D).

There is, of course, nothing especially sacred about the use of Sonine polynomials, other than possible convenience and kinetic-theory tradition, and the functions $f_{lm}(v)$ can in principle be expanded in any convenient basis set $\phi_l^{(r)}(v)$,

$$f_{lm}(v) = \sum_{r=0}^{\infty} f_{lm}^{(r)} \phi_l^{(r)}(v). \qquad (8\text{-}7\text{-}2)$$

For Sonine polynomials it happens that the coefficients $f_{lm}^{(r)}$ are conveniently related to moments of $f(\mathbf{v})$, but this may not be a compelling feature. For instance, ease of manipulation on a computer may be essential, in which case spline functions or finite-element representations might be preferred, and both have been used. Almost all investigations on electron transport differ essentially only in the choice of $\phi_l^{(r)}(v)$. The most extensive development to date of the Sonine basis expansion is that of Ness and Robson (1986).

The choice of the basis set $\phi_l^{(r)}(v)$, although not critical in principle, is in practice of crucial importance for obtaining convergence of numerical results. In this connection it is worth noting that the two indices l and r can be treated independently as far as convergence is concerned. Experience to date indicates that the convergence in l is usually much faster than in r (e.g., $l_{\max} \ll r_{\max}$); this probably just means that we have not yet been sufficiently clever in the choice of the $\phi_l^{(r)}(v)$. The optimum choice of $\phi_l^{(r)}(v)$ may also depend on the nature of the ion-neutral cross sections; for instance, Sonine polynomials are optimum if the cross sections are inversely proportional to v (Maxwell model).

Finally, we remark that Monte Carlo simulation methods have played an important role in theories of electron transport, and we mention a few selected examples. Lin et al. (1979b) used Monte Carlo calculations to check the accuracy of some of their moment solutions. Braglia (1981) used Monte Carlo simulations to resolve discrepancies between some moment results and finite-element results, and Braglia et al. (1982) used simulations to clarify some difficulties encountered in spline-function calculations. Penetrante and Bardsley (1984) have surveyed many of the numerical methods used in solving the Boltzmann equation, and checked discrepant results by means of Monte Carlo calculations. Such simulations have been an important complement to the more analytical methods based on (8-7-1).

8-8.   THERMALIZATION OF HOT ATOMS AND ELECTRONS

The success of moment methods for the treatment of steady-state swarm phenomena is an encouragement to try the same methods on other phenomena, and this last section briefly describes one such foray. Suppose we inject (or create) a dilute swarm of very energetic particles in a neutral gas that is in equilibrium at temperature T. No external fields are present. Eventually, through collisions with the neutrals, the energetic particles will also come to equilibrium, but during this approach to equilibrium they may possibly acquire peculiar energy distributions. Knowledge of these transient energy distributions is crucial to understanding the reactions that the energetic particles undergo with other species seeded into the original large amount of buffer or moderator gas.

A classic example consists of the "hot" atoms produced by nuclear reactions, which start with a very large recoil energy and subsequently undergo chemical reactions with ambient neutral molecules. Any attempt to understand this chemistry must involve the energy distribution of the hot atoms. Another similar example involves "hot" electrons produced by nuclear reactions or injected from an accelerator.

The problem of determining the distribution function of such energetic particles differs from most of the problems so far discussed, in that the result is time dependent. In the absence of an external electric field, the steady state is just the uninteresting final equilibrium state. Nevertheless, the success of the two-temperature theory when applied to relaxation in drift tubes (Section 8-4) and to ion cyclotron resonance (Section 8-5) encourages its application to the hot-atom and hot-electron problems. We give here only a few comments on the use of the two-temperature theory for these problems, which are, after all, somewhat removed from the main themes of this book.

The formulation of the hot-atom problem amounts to a combination of the time-dependent equations for relaxation (Section 8-4) with the equations for ion-molecule reactions (Section 8-6). That is, collisions with the large excess of inert moderator determine the energy distribution of the hot atoms at some time t, and the combination of this distribution with the reaction cross section of the reactants gas gives the rate coefficient at time t, $k_R(t)$. The rate equation, with this time-dependent rate coefficient, can then be integrated over time to find the "hot yield." Knierim et al. (1981) used this approach; their paper can be consulted for references to earlier work and to other methods of attacking the hot-atom problem.

The two-temperature theory works well for hot-atom problems as long as the hot atoms and the moderator atoms have rather different masses. But when the masses are close, an energetic atom can lose almost all its energy by a single nearly head-on collision. This tends to produce a bimodal energy distribution, and the two-temperature expansion based on Sonine polynomials converges poorly (if at all). There are several ways to handle this difficulty, including a bimodal or double expansion in Sonine polynomials, as was described in more detail in Section 5-3D5.

The case of hot electrons can also be handled by a time-dependent two-temperature theory, as was shown by Knierim et al. (1982), but of course the equal-mass difficulty does not arise. Unfortunately, another difficulty appears instead. The convergence of the Sonine polynomial expansion is slow when

$m \ll M$, and rather high-order approximations are needed for good results. It turns out that an expansion in speed polynomials rather than energy polynomials gives much better convergence, and this approach has been extensively exploited by Shizgal and co-workers, using a discrete ordinate method to solve the resulting equations. (Shizgal, 1983a, b; Shizgal and Blackmore, 1984; Shizgal and McMahon, 1984, 1985; McMahon and Shizgal, 1985). This undoubtedly constitutes the definitive work on the problem to date.

8-9. FINAL REMARKS

The examples taken up in this chapter are not meant to be either exhaustive or definitive. Their purpose is only to give some feel for the scope of the subject, and an entrance to some of the literature on applications. We apologize to all those friends and colleagues whose work has been given short shrift or ignored altogether.

REFERENCES

Albritton, D. L. (1979). Energy dependences of ion-neutral reactions studied in drift tubes, in *Kinetics of Ion-Molecule Reactions*, P. Ausloos, Ed., Plenum, New York, pp. 119–142.

Albritton, D. L., I. Dotan, W. Lindinger, M. McFarland, J. Tellinghuisen, and F. C. Fehsenfeld (1977). Effects of ion speed distributions in flow-drift tube studies of ion-neutral reactions, *J. Chem. Phys.* **66**, 410–421.

Albritton, D. L., I. Dotan, G. E. Streit, D. W. Fahey, F. C. Fehsenfeld, and E. E. Ferguson (1983). Energy dependence of the O^- transfer reactions of O_3^- and CO_3^- with NO and SO_2, *J. Chem. Phys.* **78**, 6614–6619.

Annis, B. K., A. P. Malinauskas, and E. A. Mason (1972). Theory of drag on neutral or charged spherical aerosol particles, *J. Aerosol Sci.* **3**, 55–64.

Annis, B. K., A. P. Malinauskas, and E. A. Mason (1973). Theory of diffusiophoresis of spherical aerosol particles and of drag in a gas mixture, *J. Aerosol Sci.* **4**, 271–281.

Baldeschwieler, J. D. (1968). Ion cyclotron resonance spectroscopy, *Science* **159**, 263–273.

Beauchamp, J. L. (1967). Theory of collision-broadened ion cyclotron resonance spectra, *J. Chem. Phys.* **46**, 1231–1243.

Beauchamp, J. L. (1971). Ion cyclotron resonance spectroscopy, *Ann. Rev. Phys. Chem.* **22**, 527–561.

Braglia, G. L. (1980). Theory of electron motion in gases: 1. Stochastic theory of homogeneous systems, *Rev. Nuovo Cimento* **3** (No. 5), 1–105.

Braglia, G. L. (1981). On the accuracy of the iterative method for swarm transport coefficients, *J. Chem. Phys.* **74**, 2990–2992.

Braglia, G. L., L. Romano, and M. Diligenti (1982). Comment on "Comparative calculations of electron-swarm properties in N_2 at moderate E/N values," *Phys. Rev.* **A26**, 3689–3694.

Burley, J. D., K. M. Ervin, and P. B. Armentrout (1987). Translational energy dependence of $O^+(^4S) + N_2 \rightarrow NO^+ + N$ from thermal energies to 30 eV c.m., *J. Chem. Phys.* **86**, 1944–1953.

Carr, T. W., Ed. (1984). *Plasma Chromatography*, Plenum, New York.

Cavalleri, G., and S. L. Paveri-Fontana (1972). Drift velocity and runaway phenomena for electrons in neutral gases, *Phys. Rev.* **A6**, 327–333.

Chapman, S., and T. G. Cowling (1970). *The Mathematical Theory of Non-Uniform Gases*, 3rd ed., Cambridge University Press, London.

Comisarow, M. B. (1971). Comprehensive theory for ion cyclotron resonance power absorption: Application to line shapes for reactive and nonreactive ions, *J. Chem. Phys.* **55**, 205–217.

Dickinson, A. S., and M. S. Lee (1978). The mobility of protons in helium, *J. Phys.* **B11**, L377–L379.

Dotan, I., F. C. Fehsenfeld, and D. L. Albritton (1978). Energy dependence of the reaction of O_2^+ with CH_4, *J. Chem. Phys.* **68**, 5665–5666.

Durup-Ferguson, M., H. Böhringer, D. W. Fahey, and E. E. Ferguson (1983). Enhancement of charge-transfer reaction rate constants by vibrational excitation at kinetic energies below 1 eV, *J. Chem. Phys.* **79**, 265–272.

Durup-Ferguson, M., H. Böhringer, D. W. Fahey, F. C. Fehsenfeld, and E. E. Ferguson (1984). Competitive reaction and quenching of vibrationally excited O_2^+ ions with SO_2, CH_4, and H_2O, *J. Chem. Phys.* **81**, 2657–2666.

Griffin, G. W., I. Dzidic, D. I. Carroll, R. N. Stillwell, and E. C. Horning (1973). Ion mass assignments based on mobility measurements, *Anal. Chem.* **45**, 1204–1209.

Heimerl, J., R. Johnsen, and M. A. Biondi (1969). Ion-molecule reactions, $He^+ + O_2$ and $He^+ + N_2$, at thermal energies and above, *J. Chem. Phys.* **51**, 5041–5048.

Hidy, G. M., and J. R. Brock (1970). *The Dynamics of Aerocolloidal Systems*, Pergamon, Elmsford, N.Y.

Howorka, F., F. C. Fehsenfeld, and D. L. Albritton (1979). H^+ and D^+ ions in He: Observations of a runaway mobility, *J. Phys.* **B12**, 4189–4197.

Huxley, L. G. H., and R. W. Crompton (1974). *The Diffusion and Drift of Electrons in Gases*, Wiley, New York.

Johnsen, R., and M. A. Biondi (1972). Reaction rates of uranium ions and atoms with O_2 and N_2, *J. Chem. Phys.* **57**, 1975–1979.

Knierim, K. D., S. L. Lin, and E. A. Mason (1981). Time-dependent moment theory of hot-atom reactions, *J. Chem. Phys.* **75**, 1159–1165.

Knierim, K. D., M. Waldman, and E. A. Mason (1982). Moment theory of electron thermalization in gases, *J. Chem. Phys.* **77**, 943–950.

Knudsen, M. (1950). *The Kinetic Theory of Gases*, 3rd ed., Methuen, London.

Kumar, K. (1984). The physics of swarms and some basic questions of kinetic theory, *Phys. Rep.* **112**, 319–375.

Kumar, K., H. R. Skullerud, and R. E. Robson (1980). Kinetic theory of charged particle swarms in neutral gases, *Aust. J. Phys.* **33**, 343–448.

Lin, S. L., and J. N. Bardsley (1975). Computation of speed distributions for ions in drift tubes, *J. Phys.* **B8**, L461–L464.

Lin, S. L., and J. N. Bardsley (1977). Monte Carlo simulation of ion motion in drift tubes, *J. Chem. Phys.* **66**, 435–445.

Lin, S. L., and J. N. Bardsley (1978). The null-event method in computer simulation, *Comput. Phys. Commun.* **15**, 161–163.

Lin, S. L., L. A. Viehland, E. A. Mason, J. H. Whealton, and J. N. Bardsley (1977). Velocity and energy relaxation of ions in drift tubes, *J. Phys.* **B10**, 3567–3575.

Lin, S. L., I. R. Gatland, and E. A. Mason (1979a). Mobility and diffusion of protons and deuterons in helium—A runaway effect, *J. Phys.* **B12**, 4179–4188.

Lin, S. L., R. E. Robson, and E. A. Mason (1979b). Moment theory of electron drift and diffusion in neutral gases in an electrostatic field, *J. Chem. Phys.* **71**, 3483–3498.

McFarland, M., D. L. Albritton, F. C. Fehsenfeld, E. E. Ferguson, and A. L. Schmeltekopf (1973a). Flow-drift technique for ion mobility and ion-molecule reaction rate constant measurements: 1. Apparatus and mobility measurements, *J. Chem. Phys.* **59**, 6610–6619.

McFarland, M., D. L. Albritton, F. C. Fehsenfeld, E. E. Ferguson, and A. L. Schmeltekopf (1973b). Flow-drift technique for ion mobility and ion-molecule reaction rate constant measurements: 2. Positive ion reactions of N^+, O^+, and N_2^+ with O_2 and O^+ with N_2 from thermal to $\sim 2\,eV$, *J. Chem. Phys.* **59**, 6620–6628.

McMahon, D. R. A., and B. Shizgal (1985). Hot-electron zero-field mobility and diffusion in rare-gas moderators, *Phys. Rev.* **A31**, 1894–1905.

Melton, L. A., and R. G. Gordon (1969). Extraction of reaction cross section from rate constant data: $D + H_2 \rightarrow HD + H$, *J. Chem. Phys.* **51**, 5449–5457.

Millikan, R. A. (1923). The general law of fall of a small spherical body through a gas, and its bearing upon the nature of molecular reflection from surfaces, *Phys. Rev.* **22**, 1–23.

Moruzzi, L. L., and Y. Kondo (1980). The mobility of H^+ ions in helium, *Jpn. J. Appl. Phys.* **19**, 1411–1412.

Ness, K. F., and R. E. Robson (1986). Velocity distribution function and transport coefficients of electron swarms in gases: 2. Moment equations and applications, *Phys. Rev.* **A34**, 2185–2209.

Penetrante, B. M., and J. N. Bardsley (1984). A critique of Boltzmann solution methods for calculating the longitudinal diffusion coefficient of electrons in gases, *J. Phys.* **D17**, 1971–1982.

Powell, C. F., and L. Brata (1932). Mobility of alkali ions in gases, *Proc. R. Soc. London* **A138**, 117–132.

Revercomb, H. E., and E. A. Mason (1975). Theory of plasma chromatography/gaseous electrophoresis—A review, *Anal. Chem.* **47**, 970–983.

Robson, R. E. (1986). Physics of reacting particle swarms in gases, *J. Chem. Phys.* **85**, 4486–4501.

Robson, R. E., and K. F. Ness (1986). Velocity distribution function and transport coefficients of electron swarms in gases: Spherical-harmonics decomposition of Boltzmann's equation, *Phys. Rev.* **A33**, 2068–2077.

Russ, C., M. V. Barnhill III, and S. B. Woo (1975). Inference of cross sections from rate constants measured in drift tubes, *J. Chem. Phys.* **62**, 4420–4427.

Rutherford, J. A., and D. A. Vroom (1971). Effect of metastable $O^+(^2D)$ on reactions of O^+ with nitrogen molecules, *J. Chem. Phys.* **55**, 5622–5624.

Shizgal, B. (1983a). Electron thermalization in gases, *J. Chem. Phys.* **78**, 5741–5744.

Shizgal, B. (1983b). Energy relaxation of electrons in helium, *Chem. Phys. Lett.* **100**, 41–44.

Shizgal, B., and R. Blackmore (1984). A discrete ordinate method of solution of linear boundary value and eigenvalue problems, *J. Comput. Phys.* **55**, 313–327.

Shizgal, B., and D. R. A. McMahon (1984). Electron distribution functions and thermalization times in inert gas moderators, *J. Phys. Chem.* **88**, 4854–4862.

Shizgal, B., and D. R. A. McMahon (1985). Electric field dependence of transient electron transport properties in rare-gas moderators, *Phys. Rev.* **A32**, 3669–3680.

Shkarofsky, I. P., T. W. Johnston, and M. P. Bachynski (1966). *The Particle Kinetics of Plasmas*, Addison-Wesley, Reading, Mass.

Thomson, J. J., and G. P. Thomson (1928). *Conduction of Electricity through Gases*, 3rd ed., Vol. 1, *General Properties of Ions. Ionization by Heat and Light*, Cambridge University Press, London. Reprinted by Dover, New York, 1969.

Thomson, J. J., and G. P. Thomson (1933). *Conduction of Electricity through Gases*, 3rd ed., Vol. 2, *Ionization by Collision and the Gaseous Discharge*, Cambridge University Press, London. Reprinted by Dover, New York, 1969.

Tyndall, A. M. (1938). *The Mobility of Positive Ions in Gases*, Cambridge University Press, London.

Viehland, L. A., and E. A. Mason (1976). Ion-molecule rate coefficients from collision-dominated ion cyclotron resonance, *Int. J. Mass Spectrom. Ion Phys.* **21**, 43–56.

Viehland, L. A., and E. A. Mason (1977). Statistical-mechanical theory of gaseous ion-molecule reactions in an electrostatic field, *J. Chem. Phys.* **66**, 422–434.

Viehland, L. A., and E. A. Mason (1979). On the choice of buffer gas mixtures for drift-tube studies of ion-neutral reactions, *J. Chem. Phys.* **70**, 2262–2265.

Viehland, L. A., E. A. Mason, W. F. Morrison, and M. R. Flannery (1975a). Tables of transport collision integrals for $(n, 6, 4)$ ion-neutral potentials, *At. Data Nucl. Data Tables* **16**, 495–514.

Viehland, L. A., E. A. Mason, and J. H. Whealton (1975b). Kinetic theory of ion cyclotron resonance collision broadening, *J. Chem. Phys.* **62**, 4715–4726.

Viehland, L. A., S. L. Lin, and E. A. Mason (1981). Kinetic theory of drift-tube experiments with polyatomic species, *Chem. Phys.* **54**, 341–364.

Waldman, M., and E. A. Mason (1981). On the density dependence of runaway mobility, *Chem. Phys. Lett.* **83**, 369–371.

APPENDIX I

INDEX OF EXPERIMENTAL DATA ON GASEOUS ION MOBILITIES AND DIFFUSION COEFFICIENTS

A large amount of experimental data has become available in the last two decades on ionic transport in gases, too much for compilation in this volume. Instead, we provide an index to data that have been screened and published in three papers appearing in the journal *Atomic Data and Nuclear Data Tables:*

Paper No. 1: H. W. Ellis, R. Y. Pai, E. W. McDaniel, E. A. Mason, and L. A. Viehland (1976), Transport Properties of Gaseous Ions over a Wide Energy Range, *At. Data Nucl. Data Tables* **17**, 177–210.

Paper No. 2: H. W. Ellis, E. W. McDaniel, D. L. Albritton, L. A. Viehland, S. L. Lin, and E. A. Mason (1978), Transport Properties of Gaseous Ions over a Wide Energy Range: Part 2, *At. Data Nucl. Data Tables* **22**, 179–217.

Paper No. 3: H. W. Ellis, M. G. Thackston, E. W. McDaniel, and E. A. Mason (1984), Transport Properties of Gaseous Ions over a Wide Energy Range: Part 3, *At. Data Nucl. Data Tables* **31**, 113–151.

The bulk of the original data dealt with here were published between 1968 and 1983; more recent results are concentrated in *The Journal of Chemical Physics, Physical Review A, The Journal of Physics (Parts B and D), Chemical Physics,* and *The Australian Journal of Physics.* The main criteria for inclusion of data in the ADNDT compilations were that (1) the identity of the ions must be well established and (2) the accuracy must be good. These criteria imply that ion-molecule reactions in the drifting and diffusing ion swarm must either be negligible or properly taken into account. Most of the mobilities and drift velocities indexed here are accurate to within ± 1 to 5%. The longitudinal and transverse diffusion coefficients are accurate to within $\pm 5\%$ at best and only to within $\pm 20\%$ in unfavorable cases.

The first part of our data index is Table I-1, which relates to measurements of standard mobilities and drift velocities at a constant gas temperature but as functions of E/N. References to the original sources of data are to be found on the indicated pages of Papers No. 1, 2, and 3 cited above. A typical table from Paper No. 1 that gives data of the type under discussion here is reproduced below (Sample Table):

Sample Table $K^+ - N_2$

Experimental Data[a] $T = 300$ K			Derived Quantities			
E/N	K_0	v_d	T_{eff}	$K_0(0)$	$\bar{\Omega}^{(1,1)}$	$\dfrac{d \ln K_0}{d \ln T_{eff}}$
4.00	2.53	0.272	300	2.53	105	0.07
6.00	2.53	0.408	400	2.58	88.8	0.06
8.00	2.53	0.544	500	2.62	78.3	0.08
10.0	2.54	0.682	600	2.66	70.4	0.08
12.0	2.54	0.819	800	2.72	59.6	0.10
15.0	2.54	1.02	1,000	2.78	52.1	0.10
20.0	2.55	1.37	1,200	2.84	46.6	0.10
25.0	2.55	1.71	1,500	2.89	41.0	0.08
30.0	2.56	2.06	1,800	2.93	36.9	0.07
40.0	2.58	2.77	2,000	2.95	34.8	0.07
50.0	2.59	3.48	2,500	2.99	30.7	0.06
60.0	2.62	4.22	3,000	3.02	27.7	0.04
80.0	2.68	5.76	3,500	3.03	25.6	0.04
100	2.76	7.42	4,000	3.05	23.8	0.01
120	2.84	9.16	5,000	3.04	21.3	-0.02
150	2.95	11.9	6,000	3.03	19.5	-0.05
200	3.04	16.3	8,000	2.96	17.3	-0.09
250	3.05	20.5	10,000	2.89	15.9	-0.12
300	3.00	24.2	12,000	2.82	14.8	-0.15
400	2.86	30.7	15,000	2.72	13.8	-0.17
500	2.71	36.4	18,000	2.63	13.0	-0.21
600	2.56	41.3	20,000	2.57	12.6	-0.22

[a]Accuracy: The total error in the experimental data is believed not to exceed $\pm 2\%$.
Source: Ellis et. al (1976), p. 204.

The energy parameter E/N is given in townsends (Td), the standard mobility K_0 in units of cm^2/V-s, and the drift velocity v_d in units of 10^4 cm/s. The tabulated numbers for K_0 and v_d were obtained from a smooth curve drawn through the experimental data and represent our best estimate of the true values.

In most cases, the precision of the data is greater than the accuracy. Often, the variation in the mobility as a function of E/N is significant even though the magnitude of the variation is within the limits of error. We therefore usually present the data to three significant figures, although the estimated error may allow only two of them to be meaningful in the absolute sense.

The right-hand part (four columns) of each entry in the sample table presents results derived from the application of theory to the experimental data on the left. The independent variable is the effective temperature T_{eff}, obtained from theory, to the first order. The quantity $K_0(0)$ is interpreted as the standard ionic mobility in the zero-field limit ($E/N \to 0$) but at the elevated ion-gas temperature T_{eff}. Calculation of the collision integral [$\bar{\Omega}^{(1,1)}(T_{eff})$ in units of 10^{-16} cm^2] is allowed by this interpretation. The logarithmic derivative of $K_0(0)$ with respect to T_{eff} is included to permit the user of the tables to make higher-order corrections to the calculated effective temperature if greater accuracy is desired, and to aid in the calculation of diffusion coefficients from values of mobility. The logarithmic derivative may contain errors of 10% or more, especially at the highest and lowest values of T_{eff}.

Our index for mobilities is completed by the much shorter Tables I-2 and I-3, which refer to zero-field mobilities as a function of gas temperature and to zero-field mobilities at single gas temperatures, respectively. Tables I-4 and I-5 relate to longitudinal and transverse diffusion coefficients, respectively. The structure of these tables as well as that of the original tables in ADNDT is sufficiently simple that no explanation is required.

Table I-1   Standard Mobilities and Drift Velocities as a Function of E/N (232 entries)

Ion	Gas	Gas Temp. (K)	Paper No.[a]	Page	E/N Range (Td)
Ar$^+$	He	300	1	185	5–130
Ar$^+$	He	82	3	124	5–70
Ar$^+$	Ne	77	3	129	6–20
Ar$^+$	Ar	300	1	196	8–2000
Ar$^+$	Ar	77	3	131	2–90
Ar^{2+}	He	300	3	125	10–96
Ar^{2+}	Ar	300	1	196	40–200
Ar$^{2+}(^1D)$	Ar	300	2	195	65–100
Ar$^{2+}(^3P)$	Ar	300	2	195	15–150
Ar$_2^+$	Ar	300	1	198	5–170
Ar$_2^+$	Ar	77	3	133	50–100
ArH$^+$	He	300	1	187	10–120
Br$^-$	He	297	2	190	2–220
Br$^-$	Ne	300	3	130	5–350
Br$^-$	Ar	297	2	194	2–425
Br$^-$	Ar	300	3	132	5–250
Br$^-$	Kr	300	3	134	5–350
Br$^-$	Xe	300	3	136	5–1000
C$^+$	He	297	3	126	3.5–100
C$^+$	CO	300	1	208	20–700
CH$^+$	He	300	3	128	15–110
CH$_2^+$	He	300	3	128	5–95

(continued)

Table I-1 (*Cont.*)

Ion	Gas	Gas Temp. (K)	Paper No.[a]	Page	E/N Range (Td)
CH_3^+	He	300	3	128	5–80
CH_3^+	Ar	298	3	132	25–110
CH_4^+	He	300	3	128	5–80
CH_5^+	He	300	1	190	5–80
$C_2H_2^-$	He	300	1	192	5–70
$CH_3O_2^+$	He	300	1	190	4–90
Cl^-	He	297	2	190	2–220
Cl^-	He	300	1	190	4–120
Cl^-	Ne	300	2	192	2–400
Cl^-	Ar	300	2	194	4–400
Cl^-	Kr	300	2	196	4–400
Cl^-	Xe	300	2	198	3–400
Cl^-	N_2	300	3	137	8–250
CO^+	He	300	1	186	5–120
CO^+	Ne	300	3	131	5–250
CO^+	Ar	300	3	133	8–300
CO^+	CO	300	1	207	15–700
CO_2^+	He	300	1	188	5–120
CO_2^+	Ne	300	2	193	8–200
CO_2^+	Ar	300	1	198	10–160
CO_2^+	N_2	298	2	203	2–260
CO_2^+	N_2	300	1	204	6–240
CO_3^-	He	300	1	192	2–120
CO_3^-	Ar	300	1	199	4–180
CO_3^-	O_2	300	1	206	3–70
CO_3^-	CO_2	300	1	209	5–150
CO_4^-	O_2	300	1	207	3–100
$C_2O_2^+$	CO	300	1	207	8–100
COH^+	He	300	1	188	5–120
COH^+	Ar	300	1	198	10–160
Cs^+	He	300	2	187	5–120
Cs^+	Ne	300	2	191	4–300
Cs^+	Ar	300	2	193	6–400
Cs^+	Kr	300	2	196	8–500
Cs^+	Xe	300	2	198	10–600
Cs^+	H_2	300	2	200	4–100
Cs^+	N_2	300	2	201	7–400
Cs^+	O_2	300	2	203	7–400
Cs^+	CO	300	2	204	7–500
Cs^+	CO_2	300	2	204	10–500
D^+	He	300	1	184	4–20
D^+	He	300	3	126	6–72
D^+	Ne	300	1	195	4–17
D^+	D_2	300	1	202	6–400
D_3^+	D_2	300	1	202	5–300
F^-	He	297	2	190	2–220

490

Table I-1 (*Cont.*)

Ion	Gas	Gas Temp. (K)	Paper No.[a]	Page	E/N Range (Td)
F^-	Ar	297	2	194	2–280
F^-	Kr	300	2	196	2–300
F^-	Xe	300	2	198	3–250
H^+	He	300	1	184	4–20
H^+	He	300	3	126	4–62
H^+	Ne	300	1	194	3–17
H^+	H_2	300	1	200	4–400
H^-	He	300	1	190	4–30
H^-	H_2	300	1	200	3–70
H_2^+	He	300	1	186	3–18
H_3^+	He	300	1	187	5–65
H_3^+	H_2	300	1	200	5–400
He^+	He	300	1	184	6–700
He^{2+}	He	303	2	189	10–70
He_2^+	He	300	1	186	1.5–24
He_2^+	He	293	3	127	2–43
HeH^+	He	300	2	189	3–60
Hg^+	He	300	1	185	4–35
Hg^+	Ne	300	1	195	9–55
Hg^+	Ar	300	1	197	35–180
H_2O^+	He	297	2	188	2–100
H_3O^+	He	297	2	188	2–100
H_3O^+	N_2	298	2	201	2–280
$H_3O^+ \cdot H_2O$	He	297	2	188	2–60
$H_3O^+ \cdot H_2O$	N_2	298	2	201	2–120
$H_3O^+ \cdot 2H_2O$	He	297	2	188	2–30
$H_3O^+ \cdot 2H_2O$	N_2	298	2	201	2–60
I^-	He	297	2	190	2–220
I^-	Ar	297	2	194	2–425
K^+	He	300	1	183	1–150
K^+	Ne	300	1	195	2–200
K^+	Ar	300	1	197	1–600
K^+	Kr	300	2	195	9–500
K^+	Xe	300	2	197	7–600
K^+	H_2	300	1	202	2–400
K^+	D_2	300	1	203	3–400
K^+	N_2	300	1	204	4–600
K^+	O_2	300	1	205	3–300
K^+	CO	300	1	207	4–600
K^+	NO	300	1	208	2–700
K^+	CO_2	300	1	209	10–700
K^+	CH_4	304	2	204	6–500
Kr^+	He	300	2	187	5–100
Kr^+	He	294	3	125	3–250
Kr^+	Ar	300	2	193	25–150
Kr^+	Kr	300	1	199	40–3000

(*continued*)

Table I-1 *(Cont.)*

Ion	Gas	Gas Temp. (K)	Paper No.[a]	Page	E/N Range (Td)
$Kr^+(^2P_{1/2})$	Kr	294	2	196	25–80
$Kr^+(^2P_{3/2})$	Kr	294	2	197	50–100
Kr^+	N_2	300	2	202	30–150
Kr_2^+	Kr	300	1	200	35–160
Kr_2^+	Kr	295	3	135	6–170
Kr^{2+}	He	300	3	125	8–90
$Kr^{2+}(A)$	Kr	301	2	197	35–150
$Kr^{2+}(B)$	Kr	301	2	197	35–150
Li^+	He	300	1	183.	2–150
Li^+	Ne	300	1	194	3–120
Li^+	Ar	300	1	196	6–200
Li^+	Kr	300	3	134	7–120
Li^+	Xe	300	3	135	10–200
Li^+	H_2	300	1	201	5–340
Li^+	D_2	300	1	203	3–300
Li^+	N_2	300	2	200	8–300
Li^+	O_2	306	2	203	3–300
N^+	He	300	1	184	6–70
N^+	N_2	300	1	204	7–500
N_2^+	He	300	1	186	8–90
N_2^+	Ne	300	3	130	7–340
N_2^+	Ar	300	3	133	12–180
N_2^+	Kr	300	3	135	14–200
N_2^+	N_2	300	1	203	6–800
N_3^+	N_2	300	1	204	2–240
N_4^+	N_2	300	2	202	2–150
N_4^+	N_2	300	1	205	2–40
Na^+	He	300	1	183	1–180
Na^+	Ne	300	1	194	4–200
Na^+	Ar	300	1	196	6–500
Na^+	Kr	300	3	134	7–500
Na^+	Xe	300	3	136	7–500
Na^+	H_2	300	1	201	4–240
Na^+	D_2	300	1	202	3–400
Na^+	CO_2	300	1	209	50–700
Ne^+	He	300	2	187	3–120
Ne^+	He	294	3	124	3–250
Ne^+	He	82	3	124	5–70
Ne^+	Ne	300	1	195	6–1500
$Ne^+(^2P_{1/2})$	Ne	78	2	191	6–70
$Ne^+(^2P_{3/2})$	Ne	78	2	191	8–70
Ne^+	Ar	294	3	131	3–220
Ne^{2+}	He	300	3	124	14–88
$Ne^{2+}(^1D)$	Ne	306	2	192	9–35
$Ne^{2+}(^3P)$	Ne	306	2	192	9–70
$Ne^{2+}(^1S)$	Ne	306	2	192	9–50

Table I-1 (*Cont.*)

Ion	Gas	Gas Temp. (K)	Paper No.[a]	Page	E/N Range (Td)
Ne_2^+	Ne	300	1	193	6–80
Ne_2^+	Ne	77	3	130	10–55
NH_3^+	He	300	1	188	10–120
NH_4^+	He	300	1	189	5–130
N_2H^+	He	295	3	129	3–140
N_2H^+	Ar	295	3	133	3–100
N_2H^+	N_2	300	1	203	6–180
N_2H^+	N_2	298	2	202	2–260
NO^+	He	300	1	187	5–130
NO^+	NO	300	1	208	10–300
$NO^+ \cdot H_2O$	He	300	1	189	4–300
N_2O^+	He	300	2	189	4–150
N_2O^+	Ne	300	2	191	12–150
N_2O^+	Ar	300	2	193	20–120
N_2O^+	N_2	300	2	202	30–200
NO_2^+	N_2	300	3	136	15–200
NO_2^-	He	300	1	192	4–120
NO_2^-	N_2	300	3	137	5–200
NO_3^-	N_2	300	3	137	5–200
$N_2O_2^+$	NO	300	1	201	8–40
N_2OH^+	He	300	1	189	5–120
N_2OH^+	Ar	300	1	199	20–190
O^+	He	300	1	185	5–130
O^+	Ar	300	1	197	5–250
O^{+*}	He	300	3	126	6–100
O^-	He	300	1	191	3–160
O^-	O_2	300	1	206	4–200
O^-	CO_2	300	1	209	25–240
O_2^+	He	300	1	187	5–130
O_2^+	Ne	300	3	131	6–380
O_2^+	Ar	300	1	198	8–240
O_2^+	Kr	300	3	135	8–170
O_2^+	O_2	300	1	205	4–500
O_2^-	He	300	1	191	3–100
O_2^-	O_2	300	1	206	3–140
O_3^-	He	300	1	191	3–60
O_3^-	Ar	300	1	199	2–120
O_3^-	O_2	300	1	206	5–120
O_4^+	O_2	300	1	205	2–20
OH^-	He	300	1	191	7–140
O_2H^+	He	300	1	188	5–120
$O_2H_2^+$	He	300	1	189	5–110
Rb^+	He	300	1	183	4–140
Rb^+	Ne	300	1	194	3–500
Rb^+	Ar	300	1	197	2–500
Rb^+	Kr	300	2	195	9–800

(*continued*)

Table I-1 (*Cont.*)

Ion	Gas	Gas Temp. (K)	Paper No.[a]	Page	E/N Range (Td)
Rb^+	Xe	300	2	198	10–800
Rb^+	H_2	300	2	200	5–200
Rb^+	N_2	300	2	200	5–600
Rb^+	O_2	300	2	203	5–400
Rb^+	CO_2	300	1	208	10–900
S^+	He	297	3	127	2.5–100
Si^+	He	300	3	127	12–100
SF_5^-	He	300	1	193	10–80
SF_6^-	He	300	1	193	2–80
SO_3^-	He	300	1	193	4–120
SO_2F^-	He	300	1	192	3–70
Tl^+	He	300	3	127	6–60
Tl^+	Ne	300	3	130	7–100
Tl^+	Ar	300	3	132	10–350
Tl^+	Kr	300	3	134	20–700
Tl^+	Xe	300	3	136	25–350
U^+	He	300	1	185	3–40
Xe^+	He	295	2	187	3–80
Xe^+	Ne	300	3	129	15–140
Xe^+	Ar	294	3	132	3–220
$Xe^+(^2P_{1/2})$	Xe	293	2	199	40–200
$Xe^+(^2P_{3/2})$	Xe	293	2	199	40–200
Xe^{2+}	He	300	3	125	10–90
Xe^{2+}	Ne	300	3	129	20–140
$Xe^{2+}(A)$	Xe	302	2	199	40–200
$Xe^{2+}(B)$	Xe	302	2	199	50–180

[a]Paper No. 1 is H. W. Ellis, R. Y. Pai, E. W. McDaniel, E. A. Mason, and L. A. Viehland (1976), Transport Properties of Gaseous Ions over a Wide Energy Range, *At. Data Nucl. Data Tables* **17**, 177–210. Paper No. 2 is H. W. Ellis, E. W. McDaniel, D. L. Albritton, L. A. Viehland, S. L. Lin, and E. A. Mason (1978), Transport Properties of Gaseous Ions over a Wide Energy Range: Part 2, *At. Data Nucl. Data Tables* **22**, 179–217. Paper No. 3 is H. W. Ellis, M. G. Thackston, E. W. McDaniel, and E. A. Mason (1984), Transport Properties of Gaseous Ions over a Wide Energy Range: Part 3, *At. Data Nucl. Data Tables* **31**, 113–151.

Table I-2 Zero-Field Mobilities as a Function of Gas Temperature (19 entries)[a]

Ion	Gas	Gas Temp. Range (K)	Page
Br^-	O_2	300–600	142
Cl^-	N_2	350–650	141
CO_3^-	N_2	225–675	141
CO_3^-	O_2	300–575	142
CO_4^-	O_2	280–475	143
Cs^+	He	75–520	140
Cs^+	Xe	190–460	141
He^+	He	170–650	140
He_2^+	He	180–680	140
He_2^{+*}	He	190–680	140
K^+	Ar	75–500	141
Li^+	He	20–500	140
NH_4^+	O_2	420–560	142
NO_2^-	N_2	300–675	141
NO_2^-	O_2	300–575	142
NO_3^-	N_2	225–575	142
NO_3^-	O_2	280–575	142
Na^+	He	90–500	140
Rb^+	Kr	85–480	141

[a]All entries are from Paper No. 3.

Table I-3 Zero-Field Mobilities (12 entries)[a]

Ion	Gas	Gas Temp. (K)	Standard Mobility $K_0(0)\,(cm^2/V\text{-}s)$
SF_6^-	He	298	12.0 ± 0.2
Ne_3^+	Ne	77	5.4 ± 0.54
Ar_3^+	Ar	77	1.65 ± 0.17
N^+	Ar	299	3.3 ± 0.2
$K^+(H_2O)$	N_2	305	2.17 ± 0.03
N_2^+	N_2	300	1.86 ± 0.18
O_2^+	O_2	294	2.14 ± 0.20
Cl^-	CH_4	298	2.12 ± 0.094
SF_5^-	SF_6	300	0.595 ± 0.007
SF_6^-	SF_6	300	0.542 ± 0.007
$SF_6^-(SF_6)$	SF_6	300	0.470 ± 0.010
$SF_6^-(SF_6)_2$	SF_6	300	0.420 ± 0.010

[a]All from Paper No. 3, p. 144.

Table I-4 Longitudinal Diffusion Coefficients as a Function of E/N (77 entries)

Ion	Gas	Gas Temp. (K)	Paper No.	Page	E/N Range (Td)
Br^-	Ne	300	3	146	5–350
Br^-	Ar	300	3	146	10–250
Br^-	Kr	300	3	147	10–450
Br^-	Xe	300	3	147	5–1000
C^+	He	300	3	145	3.5–80
CH^+	He	300	3	145	3.5–90
CH_2^+	He	300	3	145	5–90
CH_3^+	He	300	3	145	5–90
CH_4^+	He	300	3	145	5–90
CH_5^+	He	300	3	145	2.5–90
Cl^-	Ne	300	2	207	2–400
Cl^-	Ar	300	2	208	5–300
Cl^-	Kr	300	2	209	3–400
Cl^-	Xe	300	2	210	3–400
Cl^-	N_2	300	3	148	8–250
CO^+	CO	300	2	214	15–600
Cs^+	He	300	2	206	5–120
Cs^+	Ne	300	2	207	4–300
Cs^+	Ar	300	2	208	6–400
Cs^+	Kr	300	2	209	7–500
Cs^+	Xe	300	2	210	10–600
Cs^+	H_2	300	2	211	7–100
Cs^+	N_2	300	2	212	7–400
Cs^+	O_2	300	2	213	7–400
Cs^+	CO	300	2	214	7–500
Cs^+	CO_2	300	2	215	8–700
D^+	D_2	300	2	211	6–50
D^-	D_2	300	2	212	8–15
D_3^+	D_2	300	2	212	5–50
F^-	Kr	300	2	209	2–300
F^-	Xe	300	2	210	3–300
H^+	H_2	300	2	210	4–100
H^-	H_2	300	2	211	3–60
H_3^+	H_2	300	2	211	5–80
K^+	He	300	2	206	1–150
K^+	Ne	300	2	207	2–200
K^+	Ar	300	2	208	1–600
K^+	Kr	300	2	208	8–600
K^+	Xe	300	2	209	7–800
K^+	H_2	300	2	211	2–250
K^+	N_2	300	2	212	4–600
K^+	O_2	300	2	213	3–300
K^+	NO	300	2	214	30–600
K^+	CO	300	2	214	4–600
K^+	CO_2	300	2	215	12–600
Li^+	He	300	2	206	2–200

Table I-4 (*Cont.*)

Ion	Gas	Gas Temp. (K)	Paper No.	Page	E/N Range (Td)
Li^+	Ne	300	2	206	3–120
Li^+	Ar	300	2	207	6–200
Li^+	Kr	300	3	146	8–150
Li^+	Xe	300	3	147	10–200
Li^+	H_2	300	2	210	4–300
N^+	N_2	300	2	212	7–300
N_2^+	N_2	300	2	213	6–500
NO^+	NO	300	2	214	8–250
Na^+	He	300	2	206	1–200
Na^+	Ne	300	2	207	4–200
Na^+	Ar	300	2	208	6–500
Na^+	Kr	300	3	146	8–500
Na^+	Xe	300	3	147	10–400
Na^+	H_2	300	2	210	4–250
Na^+	CO_2	300	2	215	50–800
O^-	O_2	300	2	213	4–150
O_2^+	O_2	300	2	213	7–500
O_2^-	O_2	300	2	214	4–80
Rb^+	He	300	2	206	4–150
Rb^+	Ne	300	2	207	3–500
Rb^+	Ar	300	2	208	2–500
Rb^+	Kr	300	2	209	9–800
Rb^+	Xe	300	2	209	10–800
Rb^+	H_2	300	2	211	5–200
Rb^+	N_2	300	2	212	6–600
Rb^+	O_2	300	2	213	5–400
Rb^+	CO_2	300	2	215	10–800
Tl^+	Ne	300	3	146	7–110
Tl^+	Ar	300	3	146	10–350
Tl^+	Kr	300	3	147	20–800
Tl^+	Xe	300	3	147	20–450

Table I-5 Transverse Diffusion Coefficients as a Function of E/N (6 entries)[a]

Ion	Gas	Gas Temp. (K)	E/N Range (Td)
H_3^+	H_2	300	4–200
K^+	H_2	300	3–200
K^+	N_2	300	12–300
N^+	N_2	300	30–250
N_2^+	N_2	300	40–500
O_2^+	O_2	300	15–500

[a]All from Paper No. 2, p. 216.

APPENDIX II

TABLES OF TRANSPORT CROSS SECTIONS FOR MODEL POTENTIALS

These tables of cross sections can be used with any version of moment theory, as explained in Section 7-1A. Previously published tabulations give only the average collision integrals corresponding to the one- and two-temperature moment theories. The numerical results here are taken from the heretofore unpublished calculations cited.

(n, 6, 4) Models:

$$V(r) = \frac{n\varepsilon_0}{n(3 + \gamma) - 12(1 + \gamma)} \left[\frac{12}{n}(1 + \gamma)\left(\frac{r_m}{r}\right)^n - 4\gamma\left(\frac{r_m}{r}\right)^6 - 3(1 - \gamma)\left(\frac{r_m}{r}\right)^4 \right]$$

From the calculations of Viehland et al. (1975).

(n, 4) Core Models:

$$V(r) = \frac{n\varepsilon_0}{n - 4} \left[\frac{4}{n}\left(\frac{r_m - a}{r - a}\right)^n - \left(\frac{r_m - a}{r - a}\right)^4 \right]$$

From the calculations of F. J. Smith (1983).

Reduced Quantities:

$$Q^{(l)*} \equiv \bar{Q}^{(l)}/\pi r_m^2, \qquad \bar{Q}^{(l)} \text{ defined by (6-1-11);}$$

$$\varepsilon^* \equiv \varepsilon/\varepsilon_0, \qquad \varepsilon \equiv \tfrac{1}{2}\mu v_r^2;$$

$$a^* \equiv a/r_m.$$

Accuracy:

1 part in 10^4 for $\varepsilon^* > 6$, decreasing to about 1 part in 10^3 for $\varepsilon^* < 2$.

Table II-1 Cross Sections $Q^{(l)*}$ for (8, 6, 4) Potentials

ε^*	$\gamma = 0$	$\gamma = 0.2$	$\gamma = 0.4$	$\gamma = 0.6$	$\gamma = 0.8$	$\gamma = 1.0$
			$Q^{(1)*}$			
0.01	30.51	28.26	25.51	22.04	17.52	11.57
0.02	21.47	19.93	18.07	15.76	12.84	9.150
0.03	17.46	16.24	14.77	12.97	10.74	7.974
0.04	15.07	14.04	12.81	11.31	9.469	7.229
0.06	12.24	11.43	10.48	9.331	7.946	6.292
0.08	10.55	9.874	9.083	8.143	7.021	5.696
0.1	9.389	8.808	8.128	7.327	6.379	5.270
0.2	6.513	6.153	5.741	5.271	4.730	4.113
0.3	5.237	4.970	4.672	4.336	3.958	3.534
0.4	4.522	4.243	4.031	3.768	3.479	3.160
0.6	3.562	3.471	3.354	3.189	2.936	2.686
0.8	2.643	2.614	2.579	2.542	2.483	2.408
1	2.074	2.063	2.052	2.036	2.024	2.006
2	1.097	1.102	1.107	1.113	1.119	1.126
3	0.8499	0.8581	0.8657	0.8747	0.8830	0.8922
4	0.7413	0.7507	0.7601	0.7695	0.7789	0.7894
6	0.6384	0.6487	0.6590	0.6695	0.6802	0.6911
8	0.5856	0.5963	0.6069	0.6178	0.6290	0.6406
10	0.5515	0.5625	0.5732	0.5844	0.5958	0.6076
20	0.4670	0.4784	0.4898	0.5012	0.5129	0.5250
40	0.3996	0.4109	0.4225	0.4339	0.4455	0.4575
60	0.3648	0.3760	0.3873	0.3985	0.4100	0.4219
80	0.3417	0.3528	0.3638	0.3749	0.3862	0.3979
100	0.3248	0.3357	0.3465	0.3574	0.3685	0.3800
200	0.2766	0.2869	0.2969	0.3072	0.3176	0.3284
400	0.2349	0.2444	0.2537	0.2631	0.2727	0.2826
600	0.2132	0.2222	0.2311	0.2400	0.2490	0.2584
800	0.1989	0.2076	0.2161	0.2247	0.2333	0.2423
1000	0.1885	0.1969	0.2052	0.2134	0.2218	0.2305
			$Q^{(2)*}$			
0.01	32.76	30.52	27.73	24.16	19.41	12.87
0.02	23.18	21.63	19.73	17.34	14.23	10.16
0.03	18.93	17.69	16.18	14.30	11.91	8.850
0.04	16.40	15.34	14.07	12.49	10.50	8.020
0.06	13.39	12.56	11.55	10.33	8.821	6.978
0.08	11.59	10.89	10.06	9.040	7.802	6.318
0.1	10.37	9.760	9.033	8.156	7.099	5.847
0.2	7.326	6.941	6.486	5.943	5.309	4.585
0.3	5.985	5.685	5.346	4.948	4.487	3.969
0.4	5.236	4.394	4.658	4.348	3.984	3.579
0.6	4.256	4.011	3.752	3.568	3.369	3.094
0.8	3.987	3.807	3.593	3.355	3.037	2.751
1	3.455	3.361	3.249	3.099	2.942	2.719
2	1.779	1.771	1.762	1.757	1.744	1.734
3	1.248	1.251	1.252	1.258	1.259	1.264
4	1.024	1.030	1.036	1.042	1.048	1.056

Table II-1 (*Cont.*)

ε^*	$\gamma = 0$	$\gamma = 0.2$	$\gamma = 0.4$	$\gamma = 0.6$	$\gamma = 0.8$	$\gamma = 1.0$
6	0.8305	0.8389	0.8473	0.8558	0.8648	0.8743
8	0.7437	0.7532	0.7628	0.7722	0.7821	0.7925
10	0.6929	0.7029	0.7131	0.7231	0.7335	0.7444
20	0.5817	0.5928	0.6040	0.6153	0.6269	0.6389
40	0.5008	0.5126	0.5245	0.5365	0.5486	0.5612
60	0.4596	0.4715	0.4835	0.4955	0.5078	0.5205
80	0.4322	0.4440	0.4560	0.4680	0.4803	0.4930
100	0.4118	0.4236	0.4354	0.4474	0.4596	0.4723
200	0.3532	0.3648	0.3760	0.3876	0.3994	0.4116
400	0.3015	0.3125	0.3232	0.3341	0.3452	0.3567
600	0.2743	0.2849	0.2953	0.3057	0.3163	0.3273
800	0.2564	0.2666	0.2767	0.2868	0.2971	0.3077
1000	0.2432	0.2531	0.2630	0.2728	0.2828	0.2931

$$Q^{(3)*}$$

ε^*	$\gamma = 0$	$\gamma = 0.2$	$\gamma = 0.4$	$\gamma = 0.6$	$\gamma = 0.8$	$\gamma = 1.0$
0.01	35.52	32.94	29.76	25.73	20.44	13.27
0.02	25.04	23.25	21.09	18.41	14.95	10.51
0.03	20.39	18.96	17.26	15.16	12.50	9.160
0.04	17.62	16.41	14.97	13.22	11.02	8.308
0.06	14.33	13.38	12.26	10.91	9.237	7.234
0.08	12.37	11.58	10.64	9.521	8.159	6.550
0.1	11.03	10.34	9.534	8.570	7.413	6.060
0.2	7.697	7.270	6.771	6.181	5.504	4.732
0.3	6.214	5.894	5.529	5.102	4.618	4.074
0.4	5.391	5.071	4.778	4.451	4.073	3.655
0.6	4.277	4.149	3.982	3.756	3.435	3.124
0.8	3.346	3.244	3.139	3.037	2.944	2.811
1	2.901	2.819	2.722	2.619	2.499	2.389
2	1.651	1.640	1.630	1.616	1.602	1.586
3	1.178	1.181	1.183	1.185	1.188	1.190
4	0.9667	0.9728	0.9789	0.9847	0.9918	0.9984
6	0.7807	0.7893	0.7979	0.8064	0.8156	0.8253
8	0.6964	0.7060	0.7160	0.7255	0.7353	0.7459
10	0.6471	0.6572	0.6677	0.6777	0.6881	0.6991
20	0.5401	0.5511	0.5621	0.5733	0.5848	0.5966
40	0.4635	0.4749	0.4863	0.4980	0.5098	0.5221
60	0.4247	0.4362	0.4477	0.4593	0.4712	0.4835
80	0.3991	0.4105	0.4219	0.4335	0.4453	0.4575
100	0.3801	0.3915	0.4028	0.4142	0.4259	0.43.80
200	0.3257	0.3367	0.3474	0.3583	0.3695	0.3811
400	0.2778	0.2882	0.2983	0.3085	0.3190	0.3298
600	0.2527	0.2626	0.2724	0.2822	0.2921	0.3025
800	0.2361	0.2457	0.2551	0.2646	0.2742	0.2842
1000	0.2239	0.2332	0.2424	0.2516	0.2610	0.2706

$$Q^{(4)*}$$

ε^*	$\gamma = 0$	$\gamma = 0.2$	$\gamma = 0.4$	$\gamma = 0.6$	$\gamma = 0.8$	$\gamma = 1.0$
0.01	36.31	33.84	30.69	26.71	21.33	13.87
0.02	25.68	23.94	21.85	19.15	15.62	10.96
0.03	20.97	19.54	17.92	15.79	13.06	9.558

(continued)

Table II-1 *(Cont.)*

ε^*	$\gamma = 0$	$\gamma = 0.2$	$\gamma = 0.4$	$\gamma = 0.6$	$\gamma = 0.8$	$\gamma = 1.0$
0.04	18.16	16.93	15.57	13.78	11.51	8.671
0.06	14.83	13.84	12.79	11.39	9.658	7.557
0.08	12.85	12.00	11.12	9.961	8.538	6.849
0.1	11.49	10.76	9.981	8.983	7.764	6.343
0.2	8.132	7.674	7.149	6.538	5.802	4.974
0.3	6.636	6.293	5.894	5.439	4.902	4.298
0.4	5.795	5.462	5.150	4.776	4.352	3.870
0.6	4.710	4.447	4.152	3.927	3.682	3.326
0.8	4.297	4.121	3.893	3.663	3.312	2.986
1	3.733	3.605	3.471	3.299	3.137	2.910
2	2.162	2.139	2.108	2.081	2.038	1.996
3	1.497	1.494	1.490	1.487	1.479	1.475
4	1.192	1.195	1.197	1.200	1.201	1.207
6	0.9261	0.9329	0.9398	0.9463	0.9538	0.9625
8	0.8107	0.8191	0.8291	0.8369	0.8453	0.8547
10	0.7460	0.7552	0.7662	0.7750	0.7842	0.7942
20	0.6157	0.6265	0.6372	0.6483	0.6596	0.6713
40	0.5293	0.5409	0.5520	0.5640	0.5763	0.5889
60	0.4864	0.4982	0.5097	0.5219	0.5343	0.5471
80	0.4579	0.4698	0.4816	0.4937	0.5061	0.5189
100	0.4368	0.4487	0.4606	0.4726	0.4849	0.4977
200	0.3758	0.3875	0.3993	0.4109	0.4228	0.4352
400	0.3215	0.3328	0.3440	0.3551	0.3664	0.3782
600	0.2931	0.3038	0.3145	0.3253	0.3362	0.3476
800	0.2739	0.2844	0.2949	0.3053	0.3160	0.3270
1000	0.2599	0.2702	0.2804	0.2906	0.3010	0.3117

Table II-2 Cross Sections $Q^{(l)*}$ for (12, 6, 4) Potentials

ε^*	$\gamma = 0$	$\gamma = 0.2$	$\gamma = 0.4$	$\gamma = 0.6$	$\gamma = 0.8$	$\gamma = 1.0$
			$Q^{(1)*}$			
0.01	26.07	23.31	20.28	16.93	13.21	9.179
0.02	18.30	16.40	14.34	12.13	9.743	7.278
0.03	14.87	13.35	11.73	10.00	8.189	6.354
0.04	12.83	11.54	10.17	8.741	7.253	5.771
0.06	10.41	9.391	8.335	7.242	6.128	5.039
0.08	8.965	8.116	7.241	6.347	5.447	4.578
0.1	7.981	7.247	6.495	5.734	4.977	4.249
0.2	5.545	5.093	4.643	4.201	3.774	3.370
0.3	4.470	4.139	3.818	3.507	3.214	2.936
0.4	3.834	3.570	3.322	3.085	2.865	2.657
0.6	3.166	2.936	2.731	2.570	2.428	2.296
0.8	2.518	2.460	2.392	2.305	2.170	2.057
1	2.043	2.015	1.994	1.967	1.943	1.904

Table II-2 *(Cont.)*

ε^*	$\gamma = 0$	$\gamma = 0.2$	$\gamma = 0.4$	$\gamma = 0.6$	$\gamma = 0.8$	$\gamma = 1.0$
2	1.149	1.148	1.150	1.155	1.159	1.163
3	0.9148	0.9207	0.9270	0.9333	0.9385	0.9435
4	0.8125	0.8208	0.8287	0.8353	0.8412	0.8469
6	0.7165	0.7266	0.7352	0.7427	0.7495	0.7556
8	0.6676	0.6783	0.6875	0.6955	0.7026	0.7090
10	0.6362	0.6472	0.6567	0.6650	0.6723	0.6789
20	0.5596	0.5709	0.5806	0.5892	0.5969	0.6038
40	0.4985	0.5098	0.5195	0.5281	0.5357	0.5427
60	0.4666	0.4777	0.4873	0.4957	0.5033	0.5102
80	0.4451	0.4561	0.4656	0.4739	0.4814	0.4882
100	0.4292	0.4400	0.4494	0.4576	0.4650	0.4717
200	0.3831	0.3934	0.4022	0.3941	0.4170	0.4234
400	0.3420	0.3514	0.3597	0.3669	0.3735	0.3794
600	0.3200	0.3290	0.3368	0.3437	0.3499	0.3556
800	0.3052	0.3139	0.3214	0.3281	0.3341	0.3395
1000	0.2942	0.3026	0.3099	0.3164	0.3222	0.3275

$$Q^{(2)*}$$

ε^*	$\gamma = 0$	$\gamma = 0.2$	$\gamma = 0.4$	$\gamma = 0.6$	$\gamma = 0.8$	$\gamma = 1.0$
0.01	28.41	25.60	22.49	19.00	14.99	10.46
0.02	20.11	18.17	16.04	13.70	11.11	8.302
0.03	16.42	14.87	13.19	11.34	9.354	7.255
0.04	14.23	12.91	11.48	9.935	8.294	6.593
0.06	11.63	10.58	9.461	8.260	7.019	5.762
0.08	10.08	9.190	8.253	7.258	6.245	5.236
0.1	9.019	8.242	7.428	6.574	5.710	4.862
0.2	6.395	5.889	5.375	4.867	4.347	3.861
0.3	5.232	4.846	4.462	4.101	3.719	3.374
0.4	4.538	4.226	3.918	3.635	3.334	3.066
0.6	3.658	3.494	3.257	3.064	2.862	2.846
0.8	3.364	3.086	2.856	2.684	2.585	2.433
1	3.901	2.927	2.751	2.571	2.390	2.236
2	1.766	1.745	1.731	1.720	1.708	1.699
3	1.279	1.278	1.279	1.280	1.282	1.285
4	1.070	1.074	1.079	1.083	1.088	1.092
6	0.8904	0.8985	0.9055	0.9118	0.9174	0.9231
8	0.8104	0.8196	0.8277	0.8348	0.8411	0.8469
10	0.7638	0.7737	0.7822	0.7898	0.7966	0.8028
20	0.6637	0.6748	0.6843	0.6927	0.7002	0.7070
40	0.5925	0.6042	0.6142	0.6231	0.6309	0.6380
60	0.5561	0.5679	0.5780	0.5870	0.5705	0.5777
80	0.5316	0.5434	0.5535	0.5625	0.5705	0.5777
100	0.5132	0.5250	0.5351	0.5441	0.5520	0.5593
200	0.4598	0.4711	0.4810	0.4897	0.4975	0.5045
400	0.4114	0.4222	0.4315	0.4397	0.4472	0.4539
600	0.3854	0.3957	0.4047	0.4126	0.4198	0.4262
800	0.3679	0.3779	0.3866	0.3943	0.4012	0.4075
1000	0.3548	0.3645	0.3730	0.3805	0.3873	0.3934

(continued)

Table II-2 (*Cont.*)

ε^*	$\gamma = 0$	$\gamma = 0.2$	$\gamma = 0.4$	$\gamma = 0.6$	$\gamma = 0.8$	$\gamma = 1.0$
			$Q^{(3)*}$			
0.01	30.49	27.30	23.80	19.92	15.51	10.59
0.02	21.45	19.25	16.87	14.27	11.44	8.397
0.03	17.46	15.70	13.81	11.77	9.615	7.333
0.04	15.08	13.59	11.99	10.28	8.512	6.661
0.06	12.26	11.08	9.837	8.510	7.186	5.817
0.08	10.58	9.590	8.553	7.458	6.382	5.284
0.1	9.435	8.572	7.676	6.741	5.826	4.904
0.2	6.598	6.049	5.499	4.956	4.408	3.888
0.3	5.343	4.932	4.531	4.152	3.752	3.389
0.4	4.597	4.269	3.952	3.659	3.347	3.070
0.6	3.797	3.498	3.270	3.046	2.844	2.662
0.8	3.063	2.967	2.864	2.654	2.538	2.397
1	2.631	2.519	2.425	2.363	2.289	2.226
2	1.630	1.612	1.594	1.575	1.559	1.542
3	1.211	1.211	1.211	1.209	1.211	1.211
4	1.017	1.022	1.026	1.030	1.034	1.038
6	0.8452	0.8534	0.8606	0.8671	0.8729	0.8782
8	0.7677	0.7770	0.7851	0.7923	0.7988	0.8047
10	0.7225	0.7323	0.7409	0.7485	0.7554	0.7616
20	0.6253	0.6363	0.6457	0.6541	0.6615	0.6683
40	0.5570	0.5685	0.5783	0.5869	0.5946	0.6015
60	0.5224	0.5338	0.5437	0.5523	0.5600	0.5671
80	0.4992	0.5106	0.5204	0.5290	0.5367	0.5437
100	0.4819	0.4931	0.5029	0.5115	0.5192	0.5262
200	0.4315	0.4423	0.4517	0.4600	0.4674	0.4742
400	0.3859	0.3961	0.4050	0.4128	0.4199	0.4263
600	0.3615	0.3712	0.3797	0.3872	0.3941	0.4001
800	0.3450	0.3544	0.3627	0.3699	0.3765	0.3824
1000	0.3327	0.3419	0.3499	0.3570	0.3634	0.3692
			$Q^{(4)*}$			
0.01	31.50	28.37	24.89	20.98	16.49	11.27
0.02	22.26	20.12	17.75	15.10	12.21	8.946
0.03	18.16	16.47	14.58	12.48	10.27	7.817
0.04	15.72	14.29	12.68	10.92	9.093	7.104
0.06	12.83	11.71	10.44	9.068	7.676	6.207
0.08	11.11	10.17	9.101	7.964	6.817	5.641
0.1	9.940	9.115	8.186	7.212	6.223	5.238
0.2	7.059	6.510	5.915	5.342	4.720	4.159
0.3	5.789	5.354	4.905	4.503	4.035	3.634
0.4	5.029	4.661	4.301	3.990	3.617	3.302
0.6	4.055	3.855	3.556	3.351	3.106	2.884
0.8	3.706	3.399	3.131	2.932	2.805	2.621
1	3.313	3.155	2.977	2.788	2.592	2.414
2	2.074	2.032	1.993	1.955	1.917	1.883
3	1.499	1.489	1.482	1.475	1.469	1.465
4	1.221	1.221	1.223	1.224	1.226	1.227

Table II-2 (*Cont.*)

ε^*	$\gamma = 0$	$\gamma = 0.2$	$\gamma = 0.4$	$\gamma = 0.6$	$\gamma = 0.8$	$\gamma = 1.0$
6	0.9765	0.9829	0.9888	0.9940	0.9988	1.004
8	0.8714	0.8790	0.8860	0.8923	0.8982	0.9037
10	0.8123	0.8209	0.8286	0.8355	0.8418	0.8477
20	0.6938	0.7046	0.7138	0.7219	0.7292	0.7357
40	0.6174	0.6292	0.6392	0.6480	0.6558	0.6628
60	0.5799	0.5918	0.6020	0.6110	0.6189	0.6261
80	0.5549	0.5667	0.5770	0.5860	0.5940	0.6013
100	0.5362	0.5480	0.5582	0.5672	0.5752	0.5825
200	0.4813	0.4928	0.5027	0.5115	0.5195	0.5266
400	0.4312	0.4422	0.4517	0.4601	0.4678	0.4745
600	0.4041	0.4147	0.4239	0.4320	0.4394	0.4460
800	0.3858	0.3961	0.4051	0.4130	0.4202	0.4266
1000	0.3721	0.3822	0.3910	0.3987	0.4058	0.4120

Table II-3 Cross Sections $Q^{(l)*}$ for (16, 6, 4) Potentials

ε^*	$\gamma = 0$	$\gamma = 0.2$	$\gamma = 0.4$	$\gamma = 0.6$	$\gamma = 0.8$	$\gamma = 1.0$
			$Q^{(1)*}$			
0.01	24.45	21.63	18.62	15.42	12.01	8.479
0.02	17.14	15.20	13.16	11.05	8.868	6.714
0.03	13.91	12.37	10.75	9.113	7.458	5.856
0.04	11.99	10.68	9.322	7.961	6.607	5.314
0.06	9.719	8.685	7.633	6.595	5.585	4.636
0.08	8.368	7.500	6.630	5.780	4.966	4.209
0.1	7.447	6.693	5.947	5.224	4.539	3.905
0.2	5.171	4.699	4.257	3.837	3.450	3.098
0.3	4.167	3.821	3.508	3.214	2.947	2.705
0.4	3.570	3.299	3.059	2.837	2.636	2.455
0.6	2.887	2.682	2.522	2.379	2.250	2.135
0.8	2.437	2.351	2.227	2.092	2.005	1.925
1	2.013	1.981	1.948	1.909	1.846	1.785
2	1.175	1.177	1.179	1.181	1.182	1.187
3	0.9554	0.9612	0.9659	0.9700	0.9744	0.9774
4	0.8591	0.8665	0.8721	0.8770	0.8817	0.8859
6	0.7690	0.7772	0.7841	0.7898	0.7948	0.7990
8	0.7233	0.7323	0.7396	0.7458	0.7511	0.7559
10	0.6942	0.7035	0.7111	0.7176	0.7232	0.7280
20	0.6239	0.6335	0.6415	0.6484	0.6542	0.6595
40	0.5688	0.5783	0.5863	0.5932	0.5991	0.6044
60	0.5399	0.5494	0.5573	0.5641	0.5701	0.5752
80	0.5205	0.5299	0.5377	0.5444	0.5503	0.5554
100	0.5060	0.5153	0.5230	0.5297	0.5355	0.5405
200	0.4638	0.4726	0.4799	0.4863	0.4918	0.4967
400	0.4252	0.4335	0.4404	0.4464	0.4514	0.4562

(*continued*)

Table II-3   (*Cont.*)

ε^*	$\gamma = 0$	$\gamma = 0.2$	$\gamma = 0.4$	$\gamma = 0.6$	$\gamma = 0.8$	$\gamma = 1.0$
600	0.4043	0.4122	0.4188	0.4246	0.4294	0.4340
800	0.3901	0.3977	0.4042	0.4097	0.4145	0.4189
1000	0.3794	0.3869	0.3931	0.3986	0.4033	0.4075

$$Q^{(2)*}$$

ε^*	$\gamma = 0$	$\gamma = 0.2$	$\gamma = 0.4$	$\gamma = 0.6$	$\gamma = 0.8$	$\gamma = 1.0$
0.01	26.80	23.89	20.75	17.42	13.71	9.718
0.02	18.97	16.96	14.84	12.59	10.19	7.721
0.03	15.50	13.89	12.22	10.45	8.601	6.750
0.04	13.43	12.06	10.65	9.161	7.639	6.136
0.06	10.98	9.885	8.780	7.630	6.478	5.365
0.08	9.512	8.592	7.663	6.713	5.772	4.878
0.1	8.513	7.709	6.899	6.084	5.283	4.531
0.2	6.034	5.517	5.000	4.506	4.033	3.603
0.3	4.939	4.544	4.158	3.796	3.456	3.153
0.4	4.289	3.964	3.655	3.368	3.104	2.868
0.6	3.525	3.274	3.056	2.855	2.674	2.512
0.8	3.065	2.830	2.687	2.563	2.409	2.289
1	2.886	2.684	2.491	2.321	2.234	2.125
2	1.748	1.730	1.714	1.699	1.693	1.674
3	1.297	1.296	1.295	1.295	1.293	1.297
4	1.099	1.103	1.106	1.109	1.111	1.114
6	0.9294	0.9359	0.9412	0.9461	0.9506	0.9550
8	0.8537	0.8615	0.8679	0.8733	0.8782	0.8820
10	0.8099	0.8183	0.8253	0.8313	0.8365	0.8409
20	0.7178	0.7272	0.7350	0.7417	0.7472	0.7525
40	0.6542	0.6640	0.6722	0.6792	0.6853	0.6905
60	0.6219	0.6318	0.6401	0.6472	0.6536	0.6587
80	0.6002	0.6101	0.6184	0.6255	0.6321	0.6371
100	0.5839	0.5938	0.6021	0.6092	0.6157	0.6208
200	0.5360	0.5457	0.5538	0.5608	0.5667	0.5722
400	0.4922	0.5014	0.5092	0.5158	0.5211	0.5267
600	0.4683	0.4772	0.4846	0.4911	0.4961	0.5016
800	0.4520	0.4607	0.4679	0.4742	0.4793	0.4845
1000	0.4398	0.4482	0.4553	0.4615	0.4667	0.4715

$$Q^{(3)*}$$

ε^*	$\gamma = 0$	$\gamma = 0.2$	$\gamma = 0.4$	$\gamma = 0.6$	$\gamma = 0.8$	$\gamma = 1.0$
0.01	28.64	25.39	21.90	18.16	14.11	9.796
0.02	20.13	17.90	15.52	13.02	10.43	7.763
0.03	16.38	14.59	12.70	10.76	8.767	6.774
0.04	14.14	12.62	11.02	9.405	7.765	6.149
0.06	11.49	10.28	9.039	7.801	6.559	5.366
0.08	9.914	8.896	7.858	6.842	5.828	4.872
0.1	8.839	7.950	7.053	6.186	5.323	4.521
0.2	6.175	5.608	5.060	4.542	4.038	3.586
0.3	4.998	4.574	4.175	3.801	3.447	3.132
0.4	4.299	3.958	3.646	3.354	3.084	2.843
0.6	3.471	3.230	3.012	2.814	2.636	2.476
0.8	2.965	2.836	2.657	2.472	2.355	2.238

Table II-3 *(Cont.)*

ε^*	$\gamma = 0$	$\gamma = 0.2$	$\gamma = 0.4$	$\gamma = 0.6$	$\gamma = 0.8$	$\gamma = 1.0$
1	2.515	2.414	2.339	2.265	2.160	2.082
2	1.621	1.597	1.575	1.554	1.533	1.515
3	1.233	1.231	1.229	1.227	1.226	1.225
4	1.051	1.054	1.057	1.059	1.062	1.063
6	0.8883	0.8949	0.9004	0.9053	0.9097	0.9145
8	0.8150	0.8228	0.8293	0.8348	0.8397	0.8436
10	0.7726	0.7809	0.7880	0.7939	0.7992	0.8037
20	0.6831	0.6924	0.7001	0.7068	0.7122	0.7176
40	0.6218	0.6314	0.6394	0.6462	0.6523	0.6574
60	0.5908	0.6005	0.6086	0.6154	0.6218	0.6267
80	0.5701	0.5797	0.5878	0.5947	0.6011	0.6059
100	0.5545	0.5641	0.5722	0.5790	0.5854	0.5902
200	0.5090	0.5183	0.5261	0.5327	0.5384	0.5437
400	0.4673	0.4761	0.4835	0.4898	0.4947	0.5003
600	0.4445	0.4530	0.4601	0.4663	0.4709	0.4764
800	0.4291	0.4373	0.4442	0.4502	0.4549	0.4600
1000	0.4174	0.4255	0.4323	0.4381	0.4430	0.4477

$$Q^{(4)*}$$

ε^*	$\gamma = 0$	$\gamma = 0.2$	$\gamma = 0.4$	$\gamma = 0.6$	$\gamma = 0.8$	$\gamma = 1.0$
0.01	29.74	26.44	22.97	19.18	15.05	10.47
0.02	21.01	18.76	16.41	13.86	11.16	8.318
0.03	17.14	15.36	13.49	11.50	9.408	7.270
0.04	14.83	13.33	11.75	10.08	8.346	6.608
0.06	12.10	10.93	9.684	8.389	7.066	5.777
0.08	10.48	9.494	8.447	7.374	6.288	5.252
0.1	9.374	8.516	7.602	6.678	5.750	4.878
0.2	6.661	6.089	5.500	4.932	4.378	3.879
0.3	5.463	5.013	4.567	4.148	3.748	3.393
0.4	4.743	4.370	4.011	3.676	3.363	3.086
0.6	3.905	3.608	3.351	3.109	2.892	2.701
0.8	3.390	3.113	2.942	2.793	2.601	2.457
1	3.133	2.929	2.718	2.531	2.405	2.272
2	2.025	1.978	1.933	1.891	1.848	1.813
3	1.500	1.489	1.480	1.471	1.465	1.457
4	1.241	1.240	1.239	1.239	1.240	1.242
6	1.010	1.015	1.019	1.023	1.026	1.029
8	0.9097	0.9163	0.9219	0.9269	0.9312	0.9359
10	0.8541	0.8615	0.8678	0.8733	0.8784	0.8825
20	0.7449	0.7540	0.7615	0.7679	0.7729	0.7785
40	0.6767	0.6865	0.6946	0.7016	0.7077	0.7128
60	0.6434	0.6534	0.6617	0.6688	0.6757	0.6803
80	0.6213	0.6310	0.6396	0.6468	0.6539	0.6583
100	0.6047	0.6147	0.6230	0.6302	0.6372	0.6418
200	0.5558	0.5655	0.5738	0.5808	0.5866	0.5924
400	0.5107	0.5201	0.5279	0.5347	0.5392	0.5459
600	0.4860	0.4951	0.5027	0.5093	0.5136	0.5201
800	0.4692	0.4781	0.4855	0.4919	0.4966	0.5025
1000	0.4566	0.4653	0.4726	0.4788	0.4840	0.4892

Table II-4 Cross Sections $Q^{(l)*}$ for $(n, 4)$ Potentials

ε^*	$n = 6$	$n = 10$	$n = 14$	$n = 18$	$n = 300$
			$Q^{(1)*}$		
0.01	37.59	27.62	25.11	23.97	20.99
0.02	26.38	19.41	17.61	16.79	14.67
0.03	21.40	15.78	14.30	13.63	11.89
0.04	18.43	13.62	12.33	11.74	10.23
0.06	14.90	11.05	9.997	9.516	8.280
0.08	12.79	9.526	8.609	8.192	7.122
0.1	11.35	8.484	7.664	7.290	6.335
0.2	7.786	5.987	5.324	5.060	4.402
0.3	6.298	4.752	4.290	4.077	3.557
0.4	5.551	4.070	3.676	3.493	3.059
0.6	3.845	3.346	3.017	2.803	2.475
0.8	2.736	2.575	2.476	2.406	2.132
1	2.104	2.051	2.022	2.003	1.903
2	1.063	1.123	1.161	1.187	1.338
3	0.8040	0.8852	0.9365	0.9717	1.166
4	0.6903	0.7807	0.8378	0.8776	1.102
6	0.5825	0.6817	0.7450	0.7894	1.048
8	0.5273	0.6309	0.6978	0.7451	1.020
10	0.4917	0.5982	0.6677	0.7169	1.005
20	0.4038	0.5179	0.5944	0.6493	0.9746
40	0.3344	0.4536	0.5364	0.5967	0.9593
60	0.2991	0.4201	0.5060	0.5693	0.9533
80	0.2760	0.3977	0.4856	0.5509	0.9497
100	0.2592	0.3811	0.4704	0.5370	0.9472
200	0.2122	0.3335	0.4262	0.4966	0.9406
400	0.1727	0.2916	0.3863	0.4596	0.9351
600	0.1527	0.2694	0.3647	0.4394	0.9323
800	0.1398	0.2547	0.3502	0.4256	0.9304
1000	0.1305	0.2438	0.3392	0.4152	0.9290
			$Q^{(2)*}$		
0.01	39.75	29.95	27.46	26.33	23.41
0.02	28.00	21.19	19.43	18.63	16.58
0.03	22.79	17.30	15.88	15.23	13.56
0.04	19.70	14.99	13.75	13.20	11.75
0.06	16.03	12.25	11.24	10.79	9.614
0.08	13.85	10.61	9.740	9.348	8.338
0.1	12.36	9.497	8.717	8.367	7.469
0.2	8.637	6.732	6.181	5.932	5.312
0.3	7.076	5.503	5.059	4.854	4.358
0.4	5.993	4.774	4.390	4.214	3.790
0.6	5.113	3.811	3.594	3.455	3.117
0.8	4.569	3.610	3.185	2.988	2.717
1	3.820	3.244	2.972	2.814	2.443
2	1.808	1.766	1.754	1.745	1.726
3	1.229	1.264	1.288	1.305	1.378
4	0.9901	1.049	1.086	1.111	1.241

Table II-4 (*Cont.*)

ε^*	$n = 6$	$n = 10$	$n = 14$	$n = 18$	$n = 300$
6	0.7857	0.8641	0.9117	0.9444	1.131
8	0.6939	0.7814	0.8341	0.8703	1.075
10	0.6401	0.7329	0.7890	0.8277	1.048
20	0.5220	0.6274	0.6932	0.7386	0.9957
40	0.4363	0.5517	0.6261	0.6781	0.9734
60	0.3930	0.5129	0.5918	0.6475	0.9656
80	0.3645	0.4868	0.5688	0.6270	0.9611
100	0.3435	0.4674	0.5515	0.6116	0.9580
200	0.2840	0.4111	0.5009	0.5663	0.9500
400	0.2330	0.3607	0.4548	0.5246	0.9439
600	0.2068	0.3339	0.4298	0.5018	0.9410
800	0.1898	0.3159	0.4129	0.4862	0.9391
1000	0.1775	0.3026	0.4002	0.4745	0.9376

$$Q^{(3)*}$$

ε^*	$n = 6$	$n = 10$	$n = 14$	$n = 18$	$n = 300$
0.01	43.56	32.25	29.39	28.09	24.71
0.02	30.60	22.70	20.67	19.74	17.34
0.03	24.86	18.47	16.82	16.05	14.09
0.04	21.44	15.96	14.52	13.86	12.15
0.06	17.38	12.98	11.80	11.26	9.867
0.08	14.96	11.21	10.19	9.713	8.509
0.1	13.31	9.999	9.081	8.659	7.584
0.2	9.191	7.000	6.344	6.049	5.302
0.3	7.415	5.665	5.137	4.987	4.299
0.4	6.455	4.876	4.421	4.213	3.705
0.6	4.746	4.007	3.613	3.395	3.005
0.8	3.738	3.159	3.006	2.932	2.592
1	3.153	2.732	2.560	2.484	2.315
2	1.671	1.637	1.624	1.619	1.631
3	1.158	1.196	1.223	1.243	1.345
4	0.9312	0.9945	1.036	1.064	1.220
6	0.7339	0.8167	0.8686	0.9050	1.116
8	0.6449	0.7365	0.7934	0.8334	1.064
10	0.5928	0.6894	0.7497	0.7921	1.039
20	0.4799	0.5872	0.6567	0.7056	0.9892
40	0.3995	0.5149	0.5921	0.6472	0.9717
60	0.3593	0.4782	0.5594	0.6178	0.9612
80	0.3329	0.4537	0.5374	0.5981	0.9570
100	0.3135	0.4355	0.5210	0.5834	0.9541
200	0.2589	0.3828	0.4731	0.5401	0.9465
400	0.2121	0.3357	0.4295	0.5003	0.9406
600	0.1882	0.3106	0.4058	0.4785	0.9377
800	0.1727	0.2939	0.3898	0.4636	0.9358
1000	0.1615	0.2814	0.3778	0.4524	0.9344

$$Q^{(4)*}$$

ε^*	$n = 6$	$n = 10$	$n = 14$	$n = 18$	$n = 300$
0.01	44.13	33.20	30.46	29.17	25.94
0.02	31.11	23.46	21.52	20.64	18.37

(*continued*)

Table II-4 (*Cont.*)

ε^*	$n = 6$	$n = 10$	$n = 14$	$n = 18$	$n = 300$
0.03	25.33	19.14	17.55	16.87	15.02
0.04	21.89	16.57	15.19	14.62	13.02
0.06	17.80	13.52	12.39	11.95	10.64
0.08	15.37	11.71	10.73	10.35	9.224
0.1	13.71	10.47	9.604	9.267	8.258
0.2	9.614	7.432	6.828	6.568	5.863
0.3	7.848	6.090	5.599	5.370	4.805
0.4	6.706	5.299	4.857	4.656	4.175
0.6	5.596	4.235	3.972	3.828	3.430
0.8	4.896	3.926	3.519	3.302	2.986
1	4.227	3.485	3.211	3.069	2.683
2	2.249	2.109	2.047	2.009	1.901
3	1.503	1.497	1.499	1.502	1.515
4	1.172	1.208	1.232	1.249	1.338
6	0.8884	0.9547	0.9947	1.023	1.183
8	0.7658	0.8459	0.8923	0.9246	1.109
10	0.6971	0.7843	0.8351	0.8702	1.073
20	0.5583	0.6592	0.7217	0.7645	1.004
40	0.4661	0.5779	0.6497	0.6995	0.9781
60	0.4207	0.5379	0.6144	0.6680	0.9694
80	0.3908	0.5113	0.5909	0.6470	0.9646
100	0.3688	0.4915	0.5733	0.6313	0.9612
200	0.3063	0.4336	0.5215	0.5851	0.9527
400	0.2521	0.3810	0.4740	0.5424	0.9462
600	0.2241	0.3528	0.4481	0.5190	0.9433
800	0.2059	0.3339	0.4305	0.5029	0.9414
1000	0.1927	0.3199	0.4173	0.4908	0.9401

Table II-5 Cross Sections $Q^{(l)*}$ for (8, 4) Core Potentials

ε^*	$a^* = 0$	$a^* = 0.1$	$a^* = 0.2$	$a^* = 0.3$	$a^* = 0.4$	$a^* = 0.5$	$a^* = 0.6$	$a^* = 0.7$	$a^* = 0.8$
					$Q^{(1)*}$				
0.01	30.51	25.45	20.86	16.74	13.09	9.926	7.226	4.994	3.221
0.02	21.47	18.04	14.91	12.08	9.568	7.374	5.491	3.920	2.653
0.03	17.46	14.75	12.26	10.00	7.987	6.222	4.701	3.425	2.387
0.04	15.07	12.78	10.68	8.760	7.037	5.527	4.221	3.122	2.222
0.06	12.24	10.45	8.791	7.274	5.902	4.692	3.641	2.752	2.019
0.08	10.55	9.050	7.661	6.382	5.219	4.187	3.289	2.525	1.893
0.1	9.389	8.093	6.885	5.769	4.749	3.830	3.045	2.366	1.804
0.2	6.513	5.700	4.939	4.226	3.566	2.961	2.424	1.959	1.571
0.3	5.237	4.629	4.060	3.525	3.026	2.560	2.139	1.769	1.460
0.4	4.522	3.987	3.529	3.098	2.696	2.315	1.965	1.653	1.391
0.6	3.562	3.316	2.931	2.577	2.289	2.012	1.751	1.511	1.305
0.8	2.643	2.568	2.468	2.308	2.033	1.822	1.617	1.423	1.251
1	2.074	2.053	2.020	1.971	1.876	1.685	1.521	1.360	1.214
2	1.097	1.119	1.142	1.165	1.186	1.207	1.223	1.188	1.115
3	0.8499	0.8806	0.9124	0.9446	0.9762	1.008	1.039	1.069	1.070
4	0.7413	0.7761	0.8115	0.8478	0.8842	0.9208	0.9558	0.9918	1.027
6	0.6384	0.6768	0.7163	0.7568	0.7981	0.8400	0.8808	0.9252	0.9666
8	0.5856	0.6260	0.6677	0.7107	0.7548	0.8000	0.8462	0.8913	0.9367
10	0.5515	0.5932	0.6364	0.6812	0.7273	0.7747	0.8234	0.8712	0.9178
20	0.4670	0.5122	0.5595	0.6090	0.6605	0.7141	0.7698	0.8270	0.8870
40	0.3996	0.4474	0.4981	0.5516	0.6079	0.6671	0.7291	0.7938	0.8612
60	0.3647	0.4138	0.4662	0.5218	0.5807	0.6428	0.7082	0.7769	0.8489
80	0.3417	0.3915	0.4449	0.5019	0.5624	0.6266	0.6943	0.7657	0.8407
100	0.3247	0.3750	0.4292	0.4871	0.5489	0.6145	0.6840	0.7574	0.8347
200	0.2766	0.3280	0.3839	0.4445	0.5097	0.5795	0.6541	0.7334	0.8174
400	0.2349	0.2867	0.3439	0.4064	0.4745	0.5480	0.6270	0.7117	0.8020

(continued)

511

Table II-5 (*Cont.*)

ε^*	$a^* = 0$	$a^* = 0.1$	$a^* = 0.2$	$a^* = 0.3$	$a^* = 0.4$	$a^* = 0.5$	$a^* = 0.6$	$a^* = 0.7$	$a^* = 0.8$
600	0.2132	0.2650	0.3226	0.3861	0.4555	0.5309	0.6123	0.6999	0.7936
800	0.1989	0.2506	0.3084	0.3725	0.4428	0.5194	0.6024	0.6919	0.7880
1000	0.1885	0.2400	0.2979	0.3624	0.4333	0.5109	0.5950	0.6859	0.7837
					$Q^{(2)}*$				
0.01	32.76	27.65	22.92	18.72	14.89	11.48	8.524	6.001	3.904
0.02	23.18	19.70	16.51	13.60	10.98	8.603	6.525	4.731	3.207
0.03	18.93	16.16	13.63	11.30	9.211	7.294	5.609	4.141	2.878
0.04	16.40	14.05	11.91	9.918	8.142	6.500	5.050	3.778	2.674
0.06	13.39	11.54	9.851	8.269	6.855	5.541	4.370	3.332	2.420
0.08	11.59	10.05	8.615	7.279	6.076	4.958	3.955	3.057	2.262
0.1	10.37	9.023	7.767	6.600	5.538	4.554	3.665	2.864	2.151
0.2	7.327	6.470	5.646	4.987	4.176	3.521	2.918	2.360	1.856
0.3	5.985	5.323	4.696	4.126	3.557	3.045	2.570	2.122	1.713
0.4	5.236	4.628	4.124	3.656	3.182	2.754	2.354	1.974	1.624
0.6	4.256	3.722	3.441	3.078	2.727	2.398	2.089	1.790	1.511
0.8	3.987	3.546	3.061	2.695	2.449	2.180	1.924	1.676	1.440
1	3.455	3.216	2.928	2.575	2.232	2.027	1.807	1.595	1.389
2	1.779	1.765	1.746	1.723	1.691	1.635	1.529	1.380	1.255
3	1.248	1.261	1.274	1.286	1.297	1.305	1.305	1.279	1.192
4	1.023	1.046	1.068	1.091	1.113	1.134	1.146	1.168	1.143
6	0.8314	0.8606	0.8906	0.9211	0.9516	0.9817	1.010	1.038	1.068
8	0.7445	0.7775	0.8110	0.8451	0.8795	0.9139	0.9479	0.9814	1.007
10	0.6932	0.7287	0.7646	0.8010	0.8379	0.8751	0.9118	0.9466	0.9715
20	0.5811	0.6223	0.6642	0.7068	0.7502	0.7944	0.8393	0.8842	0.9315
40	0.5008	0.5458	0.5923	0.6403	0.6895	0.7400	0.7917	0.8445	0.8984
60	0.4598	0.5068	0.5556	0.6063	0.6587	0.7128	0.7684	0.8255	0.8842

80	0.4324	0.4807	0.5311	0.5836	0.6382	0.6946	0.7529	0.8130	0.8748
100	0.4120	0.4612	0.5128	0.5667	0.6228	0.6810	0.7414	0.8037	0.8680
200	0.3532	0.4050	0.4597	0.5173	0.5779	0.6414	0.7077	0.7768	0.8485
400	0.3014	0.3548	0.4118	0.4725	0.5370	0.6050	0.6767	0.7521	0.8310
600	0.2743	0.3281	0.3861	0.4484	0.5148	0.5852	0.6598	0.7385	0.8216
800	0.2563	0.3103	0.3689	0.4321	0.4997	0.5717	0.6483	0.7293	0.8151
1000	0.2431	0.2972	0.3561	0.4199	0.4884	0.5617	0.6396	0.7224	0.8102

$Q^{(3)}*$

0.01	35.52	29.67	24.37	19.64	15.41	11.67	8.504	5.862	3.739
0.02	25.04	21.05	17.44	14.18	11.28	8.676	6.456	4.591	3.061
0.03	20.39	17.23	14.36	11.74	9.418	7.322	5.523	4.003	2.743
0.04	17.62	14.95	12.52	10.27	8.298	6.503	4.956	3.643	2.546
0.06	14.33	12.23	10.32	8.530	6.956	5.519	4.271	3.203	2.303
0.08	12.37	10.61	8.997	7.487	6.147	4.923	3.854	2.932	2.152
0.1	11.03	9.503	8.091	6.773	5.590	4.512	3.565	2.743	2.046
0.2	7.697	6.732	5.819	4.985	4.188	3.469	2.827	2.253	1.767
0.3	6.214	5.491	4.798	4.174	3.551	2.991	2.486	2.025	1.633
0.4	5.391	4.744	4.185	3.677	3.163	2.700	2.276	1.884	1.549
0.6	4.277	3.949	3.466	3.059	2.689	2.342	2.018	1.711	1.445
0.8	3.346	3.126	2.956	2.662	2.393	2.119	1.856	1.604	1.379
1	2.901	2.704	2.510	2.369	2.213	1.961	1.742	1.528	1.333
2	1.651	1.633	1.611	1.581	1.542	1.481	1.410	1.323	1.210
3	1.178	1.192	1.206	1.218	1.229	1.238	1.230	1.205	1.152
4	0.9661	0.9910	1.016	1.040	1.064	1.085	1.107	1.117	1.106
6	0.7817	0.8127	0.8450	0.8784	0.9119	0.9466	0.9746	1.012	1.040
8	0.6973	0.7321	0.7679	0.8047	0.8423	0.8802	0.9198	0.9550	0.9884
10	0.6474	0.6847	0.7228	0.7619	0.8020	0.8427	0.8849	0.9226	0.9546
20	0.5394	0.5818	0.6256	0.6706	0.7170	0.7647	0.8137	0.8634	0.9165
40	0.4634	0.5091	0.5569	0.6068	0.6586	0.7123	0.7679	0.8253	0.8846
60	0.4250	0.4723	0.5222	0.5745	0.6292	0.6830	0.7456	0.8072	0.8710

(continued)

Table II-5 (*Cont.*)

ε^*	$a^* = 0$	$a^* = 0.1$	$a^* = 0.2$	$a^* = 0.3$	$a^* = 0.4$	$a^* = 0.5$	$a^* = 0.6$	$a^* = 0.7$	$a^* =$
80	0.3994	0.4478	0.4990	0.5530	0.6097	0.6690	0.7309	0.7953	0.8621
100	0.3804	0.4296	0.4818	0.5370	0.5951	0.6561	0.7199	0.7864	0.8556
200	0.3257	0.3802	0.4319	0.4905	0.5527	0.6185	0.6879	0.7608	0.8371
400	0.2777	0.3302	0.3871	0.4484	0.5141	0.5841	0.6586	0.7374	0.8206
600	0.2526	0.3053	0.3631	0.4257	0.4932	0.5654	0.6426	0.7246	0.8116
800	0.2360	0.2889	0.3471	0.4104	0.4790	0.5528	0.6317	0.7159	0.8055
1000	0.2238	0.2767	0.3351	0.3990	0.4684	0.5433	0.6236	0.7093	0.8009
					$Q^{(4)*}$				
0.01	36.31	30.59	25.35	20.59	16.40	12.57	9.270	6.480	4.165
0.02	25.68	21.80	18.22	14.96	12.09	9.393	7.079	5.089	3.412
0.03	20.97	17.89	15.04	12.42	10.13	7.950	6.076	4.444	3.058
0.04	18.16	15.55	13.13	10.90	8.947	7.076	5.464	4.047	2.837
0.06	14.83	12.77	10.85	9.075	7.516	6.022	4.720	3.562	2.564
0.08	12.85	11.11	9.485	7.983	6.650	5.382	4.265	3.263	2.394
0.1	11.49	9.971	8.550	7.236	6.052	4.939	3.948	3.054	2.273
0.2	8.132	7.134	6.214	5.372	4.545	3.810	3.131	2.510	1.953
0.3	6.636	5.871	5.165	4.530	3.864	3.290	2.750	2.254	1.798
0.4	5.795	5.117	4.531	4.013	3.454	2.971	2.516	2.094	1.700
0.6	4.710	4.121	3.781	3.366	2.956	2.582	2.228	1.896	1.577
0.8	4.297	3.861	3.358	2.946	2.650	2.344	2.050	1.772	1.500
1	3.733	3.438	3.143	2.794	2.414	2.178	1.924	1.684	1.445
2	2.162	2.102	2.035	1.958	1.865	1.756	1.608	1.449	1.302
3	1.497	1.494	1.488	1.480	1.467	1.442	1.408	1.336	1.233
4	1.191	1.205	1.217	1.230	1.241	1.250	1.245	1.236	1.183
6	0.9281	0.9517	0.9764	1.002	1.028	1.051	1.074	1.096	1.109
8	0.8125	0.8419	0.8714	0.9017	0.9325	0.9632	0.9933	1.020	1.040

ε*	a* = 0	a* = 0.1	a* = 0.2	a* = 0.3	a* = 0.4	a* = 0.5	a* = 0.6	a* = 0.7	a* = 0.8
10	0.7467	0.7796	0.8125	0.8457	0.8794	0.9135	0.9471	0.9759	0.9926
20	0.6141	0.6539	0.6942	0.7348	0.7759	0.8175	0.8596	0.9013	0.9471
40	0.5291	0.5727	0.6177	0.6639	0.7111	0.7593	0.8083	0.8581	0.9088
60	0.4870	0.5326	0.5800	0.6290	0.6795	0.7313	0.7843	0.8385	0.8938
80	0.4587	0.5058	0.5549	0.6058	0.6585	0.7128	0.7686	0.8257	0.8841
100	0.4375	0.4858	0.5361	0.5885	0.6428	0.6989	0.7569	0.8163	0.8771
200	0.3759	0.4272	0.4812	0.5377	0.5969	0.6584	0.7226	0.7889	0.8573
400	0.3213	0.3746	0.4314	0.4913	0.5548	0.6212	0.6909	0.7637	0.8396
600	0.2926	0.3466	0.4045	0.4662	0.5317	0.6007	0.6734	0.7498	0.8299
800	0.2737	0.3280	0.3865	0.4492	0.5161	0.5868	0.6616	0.7403	0.8233
1000	0.2598	0.3142	0.3732	0.4365	0.5043	0.5763	0.6526	0.7331	0.8183

Table II-6   Cross Sections $Q^{(l)*}$ for (10, 4) Core Potentials

ε*	a* = 0	a* = 0.1	a* = 0.2	a* = 0.3	a* = 0.4	a* = 0.5	a* = 0.6	a* = 0.7	a* = 0.8
					$Q^{(1)*}$				
0.01	27.62	23.10	18.98	15.28	12.00	9.145	6.705	4.676	3.058
0.02	19.41	16.35	13.55	11.03	8.779	6.808	5.111	3.686	2.532
0.03	15.78	13.36	11.14	9.134	7.334	5.752	4.385	3.229	2.285
0.04	13.62	11.58	9.703	7.998	6.465	5.115	3.943	2.949	2.132
0.06	11.05	9.465	7.991	6.645	5.427	4.349	3.410	2.608	1.943
0.08	9.526	8.201	6.967	5.833	4.803	3.886	3.085	2.399	1.825
0.1	8.484	7.337	6.266	5.276	4.374	3.568	2.860	2.252	1.742
0.2	5.897	5.184	4.513	3.880	3.297	2.764	2.289	1.876	1.525
0.3	4.752	4.223	3.725	3.250	2.810	2.399	2.207	1.701	1.422
0.4	4.069	3.647	3.249	2.870	2.513	2.177	1.867	1.593	1.358
0.6	3.346	3.035	2.674	2.409	2.150	1.905	1.672	1.461	1.279
0.8	2.575	2.490	2.365	2.120	1.924	1.736	1.552	1.380	1.230

(continued)

Table II-6 (Cont.)

ε^*	$a^* = 0$	$a^* = 0.1$	$a^* = 0.2$	$a^* = 0.3$	$a^* = 0.4$	$a^* = 0.5$	$a^* = 0.6$	$a^* = 0.7$	$a^* = 0.8$
1	2.051	2.024	1.987	1.923	1.771	1.615	1.467	1.323	1.196
2	1.123	1.143	1.162	1.182	1.200	1.217	1.225	1.172	1.105
3	0.8852	0.9133	0.9413	0.9696	0.9976	1.026	1.052	1.079	1.061
4	0.7807	0.8123	0.8444	0.8766	0.9094	0.9414	0.9731	1.004	1.032
6	0.6817	0.7168	0.7528	0.7893	0.8262	0.8639	0.8997	0.9393	0.9748
8	0.6310	0.6682	0.7063	0.7454	0.7854	0.8259	0.8664	0.9077	0.9483
10	0.5984	0.6369	0.6766	0.7174	0.7592	0.8019	0.8458	0.8881	0.9334
20	0.5179	0.5600	0.6039	0.6493	0.6965	0.7452	0.7954	0.8468	0.8996
40	0.4534	0.4986	0.5460	0.5956	0.6475	0.7015	0.7578	0.8163	0.8766
60	0.4199	0.4666	0.5158	0.5676	0.6220	0.6790	0.7386	0.8007	0.8654
80	0.3976	0.4452	0.4956	0.5489	0.6050	0.6640	0.7258	0.7904	0.8580
100	0.3810	0.4293	0.4806	0.5350	0.5923	0.6528	0.7163	0.7827	0.8525
200	0.3337	0.3836	0.4372	0.4945	0.5555	0.6202	0.6886	0.7605	0.8367
400	0.2917	0.3428	0.3983	0.4579	0.5220	0.5904	0.6633	0.7405	0.8224
600	0.2695	0.3210	0.3773	0.4382	0.5038	0.5742	0.6495	0.7296	0.8147
800	0.2547	0.3064	0.3632	0.4248	0.4915	0.5633	0.6401	0.7222	0.8094
1000	0.2438	0.2956	0.3527	0.4148	0.4823	0.5550	0.6331	0.7166	0.8054
					$Q^{(2)*}$				
0.01	29.95	25.29	21.05	17.20	13.74	10.66	7.943	5.632	3.703
0.02	21.19	18.04	15.16	12.53	10.14	7.997	6.099	4.455	3.056
0.03	17.30	14.81	12.53	10.43	8.519	6.789	5.252	3.907	2.750
0.04	14.99	12.89	10.95	9.171	7.538	6.056	4.735	3.570	2.559
0.06	12.24	10.60	9.069	7.662	6.359	5.171	4.107	3.157	2.322
0.08	10.61	9.227	7.940	6.752	5.645	4.634	3.722	2.901	2.175
0.1	9.497	8.290	7.165	6.125	5.153	4.261	3.454	2.722	2.070
0.2	6.734	5.953	5.224	4.546	3.904	3.311	2.762	2.255	1.793

516

0.3	5.502	4.913	4.354	3.832	3.335	2.875	2.440	2.034	1.660
0.4	4.771	4.292	3.833	3.401	2.990	2.607	2.241	1.897	1.577
0.6	3.811	3.543	3.212	2.884	2.572	2.281	1.997	1.727	1.473
0.8	3.610	3.181	2.795	2.584	2.319	2.079	1.845	1.621	1.407
1	3.244	2.988	2.686	2.351	2.137	1.938	1.739	1.546	1.360
2	1.766	1.752	1.731	1.704	1.666	1.599	1.465	1.349	1.236
3	1.264	1.276	1.286	1.296	1.304	1.305	1.306	1.255	1.174
4	1.049	1.069	1.089	1.109	1.126	1.145	1.155	1.170	1.133
6	0.8643	0.8909	0.9179	0.9451	0.9724	0.9987	1.022	1.047	1.067
8	0.7815	0.8113	0.8415	0.8720	0.9023	0.9324	0.9632	0.9922	1.019
10	0.7329	0.7648	0.7970	0.8299	0.8628	0.8963	0.9280	0.9597	0.9919
20	0.6272	0.6645	0.7024	0.7408	0.7799	0.8195	0.8595	0.8999	0.9397
40	0.5514	0.5928	0.6355	0.6788	0.7234	0.7689	0.8153	0.8630	0.9105
60	0.5127	0.5562	0.6012	0.6474	0.6950	0.7438	0.7939	0.8451	0.8976
80	0.4868	0.5317	0.5782	0.6264	0.6761	0.7272	0.7797	0.8334	0.8891
100	0.4675	0.5134	0.5609	0.6107	0.6619	0.7147	0.7692	0.8248	0.8829
200	0.4113	0.4601	0.5110	0.5646	0.6203	0.6783	0.7384	0.8002	0.8651
400	0.3609	0.4119	0.4659	0.5225	0.5821	0.6446	0.7099	0.7779	0.8491
600	0.3339	0.3858	0.4414	0.4995	0.5612	0.6261	0.6943	0.7657	0.8404
800	0.3159	0.3683	0.4246	0.4839	0.5469	0.6135	0.6836	0.7574	0.8345
1000	0.3025	0.3553	0.4120	0.4721	0.5362	0.6040	0.6755	0.7511	0.8300
$Q^{(3)}*$									
0.01	32.25	27.00	22.24	17.94	14.14	10.80	7.896	5.485	3.543
0.02	22.70	19.15	15.91	12.96	10.36	8.031	6.012	4.310	2.915
0.03	18.47	15.67	13.09	10.75	8.660	6.779	5.152	3.768	2.619
0.04	15.96	13.60	11.41	9.423	7.635	6.024	4.630	3.435	2.436
0.06	12.98	11.13	9.408	7.841	6.408	5.116	3.997	3.028	2.210
0.08	11.21	9.657	8.209	6.890	5.669	4.568	3.612	2.778	2.070
0.1	9.999	8.649	7.388	6.236	5.160	4.191	3.345	2.603	1.970
0.2	6.999	6.135	5.335	4.590	3.880	3.238	2.663	2.152	1.710

(continued)

517

Table II-6 *(Cont.)*

ε^*	$a^* = 0$	$a^* = 0.1$	$a^* = 0.2$	$a^* = 0.3$	$a^* = 0.4$	$a^* = 0.5$	$a^* = 0.6$	$a^* = 0.7$	$a^* = 0.8$
0.3	5.666	5.016	4.413	3.843	3.301	2.804	2.348	1.940	1.585
0.4	4.878	4.348	3.858	3.391	2.950	2.540	2.156	1.810	1.507
0.6	4.007	3.596	3.189	2.847	2.522	2.217	1.920	1.650	1.410
0.8	3.159	2.995	2.823	2.507	2.258	2.015	1.775	1.551	1.349
1	2.732	2.556	2.402	2.280	2.083	1.872	1.672	1.482	1.306
2	1.637	1.617	1.593	1.560	1.516	1.446	1.391	1.297	1.192
3	1.196	1.208	1.219	1.229	1.236	1.240	1.223	1.196	1.137
4	0.9946	1.016	1.038	1.060	1.078	1.099	1.115	1.119	1.102
6	0.8169	0.8455	0.8750	0.9048	0.9356	0.9638	0.9891	1.021	1.042
8	0.7366	0.7683	0.8009	0.8342	0.8673	0.9030	0.9349	0.9689	1.000
10	0.6893	0.7232	0.7577	0.7933	0.8294	0.8652	0.9039	0.9371	0.9751
20	0.5869	0.6259	0.6659	0.7069	0.7489	0.7921	0.8360	0.8805	0.9258
40	0.5147	0.5573	0.6018	0.6472	0.6945	0.7433	0.7935	0.8454	0.8979
60	0.4782	0.5226	0.5690	0.6172	0.6673	0.7192	0.7729	0.8284	0.8856
80	0.4538	0.4994	0.5471	0.5972	0.6492	0.7033	0.7594	0.8172	0.8775
100	0.4356	0.4821	0.5308	0.5822	0.6357	0.6914	0.7493	0.8089	0.8716
200	0.3829	0.4319	0.4835	0.5386	0.5962	0.6567	0.7199	0.7854	0.8546
400	0.3358	0.3866	0.4411	0.4987	0.5599	0.6247	0.6928	0.7642	0.8394
600	0.3106	0.3622	0.4181	0.4770	0.5401	0.6071	0.6779	0.7526	0.8312
800	0.2938	0.3458	0.4024	0.4622	0.5266	0.5952	0.6678	0.7447	0.8255
1000	0.2814	0.3336	0.3905	0.4512	0.5165	0.5861	0.6601	0.7387	0.8213

$Q^{(4)*}$

ε^*	$a^* = 0$	$a^* = 0.1$	$a^* = 0.2$	$a^* = 0.3$	$a^* = 0.4$	$a^* = 0.5$	$a^* = 0.6$	$a^* = 0.7$	$a^* = 0.8$
0.01	33.20	27.90	23.26	18.88	15.09	11.68	8.630	6.079	3.954
0.02	23.46	19.88	16.72	13.76	11.13	8.732	6.608	4.792	3.251
0.03	19.14	16.33	13.81	11.47	9.337	7.396	5.682	4.195	2.919
0.04	16.56	14.21	12.07	10.08	8.254	6.588	5.118	3.827	2.713
0.06	13.51	11.69	9.984	8.422	6.952	5.615	4.431	3.377	2.457

0.08	11.70	10.19	8.736	7.418	6.164	5.025	4.011	3.099	2.296	
0.1	10.47	9.157	7.880	6.726	5.620	4.617	3.718	2.904	2.183	
0.2	7.428	6.578	5.739	4.977	4.245	3.578	2.962	2.398	1.883	
0.3	6.088	5.422	4.779	4.186	3.621	3.098	2.610	2.159	1.739	
0.4	5.293	4.730	4.201	3.710	3.242	2.804	2.393	2.010	1.648	
0.6	4.235	3.906	3.511	3.144	2.784	2.445	2.127	1.825	1.535	
0.8	3.926	3.500	3.061	2.813	2.505	2.225	1.963	1.709	1.464	
1	3.485	3.207	2.914	2.564	2.303	2.072	1.847	1.628	1.414	
2	2.109	2.051	1.982	1.902	1.806	1.700	1.544	1.412	1.279	
3	1.497	1.493	1.485	1.474	1.457	1.428	1.391	1.309	1.213	
4	1.208	1.219	1.230	1.240	1.249	1.251	1.245	1.228	1.169	
6	0.9552	0.9768	0.9997	1.022	1.043	1.066	1.081	1.101	1.104	
8	0.8456	0.8718	0.8985	0.9257	0.9536	0.9794	1.006	1.031	1.051	
10	0.7836	0.8128	0.8419	0.8720	0.9022	0.9326	0.9621	0.9876	1.018	
20	0.6585	0.6944	0.7306	0.7669	0.8038	0.8410	0.8783	0.9156	0.9528	
40	0.5781	0.6181	0.6596	0.7008	0.7434	0.7867	0.8306	0.8756	0.9199	
60	0.5384	0.5805	0.6241	0.6686	0.7143	0.7609	0.8085	0.8569	0.9064	
80	0.5118	0.5554	0.6003	0.6471	0.6949	0.7439	0.7941	0.8448	0.8976	
100	0.4919	0.5367	0.5825	0.6311	0.6805	0.7313	0.7834	0.8360	0.8912	
200	0.4335	0.4816	0.5312	0.5840	0.6381	0.6942	0.7521	0.8110	0.8730	
400	0.3808	0.4315	0.4852	0.5406	0.5989	0.6599	0.7230	0.7885	0.8568	
600	0.3526	0.4044	0.4600	0.5168	0.5773	0.6409	0.7070	0.7761	0.8480	
800	0.3338	0.3861	0.4426	0.5006	0.5626	0.6278	0.6961	0.7676	0.8419	
1000	0.3198	0.3725	0.4293	0.4885	0.5515	0.6180	0.6878	0.7612	0.8373	

Table II-7 Cross Sections $Q^{(l)*}$ for (12, 4) Core Potentials

$Q^{(1)*}$

ε^*	$a^* = 0$	$a^* = 0.1$	$a^* = 0.2$	$a^* = 0.3$	$a^* = 0.4$	$a^* = 0.5$	$a^* = 0.6$	$a^* = 0.7$	$a^* = 0.8$
0.01	26.08	21.83	17.98	14.51	11.42	8.728	6.424	4.506	2.970
0.02	18.30	15.44	12.83	10.47	8.358	6.505	4.906	3.561	2.466
0.03	14.87	12.61	10.54	8.668	6.984	5.501	4.214	3.124	2.229
0.04	12.82	10.93	9.180	7.589	6.158	4.894	3.794	2.857	2.082
0.06	10.40	8.926	7.558	6.304	5.171	4.165	3.286	2.530	1.901
0.08	8.964	7.732	6.588	5.534	4.577	3.725	2.977	2.330	1.788
0.1	7.982	6.917	5.925	5.007	4.170	3.422	2.762	2.190	1.709
0.2	5.547	4.889	4.269	3.688	3.148	2.656	2.216	1.830	1.501
0.3	4.470	3.987	3.528	3.096	2.687	2.309	1.965	1.662	1.402
0.4	3.830	3.448	3.082	2.738	2.408	2.098	1.812	1.560	1.341
0.6	3.166	2.805	2.545	2.303	2.068	1.842	1.627	1.434	1.265
0.8	2.518	2.421	2.257	2.034	1.857	1.684	1.513	1.357	1.218
1	2.042	2.007	1.958	1.860	1.708	1.572	1.433	1.303	1.185
2	1.149	1.160	1.179	1.193	1.212	1.224	1.223	1.160	1.099
3	0.9148	0.9383	0.9642	0.9903	1.014	1.039	1.062	1.084	1.058
4	0.8125	0.8408	0.8700	0.8994	0.9286	0.9559	0.9859	1.013	1.031
6	0.7164	0.7488	0.7817	0.8151	0.8487	0.8818	0.9147	0.9491	0.9822
8	0.6677	0.7020	0.7372	0.7728	0.8094	0.8467	0.8823	0.9158	0.9562
10	0.6364	0.6721	0.7088	0.7463	0.7845	0.8234	0.8631	0.8980	0.9422
20	0.5596	0.5990	0.6398	0.6819	0.7253	0.7700	0.8157	0.8624	0.9102
40	0.4983	0.5409	0.5853	0.6315	0.6795	0.7293	0.7808	0.8343	0.8890
60	0.4663	0.5106	0.5569	0.6053	0.6558	0.7084	0.7631	0.8199	0.8787
80	0.4450	0.4903	0.5379	0.5878	0.6400	0.6945	0.7513	0.8103	0.8719
100	0.4291	0.4752	0.5237	0.5747	0.6281	0.6841	0.7425	0.8033	0.8668
200	0.3833	0.4314	0.4824	0.5365	0.5936	0.6537	0.7168	0.7828	0.8523
400	0.3421	0.3918	0.4450	0.5017	0.5620	0.6258	0.6933	0.7643	0.8392

					$Q^{(2)*}$				
600	0.3200	0.3704	0.4247	0.4828	0.5447	0.6106	0.6804	0.7542	0.8320
800	0.3055	0.3560	0.4109	0.4699	0.5329	0.6001	0.6716	0.7472	0.8271
1000	0.2941	0.3452	0.4006	0.4602	0.5241	0.5923	0.6649	0.7420	0.8234
0.01	28.42	24.03	20.05	16.39	13.11	10.20	7.629	5.432	3.594
0.02	20.11	17.14	14.44	11.95	9.689	7.666	5.870	4.306	2.973
0.03	16.42	14.08	11.93	9.950	8.143	6.514	5.061	3.782	2.679
0.04	14.23	12.26	10.42	8.750	7.210	5.815	4.568	3.459	2.496
0.06	11.63	10.08	8.633	7.312	6.087	4.970	3.966	3.062	2.268
0.08	10.08	8.778	7.561	6.446	5.408	4.457	3.598	2.817	2.126
0.1	9.019	7.888	6.828	5.850	4.940	4.102	3.341	2.645	2.025
0.2	6.397	5.668	4.999	4.350	3.752	3.196	2.678	2.197	1.759
0.3	5.232	4.680	4.179	3.673	3.211	2.779	2.368	1.985	1.631
0.4	4.535	4.091	3.683	3.263	2.882	2.524	2.178	1.854	1.551
0.6	3.658	3.395	3.077	2.771	2.483	2.212	1.944	1.691	1.450
0.8	3.364	2.961	2.697	2.474	2.241	2.020	1.799	1.590	1.387
1	3.090	2.830	2.524	2.280	2.072	1.885	1.698	1.519	1.343
2	1.766	1.740	1.719	1.698	1.642	1.567	1.429	1.329	1.223
3	1.279	1.286	1.295	1.299	1.308	1.306	1.302	1.240	1.165
4	1.070	1.087	1.105	1.121	1.138	1.148	1.163	1.169	1.127
6	0.8904	0.9147	0.9391	0.9635	0.9876	0.011	1.029	1.051	1.070
8	0.8106	0.8378	0.8652	0.8932	0.9205	0.9476	0.9747	0.9868	1.024
10	0.7640	0.7931	0.8226	0.8523	0.8821	0.9117	0.9410	0.9596	0.9965
20	0.6636	0.6978	0.7324	0.7675	0.8031	0.8391	0.8751	0.9113	0.9475
40	0.5922	0.6304	0.6696	0.7095	0.7502	0.7917	0.8339	0.8772	0.9200
60	0.5558	0.5962	0.6377	0.6803	0.7239	0.7685	0.8141	0.8607	0.9081
80	0.5314	0.5732	0.6164	0.6608	0.7063	0.7531	0.8010	0.8499	0.9003
100	0.5132	0.5561	0.6004	0.6462	0.6932	0.7416	0.7913	0.8420	0.8946
200	0.4600	0.5059	0.5537	0.6033	0.6547	0.7080	0.7631	0.8193	0.8783
400	0.4117	0.4601	0.5108	0.5638	0.6192	0.6769	0.7370	0.7990	0.8637

(continued)

Table II-7 (*Cont.*)

ε^*	$a^* = 0$	$a^* = 0.1$	$a^* = 0.2$	$a^* = 0.3$	$a^* = 0.4$	$a^* = 0.5$	$a^* = 0.6$	$a^* = 0.7$	$a^* = 0.8$
600	0.3855	0.4351	0.4873	0.5421	0.5996	0.6597	0.7225	0.7878	0.8558
800	0.3679	0.4182	0.4713	0.5273	0.5862	0.6479	0.7126	0.7801	0.8503
1000	0.3547	0.4055	0.4593	0.5162	0.5761	0.6391	0.7051	0.7742	0.8462
				$Q^{(3)*}$					
0.01	30.50	25.57	21.10	17.05	13.46	10.31	7.569	5.283	3.437
0.02	21.45	18.13	15.09	12.32	9.865	7.675	5.775	4.161	2.835
0.03	17.46	14.83	12.41	10.21	8.249	6.486	4.956	3.642	2.552
0.04	15.08	12.86	10.81	8.948	7.276	5.767	4.458	3.324	2.376
0.06	12.26	10.53	8.904	7.438	6.110	4.904	3.854	2.935	2.159
0.08	10.58	9.132	7.769	6.533	5.407	4.383	3.486	2.695	2.024
0.1	9.436	8.178	6.995	5.913	4.924	4.024	3.230	2.528	1.929
0.2	6.599	5.804	5.069	4.358	3.709	3.116	2.576	2.095	1.678
0.3	5.343	4.749	4.207	3.659	3.159	2.702	2.274	1.893	1.558
0.4	4.594	4.120	3.684	3.235	2.827	2.450	2.090	1.769	1.483
0.6	3.797	3.361	3.045	2.723	2.423	2.143	1.864	1.616	1.390
0.8	3.063	2.919	2.686	2.409	2.176	1.952	1.726	1.522	1.332
1	2.631	2.471	2.349	2.194	2.001	1.817	1.630	1.456	1.291
2	1.630	1.608	1.580	1.543	1.494	1.430	1.365	1.279	1.182
3	1.211	1.221	1.229	1.236	1.243	1.239	1.222	1.188	1.130
4	1.017	1.037	1.056	1.075	1.091	1.108	1.120	1.121	1.096
6	0.8453	0.8717	0.8985	0.9254	0.9535	0.9761	1.001	1.025	1.044
8	0.7679	0.7971	0.8270	0.8574	0.8876	0.9204	0.9478	0.9664	1.006
10	0.7226	0.7538	0.7856	0.8180	0.8510	0.8839	0.9176	0.9396	0.9816
20	0.6252	0.6612	0.6981	0.7357	0.7743	0.8137	0.8535	0.8938	0.9347
40	0.5568	0.5965	0.6376	0.6798	0.7233	0.7679	0.8138	0.8611	0.9085
60	0.5222	0.5639	0.6071	0.6518	0.6980	0.7456	0.7947	0.8453	0.8971
80	0.4991	0.5420	0.5867	0.6332	0.6812	0.7309	0.7822	0.8350	0.8897

100	0.4819	0.5258	0.5715	0.6192	0.6686	0.7199	0.7729	0.8274	0.8843
200	0.4317	0.4782	0.5271	0.5784	0.6319	0.6877	0.7459	0.8057	0.8687
400	0.3861	0.4349	0.4864	0.5408	0.5980	0.6580	0.7209	0.7862	0.8548
600	0.3615	0.4113	0.4642	0.5202	0.5793	0.6416	0.7071	0.7755	0.8472
800	0.3450	0.3953	0.4491	0.5062	0.5666	0.6304	0.6976	0.7682	0.8420
1000	0.3326	0.3834	0.4377	0.4956	0.5570	0.6219	0.6904	0.7626	0.8380

$Q^{(4)*}$

0.01	31.50	26.51	22.13	18.04	14.38	11.17	8.293	5.859	3.835
0.02	22.26	18.89	15.92	13.13	10.60	8.362	6.368	4.629	3.161
0.03	18.16	15.52	13.14	10.93	8.902	7.092	5.485	4.057	2.842
0.04	15.72	13.50	11.48	9.606	7.876	6.323	4.945	3.705	2.643
0.06	12.83	11.11	9.501	8.019	6.643	5.398	4.286	3.273	2.397
0.08	11.11	9.683	8.318	7.064	5.897	4.836	3.882	3.007	2.243
0.1	9.940	8.704	7.509	6.407	5.382	4.448	3.600	2.820	2.134
0.2	7.059	6.257	5.491	4.754	4.078	3.456	2.871	2.334	1.846
0.3	5.789	5.161	4.587	4.007	3.483	2.998	2.532	2.104	1.707
0.4	5.029	4.503	4.040	3.557	3.122	2.716	2.323	1.961	1.620
0.6	4.055	3.743	3.372	3.014	2.685	2.371	2.068	1.784	1.511
0.8	3.706	3.259	2.952	2.686	2.419	2.159	1.911	1.674	1.443
1	3.313	3.064	2.751	2.469	2.236	2.021	1.801	1.595	1.394
2	2.074	2.011	1.943	1.859	1.771	1.656	1.501	1.388	1.265
3	1.499	1.490	1.482	1.471	1.448	1.419	1.378	1.292	1.202
4	1.221	1.230	1.240	1.247	1.257	1.249	1.245	1.222	1.160
6	0.9769	0.9970	1.017	1.038	1.056	1.075	1.087	1.100	1.103
8	0.8715	0.8953	0.9197	0.9442	0.9691	0.9931	1.015	1.018	1.055
10	0.8122	0.8386	0.8654	0.8925	0.9197	0.9466	0.9734	0.9825	1.022
20	0.6933	0.7260	0.7589	0.7920	0.8255	0.8593	0.8928	0.9263	0.9600
40	0.6173	0.6542	0.6918	0.7301	0.7689	0.8081	0.8480	0.8889	0.9288
60	0.5800	0.6190	0.6591	0.7001	0.7418	0.7843	0.8276	0.8716	0.9162
80	0.5551	0.5956	0.6373	0.6802	0.7239	0.7686	0.8142	0.8604	0.9081

(continued)

Table II-7 *(Cont.)*

ε^*	$a^* = 0$	$a^* = 0.1$	$a^* = 0.2$	$a^* = 0.3$	$a^* = 0.4$	$a^* = 0.5$	$a^* = 0.6$	$a^* = 0.7$	$a^* = 0.8$
100	0.5364	0.5781	0.6210	0.6653	0.7105	0.7569	0.8044	0.8523	0.9022
200	0.4816	0.5265	0.5732	0.6215	0.6713	0.7228	0.7757	0.8293	0.8855
400	0.4313	0.4791	0.5290	0.5810	0.6350	0.6910	0.7492	0.8087	0.8707
600	0.4041	0.4532	0.5048	0.5587	0.6149	0.6734	0.7345	0.7973	0.8627
800	0.3857	0.4357	0.4883	0.5434	0.6011	0.6614	0.7243	0.7895	0.8571
1000	0.3720	0.4225	0.4759	0.5319	0.5907	0.6522	0.7166	0.7836	0.8530

Table II-8 Cross Sections $Q^{(l)*}$ for (16, 4) Core Potentials

$Q^{(1)*}$

ε^*	$a^* = 0$	$a^* = 0.1$	$a^* = 0.2$	$a^* = 0.3$	$a^* = 0.4$	$a^* = 0.5$	$a^* = 0.6$	$a^* = 0.7$	$a^* = 0.8$
0.01	24.45	20.51	16.92	13.69	10.81	8.288	6.129	4.325	2.875
0.02	17.14	14.49	12.07	9.879	7.917	6.187	4.692	3.427	2.394
0.03	13.91	11.83	9.917	8.180	6.619	5.237	4.036	3.013	2.169
0.04	11.99	10.24	8.630	7.162	5.839	4.663	3.638	2.759	2.029
0.06	9.719	8.361	7.100	5.950	4.905	3.974	3.155	2.449	1.856
0.08	8.368	7.240	6.187	5.223	4.344	3.557	2.861	2.258	1.749
0.1	7.447	6.475	5.563	4.726	3.959	3.269	2.658	2.126	1.674
0.2	5.171	4.575	4.011	3.484	2.992	2.544	2.139	1.783	1.476
0.3	4.167	3.731	3.319	2.927	2.558	2.215	1.902	1.624	1.382
0.4	3.570	3.228	2.904	2.592	2.295	2.015	1.758	1.526	1.324
0.6	2.887	2.629	2.406	2.189	1.978	1.774	1.583	1.407	1.252
0.8	2.437	2.313	2.142	1.942	1.784	1.627	1.475	1.334	1.207
1	2.013	1.971	1.915	1.776	1.648	1.524	1.401	1.283	1.175

2	1.094	1.150	1.202	1.233	1.226	1.215	1.203	1.189	1.175
3	1.056	1.084	1.078	1.059	1.039	1.019	0.9975	0.9770	0.9554
4	1.033	1.028	1.004	0.9801	0.9573	0.9332	0.9079	0.8839	0.8591
6	0.9940	0.9659	0.9392	0.9083	0.8820	0.8533	0.8247	0.7968	0.7689
8	0.9681	0.9394	0.9067	0.8765	0.8453	0.8144	0.7835	0.7532	0.7233
10	0.9547	0.9227	0.8877	0.8559	0.8223	0.7896	0.7573	0.7255	0.6943
20	0.9260	0.8860	0.8462	0.8071	0.7688	0.7312	0.6946	0.6588	0.6240
40	0.9073	0.8610	0.8155	0.7712	0.7282	0.6864	0.6459	0.6066	0.5686
60	0.8984	0.8486	0.8001	0.7530	0.7075	0.6633	0.6207	0.5795	0.5397
80	0.8926	0.8404	0.7899	0.7409	0.6936	0.6479	0.6038	0.5613	0.5204
100	0.8883	0.8343	0.7823	0.7318	0.6832	0.6363	0.5911	0.5477	0.5059
200	0.8759	0.8168	0.7600	0.7054	0.6528	0.6024	0.5541	0.5079	0.4638
400	0.8646	0.8007	0.7395	0.6809	0.6248	0.5712	0.5201	0.4715	0.4254
600	0.8584	0.7919	0.7282	0.6673	0.6093	0.5540	0.5014	0.4516	0.4044
800	0.8541	0.7859	0.7204	0.6580	0.5986	0.5422	0.4886	0.4379	0.3901
1000	0.8509	0.7812	0.7145	0.6510	0.5906	0.5333	0.4790	0.4277	0.3794

$Q^{(2)}*$

					$Q^{(2)}*$				
0.01	3.478	5.219	7.294	9.706	12.46	15.53	18.93	22.72	26.80
0.02	2.884	4.147	5.623	7.309	9.215	11.33	13.66	16.20	18.97
0.03	2.603	3.647	4.855	6.218	7.751	9.445	11.30	13.30	15.50
0.04	2.428	3.339	4.385	5.557	6.866	8.310	9.890	11.57	13.43
0.06	2.210	2.961	3.813	4.758	5.802	6.950	8.205	9.519	10.98
0.08	2.074	2.727	3.462	4.272	5.159	6.131	7.193	8.294	9.512
0.1	1.978	2.563	3.218	3.935	4.715	5.567	6.498	7.459	8.513
0.2	1.723	2.135	2.587	3.073	3.590	4.147	4.752	5.383	6.034
0.3	1.600	1.933	2.292	2.676	3.077	3.504	3.965	4.456	4.939
0.4	1.523	1.808	2.110	2.432	2.766	3.117	3.492	3.897	4.289
0.6	1.427	1.652	1.888	2.134	2.388	2.651	2.927	3.216	3.525
0.8	1.366	1.556	1.750	1.951	2.158	2.369	2.581	2.798	3.065
1	1.323	1.488	1.654	1.824	1.997	2.172	2.342	2.615	2.886

(continued)

Table II-8 (*Cont.*)

ε^*	$a^* = 0$	$a^* = 0.1$	$a^* = 0.2$	$a^* = 0.3$	$a^* = 0.4$	$a^* = 0.5$	$a^* = 0.6$	$a^* = 0.7$	$a^* = 0.8$
2	1.748	1.728	1.702	1.666	1.611	1.512	1.403	1.308	1.210
3	1.297	1.303	1.309	1.313	1.312	1.311	1.287	1.222	1.155
4	1.099	1.114	1.128	1.140	1.154	1.158	1.173	1.165	1.121
6	0.9293	0.9498	0.9705	0.9909	1.010	1.027	1.047	1.064	1.072
8	0.8538	0.8769	0.9002	0.9232	0.9461	0.9693	0.9914	1.013	1.031
10	0.8102	0.8350	0.8602	0.8853	0.9108	0.9349	0.9577	0.9860	1.006
20	0.7179	0.7473	0.7770	0.8070	0.8373	0.8678	0.8981	0.9285	0.9592
40	0.6539	0.6871	0.7209	0.7552	0.7901	0.8255	0.8613	0.8979	0.9341
60	0.6215	0.6568	0.6928	0.7295	0.7670	0.8052	0.8440	0.8836	0.9234
80	0.5999	0.6365	0.6741	0.7125	0.7518	0.7919	0.8328	0.8743	0.9168
100	0.5837	0.6214	0.6601	0.6998	0.7404	0.7820	0.8244	0.8675	0.9119
200	0.5362	0.5770	0.6191	0.6625	0.7071	0.7530	0.8001	0.8482	0.8980
400	0.4925	0.5360	0.5811	0.6279	0.6761	0.7261	0.7776	0.8309	0.8856
600	0.4685	0.5134	0.5601	0.6087	0.6590	0.7112	0.7652	0.8213	0.8788
800	0.4521	0.4979	0.5457	0.5955	0.6472	0.7009	0.7567	0.8147	0.8742
1000	0.4397	0.4862	0.5348	0.5855	0.6382	0.6931	0.7503	0.8097	0.8707
					$Q^{(3)*}$				
0.01	28.64	24.08	19.86	16.10	12.74	9.786	7.223	5.068	3.323
0.02	20.13	17.05	14.21	11.64	9.339	7.299	5.524	4.001	2.750
0.03	16.38	13.94	11.70	9.652	7.812	6.175	4.747	3.508	2.479
0.04	14.14	12.08	10.20	8.457	6.892	5.496	4.274	3.205	2.312
0.06	11.49	9.875	8.419	7.032	5.791	4.681	3.700	2.835	2.105
0.08	9.914	8.566	7.350	6.178	5.128	4.188	3.350	2.608	1.976
0.1	8.839	7.674	6.617	5.592	4.673	3.847	3.106	2.449	1.885
0.2	6.175	5.465	4.781	4.125	3.528	2.985	2.484	2.037	1.645
0.3	4.998	4.483	3.956	3.466	3.011	2.591	2.197	1.845	1.531
0.4	4.299	3.890	3.461	3.068	2.699	2.350	2.022	1.726	1.459

0.6	3.471	3.169	2.870	2.589	2.321	2.058	1.808	1.580	1.370
0.8	2.965	2.730	2.472	2.298	2.089	1.879	1.677	1.490	1.314
1	2.515	2.402	2.275	2.102	1.925	1.754	1.585	1.427	1.275
2	1.621	1.595	1.564	1.524	1.465	1.416	1.345	1.261	1.172
3	1.233	1.239	1.246	1.249	1.251	1.237	1.216	1.180	1.123
4	1.051	1.066	1.083	1.096	1.110	1.122	1.126	1.121	1.093
6	0.8883	0.9111	0.9339	0.9570	0.9782	0.9952	1.022	1.039	1.048
8	0.8151	0.8404	0.8660	0.8913	0.9185	0.9436	0.9669	0.9942	1.015
10	0.7727	0.7997	0.8272	0.8549	0.8821	0.9119	0.9360	0.9680	0.9918
20	0.6832	0.7145	0.7465	0.7790	0.8121	0.8457	0.8794	0.9136	0.9484
40	0.6215	0.6565	0.6923	0.7290	0.7665	0.8049	0.8440	0.8842	0.9243
60	0.5905	0.6274	0.6654	0.7043	0.7444	0.7854	0.8274	0.8704	0.9141
80	0.5698	0.6080	0.6474	0.6880	0.7297	0.7725	0.8166	0.8615	0.9078
100	0.5544	0.5935	0.6340	0.6758	0.7188	0.7630	0.8086	0.8550	0.9031
200	0.5092	0.5511	0.5947	0.6400	0.6867	0.7351	0.7851	0.8365	0.8898
400	0.4675	0.5120	0.5583	0.6068	0.6570	0.7093	0.7635	0.8198	0.8779
600	0.4447	0.4904	0.5382	0.5884	0.6405	0.6949	0.7516	0.8106	0.8714
800	0.4291	0.4756	0.5245	0.5757	0.6292	0.6851	0.7434	0.8043	0.8669
1000	0.4174	0.4645	0.5141	0.5661	0.6206	0.6776	0.7372	0.7994	0.8635

$Q^{(4)*}$

0.01	29.74	25.07	20.88	17.10	13.67	10.62	7.923	5.624	3.708
0.02	21.01	17.84	15.04	12.46	10.09	7.976	6.094	4.453	3.064
0.03	17.14	14.64	12.45	10.37	8.475	6.772	5.254	3.908	2.757
0.04	14.83	12.73	10.90	9.121	7.499	6.041	4.740	3.573	2.568
0.06	12.10	10.47	9.049	7.620	6.326	5.159	4.114	3.162	2.333
0.08	10.48	9.131	7.935	6.715	5.618	4.623	3.729	2.908	2.185
0.1	9.374	8.216	7.169	6.093	5.131	4.252	3.461	2.730	2.081
0.2	6.661	5.944	5.236	4.527	3.898	3.307	2.769	2.266	1.805
0.3	5.463	4.925	4.358	3.820	3.337	2.873	2.447	2.046	1.673
0.4	4.743	4.305	3.829	3.393	2.996	2.608	2.249	1.909	1.589

(continued)

Table II-8 (*Cont.*)

ε^*	$a^* = 0$	$a^* = 0.1$	$a^* = 0.2$	$a^* = 0.3$	$a^* = 0.4$	$a^* = 0.5$	$a^* = 0.6$	$a^* = 0.7$	$a^* = 0.8$
0.6	3.905	3.543	3.198	2.881	2.583	2.285	2.007	1.740	1.485
0.8	3.390	3.072	2.817	2.572	2.329	2.087	1.858	1.634	1.420
1	3.133	2.858	2.563	2.356	2.151	1.948	1.754	1.560	1.373
2	2.025	1.964	1.893	1.810	1.722	1.596	1.487	1.364	1.249
3	1.500	1.491	1.480	1.463	1.436	1.407	1.350	1.270	1.190
4	1.241	1.248	1.254	1.259	1.259	1.250	1.243	1.212	1.152
6	1.010	1.027	1.044	1.059	1.076	1.086	1.103	1.109	1.101
8	0.9099	0.9303	0.9510	0.9720	0.9914	1.012	1.029	1.050	1.059
10	0.8543	0.8768	0.8996	0.9225	0.9456	0.9679	0.9859	1.014	1.029
20	0.7448	0.7727	0.8007	0.8290	0.8574	0.8859	0.9140	0.9424	0.9710
40	0.6764	0.7083	0.7406	0.7732	0.8064	0.8399	0.8736	0.9081	0.9418
60	0.6432	0.6771	0.7118	0.7469	0.7827	0.8189	0.8557	0.8930	0.9304
80	0.6211	0.6565	0.6927	0.7295	0.7671	0.8053	0.8442	0.8834	0.9234
100	0.6046	0.6411	0.6784	0.7166	0.7556	0.7953	0.8357	0.8764	0.9184
200	0.5560	0.5957	0.6366	0.6788	0.7217	0.7659	0.8110	0.8569	0.9042
400	0.5109	0.5536	0.5978	0.6435	0.6902	0.7385	0.7881	0.8395	0.8916
600	0.4861	0.5304	0.5762	0.6238	0.6727	0.7233	0.7756	0.8298	0.8847
800	0.4692	0.5144	0.5614	0.6103	0.6606	0.7128	0.7670	0.8231	0.8800
1000	0.4565	0.5024	0.5502	0.6000	0.6514	0.7049	0.7605	0.8180	0.8764

Table II-9   Asymptotic Forms of $Q^{(l)*}$ for $\varepsilon^* \to 0$ and ∞[a,b]

	$l = 1$	$l = 2$	$l = 3$	$l = 4$
$f_0^{(l)}(4)$	2.2137[c]	2.3100[c]	2.5542[c]	2.561
$f_0^{(l)}(6)$	1.578	1.786	1.816	1.924
$f_\infty^{(l)}(6)$	1.112	1.544	1.401	1.691
$f_\infty^{(l)}(8)$	1.077	1.402	1.289	1.504
$f_\infty^{(l)}(10)$	1.058	1.319	1.226	1.398
$f_\infty^{(l)}(12)$	1.0463[d]	1.2650[d]	1.1859[d]	1.3289[d]
$f_\infty^{(l)}(14)$	1.038	1.228	1.159	1.282
$f_\infty^{(l)}(16)$	1.032	1.196	1.136	1.242
$f_\infty^{(l)}(18)$	1.026	1.171	1.117	1.210
$f_\infty^{(l)}(300)$	1.001	1.010	1.007	1.013
$f_\infty^{(l)}(\infty)$	1.0000	1.0000	1.0000	1.0000

[a] *Low-energy asymptotes, $\varepsilon^* \to 0$:*

$$Q^{(l)*} = \frac{f_0^{(l)}(4)}{(\varepsilon^*)^{1/2}} \left[\frac{3n(1-\gamma)}{n(3+\gamma) - 12(1+\gamma)} \right]^{1/2} (1 - a^*)^2 \qquad \text{for } \gamma \neq 1$$

$$Q^{(l)*} = \frac{f_0^{(l)}(6)}{(\varepsilon^*)^{1/3}} \left(\frac{n}{n-6} \right)^{1/3} (1 - a^*)^2 \qquad \text{for } \gamma = 1,$$

for $(n, 6, 4)$ core models.

High-energy asymptotes, $\varepsilon^* \to \infty$:

$$Q^{(l)*} = \frac{f_\infty^{(l)}(n)}{(\varepsilon^*)^{2/n}} \left[\frac{12(1+\gamma)}{n(3+\gamma) - 12(1+\gamma)} \right]^{2/n} \qquad \text{for } (n, 6, 4) \text{ models,}$$

$$Q^{(l)*} = a^{*2} \qquad \text{for core models.}$$

[b] All values from Higgins and Smith (1968) except as noted.
[c] Heiche and Mason (1970).
[d] Monchick and Mason (1961).

REFERENCES

Heiche, G., and E. A. Mason (1970). Ion mobilities with charge exchange, *J. Chem. Phys.* **53**, 4687–4696.

Higgins, L. D., and F. J. Smith (1968). Collision integrals for high temperature gases, *Mol. Phys.* **14**, 399–400.

Monchick, L., and E. A. Mason (1961). Transport properties of polar gases, *J. Chem. Phys.* **35**, 1676–1697.

Smith, F. J. (1983). Personal communication to E. A. Mason.

Viehland, L. A., E. A. Mason, W. F. Morrison, and M. R. Flannery (1975). Tables of transport collision integrals for $(n, 6, 4)$ ion-neutral potentials, *At. Data Nucl. Data Tables* **16**, 495–514.

APPENDIX III

TABLES OF PROPERTIES USEFUL IN THE ESTIMATION OF ION-NEUTRAL INTERACTION ENERGIES

Table III-1 Polarizabilities for atoms[a]

Atom	$\alpha_d(\text{Å}^3)$	$\alpha_q(\text{Å}^5)$
H	0.6668[b]	0.6224[b]
He	0.2050[c]	0.1014[c]
Ne	0.3946[d]	0.2665[e]
Ar	1.642[d]	2.084[e]
Kr	2.480[d]	3.965[e]
Xe	4.044[d]	8.82[e]
B	3.03[f]	
C	1.78[g]	
N	1.078[h]	
O	0.734[h]	
F	0.557[f]	
Cl	2.18[f]	
Br	3.05[f]	
I	4.9[i]	
Li	24.3[f]	57.4[e,j]
Na	23.6[f]	74.7[e,j]
K	43.4[f]	191[e,j]
Rb	47.3[f]	248[e,j]
Cs	59.6[f]	393[e,j]
Be	5.60[f]	12.6[j]
Mg	10.6[f]	34.4[j]
Ca	25.0[f]	113[j]

Table III-1 (*Cont.*)

Atom	$\alpha_d(\text{Å}^3)$	$\alpha_q(\text{Å}^5)$
Sr	27.6[f]	
Ba	39.7[f]	
Hg	5.02[k]	10.1[j]

[a]A more extensive table of α_d is given by T. M. Miller in *Handbook of Chemistry and Physics*, 67th ed., CRC Press, Orlando, Fla., 1986, pp. E-66 ff. [b]Exact value (Dalgarno, 1962). [c]Thakkar (1981). [d]Teachout and Pack (1971). [e]Summary of Standard and Certain (1985). [f]Miller and Bederson (1977). [g]Nesbet (1977). [h]Zeiss and Meath (1977). [i]Value from ref. *d* scaled by ratio of α_d's for Br from refs. *d* and *f*. [j]Maeder and Kutzelnigg (1979). [k]Stwalley and Kramer (1968).

Table III-2 Dipole Polarizabilities for Atomic Ions

Ion	$\alpha_d(\text{Å}^3)$	Ion	$\alpha_d(\text{Å}^3)$
He$^+$	0.04168[a]	H$^-$	30.54[b]
Li$^+$	0.02852[b]	O$^-$	3.2[c]
Na$^+$	0.148[d]	F$^-$	1.56[e]
K$^+$	0.811[d]	Cl$^-$	4.17[e]
Rb$^+$	1.35[d]	Br$^-$	5.40[e]
Cs$^+$	2.34[f]	I$^-$	8.10[e]
Be^{2+}	0.0080[g]	H$^+$, He^{2+}	0
Mg^{2+}	0.0072[h]		
Ca^{2+}	0.482[h]	O$^+$	0.49[c]
Sr^{2+}	0.817[i]	Ne$^+$	0.21[c]
Ba^{2+}	1.51[i]	Ne^{2+}	0.15[c]

[a]Exact value (Dalgarno, 1962). [b]Chung (1971); Glover and Weinhold (1976). [c]Dalgarno (1962). [d]Summary by Mahan (1982). [e]Coker (1976), scaled upward by 3–12% corresponding to a better value of $\alpha_d(\text{Na}^+) = 0.148$ Å^3, according to Mahan (1982). [f]Coker (1976), scaled downward by 3% according to the recommendation of Mahan (1982). [g]Average of Schmidt et al. (1979, probably low) and of Mahan (1980, probably high). [h]Eissa and Öpik (1967). [i]Schmidt et al. (1979).

Table III-3 Dipole Polarizabilities for Molecules[a]

Molecule	$\bar{\alpha}_d(\text{Å}^3)$	$\kappa = (\alpha_d^{\parallel} - \alpha_d^{\perp})/3\bar{\alpha}_d$
H_2	0.807	0.128
N_2	1.74	0.131
O_2	1.57	0.229
CO	1.97	0.090
NO	1.71	0.162
CO_2	2.93	0.266
N_2O	3.08	0.329
NH_3	2.16[b]	0.043
H_2O	1.43[b]	
H_2S	3.95	
SO_2	4.28	0.175
CH_4	2.59	0
CF_4	3.86[c]	0
CCl_4	11.16	0
SF_6	6.55[c]	0

[a]Total molecular polarizability includes both electronic and atomic contributions (Gislason et al., 1977). Values of $\bar{\alpha}_d$ are from the compilation of Maryott and Buckley (1953) unless otherwise noted. Values of κ are from Bridge and Buckingham (1966). More extensive tables of $\bar{\alpha}_d$ are given by T. M. Miller in *Handbook of Chemistry and Physics*, 67th ed., CRC Press, Orlando, Fla., 1986, pp. E-66 ff.
[b]Zeiss and Meath (1977).
[c]Tipton et al. (1964).

Table III-4 Equivalent Oscillator Numbers for Dispersion Energies

Atom	N	Isoelectronic Ions	Molecule	N
H	0.824[a]	He^+, Li^{2+}	H_2	1.63[b]
He	1.434[c]	H^-, Li^+, Be^{2+}	N_2	5.92[b]
Ne	4.45[d]	F^-, Na^+, Mg^{2+}	O_2	5.76[b]
Ar	5.90[d]	Cl^-, K^+, Ca^{2+}	CO	5.88[e]
Kr	6.70[d]	Br^-, Rb^+, Sr^{2+}	NO	5.66[b]
Xe	7.79[d]	I^-, Cs^+, Ba^{2+}		
			CO_2	8.62[f]

(*continued*)

Table III-4 *(Cont.)*
Energies

Atom	N	Isoelectronic Ions	Molecule	N
N	3.94^b	C^-, O^+, F^{2+}	N_2O	7.95^b
O	3.26^b	N^-, F^+, Ne^{2+}	NH_3	4.57^b
			H_2O	4.08^b
Cl	4.2^g	S^-, Ar^+		
Br	6.2^g	Se^-, Kr^+	CH_4	5.59^h
I	6.5^i	Te^-, Xe^+	SF_6	18.7^j
Li	0.77^k	Be^+		
Na	0.98^k	Mg^+		
K	1.05^k	Ca^+		
Rb	1.09^k	Sr^+		
Cs	1.16^k	Ba^+		
Be	1.56^k	Li^-, B^+		
Mg	1.94^k	Na^-, Al^+		
Ca	1.97^k	K^-, Sc^+		
Sr	2.05^l	Rb^-, Y^+		
Ba	1.83^l	Cs^-, La^+		
Hg	2.49^k	Tl^+		

[a]Dalgarno (1967). [b]Zeiss and Meath (1977). [c]Thakkar (1981). [d]Standard and Certain (1985). [e]Parker and Pack (1976). [f]Pack (1974). [g]LeRoy (1973, p. 169). These are spectroscopically determined values that refer to the interaction of $X(^2P_{1/2}) + X(^2P_{3/2})$ atoms. [h]Thomas and Meath (1977). [i]Danyluk and King (1977). Same comment as g. [j]Calculated from the $C^{(6)}$ for SF_6-Ne and SF_6-Ar given by Pack et al. (1982, 1984). [k]Calculated self-consistently from the α_d's and $C^{(6)}$'s of Maeder and Kutzelnigg (1979). [l]Calculated self-consistently from the α_d and $C^{(6)}$ for Sr-Kr and Ba-Xe given by Hyman (1974).

Table III-5   Dipole and Quadrupole Moments for Molecules

Molecule	$\mu^a(10^{-18}$ esu$)$	$\Theta^b(10^{-26}$ esu$)$
H_2	0	$+0.662$
N_2	0	-1.4^c
O_2	0	-0.39
CO	0.112	-2.5
NO	0.153	-1.8
CO_2	0	-4.3
N_2O	0.167	-3.0
NH_3	1.47	-1

Table III-5 (*Cont.*)

Molecule	$\mu^a(10^{-18}$ esu)	$\Theta^b(10^{-26}$ esu)
H_2O	1.85	$\sim \pm 1$
H_2S	0.97	
SO_2	1.63	± 4.4
CH_4	0	0
CF_4	0	0
CCl_4	0	0
SF_6	0	0

[a]Nelson et al. (1967).
[b]Stogryn and Stogryn (1966), unless otherwise noted.
[c]Buckingham et al. (1968); see comment by Billingsley and Krauss (1974).

Table III-6 Correlation Parameters for Homonuclear Repulsion Potentials[a]

Atom or Ion	N	α(eV-Å^2)	ρ^c
He	2	6.83	0.840
Ne	10	18.7	0.791
Ar	18	18.7	1.000
Kr	36	18.7	1.087
Xe	54	18.7	1.214
Li^+	2	17.4	0.511
Na^+	10	35.4	0.618
K^+	18	35.4	0.828
Rb^+	36	$(35.4)^b$	0.920
Cs^+	54	(35.4)	1.042
F^-	10	8.50	1.040
Cl^-	18	8.50	1.232
Br^-	36	(8.50)	1.317
I^-	54	(8.50)	1.446
Be^{2+}	2	(17.4)	0.364
Mg^{2+}	10	(35.4)	0.506
Ca^{2+}	18	(35.4)	0.703
Sr^{2+}	36	(35.4)	0.798
Ba^{2+}	54	(35.4)	0.913

[a]$V_{rep} = \alpha N(\beta/\rho)^2 \exp(-\beta r/\rho)$, $\beta = 3.66$ Å^{-1} for all systems.
[b]Values in parentheses are only inferred.
[c]From Butterfield and Carlson (1972), normalized to $\rho_{Ar} = 1.000$.

Table III-7 Potential Parameters from Mobility and Ion-Beam Measurements

System	r_m (Å)		ε_0 (eV)	
	Mobility[a,b]	Beam[c]	Mobility[a,b]	Beam[c]
Li$^+$–He	1.93	1.96	0.0735	0.0710
Ne	2.08	2.11	0.123	0.114
Ar	2.4[d]	2.42	0.27[d]	0.313
Kr	2.42	2.45	0.397	0.460
Xe	2.51	2.54	0.547	0.533
Na$^+$–He	2.44	2.22	0.0343	0.0514
Ne	2.43	2.37	0.0661	0.0762
Ar	2.69	2.75	0.190	0.163
Kr	2.91	2.87	0.220	0.210
Xe	2.98	3.03	0.258	0.259
K$^+$–He	2.86	2.56	0.0229	0.0248
Ne	2.94	2.68	0.0397	0.0473
Ar	2.94	2.96	0.127	0.125
Kr	3.24	3.16	0.144	0.131
Xe	3.37	3.39	0.210	0.164
Rb$^+$–He	—[e]		—[e]	
Ne	3.16		0.0335	
Ar	3.41		0.0857	
Kr	3.54		0.116	
Xe	3.55		0.185	
Cs$^+$–He	—[e]	2.98	—[e]	0.0158
Ne	—[e]	3.32	—[e]	0.0242
Ar	3.48	3.70	0.0852	0.0634
Kr	3.48	3.73	0.118	0.101
Xe	4.11	4.02	0.113	0.119
F$^-$–He	—[e]		—[e]	
Ne				
Ar	2.78		0.137	
Kr	2.47		0.274	
Xe	2.45		0.429	
Cl$^-$–He	—[e]		—[e]	
Ne	—[e]		—[e]	
Ar	3.14		0.104	
Kr	3.45		0.102	
Xe	3.30		0.198	

Table III-7 (*Cont.*)

System	r_m (Å)		ε_0 (eV)	
	Mobility[a,b]	Beam[c]	Mobility[a,b]	Beam[c]
Br$^-$–He	—[e]		—[e]	
Ne	—[e]		—[e]	
Ar	3.39		0.0822	
Kr	3.58		0.0887	
Xe	3.40		0.169	
I$^-$–He	—[e]		—[e]	
Ar	3.61		0.0765	

[a]Viehland (1984); alkali ions.
[b]Kirkpatrick and Viehland (1985); halide ions.
[c]Gislason and co-workers, as quoted by Viehland (1984).
[d]Gatland (1981); see comments by Viehland (1983).
[e]Potential well not accurately probed by available mobility data.

REFERENCES

Billingsley, F. P., II, and M. Krauss (1974). Quadrupole moment of CO, N$_2$, and NO$^+$, *J. Chem. Phys.* **60**, 2767–2772.

Bridge, N. J., and A. D. Buckingham (1966). The polarization of laser light scattered by gases, *Proc. R. Soc. London* **A295**, 334–349.

Buckingham, A. D., R. L. Disch, and D. A. Dunmur (1968). The quadrupole moments of some simple molecules, *J. Am. Chem. Soc.* **90**, 3104–3107.

Butterfield, C., and E. H. Carlson (1972). Ionic soft sphere parameters from Hartree–Fock–Slater calculations, *J. Chem. Phys.* **56**, 4907–4911.

Coker, H. (1976). Empirical free-ion polarizabilities of the alkali metal, alkaline earth metal, and halide ions, *J. Phys. Chem.* **80**, 2078–2084.

Chung, K. T. (1971). Dynamic polarizabilities and refractive indices of H$^-$ and Li$^+$ ions, *Phys. Rev.* **A4**, 7–11.

Dalgarno, A. (1962). Atomic polarizabilities and shielding factors, *Adv. Phys.* **11**, 281–315.

Dalgarno, A. (1967). New methods for calculating long-range intermolecular forces, *Adv. Chem. Phys.* **12**, 143–166.

Danyluk, M. D., and G. W. King (1977). Energy levels of iodine near the *B* state dissociation limit, *Chem. Phys.* **25**, 343–351.

Eissa, H., and U. Öpik (1967). The polarization of a closed-shell core of an atomic system by an outer electron: 1. A correction to the adiabatic approximation, *Proc. Phys. Soc. London* **92**, 556–565.

Gatland, I. R. (1981). Ion mobility test of Li$^+$ − Ar potentials, *J. Chem. Phys.* **75**, 4162–4163.

Gislason, E. A., F. E. Budenholzer, and A. D. Jorgensen (1977). The ion-induced dipole potential: A caveat concerning molecular polarizabilities, *Chem. Phys. Lett.* **47**, 434–435.

Glover, R. M., and F. Weinhold (1976). Dynamic polarizabilities of two-electron atoms, with rigorous upper and lower bounds, *J. Chem. Phys.* **65**, 4913–4926.

Hyman, H. A. (1974). Van der Waals constants for alkaline earth-noble gas pairs, *J. Chem. Phys.* **61**, 4063–4066.

Kirkpatrick, C. C., and L. A. Viehland (1985). Interaction potentials for the halide ion-rare gas systems, *Chem. Phys.* **98**, 221–231.

LeRoy, R. J. (1973). *Molecular Spectroscopy*, Vol. 1. *A Specialist Periodical Report*, The Chemical Society, London, pp. 113–176.

Maeder, F., and W. Kutzelnigg (1979). Natural states of interacting systems and their use for the calculation of intermolecular forces: 4. Calculation of van der Waals coefficients between one- and two-valence-electron atoms in their ground states, as well as of polarizabilities, oscillator strength sums and related quantities, including correlation effects, *Chem. Phys.* **42**, 95–112.

Mahan, G. D. (1980). Modified Sternheimer equation for polarizability, *Phys. Rev.* **A22**, 1780–1785.

Mahan, G. D. (1982). Van der Waals coefficient between closed shell ions, *J. Chem. Phys.* **76**, 493–497.

Maryott, A. A., and F. Buckley (1953). Table of dielectric constants and electric dipole moments of substances in the gaseous state, *U.S. National Bureau of Standards Circular* **537**.

Miller, T. M., and B. Bederson (1977). Atomic and molecular polarizabilities—A review of recent advances, *Adv. At. Mol. Phys.* **13**, 1–55.

Nelson, R. D., Jr., D. R. Lide, Jr., and A. A. Maryott (1967). Selected values of electric dipole moments for molecules in the gas phase. *U.S. National Bureau of Standards NSRDS-NBS 10*.

Nesbet, R. K. (1977). Atomic polarizabilities for ground and excited states of C, N, and O, *Phys. Rev.* **A16**, 1–5.

Pack, R. T. (1974). Van der Waals interactions of carbon dioxide, *J. Chem. Phys.* **61**, 2091–2094.

Pack, R. T., J. J. Valentini, and J. B. Cross (1982). Multiproperty empirical anisotropic intermolecular potentials for $ArSF_6$ and $KrSF_6$, *J. Chem. Phys.* **77**, 5486–5499.

Pack, R. T., E. Piper, G. A. Pfeffer, and J. P. Toennies (1984). Multiproperty empirical anisotropic intermolecular potentials: 2. $HeSF_6$ and $NeSF_6$, *J. Chem. Phys.* **80**, 4940–4950.

Parker, G. A., and R. T. Pack (1976). Van der Waals interactions of carbon monoxide, *J. Chem. Phys.* **64**, 2010–2012.

Schmidt, P. C., A. Weiss, and T. P. Das (1979). Effect of crystal fields and self-consistency on dipole and quadrupole polarizabilities of closed-shell ions, *Phys. Rev.* **B19**, 5525–5534.

Standard, J. M., and P. R. Certain (1985). Bounds to two- and three-body long-range interaction coefficients for S-state atoms, *J. Chem. Phys.* **83**, 3002–3008.

Stogryn, D. E., and A. P. Stogryn (1966). Molecular multipole moments, *Mol. Phys.* **11**, 371–393.

Stwalley, W. C., and H. L. Kramer (1968). Long-range interactions of mercury atoms, *J. Chem. Phys.* **49**, 5555–5556.

Teachout, R. R., and R. T. Pack (1971). The static dipole polarizabilities of all the neutral atoms in their ground states, *At. Data* **3**, 195–214.

Thakkar, A. J. (1981). The generator coordinate method applied to variational perturbation theory. Multipole polarizabilities, spectral sums, and dispersion coefficients for helium, *J. Chem. Phys.* **75**, 4496–4501.

Thomas, G. F., and W. J. Meath (1977). Dipole spectrum, sums and properties of ground-state methane and their relation to the molar refractivity and dispersion energy constants, *Mol. Phys.* **34**, 113–125.

Tipton, A. B., A. P. Deam, and J. E. Boggs (1964). Atomic polarization of sulfur hexafluoride and of carbon tetrafluoride, *J. Chem. Phys.* **40**, 1144–1147.

Viehland, L. A. (1983). Interaction potentials for Li^+–rare gas systems, *Chem. Phys.* **78**, 279–294.

Viehland, L. A. (1984). Interaction potentials for the alkali ion–rare gas systems, *Chem. Phys.* **85**, 291–305.

Ziess, G. D., and W. J. Meath (1977). Dispersion energy constants $C_6(A, B)$, dipole oscillator strength sums and refractivities for Li, N, O, H_2, N_2, O_2, NH_3, H_2O, NO and N_2O, *Mol. Phys.* **33**, 1155–1176.

AUTHOR INDEX

Page numbers in **boldface** indicate reference entry.

SUBJECT INDEX

Accuracy of measurements:
  diffusion coefficients, 9, 32, 63, 78, 104, 106, 109, 112–113, 118, 125, 269–270, 487
  mobilities and drift velocities, 9, 32, 37, 42, 44, 63, 69, 78, 156, 269–270, 487
  reaction rate coefficients, 32, 44, 46, 50, 63, 74, 78, 109
Additional residence time method, 62, 65, 67, 69
Aerosol particles, 271, 444–449
Afterglow:
  accuracy of measurements, 72, 74
  definition, 117
  diffusion controlled, 132
  flowing, 69–70, 72–78, 84, 117–118
  recombination controlled, 130–131
  stationary, 70, 117–133
Ambipolar diffusion:
  definition, 25
  ion-ion, 25–26, 76, 117, 122–125, 127, 131
Ambipolar diffusion coefficient, 26–27, 117–118, 121, 124, 128–132
Ambipolar diffusion equation, 26–27, 120
Angle of deflection:
  as function of impact parameter, 254, 385, 424, 431
  ion-neutral potential and, 327, 385, 424, 431
    orbiting collisions, 386
    rainbow scattering, 385
  relation:
    to cross section, 188, 385, 424, 429–430
    to phase shifts, 256, 334, 341, 387, 424
  small-angle approximation, 188–189, 341
Angular momentum alignment effects, 102
Anisotropic potential, *see* Noncentral interactions

Anisotropy of polarizability, 389–390
Approximate cross sections:
  capture, 189
  charge-transfer, 193, 343–344, 417
  effective-diameter, 191
  impact parameter, 254, 341–342, 385
  orbiting, 189, 385–386
  random-angle, 189, 341, 344, 430
  random-phase, 341–344
Arrival time spectra:
  nonreacting ions, 36–39, 82
  reacting ions, 40–44, 64
Attenuation method, for measuring:
  reaction rate coefficients, 37, 103, 106–109
  transverse diffusion coefficients, 37, 103–109
Attraction, ion-neutral, 188–189, 211–212, 225, 300–301, 402
Average energy of ions in gas in electric field, 2–4, 8, 33, 50, 140, 146, 148, 157, 164, 171, 174, 176–178, 181

Basis functions, 198–210, 214, 236–237, 272–273, 277, 281, 293–295, 297, 305–307, 309, 318, 323, 351–354, 359, 366–370, 459–461, 466, 472–473
Basis temperature, 203–205, 271–272, 274–278, 280–283, 293, 295–298, 302, 305–306, 323, 325, 352–359, 366–372, 457, 460–461, 467
Beam experiments, limitations, 34, 46
Beam scattering and ion-neutral interactions, 8, 34, 347, 465
Bimodal expansion, 209–210, 302, 482
Binary collision assumption, 138–142, 144, 149, 157, 173, 194, 229, 238, 349, 444–445

RETURN **PHYSICS LIBRARY**
TO➡ 351 LeConte Hall   642-3122